QC
168
.M27
cop. 1

SCIENCE

D0886616

The Chicago Public Library

Received JAN 2 1976

PHYSICAL FLUID DYNAMICS

BUSINESS/SCIENCE/TECHNOLOGY DIVISION
CHICAGO PUBLIC LIBRARY
400 SOUTH STATE STREET
CHICAGO, IL 60605

PHYSICAL FLUID DYNAMICS

P. D. McCORMACK AND **LAWRENCE CRANE**

University College
Cork, Ireland

Trinity College
Dublin, Ireland

ACADEMIC PRESS • NEW YORK AND LONDON

REF
QC
168
.M27
cop. 1

COPYRIGHT © 1973, BY ACADEMIC PRESS, INC.
ALL RIGHTS RESERVED.
NO PART OF THIS PUBLICATION MAY BE REPRODUCED OR TRANSMITTED IN ANY FORM OR BY ANY MEANS, ELECTRONIC OR MECHANICAL, INCLUDING PHOTOCOPY, RECORDING, OR ANY INFORMATION STORAGE AND RETRIEVAL SYSTEM, WITHOUT PERMISSION IN WRITING FROM THE PUBLISHER.

ACADEMIC PRESS, INC.
111 Fifth Avenue, New York, New York 10003

United Kingdom Edition published by
ACADEMIC PRESS, INC. (LONDON) LTD.
24/28 Oval Road, London NW1

LIBRARY OF CONGRESS CATALOG CARD NUMBER: 72-77330

PRINTED IN THE UNITED STATES OF AMERICA

Sci
R

CONTENTS

Chapter Three FLUID FLOW KINEMATICS

Chapter Four THE EQUATIONS OF FLUID MOTION

Chapter Five VORTEX DYNAMICS

PART II INSTABILITY OF BOUNDARY LAYER FLOWS

PART III TURBULENT FLOWS

PART IV STRUCTURE OF TURBULENCE

Chapter Nine COMPRESSIBLE FLUID FLOW

Chapter Ten PARTICLE FLUID DYNAMICS

Chapter Eleven HYDRODYNAMICS OF SUPERFLUIDS

Appendix VECTOR OPERATIONS AND IDENTITIES

PREFACE

One of the prime objectives in writing this textbook has been to provide a course in fluid dynamics which reflects its origins and the future development of the subject. Originally termed hydrodynamics, this subject was an established part of the physics curriculum. With the emphasis on nuclear physics and the development of aerodynamics and hydraulics, the subject was largely dropped and became the prerogative of the engineering curriculum. To quote G. Birkhoff:

> It seems surprising that this subject [fluid dynamics] to which so much was contributed by Stokes, Helmholtz, Rayleigh, and other physicists in the 19th century should owe all its more recent progress to mathematicians and engineers.†

Indeed, the subject has tended to become a branch of mechanics, or in the mathematical area—a branch of continuum mechanics.

The emergence of such fields as environmental physics, quantum fluids, and biophysics is now emphasizing the necessity to reestablish fluid dynamics as part of the physics curriculum (undergraduate and graduate).

Recognizing this situation, the text was formulated along the following lines:

(1) A Newtonian viscous fluid is dealt with almost exclusively, but its relationship to a general fluid is established in the introductory section of Chapter 1.

† From G. Birkhoff, "Hydrodynamics." Princeton Univ. Press, Princeton, New Jersey, 1950.

(2) To emphasize the underlying physics, Chapter 2 is devoted to the physical properties of fluids.

(3) Chapter 5 is also devoted to the important subject of vortex dynamics.

(4) An adequate course in the dynamics of real (viscous) fluids is given [kinematics (Chapter 3), equations of motion (Chapter 4), boundary-layer theory (Chapter 6), and compressible flow (Chapter 9)].

(5) Chapter 7 is devoted to slow viscous flow.

(6) The stability of flow is carefully considered in Chapter 8 and the important concept of the eddy coefficient of viscosity is introduced. Since inhomogeneous turbulence is still an area of uncertainty, it was decided to include only homogeneous isotropic turbulence.

(7) Chapter 10 is devoted to particulate fluid dynamics. This subject involves very complex problems and is very much in the development stage. The material presented has been carefully selected from recently published papers. Only suspensions of spherical, rigid, noninteracting particles are dealt with, since within these restrictions the procedures used and the results available can be taken to be reliable and likely to stand the test of time.

(8) The hydrodynamic theory of superfluid helium, based on the concepts of two mutually interacting fluids and quantized vortices, is now fairly well established and so a chapter on this subject is also included (Chapter 11).

(9) Part of the second chapter deals with the concept of similarity and dimensionality. This has proved very useful in fluid dynamics, especially in the identification of dimensionless groups such as the Reynolds and Prandtl numbers.

(10) Vector and indicial (tensor) notation has been introduced in Chapter 1 and is used throughout the test.

The first nine chapters, then, form a concise and logically developed course in contemporary Newtonian fluid dynamics, suitable for physics and engineering science students. This material could be covered in fifty one-hour lecture periods, probably at the junior or senior undergraduate level. These lectures should be accompanied by a carefully chosen set of laboratory experiments and illustrated by use of the many beautiful films on fluid dynamic topics that are currently available. The problem sets at the end of each chapter in this book should prove challenging and stimulating to the student.

The more specialized material of Chapter 10 and 11 should prove useful as a basis for graduate level instruction and reading.

The text is based on lectures presented at Trinity College, Dublin, Dartmouth College, Hew Hampshire, Oakland University, Michigan, and University College, Cork, which were given over a ten-year period.

The authors were first brought into association with each other through a research project on the dynamics of fluid jets sponsored byDrs.Wolfson and Masi of the Energetics Division, Air Force Office of Scientific Research, Washington, D.C.

Our thanks are also due to Miss Kilbride of University College, Cork, who typed most of the manuscript, and to the excellent work of the staff at Academic Press.

LIST OF NOTATION

CHAPTER 1

d_{ij}	element of deformation rate tensor
$\mathbf{F}$	vector field
$\mathbf{n}$	normal vector
n	exponent
p	hydrostatic pressure
$\mathbf{r}$	position vector
S	area
t_{ij}	element of stress tensor
$\mathbf{u}$	unit vector
$v_{i,j}$	velocity gradient tensor component
$\mathbf{v}$	velocity vector
V	volume
$\cdot\,,$	covariant derivative
δ_{ij}	Dirac delta function
ε_{ijk}	permutation symbol
Γ	circulation
φ	scalar field
φ_1	coefficient of viscosity
φ_3	coefficient of cross-viscosity
ϑ	angle
ψ	flux
ω	angular velocity

CHAPTER 2

a_x, a_y, a_z	Cartesian components of acceleration
C	molecular concentration; potential energy
C_p	specific heat at constant pressure
C_v	specific heat at constant volume
E	kinetic energy
$\mathbf{F}$	force vector
g	gravitational acceleration
Fr	Froude number

CHAPTER 2 (*cont.*)

G	universal constant of gravity
I	moment of inertia
k	Boltzmann constant; transport coefficient
Kn	Knudsen number
l, m, n	direction cosines
L	characteristic length
L (or λ)	molecular mean free path
Le	Lewis number
m	mass of molecule
N_0	Avogadro's number
Nu	Nusselt number
p	hydrostatic pressure
P	pressure
Pr	Prandtl number
R	gas constant; radius of curvature
T	temperature
u, v, w	Cartesian components of velocity
V	volume
γ	C_p/C_v
δ_{ij}	Kronecker delta
ϑ	angle
μ	coefficient of viscosity
ν	kinematic viscosity (μ/ϱ)
π	dimensionless variable
ϱ	density
σ	diagonal component of stress tensor; mean molecular cross section; surface tension
τ	off diagonal component of stress tensor
ω	angular velocity

CHAPTER 3

$\mathbf{a}$	acceleration vector
$\mathbf{A}_D$	vector potential
$\dot{e}_{ij}$	components of rate of strain tensor
Fr	Froude number V^2/gL
g	gravitational constant
i	the imaginary coefficient $(\sqrt{-1})$
L	characteristic length
$\mathbf{n}$	normal vector
p	pressure
q	flow speed
$\mathbf{r}$	source position vector
$\mathbf{r}'$	field position vector
S	radial distance
T	rate of strain tensor
u, v, w	Cartesian velocity components
U	stream velocity
$\mathbf{v}$	velocity vector
V	characteristic velocity; volume
Γ	circulation
ε_{ijk}	permutation symbol
$\boldsymbol{\zeta}$	vorticity vector
ϑ	fluid dilation; angle
$\varkappa$	doublet strength
μ	source strength
π	Bernoulli constant
ϱ	fluid density

CHAPTER 3 (*cont.*)

σ	$\mid \mathbf{r} - \mathbf{r}' \mid$
φ	velocity potential
ψ	stream function
ψ_S	Stokesian stream function
$\boldsymbol{\omega}$	angular velocity vector

CHAPTER 4

A	area (scalar)
$\mathbf{A}$	area (vector)
c	molecular concentration
C_p	specific heat at constant pressure
C_v	specific heat at constant volume
D	drag force per unit length
D_{12}	coefficient of binary diffusion
E	internal energy per unit mass
F_B	body force
F_I	inertial force
h	enthalpy per unit mass
$\mathbf{i}, \mathbf{j}, \mathbf{k}$	unit vectors (Cartesian)
k	thermal conductivity coefficient
L	length; lift force per unit length
M	turning moment
$\mathbf{n}$	normal vector
p	hydrostatic pressure
P	dimensionless pressure gradient
Q	heat added to unit mass; rate of flow
$\mathbf{r}$	position vector
R	gas constant
Re	Reynolds number
S_x, etc.	surface force components
t	time
T	temperature
u, v, w	Cartesian velocity components
v_r, v_ϑ, v_z	cylindrical velocity components
$\mathbf{v}$	velocity vector
V	volume
W	mechanical work done per unit mass
$\mathbf{X}$	body force per unit mass
Γ	circulation
ε_{ij}	components of rate of strain tensor
$\boldsymbol{\zeta}$	vorticity vector
η	dimensionless variable
ϑ	angle
$\varkappa$	vortex strength
μ	shear viscosity; line source strength
μ_B	bulk viscosity
ν	kinematic viscosity (μ/ϱ)
ϱ	fluid density
σ_{ij}	shear stress components
φ	velocity potential function
Φ	dissipation function
χ	force potential function
ψ	stream function
ω	angular velocity
Ω	vorticity function; complex potential

CHAPTER 5

a	radius of vortex core
$\mathbf{A}$	vector potential
E	energy (of a vortex ring)
f_x, f_y, f_z	components of impulsive force
g	gravitational constant
$\mathbf{i}$	current vector
I	impulse (of a vortex ring)
J_0	Bessel function of the first kind of zero order
J_1	Bessel function of the first kind of first order
p	pressure
q	tangential velocity
r_0	radius (of a vortex ring)
R	curvature; distance
$\mathbf{s}$	vector length
$\mathbf{S}$	vector area
T	temperature
V	force potential
w	specific volume of the fluid ($1/\varrho$)
$w(z)$	complex potential
z^*	the complex conjugate of z
γ	specific heat ratio
Γ, $\varkappa$	circulation or vortex strength
$\boldsymbol{\zeta}$	vorticity vector
η	dimensionless distance
ϑ	dilation in the field
μ	permeability of medium
ν	kinematic viscosity of the fluid
ϱ	mass density of the fluid
σ	cross section
φ	scalar potential
ψ	stream function
ψ_S	Stokesian stream function
$\boldsymbol{\omega}$	angular velocity vector
Ω	complex potential

CHAPTER 6

a	radius of cylinder
A	constant of integration
B	constant of integration
C	constant in the Rayleigh analogy
C_p	specific heat at constant pressure
Ei	exponential integral
f, F	functions
k	equals $U_0/2\nu$
L	length of body in x direction
M	momentum flux; Mach number
p, q	indices in the similarity variables
P	thermodynamic free energy
Pr	Prandtl number
R	gas constant
Re_L	Reynolds number based on length L
Re_x	Reynolds number based on x

CHAPTER 6 (*cont.*)

T	absolute temperature
u, v, w	velocity components in x, y, z directions
$\mathbf{u}$	velocity vector
U_0	constant velocity in x direction
x, y, z	Cartesian coordinates
X, Y	dimensionless variables in x, y directions
z	dimensionless variables in y direction
β	dimensionless variable equal to $\log(4\nu x/U_0 a^2)$
γ	Euler's constant 0.5772
δ	boundary layer thickness
δ_1	displacement thickness
ε	boundary layer thickness (order of magnitude)
ζ	z component of vorticity
$\boldsymbol{\zeta}$	vorticity vector
η	similarity variable
ϑ	cylindrical polar coordinate in angular direction
K_H	thermal diffusivity
μ	coefficient of viscosity
ν	coefficient of kinematic viscosity
ξ	similarity variable
ϱ	density
τ	skin friction
τ_{ij}	shear stresses
ψ	stream function

CHAPTER 7

a	radius
A	constant of integration
A_n	constant of integration
b	radius
B	constant of integration
B_n	constant of integration
C	constant of integration
C_n	constant of integration
d	separation distance between obstacles
D	constant of integration
D_n	constant of integration
E_n	constant of integration
f	function
F	function
G	function
h	distance between two parallel walls
H	function
i	$\sqrt{-1}$
$\mathbf{i}, \mathbf{j}, \mathbf{k}$	unit vectors in x, y, z directions
$\mathcal{I}$	imaginary part
k	$U_0/2\nu$
K_n	modified Bessel function of second kind of degree n
L	characteristic length
$\hat{\mathbf{n}}$	unit vector normal to a surface
p	pressure
p_n	solid spherical harmonic of degree n
p_∞	pressure at infinity
P_n	Legendre polynomial of degree n

CHAPTER 7 (*cont.*)

Q	volume flux
r	polar coordinate
r, ϑ	cylindrical polar coordinates
r, ϑ, φ	spherical polar coordinates
$\hat{\mathbf{r}}, \hat{\boldsymbol{\vartheta}}, \hat{\boldsymbol{\varphi}}$	unit vectors in spherical polar coordinates
$\mathcal{R}$	real part
Re_L	Reynolds number based on L
R_x, R_y	components of force in x, y directions
$\mathbf{S}$	surface area
t	independent variable
u'	perturbation velocity
$\mathbf{u}$	velocity vector
u, v, w	components of velocity in x, y, z directions
u_r, u_ϑ	cylindrical polar components of velocity
$u_r, u_\vartheta, u_\varphi$	spherical polar components of velocity
$\mathbf{u}_\mathrm{R}$	rotational component of velocity
V_n	solid spherical harmonic of degree n
W_n	solid spherical harmonic of degree n
x, y, z	rectangular Cartesian coordinates
z	$x + iy$
$\bar{z}$	$x - iy$
α_n	$[(n^2\pi^2/d^2) + k^2]^{1/2}$
Γ	circulation
ζ	component of vorticity in z direction
$\boldsymbol{\zeta}$	vorticity vector
η	$Kr\vartheta^2/2$
μ	coefficient of viscosity
ν	coefficient of kinematic viscosity
ϱ	density
φ	potential
ψ	stream function

CHAPTER 8

a	constant
A	constant of integration; function
B	function
B_{ik}, B_{ikl}	correlation tensors
c	complex wave velocity
c_i	imaginary part of c
E	energy
f	function
F	function; spectral distribution function
G	pressure gradient; differential operator
h	separation distance
H	differential operator
i	$\sqrt{-1}$
$\mathbf{i}, \mathbf{j}, \mathbf{k}$	unit vectors in x, y, z directions
$\mathcal{I}$	imaginary part
k	dimensionless constant
L	characteristic length; differential operator
$\tilde{L}$	complex conjugate of operator L
L, M, N, P	exponents

CHAPTER 8 (*cont.*)

M_0	momentum flux
n	angular frequency
p	pressure
$\bar{p}$	average pressure
p'	fluctuating pressure
p_0	mean outer pressure
p_1'	fluctuating part of outer pressure
P	pressure gradient
$\mathbf{r}$	radius vector
R	radius of curvature
Re_L	Reynolds number based on L
$\mathrm{Re}_{\mathrm{crit}}$	critical Reynolds number
t	time
T	time scale
$\mathbf{u}$	velocity vector
$\bar{\mathbf{u}}$	mean velocity
$\mathbf{u}'$	fluctuating part of velocity
u, v, w	rectangular components of velocity
u', v', w'	fluctuating components of velocity
$\bar{u}, \bar{v}, \bar{w}$	mean components of velocity
U_0	constant velocity
$U_0(x)$	outer mean velocity in x direction
U_1'	outer fluctuating velocity in x direction
x, y, z	rectangular Cartesian coordinates
X	similarity variable in x direction
y_c	critical value of y
α	wave number
β	amplification factor
Γ	shear stress
δ	boundary layer thickness
δ_0	fluctuating boundary layer thickness
δ_{ik}	unit tensor
ε	exchange coefficient
ζ	component of vorticity in z direction
$\bar{\zeta}$	mean component of vorticity in z direction
ζ'	fluctuating component of vorticity in z direction
η	similarity variable
μ	coefficient of viscosity
$\bar{\mu}$	dimensionless number
ν	coefficient of kinematic viscosity
ϱ	density
σ	dimensionless separation distance; scale factor
τ_{xy}	shear stress coefficients
φ	variable in Orr–Sommerfeld equation
$\tilde{\varphi}$	complex conjugate of φ
φ_i	solution of Orr–Sommerfeld equation
$\varphi_{1n}, \varphi_{2n}$	solutions of inviscid Orr–Sommerfeld equations
ψ	stream function
ψ'	fluctuating part of stream function

CHAPTER 9

a, c	wave (or sound) speed
$b(T)$	second virial coefficient
C_{p}	specific heat at constant pressure
C_{v}	specific heat at constant volume
e	internal energy
F	free energy
G	Gibbs free energy
h	specific enthalpy
H	enthalpy
k	coefficient of thermal conductivity
m	mass of one molecule
M	Mach number
p	pressure
Q	heat energy
R	gas constant
$\mathscr{R}$	universal gas constant
s	entropy per unit mass
S	entropy
T	temperature
T_0	stagnation temperature
T^*	throat temperature
U	stream velocity
v	volume; velocity
V	volume
α	degree of dissociation
β	coefficient of thermal expansion of the fluid
γ	specific heat ratio
ϑ_D	characteristic temperature
K_T	isothermal compressibility of the fluid
μ	coefficient of shear viscosity
ϱ_D	density
$\boldsymbol{\zeta}$	vorticity vector

CHAPTER 10

c	volume concentration of solids
C_{D}	drag coefficient
D	subscript meaning drag force
$\mathscr{D}$	drag force
e_{ij}	components of strain tensor
f	bodyforce per unit mass
f	subscript meaning fluid phase
$\mathfrak{F}_{\mathrm{L}}$	magnus lift force
$\mathfrak{F}_{\mathrm{T}}$	transverse force
I	moment of inertia
m	mass
n	particle number density
$\mathbf{n}$	normal vector
p	pressure
p	subscript meaning particle phase
Re	Reynolds number
$\mathfrak{T}$	torque force
u, v	Cartesian velocity components
α	particle concentration by volume
λ	slip relaxation length

CHAPTER 10 (*cont.*)

Λ	average velocity difference	ξ	dimensionless distance
μ	coefficient of shear viscosity	ϱ	density
ν	kinematic viscosity	τ_p	slip relaxation time for particle phase
		$\boldsymbol{\omega}, \boldsymbol{\Omega}$	angular velocity

CHAPTER 11

C_p	specific heat at constant pressure	S	entropy
E	energy	T	temperature
$\mathbf{F}$	force vector	U	velocity; internal energy density of the fluid
h	Planck's constant		
$\hbar$	$h/2\pi$		
m	mass	Γ	density ratio
M	angular momentum	ε	energy
n	subscript meaning normal fluid component	$\boldsymbol{\zeta}$	vorticity vector
p	pressure; magnitude of linear momentum vector	η	coefficient of shear viscosity
$\mathbf{p}$	linear momentum vector	$\varkappa$	vortex strength, or circulation
$\mathbf{q}$	heat flux vector	λ	source density
R	radius	μ	effective roton mass
s	subscript meaning superfluid component	ϱ	density of fluid (total)
		Ψ	wave function
		ω_0	angular velocity
		Ω	potential field function

CHAPTER ONE *INTRODUCTION AND MATHEMATICAL BACKGROUND*

1.1 Introduction

Fluid dynamics concerns itself with the investigation of the motion and equilibrium of fluids and is one of the oldest branches of physics.

The first notable works on fluids appeared in the early seventeenth century. Descartes in his "Principia Philosophiae" (1644) considered that space was filled with frictional vortices, so that the planets are carried along by the vortex motion. In Newton's "Principia," Book II, published in 1687, he deals with the influx of a fluid, the resistance of a fluid and pendulums, and the resistance of projected bodies and wave motion. He deduced that the speed of sound was 979 ft/sec, having made the momentous discovery that sound is a wave motion. He reached the conclusion that Descartes' hypotheses of vortexes in space is unreconcilable with astronomical phenomena. He showed that the motion of a sphere in an infinite vortex would give a period proportional to the square of the distance from the center rather than Kepler's law (which relates the square of the period to the cube of the distance). Moreover, he showed that the fluid resistance would cause a spiraling motion and therefore an unstable planetary system.

The name hydrodynamics is due to Daniel Bernoulli, who in 1738 used it in his famous book "Hydrodynamica" to indicate the combination of hydrostatics and hydraulics (the consideration of fluid motions). He determined the relation between pressure and velocity and formalized this in his famous theorem. But he did not have the true concept of internal pressure. In 1743 his son John Bernoulli properly introduced internal pressure and applied the momentum principle to infinitesimal elements.

Leonard Euler realized the significance of the Bernoullis' work and used it as the basis for his hydrodynamics, setting down the fundamental equations of motion and of continuity for an ideal inviscid fluid in 1755. An alternate form of these equations was stressed by Joseph Louis Lagrange in 1781 and 1789.

Up to this stage, the conditions of uniform density, incompressibility, inelasticity, and no free surfaces had been imposed. The directions for generalization of the equations were clear.

Such equations, including viscosity, were derived by Navier in 1822 and Stokes in 1845.

In 1858 Herman von Helmholtz published a famous paper on vortex motion and in 1868 another on free streamline potential flow. In 1873 he introduced the dimensional-analysis approach in which the similarity parameters are developed from the equations of motion. This really concluded the development of the classical theory of fluid dynamics.

The classical theory involved two extremes—both ideal fluids. One is a fluid with inertia and negligible viscosity, and the other is a fluid with viscosity but negligible inertia. Such ideal fluid concepts facilitate mathematical treatment. But it must be realized that such fluids *are* ideal and do not exist in nature.

Near the end of the nineteenth century, fluid flow itself began to be extensively observed and investigated. Osborne Reynolds in 1883 researched into the relation between turbulence and instability of fluid flow. It is worth noting that today the problem is to explain laminar flow, which is a much less probable state.

In 1904, Ludwig Prandtl, a German professor of mechanics, formulated the boundary layer concept. The boundary layer links potential flows and viscous fluid flows by postulating a thin region near the boundary in which the fluid velocity with respect to the boundary drops to zero. It explained the presence of drag on bodies moving through fluids.

The concept of the fluid continuum provides a theoretical model which lends itself to mathematical analysis, and so has become useful in several branches of physics. Max Born pointed out that the dense swarms of electrons in a metal could be regarded as an "electron gas." Another, recent, development is the liquid model for an atomic nucleus which consists of protons and neutrons. The absorption of neutrons causes the liquid to heat up, with the possibility of particle emission—analogous to evaporation. In statistical mechanics, phase points in the space can be conveniently regarded as a phase fluid which is subject to the laws of continuity and conservation.

The physics of real fluids is still a developing subject. A central problem is that of stability. Is fluid flow stable to infinitely small disturbances? This is a fundamental question. The subject is full of peculiarities. A small change in fluid viscosity can produce a large change in the solution of a partial differential equation describing the flow system. But a small change in compressibility normally does not do this. Such questions are still unanswered.

The problems of the viscosity of solutions and of suspensions of particles in fluids are still not resolved.

It is found that ultrasonic waves decay much faster in fluids than shear viscosity would allow. This apparently is connected with the bulk viscosity of the fluid.

There then is the question of the behavior of real fluids under extreme physical conditions. In high speed flow, compressibility becomes important. Associated with supersonic flow is the shock wave and the interactions between shock waves.

Another unusual physical condition is that of low density—for example, the case of a rarefied gas involving slip flow. This involves the relationship between fluid dynamics and kinetic theory.

Finally, there is the quantum fluid, such as liquid helium. The quantum effects are truly macroscopic ones, however. For example, the "fountain effect" is caused by the fact that the momentum process dominates the thermal effects and produces a flow from low temperature regions to high temperature ones.

The difference between solids and fluids can be understood in terms of internal stresses and strains in elastic media. The stress in a linear elastic rod is proportional to its strain, while the stress in a fluid is proportional to its true rate of strain (a Newtonian fluid). If the stress applied to a solid is below the yield stress, the deformation produced vanishes when the stress is removed; otherwise, it is left with a permanent set, or it may crack. If a shear force is applied to a fluid, the fluid will deform continuously no matter what the magnitude of the force. This continuous deformation under the action of a shear stress is manifested in the flow of the fluid.

Very strong intermolecular forces (attractive) exist in solids and give them their rigidity. These forces are weaker in liquids and very small in gases. Thus, the liquid molecules can move around relatively freely and endow the liquid with the facility to flow easily in response to a shear stress.

Although the continuum premise is widely used in fluid dynamics, on occasions the true microscopic nature of fluids must be considered. As early as 1822, C. H. Navier considered forces of molecular interaction in deriving the equations of viscous fluid flow. The explanation for the transport properties of fluids (diffusion of mass, conduction of heat, and transport of momentum) lies firmly in molecular theory. One must resort to the microscopic level also in dealing with dissociation and ionization in flowing gases.

1.2 Mathematical Background

1.2.1 Vectors and Cartesian Tensors

It is assumed that the student is already acquainted with Cartesian tensors and especially with the zero-order tensor or scalar and the first-order tensor or vector; also with vector operators such as grad and div, and tensor transformations. So this section will be largely devoted to writing these in the index, or indicial, notation.

1.2.2 Indicial Notation of a Vector

The symbol A_i represents the set of three Cartesian components

$$(A_1, A_2, A_3).$$

It designates the vector $\mathbf{A}$ in ordinary three-dimensional space, where

$$\mathbf{A} = A_1\mathbf{e}_1 + A_2\mathbf{e}_2 + A_3\mathbf{e}_3. \tag{1.1}$$

One recalls that (A_1, A_2, A_3) are the rectangular projections of **A** in the (x_1, x_2, x_3) directions, respectively, and $\mathbf{e}_1, \mathbf{e}_2, \mathbf{e}_3$ are the corresponding unit vectors (Fig. 1.1).

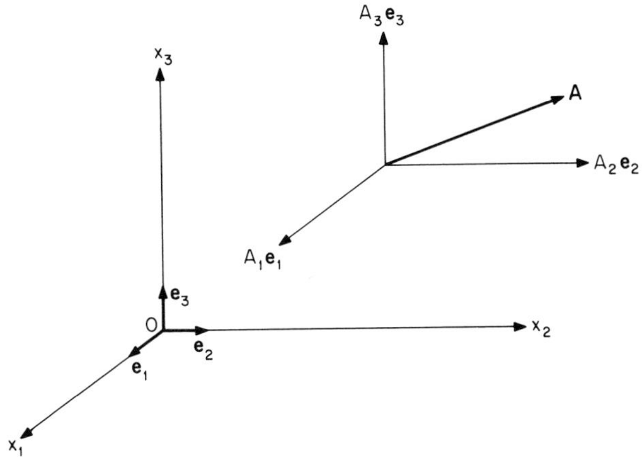

Fig. 1.1

1.2.3 Tensor Notation

It has been already mentioned that a scalar quantity is a zero-order tensor and a vector a first-order tensor. In fact, the order of a tensor is defined as the number of indices appearing with the letter,

$$\begin{aligned} A &\Rightarrow \text{a tensor of zeroth order (scalar)},\\ A_i &\Rightarrow \text{a tensor of first order (vector)},\\ A_{ij} &\Rightarrow \text{a tensor of second order (matrix)},\\ A_{ijk} &\Rightarrow \text{a tensor of third order,} \end{aligned}$$

and so on.

One notes that a tensor of order N (in three-dimensional space) has 3^N components. For example,

$$A_{ij} \Rightarrow \begin{pmatrix} A_{11} & A_{12} & A_{13} \\ A_{21} & A_{22} & A_{23} \\ A_{31} & A_{32} & A_{33} \end{pmatrix}$$

has $3^2 = 9$ components.

1.2.4 Summation Convention

Whenever a lower case Latin letter subscript occurs repeated in an expression, a summation over the range 1, 2, 3 is understood. For example, consider the vector transformation

$$A_i = a_{ij}A_j. \tag{1.2}$$

The j is repeated on the right-hand side, and so for $i = 2$, for example, this implies

$$A_2 = a_{21}A_1 + a_{22}A_2 + a_{23}A_3$$

or

$$A_iB_iC_j = (A_1B_1 + A_2B_2 + A_3B_3)C_j.$$

It is clearly meaningless to have more than two identical subscripts in any single term. As repeated subscripts get "summed out," they are called dummy subscripts.

1.2.5 Kronecker Delta

The symbol δ_{ij} is called the Kronecker delta and is defined as follows:

$$\begin{aligned} \delta_{ij} &= 1 \qquad \text{if} \quad i = j, \\ &= 0 \qquad \text{if} \quad i \neq j. \end{aligned} \tag{1.3}$$

Thus

$$A_i = \delta_{ij}A_j.$$

As i and j can take on the values 1, 2, and 3 independently, there are nine components of δ_{ij} which can be written in matrix form:

$$\delta_{ij} = \begin{pmatrix} \delta_{11} & \delta_{12} & \delta_{13} \\ \delta_{21} & \delta_{22} & \delta_{23} \\ \delta_{31} & \delta_{32} & \delta_{33} \end{pmatrix} = \begin{pmatrix} 1 & 0 & 0 \\ 0 & 1 & 0 \\ 0 & 0 & 1 \end{pmatrix}. \tag{1.4}$$

Consider the term $A_i\,\delta_{ij}$. As i is repeated,

$$\begin{aligned} A_i\,\delta_{ij} &= A_1\,\delta_{1j} + A_2\,\delta_{2j} + A_3\,\delta_{3j} \\ &= X_j \qquad \text{(a vector).} \end{aligned}$$

The components of the vector are therefore

$$X_1 = A_1, \qquad X_2 = A_2, \qquad X_3 = A_3.$$

Thus

$$X_j = A_i\,\delta_{ij} = A_j. \tag{1.5}$$

This illustrates a very important property of the Kronecker delta: That when the Kronecker delta appears in a term of an indicial equation in which one (or both) of its indices is repeated in the other symbols of the same term, the delta can be removed if the repeated index on the other symbol is removed and replaced by the remaining index of the Kronecker delta.

Examples

$$\begin{aligned} C_q\,\delta_{rq} &= C_r, \\ M_{ij}B_j\,\delta_{ik} &= M_{kj}C_j, \\ A_rB_sC_t\,\delta_{sq} &= A_rB_qC_t, \\ \delta_{ij}\,\delta_{jk} &= \delta_{ik}, \\ B_jC_kA_s\,\delta_{rk}\,\delta_{sj} &= B_sC_rA_s = B_jC_rA_j. \end{aligned} \tag{1.6}$$

If both subscripts of δ are repeated in a term, then either one (but *not* both) can be used as the one to be removed. For example,

$$A_rB_sC_t\,\delta_{st} = A_rB_tC_t \qquad \text{or} \qquad A_rB_sC_t\,\delta_{st} = A_rB_sC_s. \tag{1.7}$$

Both results are identical as the repeated indices are summed out (they are dummy indices). In fact, both results represent

$$A_r(B_1C_1 + B_2C_2 + B_3C_3).$$

Note If the repeated index in δ_{ii} is summed out, then one gets

$$\delta_{ii} = 3. \tag{1.8}$$

1.2.6 The Scalar Product of Two Vectors

The scalar product of two vectors **A** and **B** is

$$\mathbf{A}\cdot\mathbf{B} = AB\cos\vartheta.$$

If the vectors are orthogonal, then

$$\mathbf{A} \cdot \mathbf{B} = 0,$$

and if they are parallel, then

$$\mathbf{A} \cdot \mathbf{B} = AB.$$

Now the vectors can be written

$$\mathbf{A} = A_i \mathbf{e}_{(i)}, \qquad \mathbf{B} = B_j \mathbf{e}_{(j)},$$

where A_i and B_j are scalars. Thus

$$\mathbf{A} \cdot \mathbf{B} \equiv A_i B_j \mathbf{e}_{(i)} \cdot \mathbf{e}_{(j)}.$$

Now

$$\begin{aligned} \mathbf{e}_{(i)} \cdot \mathbf{e}_{(j)} &= 1 \qquad \text{if} \quad i = j, \\ &= 0 \qquad \text{if} \quad i \neq j, \end{aligned}$$

or

$$\mathbf{e}_{(i)} \cdot \mathbf{e}_{(j)} = \delta_{ij}.$$

Therefore

$$\mathbf{A} \cdot \mathbf{B} = A_i B_j \, \delta_{ij} = A_i B_i = A_j B_j = A_1 B_1 + A_2 B_2 + A_3 B_3, \tag{1.9}$$

which is just the scalar product in terms of its Cartesian components.

The component of a vector A_i in a given direction is obtained by taking the scalar product of the vector A_i with a unit vector n_i in the given direction. Thus, the component of A_i in the direction of n_i is $A_i n_i$.

It is interesting to note that the quantity $A_i B_j C_j$ in vector notation is $\mathbf{A}(\mathbf{B} \cdot \mathbf{C})$.

1.2.7 Vector Product of Two Vectors

The vector product of two vectors **A** and **B**, denoted by **A × B**, is a vector **C** which is normal to both **A** and **B** such that **A**, **B**, and **C** form a right-handed system, and

$$C = AB \sin \vartheta, \tag{1.10}$$

where ϑ is the smaller of the two angles between the vectors A and B. The Cartesian components of $\mathbf{C}$ are given by

$$C_1 = A_2B_3 - A_3B_2, \quad C_2 = A_3B_1 - A_1B_3, \quad C_3 = A_1B_2 - A_2B_1. \tag{1.11}$$

To be able to write the vector product in indicial notation requires the introduction of a new symbol ε_{ijk}, the alternating unit tensor sometimes called the permutation symbol or the Levi–Civita tensor. It is defined as follows:

$$\begin{aligned} \varepsilon_{ijk} &= +1 \quad \text{if } i, j, k \text{ are in cyclic order, 123, 231, 312,} \\ &= -1 \quad \text{if } i, j, k \text{ are in noncyclic order, 321, 132, 213,} \\ &= \;\;\, 0 \quad \text{if any two subscripts are repeated.} \end{aligned} \tag{1.12}$$

Note Interchanging two adjacent subscripts changes the sign of ε, the alternating unit tensor.

To illustrate the properties of ε_{ijk}, consider the expression $R_{ij}=\varepsilon_{ijk}T_k$. There are two floating subscripts (i and j) on the right-hand side, and so there are nine components which can be arranged in a 3×3 matrix. Summing out the k's, one obtains

$$R_{ij} = \varepsilon_{ij1}T_1 + \varepsilon_{ij2}T_2 + \varepsilon_{ij3}T_3 = \varepsilon_{ijk}T_k. \tag{1.13}$$

Now some of the terms on the right-hand side will be zero for certain components of R_{ij}. Consider R_{11}. Now $\varepsilon_{11k} = 0$ for *all* k and so $R_{11} = 0$. In the case of R_{12}, the only term on the right which survives is $\varepsilon_{123}T_3 = T_3$ since $\varepsilon_{123} = +1$. Proceeding in this way the matrix R_{ij} is completed and is

$$R_{ij} = \varepsilon_{ijk}T_k = \begin{pmatrix} 0 & T_3 & -T_2 \\ -T_3 & 0 & T_1 \\ T_2 & -T_1 & 0 \end{pmatrix}. \tag{1.14}$$

Next consider $\varepsilon_{ijk}A_iB_jC_k$. The subscripts i, j, and k are all dummy subscripts which get summed out. The result will thus be a scalar. As i, j, k all get summed from 1 to 3 independently, the result will be the sum of 27 terms. But most of these terms will vanish since $\varepsilon_{ijk} = 0$ whenever any two of the subscripts are equal. The only terms surviving in the expression $\varepsilon_{ijk}A_iB_jC_k$ are the ones in which $i\,j\,k$ are in cyclic or noncyclic order. There are only six of these, namely,

$$\varepsilon_{123} = \varepsilon_{231} = \varepsilon_{312} = +1, \qquad \varepsilon_{321} = \varepsilon_{132} = \varepsilon_{213} = -1.$$

So the sum is

$$\varepsilon_{ijk}A_iB_jC_k = A_1B_2C_3 + A_2B_3C_1 + A_3B_1C_2 - A_3B_2C_1 - A_1B_3C_2 - A_2B_1C_3. \tag{1.15}$$

Now, the components of the vector product $\mathbf{C} = \mathbf{A} \times \mathbf{B}$ are

$$C_1 = A_2B_3 - A_3B_2, \qquad C_2 = A_3B_1 - A_1B_3, \qquad C_3 = A_1B_2 - A_2B_1.$$

Consider the expression $C_i = \varepsilon_{ijk}A_jB_k$. This is a vector quantity characterized by the single floating subscript i. To find C_1 let $i = 1$. In order that ε_{ijk} is not zero, the subscripts $j\,k$ must have the values 23 or 32. Since $\varepsilon_{123} = +1$ and $\varepsilon_{132} = -1$,

$$C_1 = A_2B_3 - A_3B_2. \tag{1.16}$$

Similarly, C_2 and C_3 can be found:

$$C_2 = A_3B_1 - A_1B_3, \qquad C_3 = A_1B_2 - A_2B_1. \tag{1.17}$$

Thus $\varepsilon_{ijk}A_jB_k$ gives the components of the vector product $\mathbf{A} \times \mathbf{B}$. Thus, in indicial notation, the vector product $\mathbf{A} \times \mathbf{B}$ is written as $\varepsilon_{ijk}A_jB_k$.

Expressions such as $\varepsilon_{ijk}A_iB_k$ can always be written in vector symbols by rotating the subscripts in cyclic order until the floating subscript (in this case j) appears first. Therefore

$$\varepsilon_{ijk}A_iB_k = \varepsilon_{jki}A_iB_k = \varepsilon_{jki}B_kA_i, \tag{1.18}$$

which is the indicial notation for $\mathbf{B} \times \mathbf{A} = -(\mathbf{A} \times \mathbf{B})$.

Writing $\mathbf{C} = \mathbf{B} \times \mathbf{A}$ or $C_j = \varepsilon_{jki}B_kA_i$, one has $\mathbf{C} \cdot \mathbf{D} = (\mathbf{B} \times \mathbf{A}) \cdot \mathbf{D}$ which in the indicial notation is

$$C_jD_j = \varepsilon_{jki}B_kA_iD_j. \tag{1.19}$$

It can be shown that a very simple and useful relationship exists between the alternating unit tensor and the Kronecker delta. It is

$$\varepsilon_{ijk}\varepsilon_{irs} = \delta_{jr}\,\delta_{ks} - \delta_{js}\,\delta_{kr}. \tag{1.20}$$

Consider its use in proving the vector identity

$$\mathbf{A} \times (\mathbf{B} \times \mathbf{C}) = (\mathbf{C} \cdot \mathbf{A})\mathbf{B} - (\mathbf{B} \cdot \mathbf{A})\mathbf{C}.$$

In indicial notation,

$$\begin{aligned}\mathbf{A} \times (\mathbf{B} \times \mathbf{C}) &= \varepsilon_{rsi} A_s (\varepsilon_{ijk} B_j C_k) \\ &= \varepsilon_{irs} \varepsilon_{ijk} B_j C_k A_s \\ &= (\delta_{rj}\, \delta_{sk} - \delta_{rk}\, \delta_{sj}) B_j C_k A_s .\end{aligned}$$

The deltas are removed when the repeated indices are erased. Therefore

$$\varepsilon_{rsi} A_s (\varepsilon_{ijk} B_j C_k) = B_r C_s A_s - B_s C_r A_s = C_s A_s B_r - B_s A_s C_r .$$

In vector notation this is

$$\mathbf{A} \times (\mathbf{B} \times \mathbf{C}) = (\mathbf{C} \cdot \mathbf{A})\mathbf{B} - (\mathbf{B} \cdot \mathbf{A})\mathbf{C},$$

which was to be shown.

1.3 Vector Calculus

1.3.1 Differentiation of a Vector

The derivative of a scalar function $\varphi(t)$ with respect to t is given by

$$\frac{d\varphi(t)}{dt} = \lim_{\Delta t \to 0} \frac{\Delta\varphi}{\Delta t}, \tag{1.21}$$

where

$$\Delta\varphi = \varphi(t + \Delta t) - \varphi(t).$$

Similarly, the derivative of a vector function $\mathbf{A}(t)$ is given by

$$\frac{d\mathbf{A}}{dt} = \lim_{\Delta t \to 0} \frac{\Delta\mathbf{A}}{\Delta t}, \tag{1.22}$$

where

$$\Delta\mathbf{A} = \mathbf{A}(t + \Delta t) - \mathbf{A}(t).$$

But $\Delta\mathbf{A}$ can occur due to (i) a change in the magnitude of $\mathbf{A}$, (ii) a change in the direction of $\mathbf{A}$, or both. It is assumed that $\mathbf{A}(t)$ is a well-behaved function so that $\Delta\mathbf{A} \to 0$ as $\Delta t \to 0$.

If $\mathbf{r}$ is the displacement vector of a particle, then the velocity is given by

$$\mathbf{v} = \frac{d\mathbf{r}}{dt} = \lim_{\Delta t \to 0} \frac{\Delta \mathbf{r}}{\Delta t}. \tag{1.23}$$

For linear motion through the origin, $\Delta \mathbf{r}$ is due to change in the *magnitude* of $\mathbf{r}$. For uniform circular motion with the origin at the circle's center, $\Delta \mathbf{r}$ is due to a change in the *direction* of $\mathbf{r}$.

Let $\mathbf{u}_{(i)}$ be the unit vector in a Cartesian coordinate system so that the vector $\mathbf{A}$ can be written

$$\mathbf{A} = A_1\mathbf{u}_{(1)} + A_2\mathbf{u}_{(2)} + A_3\mathbf{u}_{(3)}. \tag{1.24}$$

Since the unit vectors are fixed,

$$\Delta \mathbf{A} = \Delta A_1\, \mathbf{u}_{(1)} + \Delta A_2\, \mathbf{u}_{(2)} + \Delta A_3\, \mathbf{u}_{(3)}.$$

Dividing by Δt and taking the limit as $\Delta t \to 0$, one obtains

$$\lim_{\Delta t \to 0} \frac{\Delta \mathbf{A}}{\Delta t} = \frac{d\mathbf{A}}{dt} = \frac{dA_1}{dt} \mathbf{u}_{(1)} + \frac{dA_2}{dt} \mathbf{u}_{(2)} + \frac{dA_3}{dt} \mathbf{u}_{(3)} \tag{1.25}$$

or, using the summation convention,

$$\frac{d\mathbf{A}}{dt} = \frac{dA_i}{dt} \mathbf{u}_{(i)}.$$

Thus the Cartesian components of $d\mathbf{A}/dt$ are given simply by dA_i/dt. Similarly, the Cartesian components of $d^2\mathbf{A}/dt^2$ are given by d^2A_i/dt^2.

In vector calculus it is advantageous to work with the indicial notation since the quantities are all really scalars and the rules of scalar algebra and calculus hold. For example, $(d/dt)(\mathbf{A} + \mathbf{B})$ can be written

$$\frac{d}{dt}(A_i + B_i) = \frac{dA_i}{dt} + \frac{dB_i}{dt};$$

therefore

$$\frac{d}{dt}(\mathbf{A} + \mathbf{B}) = \frac{d\mathbf{A}}{dt} + \frac{d\mathbf{B}}{dt}. \tag{1.26}$$

Indicially, $(d/dt)(a\mathbf{A})$ is

$$\frac{d}{dt}(aA_i) = a\frac{dA_i}{dt} + \frac{da}{dt} A_i,$$

so that

$$\frac{d}{dt}(a\mathbf{A}) = a\frac{d\mathbf{A}}{dt} + \mathbf{A}\frac{da}{dt}. \tag{1.27}$$

Similarly, $(d/dt)(\mathbf{A} \cdot \mathbf{B})$ is

$$\frac{d}{dt}(A_iB_i) = A_i\frac{dB_i}{dt} + B_i\frac{dA_i}{dt};$$

therefore

$$\frac{d}{dt}(\mathbf{A} \cdot \mathbf{B}) = \mathbf{A}\frac{d\mathbf{B}}{dt} + \mathbf{B}\frac{d\mathbf{A}}{dt}. \tag{1.28}$$

The derivative of the cross product $\mathbf{A} \times \mathbf{B}$ is more difficult. Indicially it is written $(d/dt)(\varepsilon_{ijk}A_jB_k)$. Since ε_{ijk} is constant,

$$\frac{d}{dt}(\varepsilon_{ijk}A_jB_k) = \varepsilon_{ijk}\frac{d}{dt}(A_jB_k) = \varepsilon_{ijk}\left(A_j\frac{dB_k}{dt} + B_k\frac{dA_j}{dt}\right);$$

therefore

$$\frac{d}{dt}(\mathbf{A} \times \mathbf{B}) = \mathbf{A} \times \frac{d\mathbf{B}}{dt} + \frac{d\mathbf{A}}{dt} \times \mathbf{B}. \tag{1.29}$$

1.3.2 Variable Unit Vectors

Thus far the unit vectors have been considered to be fixed. This is not always so. Consider a particle moving in uniform circular motion in the xy plane. A pair of orthogonal unit vectors $\mathbf{u}_r$ and $\mathbf{u}_\vartheta$ is introduced in the directions of increasing r and ϑ, which move around with the particle (Fig. 1.2). The position vector is $\mathbf{r} = r\mathbf{u}_r$. Since r is a constant for circular motion,

$$\text{velocity} = \mathbf{v} = \frac{d\mathbf{r}}{dt} = r\frac{d\mathbf{u}}{dt}.$$

In moving from point P_1 to point P_2, $\mathbf{r}$ changes to $\mathbf{r} + \Delta\mathbf{r}$ and $\mathbf{u}_r$ to $\mathbf{u}_r + \Delta\mathbf{u}_r$. If $\Delta\vartheta$ is small, the magnitude of $\Delta\mathbf{u}_r$ is approximately $\Delta\vartheta$ and approaches this exactly as $\Delta\vartheta \to 0$. Also, as $\Delta\vartheta \to 0$, the direction of $\Delta\mathbf{u}_r$ approaches the direction of $\mathbf{u}_\vartheta$. Hence we can write

$$\Delta\mathbf{u}_r = \Delta\vartheta\,\mathbf{u}_\vartheta$$

and

$$\frac{d\mathbf{u}_r}{dt} = \lim_{\Delta t \to 0}\frac{\Delta\mathbf{u}_r}{\Delta t} = \lim\frac{\Delta\vartheta}{\Delta t}\mathbf{u}_\vartheta = \frac{d\vartheta}{dt}\mathbf{u}_\vartheta;$$

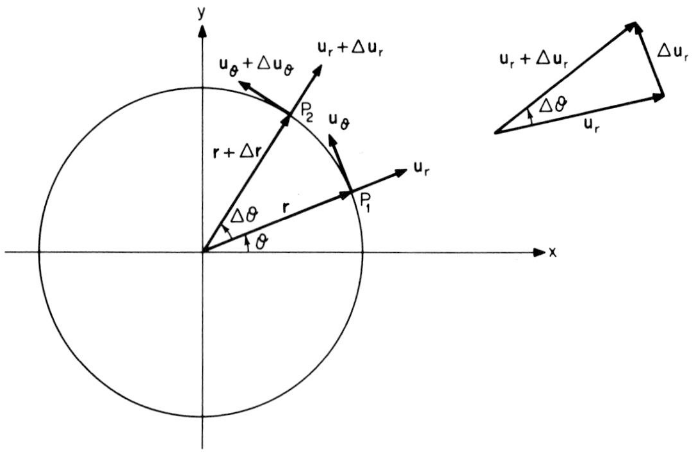

Fig. 1.2

therefore

$$\mathbf{v} = r\frac{d\mathbf{u}_r}{dt} = r\frac{d\vartheta}{dt}\mathbf{u}_\vartheta. \tag{1.30}$$

The angular velocity of the particle is defined as

$$\omega = \frac{d\vartheta}{dt};$$

therefore

$$\mathbf{v} = r\omega\mathbf{u}_\vartheta = v\mathbf{u}_\vartheta, \tag{1.31}$$

where v is a constant (speed) $= \omega r$. The acceleration of the particle is given by

$$\mathbf{a} = \frac{d\mathbf{v}}{dt} = v\frac{d\mathbf{u}_\vartheta}{dt}.$$

Now $\mathbf{u}_\vartheta$ changes to $\mathbf{u}_\vartheta + \Delta\mathbf{u}_\vartheta$ as the velocity vector is changing direction (Fig. 1.3). As $\Delta\vartheta \to 0$, the magnitude of $\Delta\mathbf{u}_\vartheta$ is $\Delta\vartheta$ and the direction is the direction of $\mathbf{u}_r$. Therefore

$$\Delta\mathbf{u}_\vartheta = -\Delta\vartheta\,\mathbf{u}_r,$$

and so

$$\mathbf{a} = v\frac{d\mathbf{u}_\vartheta}{dt} = -v\lim_{\Delta t\to 0}\frac{\Delta\vartheta}{\Delta t}\mathbf{u}_r = -v\frac{d\vartheta}{dt}\mathbf{u}_r$$

$$= -v\omega\mathbf{u}_r = -\frac{v^2}{r}\mathbf{u}_r. \tag{1.32}$$

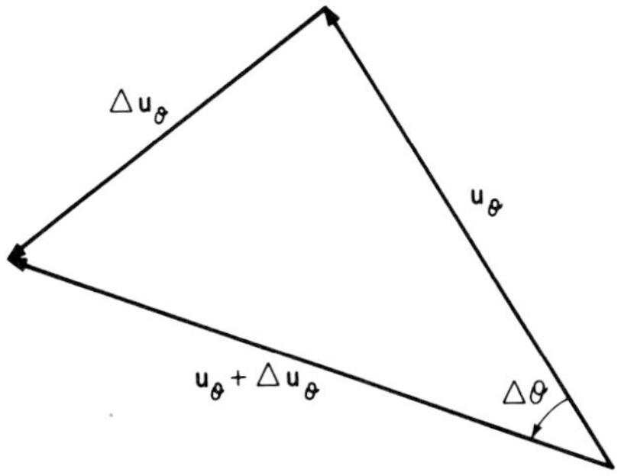

Fig. 1.3

Thus the acceleration of the particle is in the $-\mathbf{u}_r$ direction—toward the center of the circle.

Although the finite angular displacement ϑ is not a vector quantity, $\Delta\vartheta$ (in the limit as $\Delta\vartheta \to d\vartheta$) can be considered as a vector. Now

$$\mathbf{r} = r\mathbf{u}_r;$$

therefore

$$\Delta\mathbf{r} = r\,\Delta\mathbf{u}_r = r\,\Delta\vartheta\,\mathbf{u}_\vartheta.$$

Consider the vector product

$$\mathbf{r} \times \Delta\mathbf{r} = (r\mathbf{u}_r) \times (r\,\Delta\vartheta\,\mathbf{u}_\vartheta) = r^2\,\Delta\vartheta\,\mathbf{u}_r \times \mathbf{u}_\vartheta.$$

Now

$$\mathbf{u}_r \times \mathbf{u}_\vartheta = \mathbf{u}_z.$$

So one can define a vector $\Delta\boldsymbol{\vartheta} = \Delta\vartheta\,\mathbf{u}_z$ whose magnitude equals $\Delta\vartheta$ and which is directed along the axis of rotation in a right-handed sense. Therefore

$$\Delta\boldsymbol{\vartheta} = \Delta\vartheta\,\mathbf{u}_z = \frac{\mathbf{r} \times \Delta\mathbf{r}}{r^2}. \tag{1.33}$$

The angular velocity vector $\boldsymbol{\omega}$ is defined by

$$\boldsymbol{\omega} = \frac{d\boldsymbol{\vartheta}}{dt} = \lim_{\Delta t \to 0} \frac{\Delta\boldsymbol{\vartheta}}{\Delta t} = \lim_{\Delta t \to 0} \frac{\mathbf{r} \times (\Delta\mathbf{r}/\Delta t)}{r^2} = \frac{\mathbf{r} \times (d\mathbf{r}/dt)}{r^2};$$

therefore

$$\boldsymbol{\omega} = \frac{\mathbf{r} \times \mathbf{v}}{r^2}, \tag{1.34}$$

$$\boldsymbol{\omega} = \frac{(r\mathbf{u}_r) \times (\omega\mathbf{u}_\vartheta)}{r} = \omega(\mathbf{u}_r \times \mathbf{u}_\vartheta),$$

and

$$|\boldsymbol{\omega}| = \frac{v}{r}. \tag{1.35}$$

Now $\mathbf{v} = v\mathbf{u}_\vartheta$ and $v = r\omega$. Consider the cross product

$$\boldsymbol{\omega} \times \mathbf{r} = \lim_{\Delta t \to 0} \frac{\Delta \boldsymbol{\vartheta}}{\Delta t} \times \mathbf{r}.$$

Now

$$\Delta \boldsymbol{\vartheta} = \Delta \vartheta \, \mathbf{u}_z \qquad \text{and} \qquad \mathbf{r} = r\mathbf{u}_r;$$

therefore

$$\boldsymbol{\omega} \times \mathbf{r} = \lim \frac{\Delta \vartheta}{\Delta t} \mathbf{u}_z \times (r\mathbf{u}_r) = r \frac{d\vartheta}{dt} \mathbf{u}_z \times \mathbf{u}_r = r\omega \mathbf{u}_\vartheta = \mathbf{v},$$

and so

$$\mathbf{v} = \boldsymbol{\omega} \times \mathbf{r}. \tag{1.36}$$

1.3.3 Line Integrals

Suppose it is required to find the length of the curve shown in Fig. 1.4 between points A and B.

The curve can be divided into a series of short segments of length dl. It is seen that

$$dl^2 = dx^2 + dy^2 + dz^2;$$

therefore

$$dl = (dx^2 + dy^2 + dz^2)^{1/2}.$$

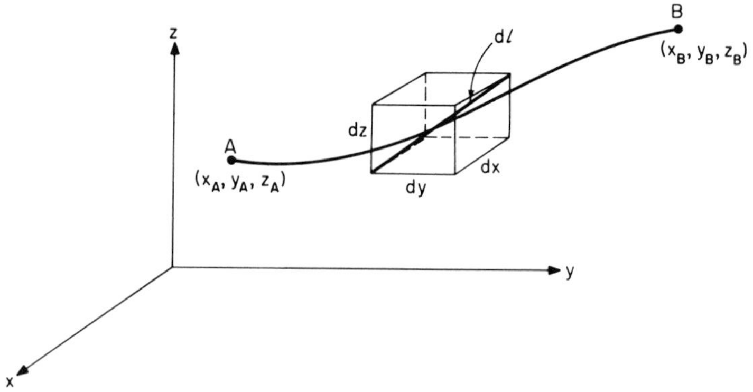

Fig. 1.4

The total path length L is found by integrating between points A and B:

$$L = \int_A^B dl = \int_A^B \left[1 + \left(\frac{dy}{dx}\right)^2 + \left(\frac{dz}{dx}\right)^2\right]^{1/2} dx. \qquad (1.37)$$

To evaluate the integral in Eq. (37), some equation representing the path along which the integration is to be carried out must be specified. The curve from A to B can be written in the form

$$f(x) = g(y) = h(z).$$

There are three equations giving the projections of the path from A to B on the three perpendicular planes xy, xz, and yz, as shown in

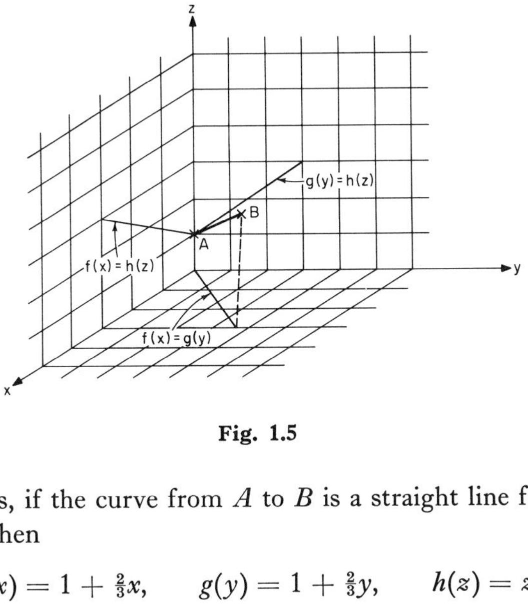

Fig. 1.5

Fig. 1.5. Thus, if the curve from A to B is a straight line from (0, 0, 1) to (3, 3, 3), then

$$f(x) = 1 + \tfrac{2}{3}x, \qquad g(y) = 1 + \tfrac{2}{3}y, \qquad h(z) = z.$$

The straight line is described by the equations

$$1 + \tfrac{2}{3}x = 1 + \tfrac{2}{3}y = z;$$

therefore $dy/dx = 1$ and $dz/dx = \frac{2}{3}$, so that the length of the curve is

$$L = \int_A^B dl = \int_0^B \left[1 + \left(\frac{dy}{dx}\right)^2 + \left(\frac{dz}{dx}\right)^2\right]^{1/2} dx$$

$$= \int_0^3 (1 + 1 + 1)^{1/2}\, dx = \sqrt{22}\,.$$

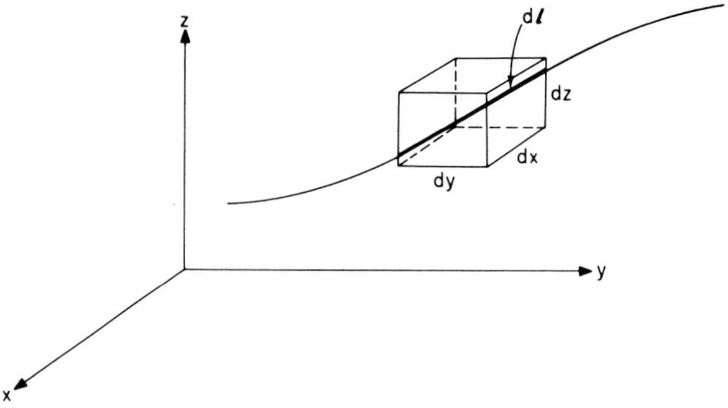

Fig. 1.6

Instead of the differential of arc length dl, it is often convenient to introduce the *vector* element $\mathbf{dl}$ (Fig. 1.6):

$$\mathbf{dl} = dx\,\mathbf{u}_x + dy\,\mathbf{u}_y + dz\,\mathbf{u}_z. \tag{1.38}$$

Obviously $dl = |\,\mathbf{dl}\,|$. The integral

$$\int_A^B \mathbf{dl} = \int_A^B dx\,\mathbf{u}_x + \int_A^B dy\,\mathbf{u}_y + \int_A^B dz\,\mathbf{u}_z \tag{1.39}$$

along a path from A to B is a *vector* quantity and is not the same as $\int_A^B dl$, which is the length of the path.

For example, the parametric equations for a circle with its center at the origin is

$$x = R\cos\vartheta, \qquad \text{thus} \quad dx = -R\sin\vartheta\,d\vartheta,$$
$$y = R\sin\vartheta, \qquad \text{thus} \quad dy = R\cos\vartheta\,d\vartheta;$$

therefore

$$dl = R\,d\vartheta.$$

The length of a semicircle is then

$$L = \int_A^B dl = \int_{-\pi/2}^{+\pi/2} R\,d\vartheta = \pi R.$$

But the *vector* $\mathbf{L}$ is (Fig. 1.7)

$$\mathbf{L} = \int_A^B \mathbf{dl} = \int_{x_A}^{x_B} dx\,\mathbf{u}_x + \int_{y_A}^{y_B} dy\,\mathbf{u}_y = 2R\mathbf{u}_y.$$

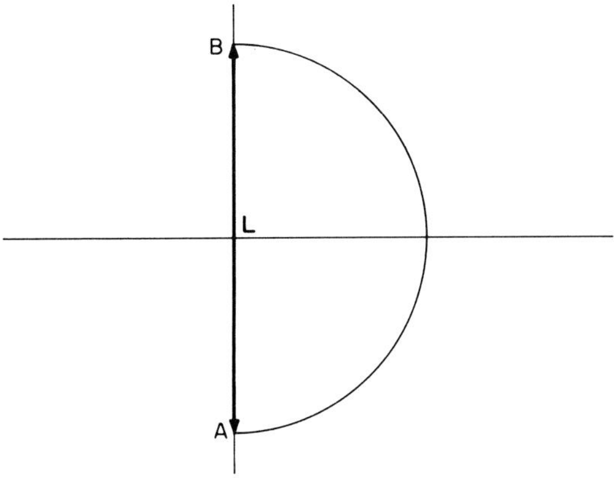

Fig. 1.7

It is interesting to note that if $A = B$—that is, the integration is around a closed path (designated by $\oint$)—then

$$\oint dl = 0.$$

But $\oint dl$ is the length of the closed path.

Consider the scalar field (for example, temperature) defined by the scalar function $f(x, y, z)$, and a path from A to B in the region in which f is defined (Fig. 1.8). The path is broken up into a series of N small segments Δl, small enough that one can assign some average value of $f(x, y, z)$ to each Δl.

Multiplying each Δl by the value of the field at this point, adding the products, and taking the limit $\Delta l \to 0$, one obtains

$$\lim_{\substack{\Delta l \to 0 \\ N \to \infty}} \sum_{n=1}^{N} f_n(x, y, z)\, \Delta l_n = \int_A^B f(x, y, z)\, dl. \tag{1.40}$$

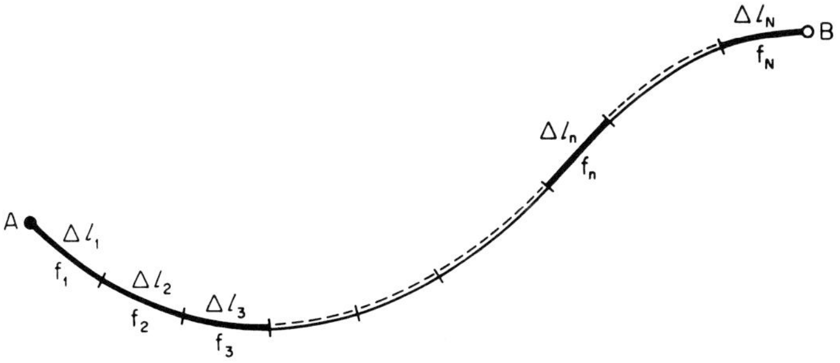

Fig. 1.8

If there is a vector field $\mathbf{F}(x, y, z)$ and if one takes the component of $\mathbf{F}$ that is parallel to the curve from $A \to B$ (that is, $F \cos \vartheta$), then one can write the line integral

$$\int_A^B F \cos \vartheta \, dl = \int_A^B \mathbf{F} \cdot d\mathbf{l} = \int_A^B (F_x \, dx + F_y \, dy + F_z \, dz). \quad (1.41)$$

For example, if the vector field is $\mathbf{F} = 3xy\mathbf{u}_x - y^2\mathbf{u}_y$, then the line integral along the curve $y = 2x^2$ in the xy plane from $(0, 0)$ to $(1,2)$ evaluates to $-7/6$.

Note The line integral $\int_A^B \mathbf{F} \cdot d\mathbf{l}$ can be written indicially as $\int_A^B F_i \, dl_i$, and summing out, one has

$$\int_A^B F_i \, dl_i = \int_A^B (F_1 \, dx_1 + F_2 \, dx_2 + F_3 \, dx_3). \quad (1.42)$$

The line integral of $\mathbf{F}$ around a closed path is called the *circulation* and is written as $\oint \mathbf{F} \cdot d\mathbf{l}$.

If the circulation is zero, the vector field $\mathbf{F}$ is said to be *irrotational.* For example, if $\mathbf{F} = 2xy\mathbf{u}_x + x^2\mathbf{u}_y$ and the closed path is a circle of radius R in the xy plane centered at the origin and starting at $(R, 0)$, then

$$\oint \mathbf{F} \cdot d\mathbf{l} = 0,$$

and so $\mathbf{F}$ is an irrotational field.

1.3.4 Surface Integrals

The area of the parallelogram formed by $\mathbf{A}$ and $\mathbf{B}$ and shown in Fig. 1.9 is $AB \sin \varphi$. Now

$$\mathbf{A} \times \mathbf{B} = AB \sin \varphi \, \mathbf{n},$$

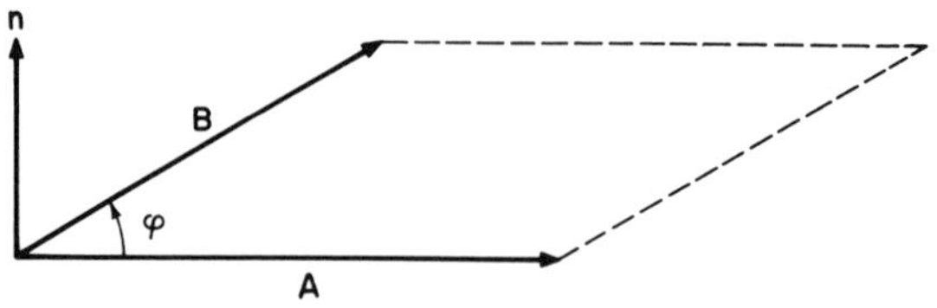

Fig. 1.9

where **n** is a unit vector normal to the plane formed by **A** and **B**. Thus the area of the parallelogram formed by **A** and **B** can be represented by the vector $\mathbf{A} \times \mathbf{B}$ (Fig. 1.9). Using the above result, the differential surface element dS can be written as a vector in terms of $d\mathbf{l}_1$ and $d\mathbf{l}_2$ in the form

$$d\mathbf{S} = dS\,\mathbf{n} = d\mathbf{l}_1 \times d\mathbf{l}_2,$$

where **n** is a unit vector normal to the surface element dS (Fig. 1.10).

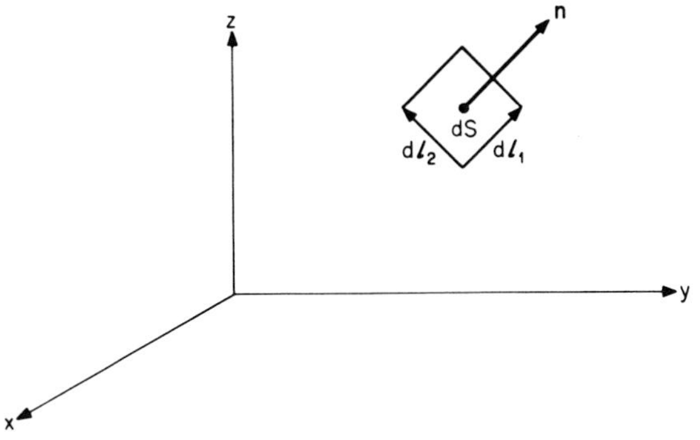

Fig. 1.10

Consider a surface S in a vector field $\mathbf{A}(x, y, z)$. At the point of the differential surface element dS, let A_n be the component of **A** normal to dS (Fig. 1.11).

The product $A_n\,dS$ is called the flux of **A** through the surface element dS. The total flux through the surface S is then

$$\int_S A_n\,dS = \int_S \mathbf{A} \cdot d\mathbf{S} = \int_S \mathbf{A} \cdot \mathbf{n}\,dS$$

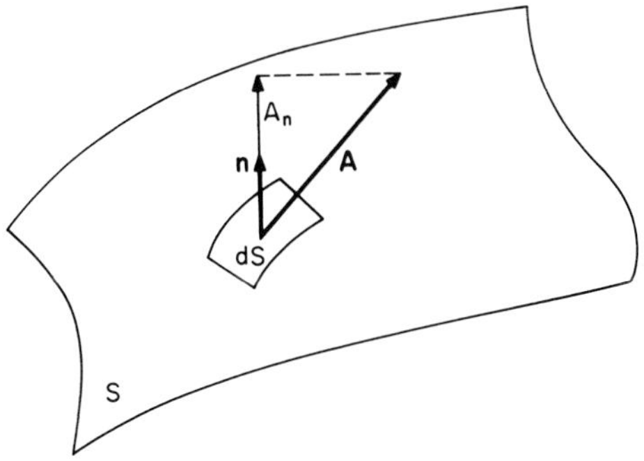

Fig. 1.11

or, in indicial notation,

$$\int_S A_i n_i \, dS = \int_S (A_1 n_1 + A_2 n_2 + A_3 n_3) \, dS. \tag{1.43}$$

If the surface integral is integrated over a closed surface, it is written

$$\oint \mathbf{A} \cdot \mathbf{n} \, dS,$$

where $\mathbf{n}$ is the outward normal.

If $\oint_s \mathbf{A} \cdot \mathbf{n} \, dS = 0$, the vector field $\mathbf{A}$ is said to be solenoidal.

1.3.5 Gradient of a Scalar Field

Consider the scalar field $\varphi(x_1, x_2, x_3)$. At some point P specified by the displacement vector $\mathbf{r} = x_1\mathbf{u}_{(1)} + x_2\mathbf{u}_{(2)} + x_3\mathbf{u}_{(3)}$, the value of φ is $\varphi(x_1, x_2, x_3)$. At a nearby point Q, specified by $\mathbf{r} + \Delta\mathbf{r}$, the value of φ has changed to $\varphi(x_1 + \Delta x_1, x_2 + \Delta x_2, x_3 + \Delta x_3)$. One can write

$$\Delta\varphi = \frac{\partial\varphi_1}{\partial x_1}\Delta x_1 + \frac{\partial\varphi_2}{\partial x_2}\Delta x_2 + \frac{\partial\varphi_3}{\partial x_3}\Delta x_3 = \frac{\partial\varphi_i}{\partial x_i}\Delta x_i; \tag{1.44}$$

$\Delta\varphi$ must be the scalar product of two vectors, the Cartesian components of one being given by $\partial\varphi/\partial x_i$ and of the other by Δx_i. The latter vector is written in terms of the unit vectors as

$$\frac{\partial\varphi}{\partial x_i}\mathbf{u}_{(i)} = \frac{\partial\varphi}{\partial x_1}\mathbf{u}_{(1)} + \frac{\partial\varphi}{\partial x_2}\mathbf{u}_{(2)} + \frac{\partial\varphi}{\partial x_3}\mathbf{u}_{(3)} \tag{1.45}$$

and is called the gradient of φ, written as grad φ. So the change in φ is

$$\Delta\varphi = \mathbf{grad}\,\varphi \cdot \Delta\mathbf{r} \tag{1.46}$$

or, in differentials,

$$d\varphi = \mathbf{grad}\,\varphi \cdot d\mathbf{r}. \tag{1.47}$$

Indicially, the gradient of φ is

$$\text{grad}\,\varphi = \frac{\partial\varphi}{\partial x_i}. \tag{1.48}$$

It is convenient to introduce a comma notation to represent differentiation with respect to a spatial coordinate. Let

$$\frac{\partial}{\partial x_i} \equiv {}_{,i}.$$

Thus the gradient of φ can be written as

$$\frac{\partial \varphi}{\partial x_i} = \varphi_{,i}. \tag{1.49}$$

The "del operator" ∇ is often used instead of grad, where

$$\nabla = \frac{\partial}{\partial x_i}\mathbf{u}_{(i)} = \frac{\partial}{\partial x_1}\mathbf{u}_{(1)} + \frac{\partial}{\partial x_2}\mathbf{u}_{(2)} + \frac{\partial}{\partial x_3}\mathbf{u}_{(3)}. \tag{1.50}$$

A scalar field can be represented graphically in the form of a contour map where the contour lines traverse points with equal values of φ (Fig. 1.12). If the lines are drawn for equal increments in φ, then when the

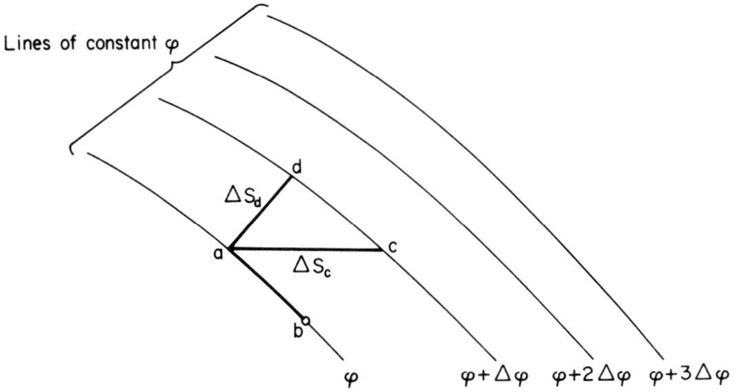

Fig. 1.12

contour lines are crowded, φ is changing rapidly. Moving from a to c produces a change in φ of $\Delta\varphi$, and the rate of change of φ with distance is $\Delta\varphi/\Delta S_c$. Similarly, moving from a to d, the spatial rate of change of φ is $\Delta\varphi/\Delta S_d$. It is clear that the maximum spatial rate of φ will occur when for a given $\Delta\varphi$ the distance ΔS is a minimum. This will clearly occur in a direction that is normal to the lines of constant φ. Now

$$\Delta\varphi = \nabla\varphi \cdot \Delta\mathbf{r}.$$

If $\Delta\mathbf{r}$ is taken along one of the constant φ lines, then $\Delta\varphi = 0$ and

$\nabla\varphi \cdot \Delta\mathbf{r} = 0$. Then $\nabla\varphi$ is normal to $\Delta\mathbf{r}$. Hence $\nabla\varphi$ is a vector which points in the direction of the maximum spatial rate of change of φ.

The directional derivative of φ along a path of length ΔS (Fig. 1.13) is given by

$$\frac{d\varphi}{dS} = \lim_{\Delta S\to 0} \frac{\Delta\varphi}{\Delta S} = \nabla\varphi \cdot \frac{d\mathbf{r}}{dS}. \tag{1.51}$$

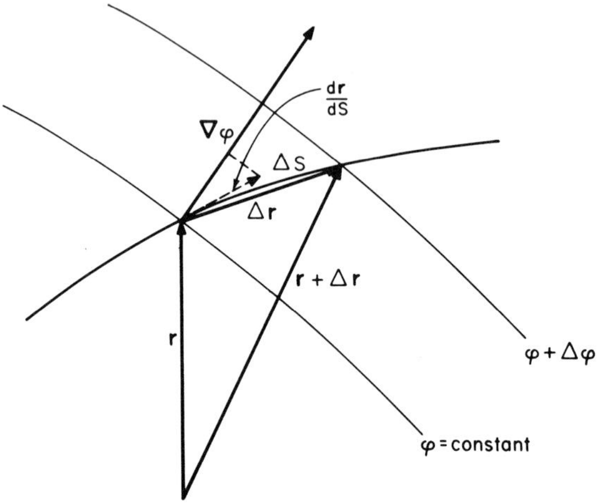

Fig. 1.13

It is clear that $|\, d\mathbf{r}/ds \,| = 1$, and the direction of $d\mathbf{r}/ds$ is tangent to the curve along which the change in φ is being measured. Thus the maximum of $d\varphi/ds$ occurs when $d\mathbf{r}/ds$ is perpendicular to the constant φ lines. Therefore

$$\left.\frac{d\varphi}{dS}\right|_{\max} = \left|\nabla\varphi \cdot \frac{d\mathbf{r}}{dS}\right|_{\max} = |\,\nabla\varphi\,|.$$

Thus the *magnitude* of the vector $\nabla\varphi$ equals the maximum spatial rate of change of φ.

If $\mathbf{F} \cdot d\boldsymbol{l}$ is a perfect differential, say $d\varphi$, then the integral of $\mathbf{F} \cdot d\boldsymbol{l}$ between any two points along the path depends *only* on these two points and not on the path. Therefore

$$\oint_a^b \mathbf{F} \cdot d\boldsymbol{l} = \oint_a^b d\varphi = \varphi(b) - \varphi(a). \tag{1.52}$$

Since $d\varphi = \nabla\boldsymbol{\varphi} \cdot d\boldsymbol{l}$, $\mathbf{F}$ will be a conservative field if $\mathbf{F} = \nabla\varphi$. Thus any vector field which is the gradient of a scalar field is a conservative field.

Consider two paths from a to b in a vector field $\mathbf{F}$ (Fig. 1.14).

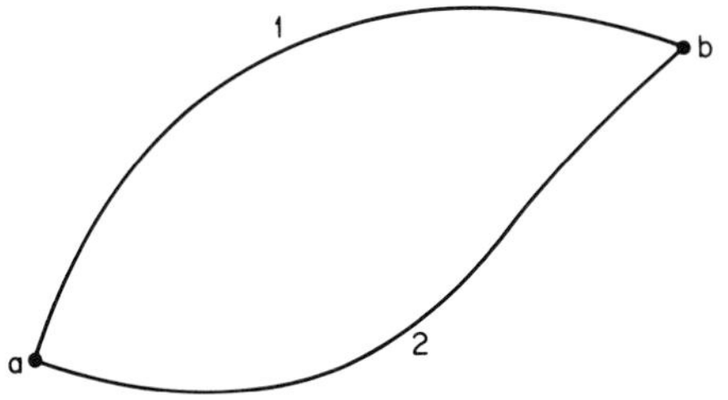

$$\int_b^a \mathbf{F} \cdot d\mathbf{l} \bigg|_1 = -\int_a^b \mathbf{F} \cdot d\mathbf{l} \bigg|_2$$

Fig. 1.14

If $\mathbf{F}$ is a conservative field, the line integral from a to b is the same for paths 1 and 2:

$$\int_a^b \mathbf{F} \cdot d\mathbf{l} \bigg|_1 = \int_a^b \mathbf{F} \cdot d\mathbf{l} \bigg|_2 .$$

If the closed line integral is now taken by going from a to b along 1 and returning to a along 2, then

$$\begin{aligned} \int_{cba} \mathbf{F} \cdot d\mathbf{l} &= \int_a^b \mathbf{F} \cdot d\mathbf{l} \bigg|_1 + \int_b^a \mathbf{F} \cdot d\mathbf{l} \bigg|_2 \\ &= \int_a^b \mathbf{F} \cdot d\mathbf{l} \bigg|_1 - \int_a^b \mathbf{F} \cdot d\mathbf{l} \bigg|_2 \\ &= \int_a^b \mathbf{F} \cdot d\mathbf{l} \bigg|_1 - \int_a^b \mathbf{F} \cdot d\mathbf{l} \bigg|_1 = 0. \end{aligned}$$

Now $\oint \mathbf{F} \cdot d\mathbf{l}$ is the *circulation*, and since the circulation is zero if the field is conservative, a conservative field is also irrotational. Hence, any vector field which is the gradient of a scalar field is irrotational.

1.3.6 Divergence of a Vector Field

The gradient of a scalar field describes the maximum spatial rate of change of the field. Suppose there is a vector field and one wishes to describe its spatial rate of change. The vector field can change both in magnitude and direction. To describe the vector field it is found that both a scalar function—the divergence of the vector field—and a vector function—the curl of the vector field—are required.

The divergence will be discussed in this section.

Consider a closed surface S which completely encloses a volume V. Now, as has been shown earlier, the flux of $\mathbf{F}$ through the surface element dS with an outward normal $\mathbf{n}$ is $\mathbf{F} \cdot \mathbf{n}\, dS$. The total outward flux ψ through the closed surface S is then given by

$$\psi = \iint_S \mathbf{F} \cdot \mathbf{n}\, dS. \tag{1.53}$$

Let the volume V be divided into two volumes V_1 and V_2 ($V = V_1 + V_2$) by the common surface S_C, as shown in Fig. 1.15. The original surface S is split into two open surfaces S_A and S_B, ($S = S_A + S_B$). The closed surface around V_1 is then $S_1 = S_A + S_C$, and the closed surface around V_2 is $S_2 = S_B + S_C$.

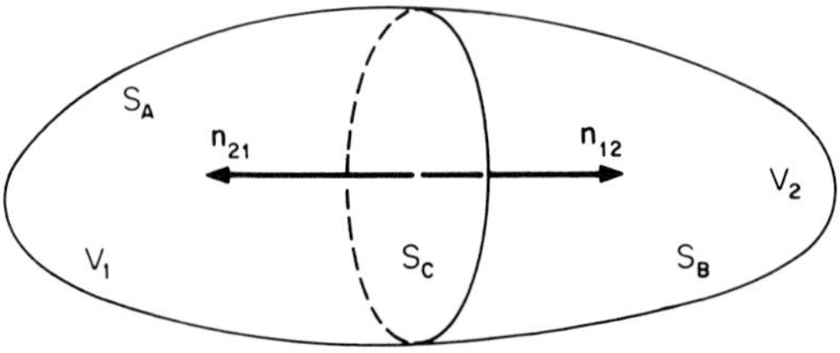

Fig. 1.15

The total flux leaving V_1 is

$$\psi_1 = \iint_{S_1} \mathbf{F} \cdot \mathbf{n}\, dS = \iint_{S_A} \mathbf{F} \cdot \mathbf{n}\, dS + \iint_{S_C} \mathbf{F} \cdot \mathbf{n}\, dS = \psi_{1A} + \psi_{1C}$$

and is the sum of the flux through $S_A(\psi_{1A})$ and the flux through $S_C(\psi_{1C})$ from V_1 into V_2. Similarly, the total flux leaving V_2 is

$$\psi_2 = \psi_{2B} + \psi_{2C},$$

which is the sum of the flux through $S_B(\psi_{2B})$ and the flux through $S_C(\psi_{2C})$ from V_2 into V_1.

But all the flux which leaves V_1 through S_C enters V_2; that is, $\psi_{1C} = -\psi_{2C}$. Thus, the sum of the flux leaving V_1 and V_2 is

$$\psi_1 + \psi_2 = \psi_{1A} + \psi_{2B} = \iint_{S_A} \mathbf{F} \cdot \mathbf{n}\, dS + \iint_{S_B} \mathbf{F} \cdot \mathbf{n}\, dS$$
$$= \oint\!\!\oint_S \mathbf{F} \cdot \mathbf{n}\, dS = \psi.$$

So the total flux leaving V equals the sum of the flux leaving V_1 and the flux leaving V_2.

If one continues to subdivide the volume V into a large number of small volumes, the total flux leaving V will still be the sum of the flux leaving all the small volumes.

The flux leaving a given small volume will decrease as the volume is decreased. In the limit of *zero* volume, the ratio of flux leaving the closed volume to the volume approaches a constant value which is called the divergence of the vector field at that point and is written as div $\mathbf{F}$; that is,

$$\operatorname{div} \mathbf{F} = \lim_{\Delta V \to 0} \frac{\psi}{\Delta V} = \lim_{\Delta V \to 0} \frac{1}{\Delta V} \int_S \mathbf{F} \cdot \mathbf{n} \, dS. \tag{1.54}$$

The definition of div $\mathbf{F}$ will now be applied to the cubical volume V shown in Fig. 1.16. Here $\Delta V = \Delta x \, \Delta y \, \Delta z$ and the point P is located at the center of the cube.

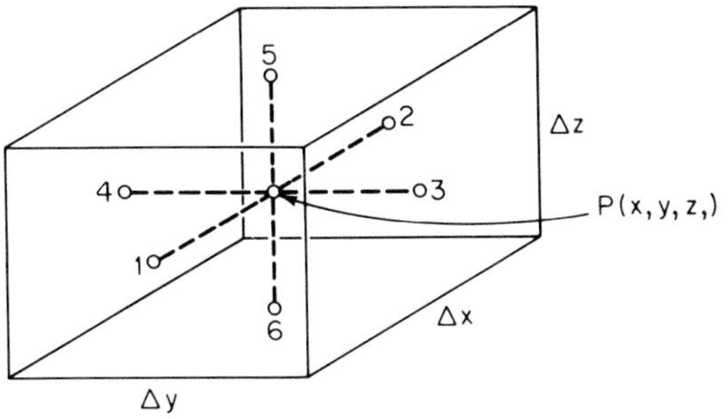

Fig. 1.16

Consider side 1 whose area is $\Delta y \, \Delta z$. The outward normal from this side is $\mathbf{n} = (n_x, 0, 0) = (1, 0, 0)$ so that

$$\mathbf{F} \cdot \mathbf{n} = F_x.$$

Since $\Delta y \, \Delta z$ is a small surface element, one can take the average value of F_x over the entire surface to be the value of F_x at the center point 1.

The value $F_x(x + (\Delta x/2), y, z)$ is different from the value of $F_x(x, y, z)$ at point P. Since the small distance between P and 1 is $\Delta x/2$, one can find F_x at 1 by expanding F_x at P in a Taylor series and keep only the first two terms in the series:

$$F_{x1}\left(x + \frac{\Delta x}{2}, y, z\right) = F_x(x, y, z) + \frac{\partial F_x}{\partial x} \frac{\Delta x}{2}, \tag{1.55}$$

where $\partial F_x/\partial x$ is evaluated at $P(x, y, z)$.

The outward flux from side 1 is therefore

$$\psi_1 = F_{x1}\,\Delta y\,\Delta z = F_x(x, y, z)\,\Delta y\,\Delta z + \frac{\partial F_x}{\partial x}\frac{\Delta x}{2}\,\Delta y\,\Delta z. \quad (1.56)$$

Consider now the flux leaving surface 2. Since the outward normal drawn from surface 2 is $-\mathbf{n}_x$, the flux leaving the volume element through surface 2 is

$$\mathbf{F}\cdot\mathbf{n}\,dS = -F_{x2}\,\Delta y\,\Delta z,$$

where $F_{x2}(x - (\Delta x/2), y, z)$ is the value of F_x at the center point 2 and is taken as the average of F_x on side 2. One can find F_{x2} in terms of F_x at point P by expanding F_x in a Taylor series about P. Thus

$$F_{x2} = F_x\!\left(x - \frac{\Delta x}{2}, y, z\right) = F_x(x, y, z) - \frac{\partial F_x}{\partial x}\frac{\Delta x}{2}.$$

The outward flux from side 2 is then

$$\psi_2 = (\mathbf{F}\cdot\mathbf{n}\,\Delta S)_2 = -F_x(x, y, z)\,\Delta y\,\Delta z + \frac{\partial F_x}{\partial x}\frac{\Delta x}{2}\,\Delta y\,\Delta z. \quad (1.57)$$

The *net* outward flux from sides 1 and 2 is therefore

$$\psi_1 + \psi_2 = \frac{\partial F_x}{\partial x}\,\Delta x\,\Delta y\,\Delta z. \quad (1.58)$$

Similarly, the net outward flux from sides 3 and 4 can be deduced to be

$$\psi_3 + \psi_4 = \frac{\partial F_y}{\partial y}\,\Delta x\,\Delta y\,\Delta z, \quad (1.59)$$

and from sides 5 and 6,

$$\psi_5 + \psi_6 = \frac{\partial F_z}{\partial z}\,\Delta x\,\Delta y\,\Delta z. \quad (1.60)$$

Combining these results, one gets the total flux leaving the volume ΔV:

$$\psi = \psi_1 + \psi_2 + \psi_3 + \psi_4 + \psi_5 + \psi_6 = \left(\frac{\partial F_x}{\partial x} + \frac{\partial F_y}{\partial y} + \frac{\partial F_z}{\partial z}\right)\Delta x\,\Delta y\,\Delta z.$$

If ψ is divided by the volume element $\Delta V = \Delta x\,\Delta y\,\Delta z$, then the divergence of $\mathbf{F}$ is given by

$$\operatorname{div}\mathbf{F} = \lim_{\Delta V\to 0}\frac{\psi}{\Delta V} = \frac{\partial F_x}{\partial x} + \frac{\partial F_y}{\partial y} + \frac{\partial F_z}{\partial z}. \quad (1.61)$$

If the axes had been labeled x_1, x_2, x_3, then the vector **F** would have Cartesian components (F_1, F_2, F_3). The divergence would then be written as

$$\operatorname{div} \mathbf{F} = \frac{\partial F_1}{\partial x_1} + \frac{\partial F_2}{\partial x_2} + \frac{\partial F_3}{\partial x_3} = \frac{\partial F_i}{\partial x_i}. \tag{1.62}$$

In the comma notation,

$$\operatorname{div} \mathbf{F} = \frac{\partial F_i}{\partial x_i} = F_{i,i}. \tag{1.63}$$

Now the "del" operator is given by

$$\boldsymbol{\nabla} = \frac{\partial}{\partial x_i} \mathbf{u}_{(i)},$$

and if **F** is written as $\mathbf{F} = F_j \mathbf{u}_{(j)}$, then

$$\boldsymbol{\nabla} \cdot \mathbf{F} = \frac{\partial}{\partial x_i} \mathbf{u}_{(i)} \cdot F_j \mathbf{u}_{(j)} = \frac{\partial F_j}{\partial x_i} \mathbf{u}_{(i)} \cdot \mathbf{u}_{(j)} = \frac{\partial F_j}{\partial x_i} \delta_{ij} = \frac{\partial F_i}{\partial x_i}. \tag{1.64}$$

Thus, in vector notation the divergence of F is written as $\boldsymbol{\nabla} \cdot \mathbf{F}$.

Note The expression $\boldsymbol{\nabla} \cdot (a\mathbf{F})$ can be written indicially as $(aF_i)_{,i}$. Now

$$(aF_i)_{,i} = a_{,i}F_i + aF_{i,i} = (\boldsymbol{\nabla} a) \cdot \mathbf{F} + a\boldsymbol{\nabla} \cdot \mathbf{F}.$$

1.3.7 The Curl of a Vector Field

In Fig. 1.17 the line *ca* cuts the surface S bounded by the curve *abcda* into two parts S_1 and S_2.

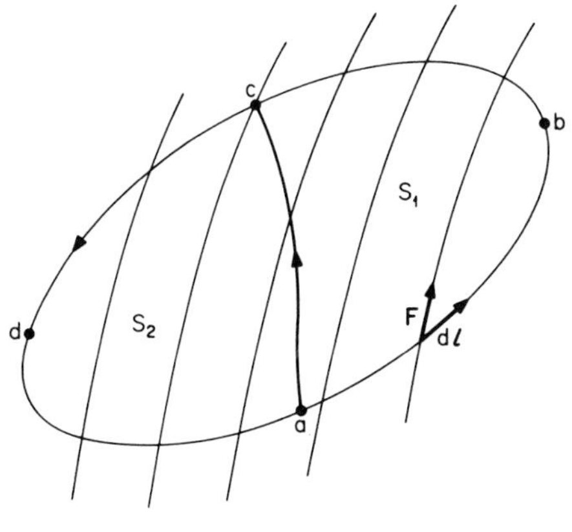

Fig. 1.17

Let the circulation of **F** around the path *abcda* be Γ. Then

$$\Gamma = \oint_{abcda} \mathbf{F} \cdot d\mathbf{l}.$$

Let Γ_1 be the circulation around *abca* enclosing S_1 and that around *acda* enclosing S_2 be Γ_2. Now

$$\Gamma_1 = \oint_{abca} \mathbf{F} \cdot d\mathbf{l} = \int_{abc} \mathbf{F} \cdot d\mathbf{l} + \int_{ca} \mathbf{F} \cdot d\mathbf{l},$$

$$\Gamma_2 = \oint_{acda} \mathbf{F} \cdot d\mathbf{l} = \int_{acda} \mathbf{F} \cdot d\mathbf{l} + \int_{cda} \mathbf{F} \cdot d\mathbf{l},$$

and since

$$\int_{ca} \mathbf{F} \cdot d\mathbf{l} = -\int_{ac} \mathbf{F} \cdot d\mathbf{l},$$

one therefore has

$$\Gamma_1 + \Gamma_2 = \oint_{abcda} \mathbf{F} \cdot d\mathbf{l} = \Gamma. \tag{1.65}$$

It is clear that one could continue to subdivide the surface area S into N small surfaces ΔS_n such that

$$\sum_{n=1}^{N} \Delta S_n = S.$$

Then the circulation Γ around the original path *abcda* will be the sum of the circulations around all the small surface elements ΔS_n. As the surface element ΔS_n gets smaller, the circulation around it also gets smaller. If, however, $\Delta S_n \to 0$ in such a way that the length of the path enclosing ΔS_n also approaches zero, then the ratio of the circulation Γ_n to the surface element ΔS_n approaches a constant value. That is,

$$\lim_{\Delta S \to 0} \frac{\Gamma}{\Delta S_n} = \lim_{\Delta S \to 0} \frac{1}{\Delta S} \oint \mathbf{F} \cdot d\mathbf{l} = \text{a constant}. \tag{1.66}$$

Equation (1.66) gives the circulation per unit area at a point P in the field. This depends in general on the orientation of the surface element ΔS. The orientation is specified by the unit normal **n** where ***dl*** and **n** are related in the right-hand sense as shown in Fig. 1.18. If **n** is oriented along the x, y, and z axes in turn, then three different values are obtained for the circulation per unit area. The three values form the components of a vector called the curl of **F**. If **n** points in some arbitrary direction,

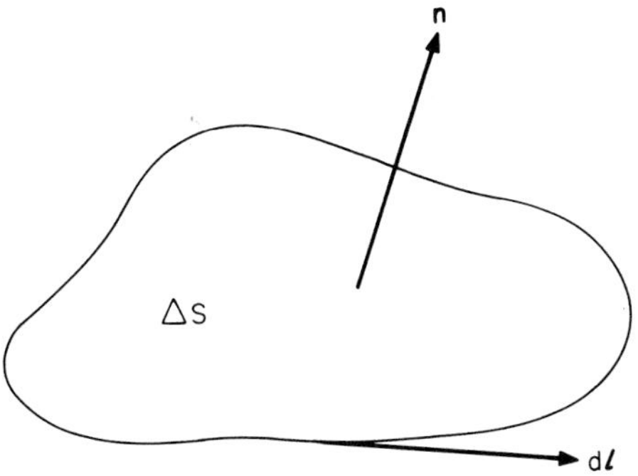

Fig. 1.18

then the circulation per unit area is equal to the component of the curl in the **n** direction. Thus,

$$\mathbf{n} \cdot \operatorname{curl} \mathbf{F} = \lim_{\Delta S \to 0} \frac{1}{\Delta S} \oint_C \mathbf{F} \cdot d\mathbf{l} \tag{1.67}$$

defines the curl of **F**. This expression is a maximum when **n** and curl **F** are parallel. Thus, the curl of **F** is a vector in the direction of **n** when **n** is oriented so that the circulation per unit area is a maximum. The magnitude of **F** is equal to this maximum circulation per unit area.

To find the x component of curl **F**, one calculates the circulation per unit area around the surface element $\Delta S = \Delta y\, \Delta z$ which lies in the yz plane as shown in Fig. 1.19. Since the element is small, the average of **F** along each side can be taken as the value of **F** at the center point. Therefore

$$(\operatorname{curl} \mathbf{F})_x = \lim_{\Delta S \to 0} \frac{1}{\Delta S} \oint \mathbf{F} \cdot d\mathbf{l}$$

$$= \lim_{\Delta S \to 0} \frac{1}{\Delta y\, \Delta z} [F_{z1}\, \Delta z - F_{y2}\, \Delta y - F_{z3} \Delta z + F_{y4} \Delta y]. \tag{1.68}$$

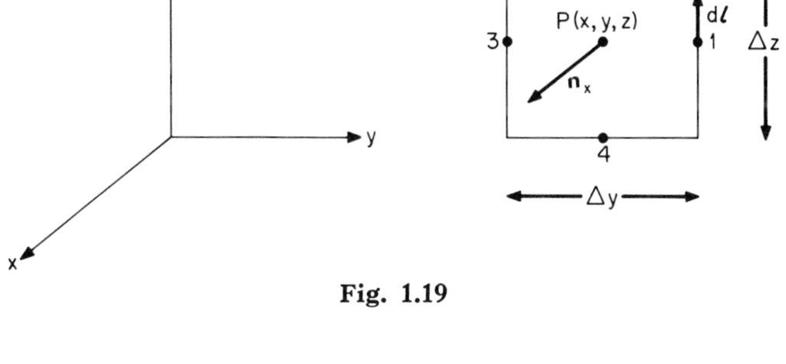

Fig. 1.19

The values of F_{z1}, F_{y2}, F_{z3}, and F_{y4} can be found in terms of **F** at the point P from the first terms of a Taylor series expansion. Thus,

$$F_{z1} = F_z\left(x, y + \frac{\Delta y}{2}, z\right) = F_z(x, y, z) + \frac{\partial F_z}{\partial y}\frac{\Delta y}{2},$$

$$F_{y2} = F_y\left(x, y, z + \frac{\Delta z}{2}\right) = F_y(x, y, z) + \frac{\partial F_y}{\partial z}\frac{\Delta z}{2},$$

$$F_{z3} = F_z\left(x, y - \frac{\Delta y}{2}, z\right) = F_z(x, y, z) - \frac{\partial F_z}{\partial y}\frac{\Delta y}{2},$$

$$F_{y4} = F_y\left(x, y, z - \frac{\Delta z}{2}\right) = F_y(x, y, z) - \frac{\partial F_y}{\partial z}\frac{\Delta z}{2}.$$

Substituting these values into Eq. (1.68), one obtains

$$(\text{curl } \mathbf{F})_x = \lim_{\Delta S \to 0} \frac{1}{\Delta y\, \Delta z}\left\{\left[F_z + \frac{\partial F_z}{\partial y}\frac{\Delta y}{2}\right]\Delta z - \left[F_y + \frac{\partial F_y}{\partial z}\frac{\Delta z}{2}\right]\Delta y - \left[F_z - \frac{\partial F_z}{\partial y}\frac{\Delta y}{2}\right]\Delta z + \left[F_y - \frac{\partial F_y}{\partial z}\frac{\Delta z}{2}\right]\Delta y\right\}$$
$$= \frac{\partial F_z}{\partial y} - \frac{\partial F_y}{\partial z}. \tag{1.69}$$

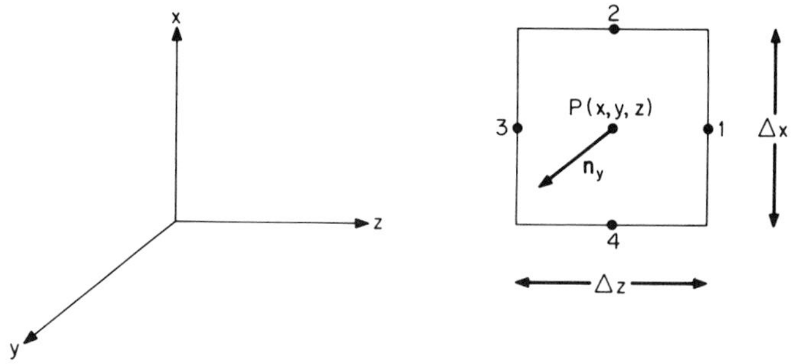

Fig. 1.20

Similarly, the y component of curl **F** can be found by calculating the circulation per unit area around a square path in the zx plane, as shown in Fig. 1.20:

$$(\text{curl } \mathbf{F})_y = \lim_{\Delta S \to 0} \frac{1}{\Delta S}\oint \mathbf{F} \cdot d\mathbf{l}$$
$$= \lim_{\Delta S \to 0} \frac{1}{\Delta x\, \Delta z}\,[F_{x1}\,\Delta x - F_{z2}\,\Delta z - F_{x3}\,\Delta x + F_{z4}\,\Delta z].$$

Expanding **F** in a Taylor series about P, one has

$$F_{x1} = F_x\left(x, y, z + \frac{\Delta z}{2}\right) = F_x(x, y, z) + \frac{\partial F_x}{\partial z}\frac{\Delta z}{2},$$

$$F_{z2} = F_z\left(x + \frac{\Delta x}{2}, y, z\right) = F_z(x, y, z) + \frac{\partial F_z}{\partial y}\frac{\Delta y}{2},$$

$$F_{x3} = F_x\left(x, y, z - \frac{\Delta z}{2}\right) = F_x(x, y, z) - \frac{\partial F_x}{\partial z}\frac{\Delta z}{2},$$

$$F_{z4} = F_z\left(x - \frac{\Delta z}{2}, y, z\right) = F_z(x, y, z) - \frac{\partial F_z}{\partial x}\frac{\Delta x}{2}.$$

Substituting the above into the expression for $(\text{curl}\,\mathbf{F})_y$, one gets

$$(\text{curl}\,\mathbf{F})_y = \lim_{\Delta S \to 0} \frac{1}{\Delta x\,\Delta z}\left\{\left[F_x + \frac{\partial F_x}{\partial z}\frac{\Delta z}{2}\right]\Delta x - \left[F_z + \frac{\partial F_z}{\partial x}\frac{\Delta x}{2}\right]\Delta z - \left[F_x - \frac{\partial F_x}{\partial z}\frac{\Delta z}{2}\right]\Delta x + \left[F_z - \frac{\partial F_z}{\partial x}\frac{\Delta x}{2}\right]\Delta z\right\};$$

thus

$$(\text{curl}\,\mathbf{F})_y = \frac{\partial F_x}{\partial z} - \frac{\partial F_z}{\partial y}. \tag{1.70}$$

Finally the z component of curl **F** is found by calculating the circulation per unit area around the square path in the xy plane (Fig. 1.21). Proceeding as before, one finds

$$(\text{curl}\,\mathbf{F})_z = \frac{\partial F_y}{\partial x} - \frac{\partial F_x}{\partial y}. \tag{1.71}$$

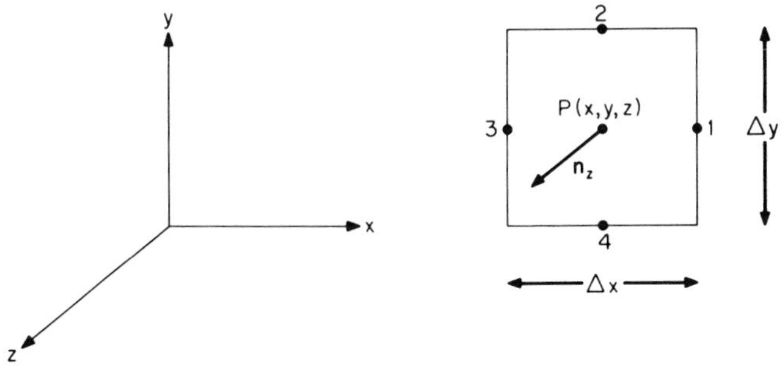

Fig. 1.21

Thus the vector curl **F** in terms of its Cartesian components is

$$\text{curl}\,\mathbf{F} = \left(\frac{\partial F_z}{\partial y} - \frac{\partial F_y}{\partial z}\right)\mathbf{u}_x + \left(\frac{\partial F_x}{\partial z} - \frac{\partial F_z}{\partial x}\right)\mathbf{u}_y + \left(\frac{\partial F_y}{\partial x} - \frac{\partial F_x}{\partial y}\right)\mathbf{u}_z. \tag{1.72}$$

A convenient way to remember the components of curl **F** is to note that curl **F** can be written as the determinant

$$\operatorname{curl} \mathbf{F} = \begin{vmatrix} \mathbf{u}_x & \mathbf{u}_y & \mathbf{u}_z \\ \dfrac{\partial}{\partial x} & \dfrac{\partial}{\partial y} & \dfrac{\partial}{\partial z} \\ F_x & F_y & F_z \end{vmatrix}$$

Since $\partial/\partial x$, $\partial/\partial y$, and $\partial/\partial z$ are components of the "del" operator, one can write

$$\operatorname{curl} \mathbf{F} = \nabla \times \mathbf{F}. \tag{1.73}$$

Now indicially $\mathbf{A} \times \mathbf{B}$ is written as $\varepsilon_{ijk} A_j B_k$; thus it seems likely that $\nabla \times \mathbf{F}$ should be written

$$\varepsilon_{ijk} \frac{\partial F_k}{\partial x_j}.$$

Summing out the repeated indices j and k and using the definition of ε_{ijk}, one finds that

$$i = 1; \qquad \varepsilon_{1jk} \frac{\partial F_k}{\partial x_j} = \frac{\partial F_3}{\partial x_2} - \frac{\partial F_2}{\partial x_3},$$

$$i = 2; \qquad \varepsilon_{2jk} \frac{\partial F_k}{\partial x_j} = \frac{\partial F_1}{\partial x_3} - \frac{\partial F_3}{\partial x_1},$$

$$i = 3; \qquad \varepsilon_{3jk} \frac{\partial F_k}{\partial x_j} = \frac{\partial F_2}{\partial x_1} - \frac{\partial F_1}{\partial x_2},$$

which are recognized as the three components of curl $\mathbf{F} \equiv \nabla \times \mathbf{F}$. Using the comma notation, one has

$$\nabla \times \mathbf{F} = \varepsilon_{ijk} \frac{\partial F_k}{\partial x_j} = \varepsilon_{ijk} F_{k,j}. \tag{1.74}$$

1.4 Vector Integral Theorems

1.4.1 The Divergence Theorem

The definition of div **F** or $\nabla \cdot \mathbf{F}$ is

$$\nabla \cdot \mathbf{F} = \lim_{\Delta V \to 0} \frac{1}{\Delta V} \oint\!\!\oint_S \mathbf{F} \cdot \mathbf{n}\, dS. \tag{1.75}$$

From the definition of divergence, the net outward flux from a small volume element ΔV can be written in terms of $\nabla \cdot \mathbf{F}$ as

$$\psi_{\Delta V} = \lim_{\Delta V \to 0} \nabla \cdot \mathbf{F}\, dV. \tag{1.76}$$

If a volume V is divided into N small elements of volume, the total flux leaving V is

$$\psi = \oint\!\!\oint \mathbf{F} \cdot \mathbf{n}\, dS \tag{1.77}$$

and is equal to the sum of the flux leaving all of the small volumes. Thus one can write

$$\psi = \sum_{n=1}^{N} \psi_{\Delta V}$$

or

$$\oint\!\!\oint \mathbf{F} \cdot \mathbf{n}\, dS = \lim_{\substack{\Delta V \to 0 \\ N \to \infty}} \sum_{n=1}^{N} \nabla \cdot \mathbf{F}\, \Delta V.$$

But the right-hand term is just the volume integral $\iiint_V \nabla \cdot \mathbf{F}\, dV$. Therefore

$$\oint\!\!\oint \mathbf{F} \cdot \mathbf{n}\, dS = \iiint_V \nabla \cdot \mathbf{F}\, dV. \tag{1.78}$$

This is called the *divergence theorem.* It converts a volume integral into a surface integral (or vice versa) where S is the closed surface which bounds the volume V.

In indicial notation the divergence theorem can be written as

$$\iiint_V F_{i,i}\, dV = \oint\!\!\oint_S F_i n_i\, dS. \tag{1.79}$$

Thus, if one uses the indicial notation, a volume integral can be converted into a surface integral by replacing $_{,i}$ by n_i.

Note The , operates on the entire integrand of the volume integral. For example,

$$\iiint_V (A_i B_j C_j)_{,i}\, dV = \oint\!\!\oint_S A_i B_j C_j n_i\, dS.$$

This theorem is also referred to as *Gauss's theorem.* In general (that is, for a general tensor field), it can be written as

$$\oiint F_{ijk\ldots}\, n_r\, dS = \iiint_V F_{ijk\ldots,r}\, dV. \tag{1.80}$$

1.4.2 Stokes's Theorem

The definition of curl $\mathbf{F}$ or $\nabla \times \mathbf{F}$ is

$$(\nabla \times \mathbf{F}) \cdot \mathbf{n} = \lim_{\Delta S \to 0} \frac{1}{\Delta S} \oint_C \mathbf{F} \cdot dl. \tag{1.81}$$

From the definition of curl, the circulation around a small surface element can be written in terms of $\nabla \times \mathbf{F}$ as

$$\Gamma_{\Delta S} = \lim_{\Delta S \to 0} (\nabla \times \mathbf{F}) \cdot \mathbf{n}\, dS. \tag{1.82}$$

If a surface area S which is bounded by the curve C is divided into N small surface elements ΔS, then the total circulation around S is

$$\Gamma = \oint_C \mathbf{F} \cdot dl \tag{1.83}$$

and equals the sum of the circulations $\Gamma_{\Delta S}$ around all the small surface elements ΔS. Therefore

$$\Gamma = \sum_{n=1}^{N} \Gamma_{\Delta S}$$

or

$$\oint_C \mathbf{F} \cdot dl = \lim_{\substack{\Delta S \to 0 \\ N \to \infty}} \sum_{n=1}^{N} (\nabla \times \mathbf{F}) \cdot \mathbf{n}\, \Delta S.$$

But the term on the right-hand side is just the surface integral

$$\iint_S (\nabla \times \mathbf{F}) \cdot \mathbf{n}\, dS;$$

therefore

$$\oint \mathbf{F} \cdot dl = \iint_S (\nabla \times \mathbf{F}) \cdot \mathbf{n}\, dS. \tag{1.84}$$

This is called *Stokes's theorem.* It converts a surface integral into a line integral (or vice versa) where C is the closed curve that bounds the surface S. In indicial notation the Stokes theorem is

$$\oint_C F_i \, dl_i = \iint_S \varepsilon_{ijk} F_{k,j} n_i \, dS. \tag{1.85}$$

For example, one can evaluate the integral

$$\int_C (3x + 4y)\, dx + (2x + 3y)\, dy,$$

where C is a circle of radius 3 with its center at the origin of the xy plane and is traversed in the positive sense, by using Stokes's theorem as having the value zero.

1.4.3 Related Integral Theorems

Now

$$\iiint_V (\psi\varphi_{,i})_{,i}\, dV = \oiint_S \psi\varphi_{,i} n_i \, dS.$$

If the integrand of the volume integral is expanded and one writes $\varphi_{,i} n_i = \partial\varphi/\partial n$ (the normal derivative of φ at the surface S), then

$$\iiint_V (\psi_{,i}\varphi_{,i} + \psi\varphi_{,ii})\, dV = \oiint_S \psi \frac{\partial\varphi}{\partial n}\, dS. \tag{1.86}$$

This is called *Green's first identity*.

If one interchanges φ and ψ in expression (1.86), then

$$\iiint_V (\varphi\psi_{,i})_{,i}\, dV = \oiint_S \varphi\psi_{,i} n_i \, dS$$

and

$$\iiint_V (\varphi_{,i}\psi_{,i} + \varphi\psi_{,ii})\, dV = \oiint_S \varphi \frac{d\psi}{dn}\, dS.$$

Subtracting this from Eq. (1.86), one thus obtains

$$\iiint_V (\psi\varphi_{,ii} - \varphi\psi_{,ii})\, dV = \oiint_S \left(\psi \frac{\partial\varphi}{\partial n} - \varphi \frac{\partial\psi}{\partial n}\right) dS, \tag{1.87}$$

which is called *Green's second identity*, or *theorem*.

1.5 The Rheological Behavior of Fluids

The classical theory of the hydrodynamics of viscous fluids is based on the constitutive equations

$$t_{ij} = -p\,\delta_{ij} + \varphi_1\, d_{ij}, \tag{1.88}$$

where t_{ij} is the stress tensor, p the hydrostatic pressure, d_{ij} the deformation rate tensor $[= \frac{1}{2}(v_{i,j} + v_{j,i})]$, $v_{i,j}$ the velocity gradient, and ., denotes the covariant derivative.

The fluids which obey Eq. (1.88) are called Newtonian fluids. In this textbook it is this class of fluids which will be dealt with almost exclusively. But a review of the non-Newtonian fluids is not out of place in this introduction.

The coefficient of viscosity φ_1 is generally a function of the material properties such as density, temperature, and molecular properties. Common gases such as air and liquids similar to water and mercury all belong to the class of Newtonian fluids.

Experimental results on thick liquids show a considerable departure from Eq. (1.88). Liquids which do not obey Eq. (1.88) are called non-Newtonian liquids. Examples are lubricants, colloids, polymer solutions, and pastes.

Several nonlinear constitutive equations have been proposed by introducing nonlinearity and elasticity or memory to explain the behavior of such highly viscous fluids.

The non-Newtonian fluids are subdivided into three classes: (a) viscoinelastic fluids, (b) time-dependent fluids, (c) viscoelastic fluids.

(a) *Viscoinelastic fluids* are Stokesian fluids. They are isotropic and homogeneous at rest. The resultant stress depends only on the rate of shear when they are subjected to a shear. These fluids are further classified into the following types.

(i) *Bingham plastics* These obey Eq. (1.88) but differ from Newtonian fluids in being able to sustain a certain finite stress—the "yield stress"—before flow begins. The stress–shear rate characteristics of the various types are shown in Fig. 1.22.

(ii) *Pseudoplastic and dilatant fluids* These obey Ostwald's law

$$t = \mu\left(\frac{\partial u}{\partial y}\right)^n, \tag{1.89}$$

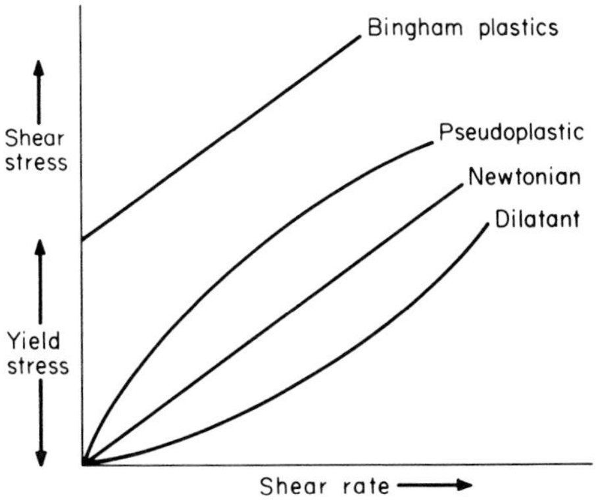

Fig. 1.22

where μ and n are constants, $n < 1 \Rightarrow$ pseudoplasticity, $n > 1 \Rightarrow$ dilatancy, and $n = 1 \Rightarrow$ Newtonian law. The main difference between the pseudoplastic and dilatant fluids is that in the former class of fluids the apparent viscosity decreases with the increase of rate of shear, while the opposite is true for dilatant fluids.

(iii) *Reiner–Rivlin fluids* Reiner and Rivlin assumed that the stress tensor t_{ij} can be expressed in terms of a power series in d_{ij} and obtained

$$t_{ij} = -p\,\delta_{ij} + \varphi_1\,d_{ij} + \varphi_3\,d_{im}\,d_{mj} \tag{1.90}$$

using the Cayley–Hamilton theorem on matrices. Here φ_1 and φ_3 besides being functions of material properties are also functions of the three invariants of d_{ij}. The φ_1 has the dimensions of viscosity—$ML^{-1}T^{-1}$—and is called the coefficient of viscosity, and φ_3 is called the coefficient of cross viscosity, although its dimensions are ML^{-1}.

The Weissenberg effect, namely the climbing of the fluid along the inner fixed cylinder against the centrifugal force when the fluid is sheared between two coaxial cylinders by rotating the outer cylinder, has been attributed to the nonlinear terms involving the cross viscosity in Eq. (1.90). It has also been regarded as the source of the Merrington effect which involves the swelling of the stream of fluid at the exit of a circular pipe through which it has been flowing under a pressure gradient.

(b) *Time-dependent fluids* are those that show either an increase or decrease in viscosity as time passes, when subjected to a steady rate of shear under isothermal conditions. The fluids which show an increase in viscosity under the above conditions are called "rheopectic," while those which show the opposite effect are called "thixotropic."

(c) *Viscoelastic fluids* are solid materials that possess fluid characteristics. These materials possess a certain amount of rigidity that is characteristic of solid bodies, and at the same time they flow and dissipate energy by frictional losses as do some fluids. Solids having internal friction belong to this class. With the recent explosive increase in the use of plastic and high polymers, the field of viscoelasticity has become a very active one both in engineering techniques and mathematical analysis.

In a normal inelastic fluid one is concerned only with the *rate* of strain, but in elastic fluids the strain cannot be neglected however small it is. The strain is responsible for the recovery of the original state and for the reverse flow that follows the removal of stress.

As mentioned previously, only the dynamics of inelastic Newtonian fluids will be considered. The student must be aware of the fact, however, that this is now but one branch of the dynamics of a wide variety of fluids having almost as wide a variety of constitutive equations. They *all*, however, must obey the laws of conservation of mass, momenta, and energy.

PROBLEMS

1.1 Use Eq. (1.15) to show that

$$\varepsilon_{ijk}A_iB_jC_k = \begin{vmatrix} A_1 & A_2 & A_3 \\ B_1 & B_2 & B_3 \\ C_1 & C_2 & C_3 \end{vmatrix}.$$

1.2 Write the following expressions in indicial notation:

$$\mathbf{A} \cdot \mathbf{B} \times \mathbf{C}, \qquad (\mathbf{C} \cdot \mathbf{A})\mathbf{B}.$$

1.3 Write the following indicial expressions in vector notation:

$$\varepsilon_{ijk}A_iB_jC_k, \qquad \varepsilon_{ijk}F_kG_j.$$

1.4 Use the relation given in Eq. (1.20) to write the following expressions in terms of the Kronecker deltas:

$$\varepsilon_{kij}\varepsilon_{ipq}, \qquad \varepsilon_{mnp}\varepsilon_{rps}.$$

1.5 Use Eq. (1.20) to prove the vector identity

$$(\mathbf{A} \times \mathbf{B}) \times \mathbf{C} = \mathbf{B}(\mathbf{A} \cdot \mathbf{C}) - \mathbf{A}(\mathbf{B} \cdot \mathbf{C}),$$

and verify that

$$(\mathbf{A} \times \mathbf{B}) \times \mathbf{C} \neq \mathbf{A} \times (\mathbf{B} \times \mathbf{C}).$$

1.6 Show that

$$\varepsilon_{ijr}\varepsilon_{ijs} = 2\,\delta_{rs} \qquad \text{and} \qquad \varepsilon_{ijk}\varepsilon_{ijk} = 6.$$

1.7 If $\varphi(x, y, z) = 3x^2 - y^3z^2$, then evaluate $\nabla\varphi$ at the point (1, −2, −1).

1.8 If $\varphi(x_i) = x_1{}^2 + x_2{}^2 + x_3{}^2$, then evaluate the x_1 component of grad φ.

1.9 Write the expression $\nabla(\mathbf{A} \cdot \mathbf{B})$ in indicial notation.

1.10 If $\varphi = x^2 + y + yz$, then at the point (1, 1, 1) the maximum rate of change of φ is in the direction of the unit vector **n**. Find **n**.

1.11 If $\mathbf{F} = x^2z\mathbf{u}_x - 2y^3z^2\mathbf{u}_y + xy^2\mathbf{u}_z$, then find $\nabla \cdot \mathbf{F}$ at the point (1, −1, 1).

1.12 If $F_i = (x_1\,x_2, x_2{}^2x_3, x_1\,x_2)$, then determine the divergence of **F**.

1.13 Express $\nabla \cdot (a\mathbf{F})$ in indicial notation.

1.14 Write the following equation in vector notation:

$$(aF_i)_{,i} = a_{,i}F_i + aF_{i,i}.$$

1.15 Verify the following vector identities by writing each left-hand expression in indicial notation and carrying out the indicated operations:

(i) $\nabla \cdot (\mathbf{A} + \mathbf{B}) = \nabla \cdot \mathbf{A} + \nabla \cdot \mathbf{B}$,
(ii) $\nabla \cdot \nabla\varphi = \nabla^2\varphi$,

where

$$\nabla^2 \equiv \frac{\partial^2}{\partial x^2} + \frac{\partial^2}{\partial y^2} + \frac{\partial^2}{\partial z^2} \equiv {}_{,jj}.$$

1.16 If $\mathbf{F} = xz^3\mathbf{u}_x - 2x^2yz\mathbf{u}_y + 2yz^4\mathbf{u}_z$, then determine $\nabla \times \mathbf{F}$ at the point (1, −1, 1).

1.17 If $F_i = (x_1x_2, x_2{}^2x_1x_3, x_2x_3{}^2)$, then evaluate the three components of

$$\nabla \times \mathbf{F} = \varepsilon_{ijk}F_{k,j}.$$

1.18 (i) Write the following expression in indicial notation:

$$\nabla \times \nabla \times \mathbf{A}.$$

(ii) Write the following expression in vector notation:

$$\varepsilon_{rsi}\varepsilon_{ijk}(A_jB_k)_{,s}$$

1.19 Use the divergence theorem to evaluate $\oiint_S \mathbf{F} \cdot \mathbf{n}\, dS$, where

$$\mathbf{F} = 4xz\mathbf{u}_x + y^2\mathbf{u}_y + yz\mathbf{u}_z$$

and S is the surface of a cube bounded by

$$x = 0, \quad x = 1, \quad y = 0, \quad y = 1, \quad z = 0, \quad z = 1.$$

1.20 Show that if the total outward flux of a vector **F** through an arbitrary closed surface S is zero, then the divergence of **F** is zero.

1.21 Use Stokes's theorem to evaluate

$$\oint_C (3x + 4y)\, dx + (2x + 3y)\, dy,$$

where C is a circle of radius 3 with center at the origin of the xy plane and is traversed in the positive sense.

BIBLIOGRAPHY

I. Newton, "Philosophiae Naturalis Principia Mathematica." London, 1687.

D. Bernoulli, "Hydrodynamica, Sive de Viribus et Motibus Fluidorum Commentarii." Strasbourg, 1738.

L. Euler, Principes generaux de l'etat d'equilibre des fluides, *Mem. Acad. Roy. Sci. Berlin* **11**, 217–273 (1755).

L. Euler, Principes generaux du mouvement des fluides, *Mem. Acad. Roy. Sci. Berlin* **11**, 274–315 (1755).

L. Euler, Continuation des recherches sur la theorie du mouvement des fluides, *Mem. Acad. Roy. Sci. Berlin* **11**, 316–361 (1755).

J. L. Lagrange, Memoire sur la theorie du mouvement des fluides, *Nouv. Mem. Acad. Sci. Berlin* 151–198 (1781).

L. M. NAVIER, Memoire sur les lois du mouvement des fluides, *Mem. Acad. Roy. Sci.* **6** (1823).

G. G. STOKES, On the theories of the internal friction of fluids in motion and of the equilibrium and motion of elastic solids, *Trans. Cambridge Phil. Soc.* **8** (1845).

H. VON HELMHOLTZ, Über Integrale der hydrodynamischen Gleichungen, welche den Wirbelwegung Entstrechen, *J. Reine Angew. Math.* **55**, 25–55 (1858). [Translated by P. G. Tait in *Phil. Mag. Ser.* 4, **33** 485–510 (1867).]

H. VON HELMHOLTZ, Über Discontinuichiche Flussigkeitsbewegungen, *Phil. Mag. Ser.* 4, **36**, 337–345 (1868).

O. REYNOLDS, *Trans. Roy. Soc.* (*London*) **174A**, 935–982 (1883).

O. REYNOLDS, *Trans. Roy. Soc.* (*London*) **186A**, 123–164 (1895).

L. PRANDTL, Verbandl d. III, *Intern. Math. Kongr., Heidelberg* (1904). [Translated in NACA TM 452 (1928).]

G. E. HAY, "Vector and Tensor Analysis." Dover, New York, 1953.

C. TRUESDELL, The mechanical foundations of elasticity and fluid dynamics, *J. Ratl. Mech. and Anal.* **1**, 125–300 (1952).

J. L. SYNGE, and A. SCHILD, "Tensor Calculus." Univ. of Toronto Press, 1956.

A. C. ERINGEN, "Nonlinear Theory of Continuous Media." McGraw-Hill, New York, 1962.

B. COLEMAN and W. NOLL, Foundations of linear viscoelasticity, *Rev. Mod. Phys.* **33**, 239–249 (1961).

H. GREENSMITH and R. RIVLIN, The hydrodynamics of non-Newtonian fluids. III. The normal stress effect in high-polymer solutions. *Phil. Trans. Roy. Soc.* **A245**, 399–428 (1954).

J. G. OLDROYD, *Proc. Roy. Soc.* **A200**, 523 (1950).

R. S. RIVLIN, in "Viscoelastic Fluids in Research Frontiers in Fluid Dynamics" (G. Seeger and G. Temple, eds.), pp. 144–170. Wiley (Interscience), New York, 1965.

J. EISILE and R. MASON, "Applied Matrix and Tensor Analysis." Wiley (Interscience), New York, 1970.

CHAPTER TWO *THE PHYSICAL PROPERTIES OF FLUIDS*

According to the molecular theory of matter, the three states of matter—solid, liquid, and gaseous—differ in their average spacings of molecules.

The liquid and gas states are referred to generally as fluids. The fluid state is distinguished by the relative ease of mobility of the molecules. In a solid, vibrational motion about equilibrium positions is the predominant motion. In the fluid state, rotational and translational motion are predominant.

Thus, fluids have the unique property that they are easily deformed—they do not have preferred shapes. In fact, it is the relative motion possible in a fluid under the influence of external forces which gives rise to the science of fluid dynamics.

In spite of the molecular structure of matter, in fluid dynamics, fluids are normally regarded as totally continuous and without voids. This simplifies the mathematical analysis of fluid motion.

It is, however, the nature of the intermolecular forces which is responsible for the fact that fluids can exist in one of two stable forms—gases or liquids. Although liquids are generally much more dense than gases, this is not the most basic difference. The basic difference lies in their bulk elasticity or, in other words, in their compressibility. Gases are much more compressible than liquids. So any motion of a gas involving appreciable variations in pressure will result in significant changes in specific volume. In some circumstances, the motions of a fluid are accompanied only by slight variations in pressure, and in this case gases and liquids behave similarly and are then classed as incompressible fluids.

2.1 Intermolecular Forces

The form of the interaction between molecules can be superficially recognized. The force must be repulsive at very small intermolecular distances when the molecules actually come into contact. This is at an intermolecule center distance of about 10^{-8} cm for simple molecules. At larger separations, the force must be attractive or else a gas would never condense since there would be no force holding it together. So one can draw a smooth potential energy curve (Fig. 2.1) which gives an attractive force at large distances and a repulsive one at small distances. If the atoms were hard spheres, the force should become infinite when they touched—as shown by the dotted curve in Fig. 2.1. The solid curve is for the more realistic case of "soft spheres."

For given molecular mass and fluid density, the average distance between adjoining molecule centers can be computed. For simple molecules,

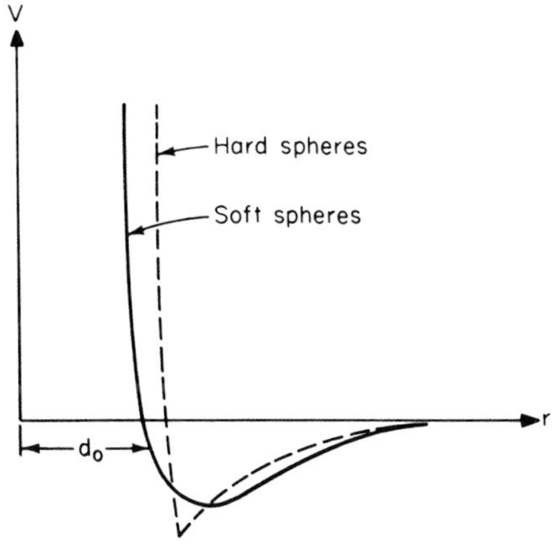

Fig. 2.1

this shows that the average spacing between molecules in the gaseous phase at normal temperature and pressure is about 10 d_0, whereas in liquids it is about d_0. So in gases under normal conditions the molecules are so far apart that only weak cohesive forces act, except when two molecules come close to each other. Thus, the concept of an ideal, or perfect, gas is commonly used. In such a gas, the potential energy of a molecule in the force field of the surrounding molecules is much less than its kinetic energy. In other words, intermolecular effects are negligible and the molecules move independently.

In liquids, the molecules are well within the strong force fields of their neighbors at all times. The theory of the liquid state is still not satisfactory. But it appears that the molecules are partially ordered at least in groups, which can form arrays and are periodically broken into smaller groups. Any shear force applied to a liquid will thus produce a deformation which continually increases in magnitude as long as the force is applied.

2.1.1 Van der Waals' Equation of State

The kind of effect which intermolecular forces have on the equations of state of a perfect gas can be illustrated by examining this modified equation of state.

In the hard sphere model of a gas, each molecule has a definite volume so that if V is the volume of the vessel containing 1 mole (N_0 molecules) of the gas, then the volume which is not occupied by molecules is $V - b$, where b is approximately N_0 times the volume of one molecule. The pressure is thereby increased, so that instead of $P = kT/V$ one should write

$$P = RT/(V - b).$$

This is called the *Clausius equation of state.*

To relate b more carefully to the molecular diameter, note that any given molecule excludes the centers of other molecules from a sphere circumscribed about the center of the given molecule (Fig. 2.2). The radius of this sphere is equal to the diameter d of the molecule, so that its volume of exclusion is eight times the molecular volume. Each one of a pair of molecules excludes the other from a volume of this size. So

Fig. 2.2

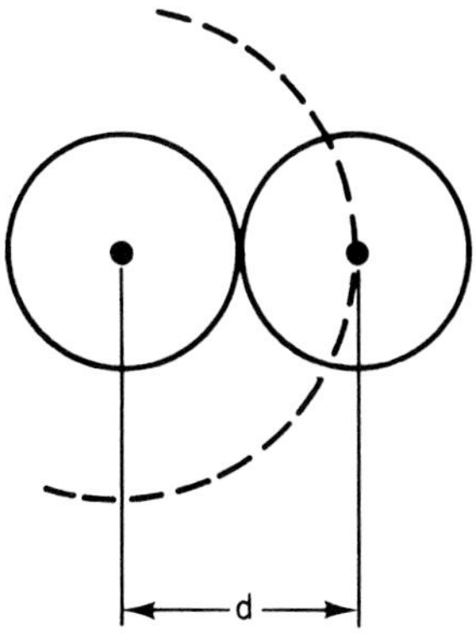

the total excluded volume in 1 mole is $\frac{1}{2}N_0$ times the molecular volume,

$$b = \tfrac{1}{2}N_0\tfrac{4}{3}\pi\, d^3.$$

To include attractive forces, one must consider the reduced velocity with which the molecules on the average strike the walls of the container. The reduction in pressure will be proportional both to the number of molecules striking the wall and to the number of molecules in the interior of the container. Both are proportion to the number density of the molecules, n. Then P will be decreased proportionally to n^2, or $1/V^2$. Thus,

$$P = \frac{RT}{V - b} - \frac{a}{V^2},$$

where a is a constant.

This is called the *van der Waals equation of state*. It describes the departure to be expected from the ideal gas law due to the effect of intermolecular forces.

It is found that if a and b are made functions of temperature the agreement with experiment is better. The best approach is to use a power series expansion. The form

$$\frac{PV}{RT} = 1 + \frac{B(T)}{V} + \frac{C(T)}{V^2} + \frac{D(T)}{V^3} + \cdots$$

is called the *virial equation of state*.

This form also holds for dense gases or liquids. The analytical form of the expressions for the first, second, etc. virials are very complex and obtaining them is a major objective in statistical thermodynamics.

2.2 The Continuum Concept

In many cases problems involve systems in which the dimensions are very large compared with molecular distances. One is interested in the statistical average properties and the behavior of large numbers of molecules, and not in that of individual molecules (that is, macroscopic, and not microscopic, properties are of interest).

As individual molecules are not being considered, the fluid can be regarded as a continuous substance. A continuum model of the fluid is adopted. Physical quantities such as the mass and momentum of the matter contained in a very small volume are regarded as being spread uniformly throughout that volume.

With normal measuring instruments (transducers, hot-wire anemometers, etc.), the continuous and smoothly varying properties of fluids are easily demonstrated and support the continuum hypothesis.

The sensitive volume of the instrument is usually chosen so that the property being measured does not change with the volume (the measurement is "local"). If the sensitive volume is reduced so much that it contains only a few molecules at the time of observation, then the measurement will vary irregularly from time to time. This is due to the statistical fluctuations in the number and kind of molecules in the sensitive volume. Figure 2.3 illustrates how the measured property, for example, density, might vary.

Under normal conditions, a cubic millimeter of air contains 2.7×10^{16} molecules. One is usually involved with dimensions of 1 cm or more, and

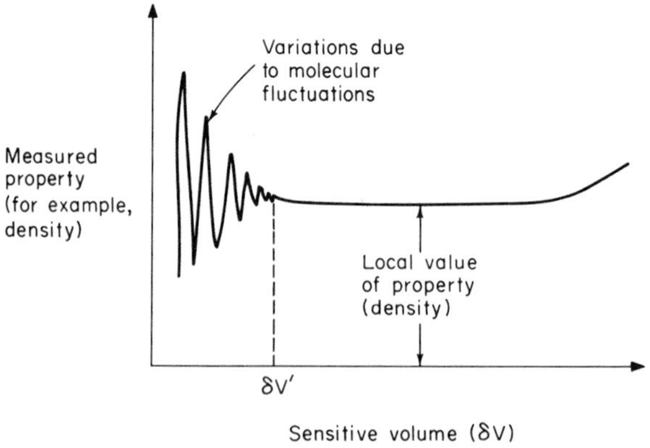

Fig. 2.3

very little variation in the physical and dynamical properties of the fluid occurs over a distance of 10^{-3} cm (except perhaps in a shock wave). Thus, an instrument with a sensitive volume of 10^{-9} cm^3 would still give a measure of a local property. This volume still contains more than 10^{10} molecules of air, say at NTP, and a property average over such a number is independent of the actual number (law of large numbers). In dealing with the structure of shock fronts, or with the flow of rarefied gases, the continuum approach of classical fluid dynamics and thermodynamics must be abandoned and replaced by the microscopic approach of kinetic theory and statistical mechanics.

In continuum mechanics one assumes that the macroscopic fluid properties, for example mean density, mean pressure, and mean viscosity, vary continuously with (a) the size of the lump of fluid considered, (b) the position in the fluid system, and (c) the time. In (a), the variation becomes imperceptible when the element, or lump, is very small but still large enough to satisfy the continuum criterion. Such an element is called a *fluid particle*. The mean properties of the fluid particle are assigned to a point in space, so that a field representation may be used for continuum properties. Thus, fluid properties, for example density, pressure, and velocity are expressed as continuous functions of position and time only. On this basis, it is possible to establish equations governing the motion of a fluid, which are independent, in their form, of the nature of the particle structure. So gases and liquids may be treated together.

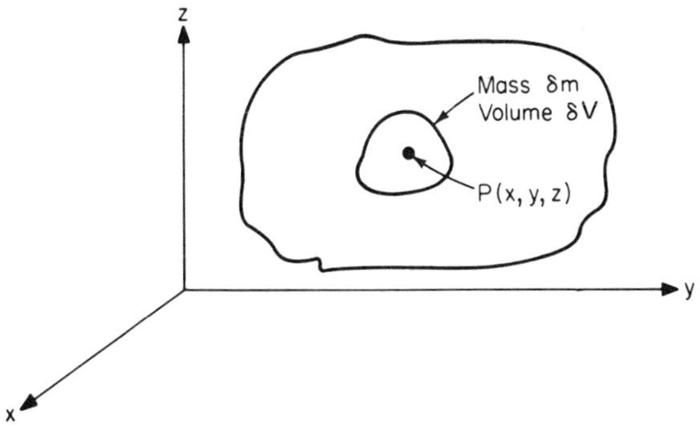

Fig. 2.4

Consider, as an illustration, the definition of the density of a fluid at a given point. Figure 2.4 shows a fluid mass δm in a small volume δV around the point $P(x, y, z)$ in a continuous fluid. The mean density of the fluid in this volume is defined as $\delta m/\delta V$. As the volume V is allowed

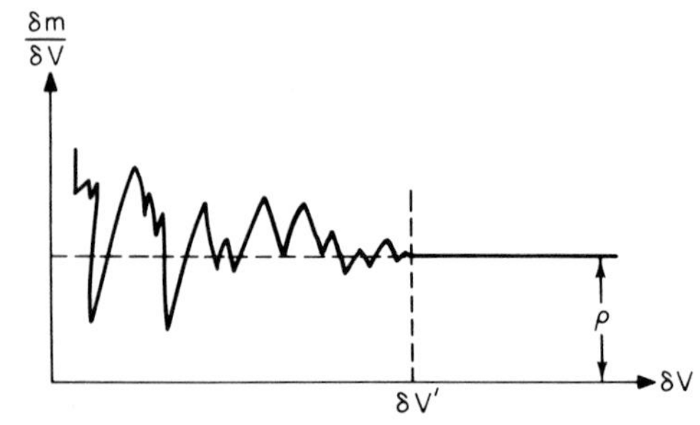

Fig. 2.5

to shrink about P, Fig. 2.5 shows how $\varrho = \delta m/\delta V$ varies with δV. When δV is shrunk below $\delta V'$, the mean density starts to fluctuate wildly due to the fluctuation of the small number of molecules in the volume. So one cannot fix a definite value of ϱ when $\delta V \leq \delta V'$. The density at P is defined as

$$\varrho \equiv \lim_{\delta V \to \delta V'} \frac{\delta m}{\delta V}. \tag{2.1}$$

The field representation for ϱ is written as

$$\varrho = \varrho(x, y, z, t). \tag{2.2}$$

This, of course, is a scalar density field. There are also vector fields such as velocity, and tensor fields such as stress.

2.2.1 Stress in a Continuum

There are two kinds of forces which act on matter in bulk. Volume forces, which are long range, can penetrate into the interior of a fluid, acting on all fluid elements. Such forces are those of gravity, centrifugal force, and electromagnetic forces in a charged fluid. Within a small element of volume, volume forces, due to their long range, act equally on all the elements within the volume, and the total force is proportional to the size of the volume element. Thus, the name. They are also called *body forces*. The total body force acting at time t on a fluid element of volume δV with position vector $\mathbf{r}$ will be designated as

$$\mathbf{F}(\mathbf{r}, t)\varrho\, \delta V, \tag{2.3}$$

where ϱ is the fluid density. In the case of the earth's gravitational field,

$$\mathbf{F} = \mathbf{g}.$$

The second kind of forces are stress, or surface, forces, and these are characterized by being short ranged. These forces are molecular in origin and are appreciable only when the distance between interacting elements is of the order of the intermolecular distance. Thus, these forces are negligible unless the interacting elements are in actual mechanical contact.

The force at a common boundary between two gaseous fluid masses is due to transport of momentum across that boundary by migrating molecules. In the case of liquids, intermolecular forces across contact surfaces are significant. But as mentioned previously, the laws of continuum mechanics do *not* depend on the molecular origin of these contact forces.

These surface (contact) forces between a fluid element and its surroundings can act only on a thin layer next to the boundary of the element. So the total surface force acting on the element is determined by its surface area. It is convenient to consider a plane surface fluid element, so that the total surface (short-range, or contact) force is the total force exerted on the fluid on one side of the element by that on the other.

Let the elemental area be δA and the total force exerted across it be $\mathbf{F}(\mathbf{n}, \mathbf{r}, t)$, where $\mathbf{n}$ is the normal to the surface element. Then the force per unit area, or the stress, at a point in the fluid is defined as

$$\mathbf{t} = \lim_{\delta A \to 0} \frac{\mathbf{F}}{\delta A}; \tag{2.4}$$

$\mathbf{n}$ is normally taken as the outward pointing unit normal vector (Fig. 2.6).

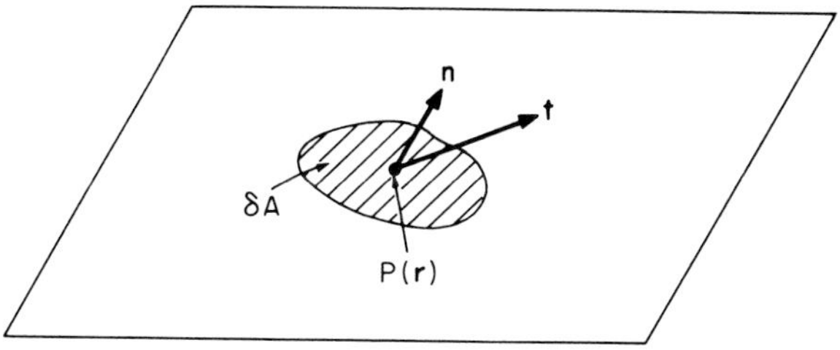

Fig. 2.6

The stress component in the direction of **n** is called *tensile stress*. Negative normal stress is known as *compressive stress*. As fluids in stable thermodynamic equilibrium do not support tension, it is the compressive, or negative, normal stress which is of most concern.

The component of **t** tangent to the surface at the point in question is known as the *tangential, or shear, stress*.

2.2.2 Stress Tensor

The stress at a point, as given by Eq. (2.4), depends in fact on two vector quantities **F** and ∂A. The elemental area is also a vector quantity. Thus, nine scalar quantities are required to completely specify the state of stress at a given point. Hence, the stress at a point is a tensor quantity of second order.

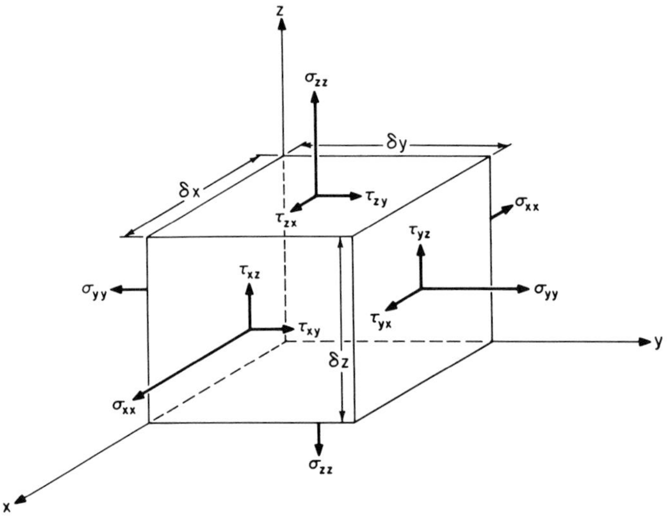

Fig. 2.7

The volume element, shown in Fig. 2.7, represents a fluid element isolated as a free body. The surface force per unit area, or stress vector, on each surface is resolved into a normal component and two tangential (shear) components parallel to the three Cartesian coordinates.

The surface stress components shown are average values on each surface that passes through the fluid element.

Double subscripts are used to identify the component stresses. The leading subscript indicates the direction of the normal to the surface on which the component stress acts. The second subscript denotes the direction of the stress component itself. Thus, the normal stress components have repeated subscripts, for example σ_{xx}. The σ_{xy} indicates the shear stress acting in the direction of the y axis, along the surface perpendicular to the x axis.

Arbitrarily it will be agreed that area vectors pointing out of a volume element are positive. Therefore, a stress component is positive if the stress itself, and the area vector of the surface on which it acts, both have the same direction. If they point in opposite directions, the stress component is negative.

Tensile stresses are thus positive. Shear stresses on faces farthest from the reference planes are positive if they point in the positive direction of the reference axes, while those on the faces nearest the reference planes are negative if they point in the positive directions of the reference axes. All the stress components shown in Fig. 2.7 are positive.

A stress matrix of the scalar components of the stress tensor acting on a fluid particle can be constructed as follows:

$$\begin{pmatrix} \sigma_{xx} & \tau_{xy} & \tau_{xz} \\ \tau_{yx} & \sigma_{yy} & \tau_{yz} \\ \tau_{zx} & \tau_{zy} & \sigma_{zz} \end{pmatrix}.$$

It can be shown that this is a symmetric matrix.

Taking moments about the x, y, and z axes, respectively, one obtains the moment equations

$$\begin{aligned} (\tau_{yz}\,\delta z\,\delta x)\,\delta y &= (\tau_{zy}\,\delta y\,\delta x)\,\delta z\,, \\ (\tau_{xz}\,\delta z\,\delta y)\,\delta x &= (\tau_{zx}\,\delta x\,\delta y)\,\delta z\,, \\ (\tau_{xy}\,\delta y\,\delta z)\,\delta x &= (\tau_{yx}\,\delta x\,\delta z)\,\delta y\,. \end{aligned} \tag{2.5}$$

Then

$$\tau_{yz} = \tau_{zy}\,, \qquad \tau_{xz} = \tau_{zx}\,, \qquad \tau_{xy} = \tau_{yx}\,.$$

There are only three independent shear stresses, and the stress tensor becomes a symmetric matrix:

$$\begin{pmatrix} \sigma_{xx} & \tau_{xy} & \tau_{xz} \\ \tau_{xy} & \sigma_{yy} & \tau_{yz} \\ \tau_{xz} & \tau_{yz} & \sigma_{zz} \end{pmatrix}.$$

An element of the tensor can be written as σ_{ij}, where σ_{ij} is the ith component of the force per unit area exerted across a plane surface element normal to the j direction, at position $\mathbf{r}$ in the fluid and at time t.

To show that the six scalar stress components are enough to determine the state of stress at a given point, consider the volume element in the shape of a tetrahedron with three orthogonal faces (Fig. 2.8). In other words, an oblique plane is passed through the previous volume element.

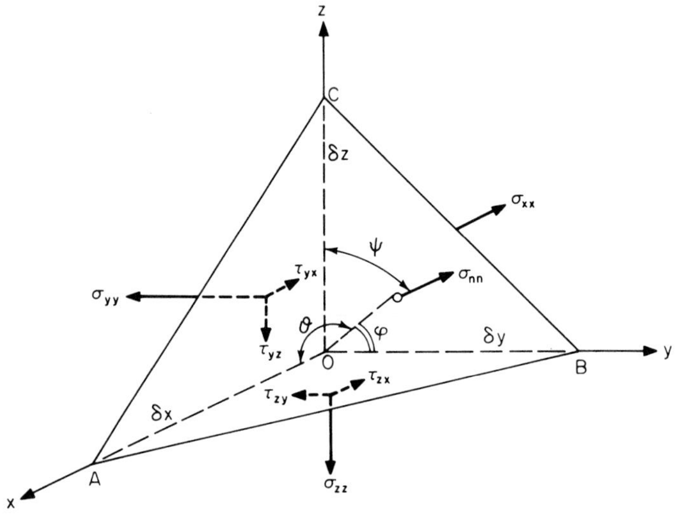

Fig. 2.8

Let the area of the oblique face be δA_0. The direction of this surface is defined by the angle which its outward normal makes with the three axes ϑ, φ, and ψ, respectively. The direction cosines of the surface ABC are

$$\cos\vartheta = l, \qquad \cos\varphi = m, \qquad \cos\psi = n. \tag{2.6}$$

The areas of the three orthogonal faces of the tetrahedron are related as follows:

$$OBC = \frac{\delta y\, \delta z}{2} = l\, \delta A_0, \qquad OAC = \frac{\delta x\, \delta z}{2} = m\, \delta A_0,$$
$$OAB = \frac{\delta x\, \delta y}{2} = n\, \delta A_0. \tag{2.7}$$

The normal stress σ_{nn} is perpendicular to δA_0, and the shear stress σ_{ss} lies in the plane of δA_0. The l, m, and n are also the direction cosines of σ_{nn}.

Application of the force/acceleration relationship of Newton, in the direction of σ_{nn}, leads to the equation

$$\begin{aligned}\sigma_{nn}\,\delta A_0 &- \sigma_{xx}(OBC)l - \tau_{xy}(OBC)m - \tau_{xz}(OBC)n \\ &- \tau_{yx}(OAC)l - \sigma_{yy}(OAC)m - \tau_{yz}(OAC)n - \tau_{zx}(OAB)l \\ &- \tau_{zy}(OAB)m - \sigma_{zz}(OAC)n + \varrho\,\frac{\delta x\,\delta y\,\delta z}{6}\,g_n = \varrho\,\frac{\delta x\,\delta y\,\delta z}{6}\,a_n,\end{aligned} \tag{2.8}$$

where g is the gravitational acceleration component in the σ_{nn} direction, and a_n is the acceleration component. As the volumes are an order of magnitude less than the areas, the acceleration terms may be dropped. So, the equation becomes, on using Eq. (2.7),

$$\sigma_{nn} = \sigma_{xx}l^2 + \sigma_{yy}m^2 + \sigma_{zz}n^2 + 2\tau_{xy}lm + 2\tau_{yz}mn + 2\tau_{zx}nl. \tag{2.9}$$

Note The last term on the right-hand side of Eq. (2.8) is, of course, a body force term. In this case the body force is taken as much smaller than the surface force and so is neglected. This is not always justifiable.

2.2.3 Stress Tensor in a Fluid at Rest

When a fluid is at rest, or in uniform motion, all the shear stress components are zero everywhere. Then the force/acceleration law can be written for the x, y, and z axes separately. Dropping the acceleration terms as before, one gets the result

$$\sigma_{xx} = \sigma_{nn}, \qquad \sigma_{yy} = \sigma_{nn}, \qquad \sigma_{zz} = \sigma_{nn}. \tag{2.10}$$

Hence, in the absence of shear stresses, the normal stress at a point is the same in all directions:

$$\sigma_{xx} = \sigma_{yy} = \sigma_{zz} = \sigma_{nn} = -p. \tag{2.11}$$

Thus, for a fluid at rest, or in uniform motion, the normal stress is simply the negative of the pressure. In the absence of shear stresses, the tensor stress field degenerates into a scalar pressure field.

It can be shown generally that it is always possible to choose the directions of the orthogonal axes of reference so that the off-diagonal terms of a symmetric second-order tensor are zero. These are called the *principal axes* of the stress tensor σ_{ij} at the given point $\mathbf{r}$, the diagonal elements then being called *principal stresses*. The trace of this stress matrix is invariant with respect to transformation:

$$\sigma_{xx} + \sigma_{yy} + \sigma_{zz} = \sigma_{ii} = \text{constant}.$$

Thus, in a fluid at rest, the principal stresses are all the same and equal to $\frac{1}{3}\sigma_{ii}$ at all points in the fluid. The stress tensor is therefore everywhere isotropic and *all* orthogonal axes are principal axes for the stress tensor.

Fluids at rest are normally in a state of compression, and it is usual to write the stress tensor for a fluid at rest as

$$\sigma_{ij} = -p\delta_{ij},$$

where, as has been indicated before,

$$p = -\tfrac{1}{3}\sigma_{ii}$$

is the static fluid pressure (in general, a function of position).

2.3 Fluid Statics

The forces in a fluid at rest and the forces which act at fluid interfaces now will be considered in some detail.

Consider the three-dimensional fluid element shown in Fig. 2.9. Pressure (force per unit area) is assumed to vary continuously throughout

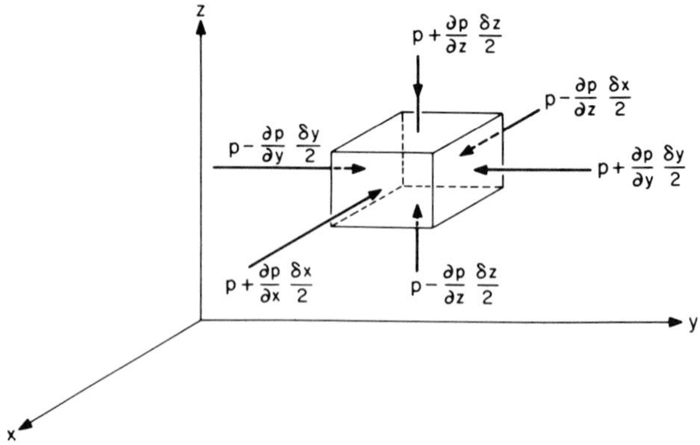

Fig. 2.9

the element. The center of the element is at (x, y, z) and the pressure there is $p(x, y, z)$.

Consider the pressure variation along a line parallel to the x axis and passing through the point (x, y, z) (Fig. 2.10). The slope of the tangent

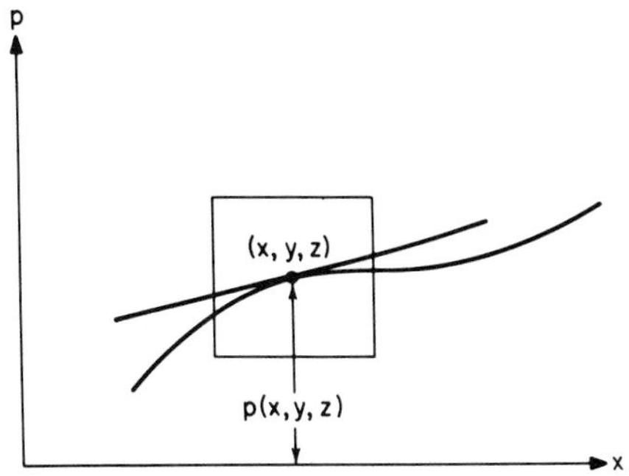

Fig. 2.10

to this curve at this point is $\partial p/\partial x$. To a first approximation, the average pressure at the left side is

$$p - \frac{\partial p}{\partial x}\frac{\delta x}{2},$$

and that at the right side is

$$p + \frac{\partial p}{\partial x}\frac{\delta x}{2}.$$

As the element is allowed to shrink toward the point (x, y, z), these values become more exact.

The resultant pressure force in the positive x direction is

$$\delta P_x = \left(p - \frac{\partial p}{\partial x}\frac{\delta x}{2}\right)\delta y\,\delta z - \left(p + \frac{\partial p}{\partial x}\frac{\delta x}{2}\right)\delta y\,\delta z = -\frac{\partial p}{\partial x}\,\delta x\,\delta y\,\delta z. \tag{2.12}$$

Similarly,

$$\delta P_y = -\frac{\partial p}{\partial y}\,\delta x\,\delta y\,\delta z, \tag{2.13}$$

$$\delta P_z = -\frac{\partial p}{\partial z}\,\delta x\,\delta y\,\delta z. \tag{2.14}$$

The total pressure force on the element is then

$$\delta\mathbf{P} = \delta P_x\,\mathbf{i} + \delta P_y\,\mathbf{j} + \delta P_z\,\mathbf{k} = -\left(\frac{\partial p}{\partial x}\mathbf{i} + \frac{\partial p}{\delta y}\mathbf{j} + \frac{\partial p}{\partial z}\mathbf{k}\right)\delta x\,\delta y\,\delta z. \tag{2.15}$$

The force per unit volume is therefore

$$\frac{\delta\mathbf{P}}{\delta x\,\delta y\,\delta z} = \mathbf{F} = -\left(\frac{\partial p}{\partial x}\mathbf{i} + \frac{\partial p}{\delta y}\mathbf{j} + \frac{\partial p}{\delta z}\mathbf{k}\right) = -\boldsymbol{\nabla} p, \tag{2.16}$$

where $\boldsymbol{\nabla}$ is the gradient operator. This equation is the necessary and sufficient condition for a fluid to be in mechanical equilibrium. It is an important equation in the mechanics of continuum in that it relates a vector force and a scalar pressure field.

Note The two common types of body force per unit volume are gravity and centrifugal. In the case of gravity,

$$\mathbf{F} \equiv \varrho\mathbf{g}, \tag{2.17}$$

where g is a constant vector and points vertically down, and for centrifugal force,

$$\mathbf{F} \equiv \varrho(\textbf{centrifugal acceleration}). \tag{2.18}$$

2.3.1 Fluid at Rest under Gravity

The case in which gravity is the only volume force acting is important. There are two specific cases of particular importance. In the first, the fluid mass involved is large and isolated, and the gravitational attraction of *other* parts of the fluid provides the volume force on any particular fluid element; for example, a gaseous star. In the second case, the fluid mass is much smaller than nearby matter and the gravitational field is approximately constant over the region occupied by the fluid.

The first case can be referred to as a self-gravitating fluid. If Ψ is the gravitational potential, then

$$\mathbf{F} = -\varrho\,\boldsymbol{\nabla}\Psi, \tag{2.19}$$

where Ψ is related to the distribution of density by the equation

$$\nabla^2\Psi = 4\pi G\varrho, \tag{2.20}$$

where G is the constant of gravitational force. Using Eq. (2.16), one therefore has

$$\nabla \cdot \left(\frac{\nabla p}{\varrho}\right) = -4\pi G\varrho. \tag{2.21}$$

In the case of a spherically symmetric distribution of density and pressure, Eq. (2.21) becomes

$$\frac{d}{dr}\left(\frac{r^2}{\varrho}\frac{dp}{dr}\right) = -4\pi G r^2 \varrho. \tag{2.22}$$

One cannot develop this further without knowing the distribution of density in, for example, the gaseous star. As a simple model one can take the density to be a function of pressure only. For example, assume that

$$p \propto \varrho^{1+(1/n)}, \tag{2.23}$$

where $n \geq 0$. Then for any given n, Eq. (2.22) can be integrated numerically. For $n = 0$ and $n = 5$, analytical solutions are available.

(i) $n = 0$, which implies a fluid of uniform density ϱ. Then,

$$p = \tfrac{2}{3}\pi G \varrho_0^2 (a^2 - r^2), \tag{2.24}$$

where $r = a$ can be interpreted as the outer edge of the star.

(ii) $n = 5$. It is found that

$$p = C\varrho^{6/5} = \frac{27 a^3 C^{5/2}}{(2\pi G)^{3/2}(a^2 + r^2)^3}, \tag{2.25}$$

where p and ϱ are nonzero for *all* r in this case and there is no definite outer boundary. However, the total mass of fluid is finite.

In the second case, there is a constant body force due to gravity. Then

$$\mathbf{F} = \varrho\mathbf{g}, \qquad \Psi = -\mathbf{g} \cdot \mathbf{r}, \tag{2.26}$$

and the equation for the pressure in a fluid at rest is

$$\nabla p = \varrho\mathbf{g}. \tag{2.27}$$

If the z axis of the Cartesian coordinate system is the vertical axis (positive upward) so that $\mathbf{g} \cdot \mathbf{r} = -gz$, then Eq. (2.27) becomes

$$\frac{dp}{dz} = -g\varrho(z). \tag{2.28}$$

Again one needs information on the density distribution to proceed further.

Case (i) Fluid of uniform density Integration of (2.28) leads to

$$p = p_0 - \varrho g z, \tag{2.29}$$

the well-known pressure/height relation of hydrostatics.

Case (ii)

$$p/\varrho = \text{constant} = gZ,$$

where Z is a height factor. This law holds approximately in the earth's atmosphere (ignoring thermal effects). It leads to a pressure/height relation in the atmosphere of the form

$$p = p_0 e^{-z/H}, \tag{2.30}$$

where p_0 is the pressure at the earth's surface ($z = 0$). Thus p and ϱ decrease by a factor e^{-1} over a height interval H, and so H can be called a "scale height" for the atmosphere. For air at a temperature of 0°C, H is 8 km.

A more realistic pressure/density relation in the atmosphere is

$$\frac{p(z)}{\varrho^n(z)} = \frac{p_0}{\varrho_0{}^n} = \text{constant}, \tag{2.31}$$

where the subscript 0 refers to the earth's surface and n is taken as having the value 1.235.

Now the gravitational acceleration $g(z)$ at any height z varies inversely as the square of the distance from the earth's center:

$$g(z) = \left(\frac{r}{r+z}\right)^2 g_0 = \frac{g_0}{[1 + (z/r)]^2}, \tag{2.32}$$

where g_0 is the gravitational acceleration at the earth's surface and r is the mean value of the earth's radius (about 3960 miles). From Eq. (2.31),

$$\varrho(z) = \varrho_0 \left(\frac{p(z)}{p_0}\right)^{1/n}. \tag{2.33}$$

Then,

$$\frac{dp(z)}{dz} = -\varrho(z) g(z) = -g(z)\varrho_0 \left(\frac{p(z)}{p_0}\right)^{1/n}. \tag{2.34}$$

Assuming that z/r is small, the expression for $g(z)$ can be simplified as follows:

$$g(z) = g_0\left(1 + \frac{z}{r}\right)^{-2} = g_0\left(1 - \frac{2z}{r} + \cdots\right). \tag{2.35}$$

Using this in Eq. (2.34) and rearranging, one obtains

$$p(z)^{-1/n}\, dp(z) = -\varrho_0 g_0 p^{-1/n}\left(1 - \frac{2z}{r}\right) dz.$$

Integrating this from the earth's surface (z_0) to the height z, one gets the relation

$$\frac{n}{n-1}\left[p(z)^{((n-1)/n)} - p^{((n-1)/n)}\right] = -\varrho_0 g_0 p_0^{-1/n}\left[(z - z_0) - \frac{(z - z_0)^2}{r}\right]. \tag{2.36}$$

The last term on the right-hand side is the correction for the variation of gravitational acceleration.

2.3.2 Pressure in an Incompressible Fluid with a Free Surface

Integrating Eq. (2.28), one obtains

$$p - p_0 = -g\varrho z, \tag{2.37}$$

where p_0 is the (static) pressure at $z = 0$. When one is dealing with a body of liquid with a free surface, the $z = 0$ plane is taken as the free surface, and with p_a denoting the atmospheric pressure on the free surface, one gets

$$p = p_a + \varrho g h, \tag{2.38}$$

where $h = -z$ is the vertical distance below the free surface, as shown in Fig. 2.11. The term ϱgh is the pressure at a level in the liquid in excess of the atmospheric pressure.

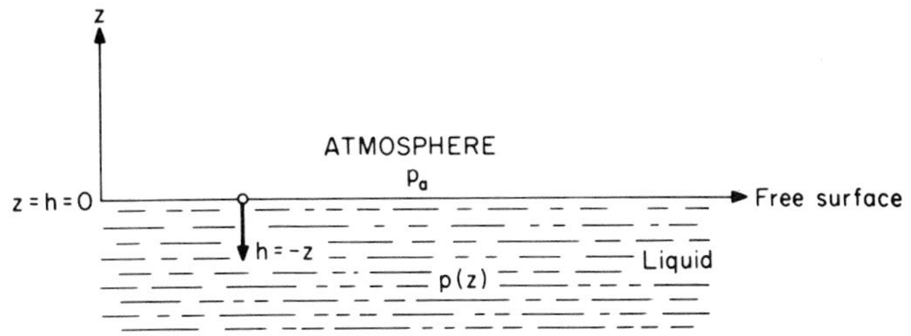

Fig. 2.11

2.4 Boundary Conditions between Two Fluid Media

2.4.1 Surface Tension

The free surfaces of a liquid act as though they were elastic membranes of constant force per unit length. This force is also evidenced by the fact that small liquid drops in air and bubbles in water assume a spherical form.

The origin of the phenomenon of surface tension lies in the intermolecular cohesive forces.

The total free energy of a system consisting of two uniform media of densities ϱ_1 and ϱ_2 and volumes V_1 and V_2 with interfacial area A is of the form

$$H = \varrho_1 V_1 F_1 + \varrho_2 V_2 F_2 + A\sigma, \tag{2.39}$$

where σ is the constant of proportionality, A the surface energy, and F_1 and F_2 the free energies per unit mass of the two media.

In any small reversible isothermal change in the fluid system, the total work done on the system equals the gain in total free energy. If the densities also remain unchanged, then the total work done on the system is $\sigma\ \delta A$. The work is done in *stretching* the interface—just as if it was a uniformly stretched membrane. So σ may be interpreted both as the free energy per unit area of the interface and as the *surface tension*. Across any line drawn on the interface there is exerted a force of magnitude σ per unit length in a direction normal to that line and tangential to the interface.

The mean free energy of a molecule in the bulk of the liquid is independent of its position. Within a distance from the interface less than the range of the cohesive molecular forces (about 10^{-7} cm), the free energy of a molecule changes with its position. Since this layer is very thin, all parts of the surface contribute equally to the term in Eq. (2.39), which corrects the total free energy for the presence of the interface.

The molecules which are near an interface with a gas experience an unbalanced cohesive force in the direction away from the interface. Thus, the interface tends to contract. Of course, σ is positive in this case. When an interface separates two liquids, the sign of σ can be positive or negative. For some pairs of liquids, such as alcohol and water, an interface cannot normally be observed, because these liquids are miscible.

The value of σ at a liquid–fluid interface in equilibrium may be greatly affected by the presence of adsorbed material at the liquid surface. Since the concentration of adsorbed material may vary from point to point on the interface, the surface tension varies from point to point and leads to unbalanced forces on an element of the surface. This can set up fluid motion in the interface. It also explains the motion on a water surface of a small boat with a piece of camphor attached.

2.4.2 Shape of the Interface between Stationary Fluids

It is assumed that the surface tension σ is constant over the interface, with the fluids stationary and in thermodynamic equilibrium. The problem of finding the geometrical shape of the interface consistent with mechanical equilibrium is generally difficult.

Consider a point O on the surface with a tangent plane (x, y), the origin of the coordinate system being at O. The equation of the surface can be written

$$z = F(x, y) = 0, \tag{2.40}$$

where the function F and its derivatives are zero at O. In the neighborhood of O the unit vector $\mathbf{n}$ normal to the surface has, to a first approximation, the components

$$-\frac{\delta F}{\delta x}, \quad -\frac{\delta F}{\delta y}, \quad \text{and} \quad 1.$$

Suppose δl is a line element of a closed curve around O. Then the resultant tensile force on this element of surface is

$$-\sigma \oint \mathbf{n} \times dl.$$

To second order, this is a force parallel to the normal at O (the z axis), of magnitude

$$-\sigma \oint \left(-\frac{\partial F}{\partial x}\,dy + \frac{\partial F}{\partial y}\,dx\right) = \sigma\left(\frac{\partial^2 F}{\partial x^2} + \frac{\partial^2 F}{\partial y^2}\right)_0 \delta A. \tag{2.41}$$

The tension, therefore, across the curve bounding the surface element has the same effect on the element as a *pressure* on the surface, of magnitude

$$\sigma\left(\frac{\delta^2 F}{\delta x^2} + \frac{\delta^2 F}{\delta y^2}\right)_0 = \sigma\left(\frac{1}{R_1} + \frac{1}{R_2}\right), \tag{2.42}$$

where R_1 and R_2 are the radii of curvature in mutually perpendicular

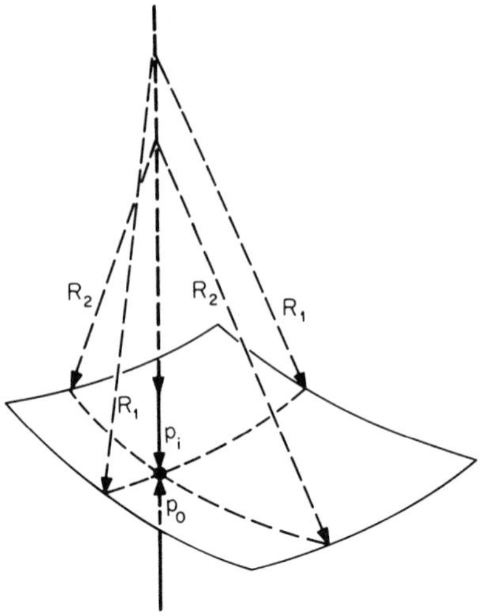

Fig. 2.12

planes, which are also perpendicular to the tangent plane, of the element of the free surface (Fig. 2.12). Since the interface has zero mass, a curved interface can be in equilibrium only if the surface tension pressure is balanced by a pressure difference between the fluids on the two sides of the interface, p_i and p_o, respectively:

$$\Delta P = P_i - P_o = \sigma\left(\frac{1}{R_1} + \frac{1}{R_2}\right). \tag{2.43}$$

Note ΔP is positive in the direction of the center of curvature of the interface.

2.4.3 Shape of a Liquid Surface near a Solid Wall

Suppose the interface separates a gas (pressure constant) and a liquid of density ϱ, in which the pressure variation with height is given by

$$p = p_0 - \varrho g z, \tag{2.44}$$

the relation for an incompressible fluid. The condition for equilibrium at any point of the interface is then

$$\varrho g z - \sigma\left(\frac{1}{R_1} + \frac{1}{R_2}\right) = \text{constant}, \tag{2.45}$$

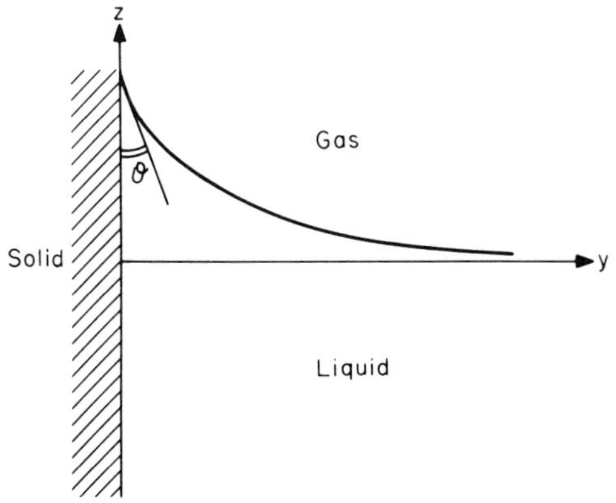

Fig. 2.13

where R_1 and R_2 are positive when the centers of curvature are on the gas side. Now consider a free liquid meeting a plane, vertical rigid wall, as shown in Fig. 2.13. The equation of the interface in this two-dimensional case is

$$z = F(y), \tag{2.46}$$

and the curvatures of the interface are

$$\frac{1}{R_1} = 0, \qquad \frac{1}{R_2} = \frac{F''}{(1 + F'^2)^{3/2}}, \tag{2.47}$$

where $F' \equiv \partial F/\partial y$. Using (2.46) and (2.47) in (2.45), one therefore obtains

$$\frac{\varrho g}{\sigma} F - \frac{F''}{(1 + F'^2)^{3/2}} = 0. \tag{2.48}$$

The constant is zero since $F = 0$ at $y = \infty$. Integrating once, one gets

$$\frac{1}{2} \frac{\varrho g}{\sigma} F^2 + \frac{1}{(1 + F'^2)^{1/2}} = C, \tag{2.49}$$

and the same boundary condition shows that $C = 1$. Now $\partial z/\partial y = F'$ and the second term in (2.49) can be rearranged as

$$\frac{dy}{(dy^2 + dz^2)^{1/2}} = \sin \vartheta,$$

where ϑ is the contact angle (known from the properties of the system).

So the height to which the liquid climbs up the wall is

$$h^2 = 2\frac{\sigma}{\varrho g}(1 - \sin\vartheta). \tag{2.50}$$

The boundary condition at $y = 0$, $F(0) = h$, can be used to fix the constant in the integration of Eq. (2.49). The result is

$$\frac{y}{k} = \cosh^{-1}\frac{2d}{F} - \cosh^{-1}\frac{2d}{h} + \left(4 - \frac{h^2}{d^2}\right)^{1/2} - \left(4 - \frac{F^2}{d^2}\right)^{1/2}, \tag{2.51}$$

where $k^2 = \sigma/\varrho g$.

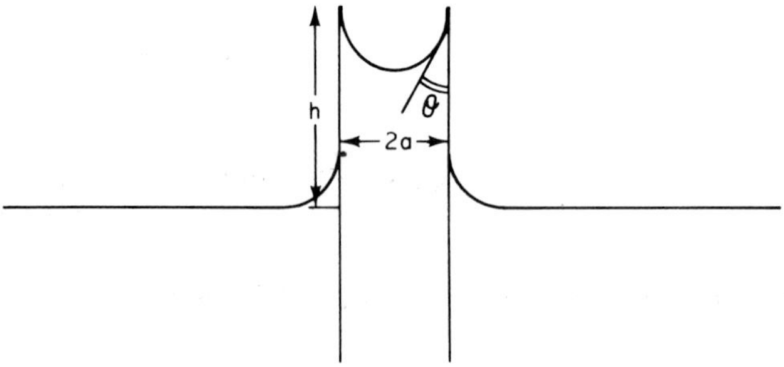

Fig. 2.14

This rise or fall of a liquid in contact with a wall is known as *capillarity*. In the case of the capillary rise in a small tube, the radius of curvature is $a/\cos\vartheta$ (Fig. 2.14). Now

$$\Delta p = \sigma\left(\frac{1}{R_1} + \frac{1}{R_2}\right) = \sigma\left(\frac{\cos\vartheta}{a} + \frac{\cos\vartheta}{a}\right) = \frac{2\sigma\cos\vartheta}{a}. \tag{2.52}$$

The condition for equilibrium of a column of height h is approximately

$$\varrho g h = \Delta p = \frac{2\sigma\cos\vartheta}{a}, \tag{2.53}$$

that is,

$$h = \frac{2k^2\cos\vartheta}{a}. \tag{2.54}$$

Thus, in very small tubes, the liquid may rise substantially. In the case of a liquid which does not "wet" the walls, $\vartheta > \pi/2$ and so h is negative, indicating a depression of the free surface in a tube.

2.5 Fluid Motion

The fundamental property of a fluid is that it cannot be in equilibrium in a state of shear stress. A shear stress is one that has a component tangential to a common surface in the fluid. As long as these tangential or shear stresses do not come into play, the fluid moves as a solid body. Normal stresses are, of course, not incompatible with equilibrium. In this first instance, only the normal stresses, which appear as the pressure in the fluid, will be considered. One can either have the objective of acquiring a knowledge of the velocity, pressure, and density at all points of space occupied by the fluid at all times, or one can determine the history of every particle. The equations of motion in the former case are called the Euler equations, and in the latter the Lagrangian form of the equations.

2.5.1 The Lagrange Equations

Each fluid particle is identified by its initial position (a, b, c) relative to the origin of the coordinate system at some arbitrary initial time t_0. The paths of fluid particles of fixed identity are followed. At each instant the position (x, y, z), density, temperature, state of stress, etc. associated with each fluid particle are functions of the initial position and of time. The coordinates of a fluid particle may be written

$$x = x(a, b, c, t), \qquad y = y(a, b, c, t), \qquad z = z(a, b, c, t), \tag{2.55}$$

and the velocity and acceleration components then are

$$\begin{aligned} u &= \frac{\partial x}{\partial t} & \qquad & a_x = \frac{\partial^2 x}{\partial t^2}, \\ v &= \frac{\partial y}{\delta t} & \text{and} \qquad & a_y = \frac{\partial^2 y}{\partial t^2}, \\ w &= \frac{\partial z}{\partial t} & \qquad & a_z = \frac{\partial^2 z}{\partial t^2}. \end{aligned} \tag{2.56}$$

So in solving a problem in this mode one seeks solutions in the form of parametric equations such as Eq. (2.55).

2.5.2 The Euler Equations

In the Eulerian method, the history of each fluid particle is not considered. The problem in this method is to determine the velocity components of the fluid motion at various positions (x, y, z) in the space at any time t, where

$$u = u(x, y, z, t), \qquad v = v(x, y, z, t), \qquad w = w(x, y, z, t). \tag{2.57}$$

It is assumed that u, v, w are finite and continuous functions of x, y, z and also of their first-order spatial derivatives ($\partial u/\partial x$, etc.). This is what is implied by continuous motion.

Let $F(x, y, z, t)$ be any continuous property (for example, velocity, temperature, density) which is a function of position and time. Now, at the time $t + \delta t$ the particle which was originally at the point (x, y, z) is in the position $(x + u\,\delta t, y + v\,\delta t, z + w\,\delta t)$, so that the corresponding value of F is

$$\begin{aligned} &F(x + u\,\delta t,\ y + v\,\delta t,\ z + w\,\delta t,\ t + \delta t) \\ &\quad = F + u\,\delta t\,\frac{\partial F}{\partial x} + v\,\delta t\,\frac{\partial F}{\partial y} + w\,\delta t\,\frac{\partial F}{\partial z} + \delta t\,\frac{\partial F}{\partial t}. \end{aligned} \tag{2.58}$$

Following Stokes, the term D/Dt will be used to denote a differentiation following the motion of the fluid. The new value of F is

$$F + \frac{DF}{Dt}\,\delta t,$$

and so

$$\frac{DF}{Dt} = \frac{\partial F}{\partial t} + u\,\frac{\partial F}{\partial x} + v\,\frac{\partial F}{\partial y} + w\,\frac{\partial F}{\partial z}. \tag{2.59}$$

The derivative DF/Dt is called the material or substantive derivative (or the derivative following the fluid). The last three terms on the right-hand side of (2.59) are called the convective derivatives, and they represent the change in F due to the convection of a fluid particle from one position to a second position, where the value of F is different.

In vector notation, Eq. (2.59) can be written compactly as

$$\frac{DF}{Dt} = \frac{\partial F}{\partial t} + (\mathbf{V} \cdot \boldsymbol{\nabla})F, \tag{2.60}$$

where

$$\mathbf{V} \equiv \mathbf{i}u + \mathbf{j}v + \mathbf{k}w \qquad \text{and} \qquad \boldsymbol{\nabla} \cdot \equiv i\frac{\partial}{\partial x} + j\frac{\partial}{\partial y} + k\frac{\partial}{\partial z},$$

if the Cartesian coordinate system is used.

This equation relates the material and spatial time derivatives of a scalar, or vector-field, function F. To establish the acceleration field of flow, suppose the velocity $\mathbf{V}$ is the continuum property F. The substantive derivative is then the acceleration **a**, and

$$\mathbf{a} = \frac{D\mathbf{V}}{Dt} = \frac{\partial V}{\partial t} + u\frac{\partial V}{\partial x} + v\frac{\partial V}{\partial y} + w\frac{\partial V}{\partial z} = \frac{\partial V}{\partial t} + (\mathbf{V} \cdot \boldsymbol{\nabla})\mathbf{V}. \tag{2.61}$$

This equation is known as the *d'Alembert–Euler acceleration formula.* The $\partial V/\partial t$ is the local acceleration and represents the acceleration felt by an observer fixed at a location in space. The local acceleration term exists only in unsteady flow. The term $(\mathbf{V} \cdot \boldsymbol{\nabla})\mathbf{V}$ is the acceleration felt by an observer at a given instant due to a change in position in space. It is the convective acceleration.

2.6 Transport Phenomena in Fluids

It is found normally that when two portions of matter are contiguous, any differences in material properties between them tend to get smoothed out in time. This involves some exchange or transfer of mechanical or thermal properties. It appears that interacting portions of matter generally tend to come to some equilibrium condition.

The total amount of the property involved (for example, mass) is generally conserved, and so the amount in one decreases while that in the other increases (or vice versa) as the equilibrium state is approached. Such exchanges come under the general heading of transport phenomena. The three most important basic properties which are exchanged are (a) matter (mass), (b) energy, and (c) momentum.

2.6.1 Transport of Matter

This occurs in a fluid mixture in which the composition varies with position. All the molecules are in continual random motion. If there is a larger proportion of a certain type of molecule on one side of an element of surface drawn in the fluid, a net flux of these molecules ($\otimes$) crosses this surface from the higher concentration side to the lower (Fig. 2.15). This nonzero flux J of a molecular constituent of the fluid is called *diffusion*. The concentration gradient is dC/dx. As long as the concentration gradient persists, the flux continues.

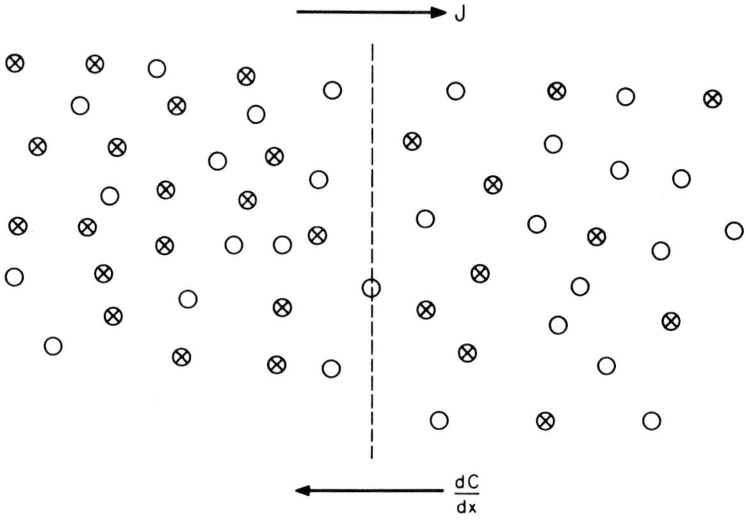

Fig. 2.15

2.6.2 Transport of Kinetic Energy

When two fluid regions are separated by a thin wall permeable to heat, kinetic energy will be transferred from the hotter to the colder region. When the temperatures are equalized, this flux of kinetic energy ceases and the system is in thermal equilibrium. The thin wall can be just an imaginary plane in a gas or liquid.

Now the average kinetic energy of a molecule is proportional to the temperature. So the molecules which randomly migrate from the hotter to the colder region carry more kinetic energy with them than those which cross in the opposite direction. There is thus a net transport of energy from the hot to the cold side. Through molecular collisions the "hot" molecules share their extra energy with those on the cold side.

Thus, the cold gas gets hotter and the hot gas colder, until thermal equilibrium is reached.

This net flux of molecular energy which occurs when the temperature is nonuniform is called *conduction* of heat.

2.6.3 Transport of Momentum

Consider an element of surface moving with the local continuous velocity, which separates slow-moving fluid on one side from faster-moving fluid on the other. As a molecule crosses the surface it carries with it a flow momentum. The molecule crossing to the slow side will carry more momentum with it than a molecule crossing to the fast side. Thus, while there is a velocity difference at the surface, there will be a flux of momentum across it. This flux of momentum appears as a tangential, or shear, stress in the surface. Transport of momentum then constitutes internal frictions in the fluid. A fluid which exhibits such internal friction is termed a *viscous fluid.*

These three kinds of transport involve a flux of some quantity and a gradient in some associated intensity. It will be noted that of these intensities, molecular concentration and kinetic energy are scalar in nature, while the third one, linear momentum, is a vector.

2.7 Scalar Intensities

Suppose $\mathbf{F}$ is a flux vector. Then the net transfer of the associated intensity per second across a surface element of area δA and with a unit normal $\mathbf{n}$ is

$$\mathbf{F} \cdot \mathbf{n} \, \delta A,$$

where the flux vector $\mathbf{F}$ is a function, generally, of the position in the fluid, $\mathbf{r}$. This expression is, of course, a scalar. Let the scalar intensity be denoted by W, where $W = W(\mathbf{r})$. Both W and $\mathbf{F}$ can also be functions of time. The objective is to deduce a relation between W and $\mathbf{F}$. A deduction in terms of molecular kinetics is extremely difficult. So based on experience, it is postulated that there is a *linear* relation between the two quantities. This hypothesis may be expressed as

$$F_i = k_{ij} \frac{\partial W}{\partial r_j}; \tag{2.62}$$

F_i and $\partial W/\partial r_j$ are vectors, and if (2.62) is to be true for all coordinate systems, then the transport coefficient k_{ij} must be a second-order tensor. In effect, Eq. (2.62) is the first term in a Taylor expansion in the components of ∇W. Experimental results indicate that this linear relation is very accurate for normal values of ∇W.

In an isotropic material, then, k_{ij} (which depends on local material properties) must have a form in which directional dependence does not exist. In this situation, all sets of orthogonal axes must be principal axes of the coefficient k_{ij}, and thus

$$k_{ij} = -k\,\delta_{ij}. \tag{2.63}$$

Then Eq. (2.62) can be written

$$\mathbf{F} = -k\,\nabla W, \tag{2.64}$$

which applies at all points of the fluid.

Now, the total transfer per second of fluid out of the volume enclosed by a closed surface A with outward normal $\mathbf{n}$ is

$$-\int \mathbf{F}\cdot\mathbf{n}\,dA = -\int k\mathbf{n}\cdot\nabla W\,dA = -\int \nabla\cdot(k\,\nabla W)\,dV, \tag{2.65}$$

where V is the enclosed volume. At present the fluid will be assumed to be at rest.

2.7.1 Equation of Diffusion

In this case, W represents the fraction of a certain type A of molecule in a mixture. A simple conservation law holds. The number of type-A molecules in a volume V of the fluid is $\int WN\,dV$, where N is the total number of molecules per unit volume. This number can change only by molecular transport across the surface. Therefore

$$\frac{\partial}{\partial t}\int WN\,dV = \int \nabla\cdot(k_{\mathrm{D}}\,\nabla W)\,dV,$$

and so

$$\int \left\{ \frac{\partial(WN)}{\partial t} - \nabla \cdot (k_D \nabla W) \right\} dV = 0, \tag{2.66}$$

where k_D is the value of k for the diffusion of type-A molecule. Now the total number of molecules is conserved—that is, N is a constant. Moreover, the above relation is true for all choices of a volume V situated entirely in the fluid, and so the integrand must be zero everywhere. Therefore

$$N \frac{\partial W}{\partial t} = \nabla \cdot (k_D \nabla W). \tag{2.67}$$

Normally, k_D is a function of position in the fluid (being affected by the local fluid state and the concentration W). However, if the gradient of k_D is small, then k_D can be taken as a constant, and Eq. (67) assumes the approximate form

$$N \frac{\partial W}{\partial t} = k_D \nabla^2 W. \tag{2.68}$$

This is known as the *diffusion equation*. It is usually written in the form

$$\frac{\partial W}{\partial t} = \varkappa_D \nabla^2 W, \tag{2.69}$$

where $\varkappa_D = k_D/N$ is the coefficient of diffusion of type-A molecules in the surrounding fluid. It has the dimensions L^2T^{-1}.

2.7.2 Equation of Heat Conduction

In this case, W represents the temperature of the fluid. The law of conservation of energy can be used. The quantity transported is heat and the rate of transport of heat into a small volume δV across the bounding surface is, according to Eq. (2.65),

$$\nabla \cdot (k_H \nabla T)\, \delta V,$$

where k_H is the appropriate k for a heat conductor. It is called the *thermal conductivity*.

Now,

$$\varrho = \frac{m}{\delta V},$$

where ϱ is the fluid density and m the mass of the volume δV. So the above expression can be written

$$\nabla \cdot (k_{\mathrm{H}} \nabla T) \frac{m}{\varrho}.$$

Due to this heat flux, the thermodynamic state of the fluid changes. If the change is slow, however, the change from one state to another can be regarded as reversible. If the rate of heat addition, per unit mass, is $\delta Q/\delta t$, then

$$\delta Q = \frac{\delta t}{\varrho} \nabla \cdot (k_{\mathrm{H}} \nabla T). \tag{2.70}$$

Some of this heat addition may go into work done by the fluid (in expansion against the pressure of the surrounding fluid) and some into increasing the internal energy of the fluid per unit mass. Both these effects are combined in the increase of entropy per unit mass, $-\delta Q/T$. In fact,

$$T\,\delta S = \delta Q = C_{\mathrm{p}}\,\delta T - Tv\,\delta p = C_{\mathrm{p}}\,\delta T - \frac{\beta T}{\varrho}\,\delta p \tag{2.71}$$

from classical thermodynamics, where C_{p} is the specific heat of the fluid at constant pressure, and

$$\beta = \frac{1}{v}\left(\frac{\partial v}{\partial T}\right)_{\mathrm{p}}$$

is the coefficient of thermal expansion of the fluid. Converting to rates of change, and using Eqs. (2.70) and (2.71), one has

$$T\frac{\partial S}{\partial t} = C_{\mathrm{p}}\frac{\partial T}{\partial t} - \frac{\beta T}{\varrho}\frac{\partial p}{\partial t} = \frac{1}{\varrho}\nabla \cdot (k_{\mathrm{H}} \nabla T). \tag{2.72}$$

This is a general equation and represents the effect of conduction of heat in a fluid at rest. For a fluid at rest and which is free to expand, p is constant. Also, for a confined fluid at rest in which the average temperature and pressure are approximately constant, Eq. (2.72) assumes the form

$$T\frac{\partial S}{\partial t} = C_{\mathrm{p}}\frac{\partial T}{\partial t} = \frac{1}{\varrho}\nabla \cdot (k_{\mathrm{H}} \nabla T). \tag{2.73}$$

If k_H can be taken as approximately uniform throughout the fluid, then Eq. (2.73) can be written as

$$\frac{\partial T}{\partial t} = \varkappa_H \nabla^2 T, \qquad (2.74)$$

where

$$\varkappa_H = \frac{k_H}{\varrho C_p} \qquad (2.75)$$

Equation (2.74) is the heat conduction equation and, for a fluid at rest, has the same form as the diffusion equation. The parameter $\varkappa_H$ is called the *thermal diffusivity*.

2.7.3 Molecular Transport of Momentum

Since momentum is a vector quantity, the transport of momentum has to be handled in a separate way.

Consider the fluid velocity in a simple shearing motion with components $u(y)$ and 0 with respect to the rectilinear axes (x, y) (Fig. 2.16). With the

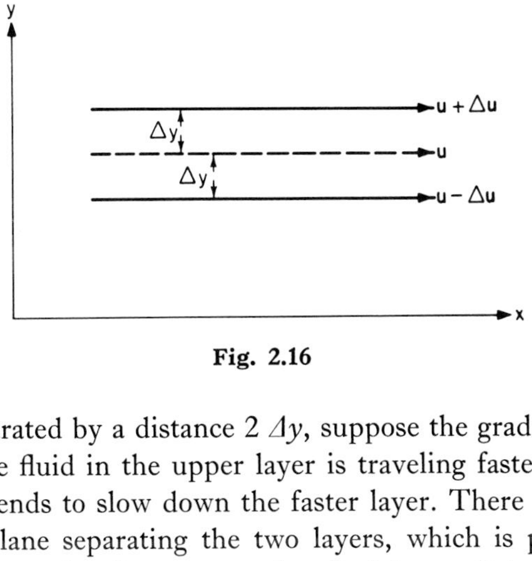

Fig. 2.16

two layers separated by a distance $2\,\Delta y$, suppose the gradient of velocity is such that the fluid in the upper layer is traveling faster. The slower-moving layer tends to slow down the faster layer. There is thus a shear stress in the plane separating the two layers, which is proportional to $\Delta u/\Delta y$. The proportionality constant is called the *coefficient of viscosity* η. In terms of the indicial notation used in discussing the stress tensor, one can write

$$\sigma_{12} = \eta \frac{du}{dy}. \qquad (2.76)$$

Because the molecules which cross from the fast side to the slow side carry more momentum than those which cross from the opposite side, there is a net transfer of momentum across the plane. By Newton's law, the time rate of this momentum transfer is equal to the force across the plane.

The linear relation, Eq. (2.76), is found to be accurate over a large range of values of $|du/dy|$. It is evident that the quantity which, like the diffusivities $\varkappa_D$ and $\varkappa_H$, measures the ability of molecular transport to eliminate the gradient of velocity, and which is the source of the transport, is

$$\nu = \frac{\eta}{\varrho}, \tag{2.77}$$

where ν is called the *kinematic viscosity* and is in fact the diffusivity for momentum.

2.8 Special Properties of Liquid Fluids

A liquid is known as a condensed phase in which a molecule is always within the strong cohesive force field of neighboring molecules. In spite of having this property in common with solids, liquids have the ability to change shape easily.

It is found that the density of water increases by only $\frac{1}{2}\%$ when the pressure is increased from 1 to 100 atmospheres at constant temperature. From the point of view of fluid dynamics, this great resistance to compression means that liquids can be regarded for most purposes as being incompressible.

When the pressure in a flowing liquid falls below the vapor pressure (corresponding to the liquid temperature), vapor bubbles form in it. The process of formation of the vapor phase in a flowing liquid is known as *cavitation.*

In a liquid, transport of energy and momentum occurs through the action of intermolecular forces (whereas in a gas, random motion of the molecules is the mode). Diffusion of specific molecules through a liquid is much less than that of molecules in a gas. The coefficient of diffusion for NaCl molecules in water is found to be of the order of 10^{-5} cm^2/sec (it is roughly 0.2 for the coefficient of self-diffusion of nitrogen molecules).

Transport of heat in a liquid is primarily by direct exchange of translational energy between molecules lying within each other's force fields. It is thus a stronger process than material diffusion. The thermal diffusivity for water is 1.4×10^{-3} cm^2/sec at 15^{0}C, and for most liquids at normal temperatures, the value is of the same order. In the case of liquid metals, $\varkappa_H$ is about 10 times larger, due to the effect of the heat transported by free electrons which can move virtually unrestrained through the liquid.

In the case of momentum transport in liquids, the transfer is not primarily by migration of molecules across hypothetical surfaces in the liquid. The actual mechanism is obviously a complex one about which little is known. It appears that coherent groups of molecules can resist deformation in some way involving intermolecular forces. The effect of a shear motion is to break up these coherent groups—which continually reform. Energy is thus dissipated into disordered molecular motion, or heat. The linear relation between tangential stress and velocity gradient is based mainly on the abundant experimental evidence that is available for nearly homogeneous liquids not containing macromolecules in solution. The most common unit of the shear coefficient of viscosity is the centipoise ($= 10^{-2}$ gm $\cdot$ cm^{-1} sec^{-1}). In general, η depends on both temperature and pressure. There is wide variation in the values of kinematic viscosity for liquids; for example, ν for mercury is 0.0012 cm^2/sec, and for olive oil it is 1.0 cm^2/sec, at normal temperature.

A time lag exists in the distribution of molecular energy between the translational modes of molecular motion on the one hand and the vibrational and rotational modes on the other. This leads to what are called relaxation effects in liquids. Such effects are discussed in length in more detailed texts.

2.9 Special Properties of Gases

The distinctive characteristic of a gaseous fluid is its low density and the consequent wide separation of the molecules. At 0^{0}C and 1 atmosphere pressure, there are about 2.7×10^{19} molecules in 1 cubic centimeter. If the molecules were spaced in a cubic lattice array, they would be a distance 3.3×10^{-7} cm apart. With a mean molecular diameter of about 3.5×10^{-8} cm, the average separation in a gas is about 10 times the mean molecular diameter. Thus, for most of the time a molecule travels freely in a straight line at constant speed.

Elementary kinetic theory shows that the mean distance between collisions is given approximately by the relation

$$L = \frac{1}{N\sigma}, \tag{2.78}$$

where L is called the *mean free path*, N is the number of molecules per cubic centimeter (2.7×10^{19}), and σ is roughly the mean molecular cross section (based on defining a collision as occurring when molecules come close enough to bring repulsive forces into play).

It is found that if d_0 is the mean molecular diameter, then the mean free path is about $200d_0$.

If it is assumed that the molecules collide only with the wall, then elementary kinetic theory shows that the pressure exerted by the gas is given by the expression

$$p = \tfrac{1}{3}\varrho\bar{u}^2, \tag{2.79}$$

where $\bar{u}^2$ is the mean square molecular velocity.

2.10 The Maxwell–Boltzmann Distribution

Since it is impossible to specify the velocities of all the molecules in a gas, it is necessary to define a distribution function for velocities.

The number of molecules in the velocity interval dv is defined as

$$dN(v_x) = Nf(v_x)\,dv_x, \tag{2.80}$$

where N is the total number of molecules. The quantity $f\,dv_x$ equals the fraction of molecules whose x component of velocity lies in the interval between v_x and $v_x + dv_x$.

The distribution function f represents the fraction of molecules with x velocity between v_x and $v_x + dv_x$ per unit velocity interval. One notes that

$$\int_{-\infty}^{+\infty} Nf(v_x)\,dv_x = N.$$

Therefore

$$\int_{-\infty}^{+\infty} f(v_x)\, dv_x = 1. \tag{2.81}$$

The function f is said to be normalized to unity.

The exact form of the function f can be deduced as follows. If the exchange of velocities among the molecules is random, then one can expect a normal, or Gaussian, distribution,

$$f(v_x) = A_x \exp(-\beta v_x^2). \tag{2.82}$$

This function is centered about $v_x = 0$ and vanishes as v_x goes to ∞ (as required). The constant A_x is determined by the normalization condition, Eq. (2.81):

$$\int_{-\infty}^{+\infty} f(v_x)\, dv_x = 1 = \int_{-\infty}^{+\infty} A_x \exp(-\beta v_x^2)\, dv_x$$

$$= A_x \frac{1}{\beta^{1/2}} \int_{-\infty}^{\infty} \exp(-\xi^2)\, d\xi = A_x \frac{1}{\beta^{1/2}} \pi^{1/2}.$$

So

$$A_x = \left(\frac{\beta}{\pi}\right)^{1/2}. \tag{2.83}$$

The constant β determines the width of the distribution. The larger the β the narrower the distribution.

For a three-dimensional distribution function,

$$f(v_x, v_y, v_z) = A \exp[-\beta(v_x^2 + v_y^2 + v_z^2)] \tag{2.84}$$

and

$$A = \left\{\iiint \exp[-\beta(v_x^2 + v_y^2 + v_z^2)]\, dv_x\, dv_y\, dv_z\right\}^{-1}. \tag{2.85}$$

2.10.1 Averages and the Concept of Temperature

The distribution function can be used to calculate the average value of some quantity that depends on the velocity.

If $g(\mathbf{v}) = g(v_x, v_y, v_z)$ is such a quantity, then its average value is

$$\bar{g} = \int_{-\infty}^{\infty}\int_{-\infty}^{\infty}\int_{-\infty}^{\infty} g(v_x, v_y, v_z) f(v_x, v_y, v_z)\, dv_x\, dv_y\, dv_z. \tag{2.86}$$

For example, the kinetic energy is

$$E = \tfrac{1}{2}mv^2 = \tfrac{1}{2}m(v_x^2 + v_y^2 + v_z^2). \tag{2.87}$$

Then,

$$\bar{E} = \frac{\iiint E \exp[-\beta(v_x^2 + v_y^2 + v_z^2)]\, dv_x\, dv_y\, dv_z}{\iiint \exp[-\beta(v_x^2 + v_y^2 + v_z^2)]\, dv_x\, dv_y\, dv_z}. \tag{2.88}$$

Using Eq. (2.87), one obtains

$$\bar{E} = \frac{\int_{-\infty}^{\infty} \tfrac{1}{2}mv_x^2 \exp(-\beta v_x^2)\, dv_x \int_{-\infty}^{\infty}\int_{-\infty}^{\infty} \exp(-\beta(v_y^2 + v_z^2))\, dv_y\, dv_z}{\int_{-\infty}^{\infty} \exp(-\beta v_x^2)\, dv_x \int_{-\infty}^{\infty}\int_{-\infty}^{\infty} \exp(-\beta(v_y^2 + v_z^2))\, dv_y\, dv_z}.$$

The integrals in v_y and v_z cancel in the numerator and denominator, and the remainder

$$\frac{\tfrac{1}{2}m \int_{-\infty}^{\infty} v_x^2 \exp(-\beta v_x^2)\, dv_x}{\int_{-\infty}^{+\infty} \exp(-\beta v_x^2)\, dv_x} = \frac{2(m/\beta) \int_{-\infty}^{\infty} \xi^2 \exp(-\xi^2)\, d\xi}{\int_{-\infty}^{\infty} \exp(-\xi^2)\, d\xi}, \tag{2.89}$$

where

$$\xi^2 = \beta v_x^2.$$

Now

$$\int_{-\infty}^{\infty} \exp(-\xi^2)\, d\xi = \pi^{1/2}.$$

It is easy to show that

$$\int_{-\infty}^{\infty} \xi^2 \exp(-\xi^2)\, d\xi = \tfrac{1}{2}\,\pi^{1/2},$$

and so Eq. (2.89) has the value $\tfrac{1}{2}(m/2\beta)$.

There are two equal contributions to $\bar{E}$ from $\tfrac{1}{2}mv_y^2$ and $\tfrac{1}{2}mv_z^2$, and so

$$\bar{E} = \frac{3}{2}\,\frac{m}{2\beta}. \tag{2.90}$$

Now kinetic theory shows that the mean energy of a molecule is also given by

$$\bar{E} = \tfrac{3}{2}kT. \tag{2.91}$$

Hence

$$\beta = \frac{m}{2kT}. \tag{2.92}$$

This relation introduces temperature into the statistical theory, and

$$f(\mathbf{v}) = \left(\frac{m}{2\pi kT}\right)^{3/2} \exp[-m(v_x^2 + v_y^2 + v_z^2)/2kT]. \qquad (2.93)$$

This function is called the *Maxwell distribution of velocities*. The higher the temperature the wider the distribution and the more molecules with high velocities. There is another useful form in which to express $f(v)$. The vector $\mathbf{v}$ can be represented by a point in so-called "velocity space." It is preferable to use a generalized coordinate system for velocity space. This system can consist of the length of the velocity vector and the spherical coordinate angles ϑ and φ. A volume element in this space is

$$v^2\, dv \sin\vartheta\, d\vartheta\, d\varphi,$$

and the fraction of molecules with velocities in this element is

$$f_1(v, \vartheta, \varphi) = \left(\frac{m}{2\pi kT}\right)^{3/2} \exp\left(-\frac{mv^2}{2kT}\right). \qquad (2.94)$$

The new function $f_1(v, \vartheta, \varphi)$ is obtained from $f(v_x, v_y, v_z)$ and is gotten by substituting the coordinate transformation

$$v^2 = v_x^2 + v_y^2 + v_z^2,$$

where v is now a (scalar) speed.

Since the distribution is isotropic (the probability of moving in any direction in space is the same), f_1 does not depend on ϑ and φ. So these can be integrated out to give the total solid angle 4π. Then

$$f_1(v) = 4\pi\left(\frac{m}{2\pi kT}\right)^{3/2} v^2 \exp\left(-\frac{mv^2}{2kT}\right). \qquad (2.95)$$

This is called the *Maxwell distribution of speeds*.

2.10.2 Generalization of the Distribution Function

Suppose there is an external (conservative) force acting on the molecules. In this case the total energy of a molecule is

$$E = \tfrac{1}{2}m(v_x^2 + v_y^2 + v_z^2) + V(x, y, z). \qquad (2.96)$$

The probability of finding a molecule then depends on position as well as velocity:

$$f(v_x, v_y, v_z, x, y, z)\, dv_x\, dv_y\, dv_z\, dx\, dy\, dz = A'e^{-E/kT}\, dv_x\, dv_y\, dv_z\, dx\, dy\, dz. \tag{2.97}$$

This is the probability that a molecule has a velocity in the element of velocity space $dv_x\, dv_y\, dv_z$, and a position in the element of configuration space $dx\, dy\, dz$.

This generalized distribution function is called the *Boltzmann distribution.*

2.10.3 Heat Capacity of Polyatomic Gases

As a second example of the Boltzmann distribution, consider again an ideal gas with no forces acting on it, but with polyatomic molecules. A rotational term must now be added to the kinetic energy:

$$E = \tfrac{1}{2}mv^2 + \tfrac{1}{2}I\omega^2,$$

where I is the molecular moment of inertia and ω is the angular velocity. The molecules can rotate freely about any of the three coordinate axes, and so

$$E = \tfrac{1}{2}m(v_x^2 + v_y^2 + v_z^2) + \tfrac{1}{2}I_x\omega_x^2 + \tfrac{1}{2}I_y\omega_y^2 + \tfrac{1}{2}I_z\omega_z^2. \tag{2.98}$$

The distribution function is

$$f(v_x, v_y, v_z, \omega_x, \omega_y, \omega_z) = A' \exp\{[-m(v_x^2 + v_y^2 + v_z^2)/2kT] - [(I_x\omega_x^2 + I_y\omega_y^2 + I_z\omega_z^2)/kT]\}. \tag{2.99}$$

The average energy of a molecule can again be calculated from

$$\bar{E} = \int E f\, dv_x\, dv_y\, dv_z\, d\omega_x\, d\omega_y\, d\omega_z. \tag{2.100}$$

As before, one gets three terms of the form

$$\frac{kT \int_{-\infty}^{\infty} \xi^2 \exp(-\xi^2)\, d\xi}{\int_{-\infty}^{\infty} \exp(-\xi^2)\, d\xi} = \frac{1}{2}kT. \tag{2.101}$$

Now there are also three terms for the components of angular velocity. The average energy per molecule for an extended (polyatomic) molecule is thus

$$\bar{E} = 3kT.$$

If the molecule is diatomic, the moment of inertia about one axis is zero and so there are only five terms. Therefore

$$E = \tfrac{5}{2}kT.$$

Thus, each degree of freedom contributes $\frac{1}{2}kT$ to the average energy. For n degrees of freedom,

$$\bar{E} = \frac{n}{2}\,kT.$$

Since in 1 mole there are N_0 molecules, the heat energy in 1 mole is

$$U_T = N_0\bar{E} = \frac{n}{2}\,N_0kT = \frac{n}{2}\,kT. \qquad (2.102)$$

The heat capacity per mole of the gas is the rate of change of this internal energy with temperature (for constant volume):

$$C_v = \frac{dU_T}{dT} = \frac{n}{2}\,k. \qquad (2.103)$$

At normal temperatures the predicted values agree well with experiment. But at low and high temperatures, deviations occur and are indicative of the breakdown of the classical approach and the presence of quantum effects.

It must be emphasized that in the above treatment, although an external force was admitted, no intermolecular forces were admitted. The reason for this is the extremely complex mathematical problem which results. In the presence of such forces, the potential energy of a molecule depends not only on the position of that molecule but also on that of every other molecule in the gas. In other words, there is a correlation between the molecules. Instead of the six variables, there are a very large number of variables (the distances from all the other molecules). One then has a "many-particle distribution function" for all the particles in the system, giving the probability of finding particle 1 at point 1 in the combined velocity and configuration space (phase space), particle 2 at point 2, particle 3 at point 3, and so on.

A Boltzmann distribution function can then, in principle, be derived. But the computation of properties from it is exceedingly difficult. Up to now, treatment of dense gases or liquids has been only modestly successful.

2.10.4 Transport Coefficients in a Perfect Gas

It has already been pointed out that if J_A is the net flux of molecules of type A per unit area in the presence of a concentration gradient dc_A/dx, then

$$J_A = -D\frac{dc_A}{dx}, \tag{2.104}$$

where D is the diffusion coefficient. This equation is called *Fick's law*.

Let the concentration of type-A molecules to the left of a plane be $c_A - dc_A$ and $c_A + dc_A$ to the right. Now, in general, if c is the concentration and $\bar{v}$ is the mean molecular velocity for a stationary gas, the flux of molecules across a plane is $\frac{1}{2}c\bar{v}$. Thus, in this case, the net flux across the plane in one second is

$$J = \tfrac{1}{2}(c_A - dc_A)\bar{v} - \tfrac{1}{2}(c_A + dc_A)\bar{v} = -\bar{v}\,dc_A. \tag{2.105}$$

If L is the mean free path between collisions, then the change in concentration across the plane is given approximately by

$$\Delta c_A = L\left(\frac{dc_A}{dx}\right). \tag{2.106}$$

Therefore

$$J = -\bar{v}L\left(\frac{dc_A}{dx}\right), \tag{2.107}$$

and so

$$D = \bar{v}L. \tag{2.108}$$

With respect to the transport of momentum in the presence of a velocity gradient, it has been seen that the tangential stress acting in a plane across which there is a change in velocity from $u + du$ to $u - du$ is given by

$$\sigma_{12} = -\eta\frac{du}{dx}.$$

Now, as above, the flux of molecules across the surface is $\frac{1}{2}c_A\bar{v}$. Those going from left to right carry momentum $m(u - \Delta u)$, and from right to left $m(u + \Delta u)$. The net momentum transferred across unit area per second, or the stress, is given by

$$\sigma_{12} = (2m\,\Delta u)(\tfrac{1}{2}c_A\bar{v}) = -c_A\bar{v}m\,\Delta u, \tag{2.109}$$

and, as before, if the plane is taken as one mean free path across, one obtains

$$\Delta u = L\left(\frac{du}{dx}\right).$$

Now,

$$c_A m = \varrho,$$

and so

$$\sigma_{12} = -\varrho\bar{v}L\left(\frac{du}{dx}\right). \tag{2.110}$$

Therefore

$$\eta = \varrho\bar{v}L. \tag{2.111}$$

Finally, the transport of heat will be considered again. Consider an imaginary plane in the gas (Fig. 2.17). Suppose the average energy of

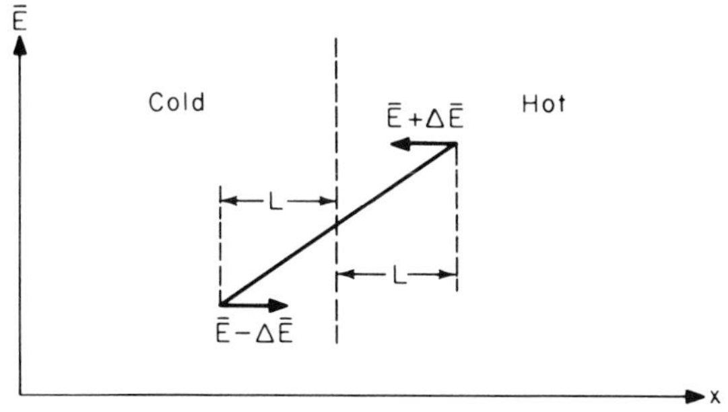

Fig. 2.17

a molecule at the plane is $\bar{E}$. Each molecule crossing from the cold to the hot side has a kinetic energy $\bar{E} - d\bar{E}$ (on the average) and each crossing from the hot side to the cold has a kinetic energy $\bar{E} + d\bar{E}$. So the net heat flow from the hot to the cold side is

$$H = \tfrac{1}{2}c_A\bar{v}(\bar{E} - \Delta\bar{E}) - \tfrac{1}{2}c_A\bar{v}(\bar{E} + \Delta\bar{E}) = -c_A\bar{v}\,\Delta\bar{E}. \tag{2.112}$$

As before,

$$\Delta\bar{E} = L\left(\frac{d\bar{E}}{dx}\right) \quad \text{and} \quad \frac{d\bar{E}}{dx} = \frac{d\bar{E}}{dT}\,\frac{dT}{dx}. \tag{2.113}$$

Therefore

$$H = -c_A\bar{v}L\left(\frac{d\bar{E}}{dT}\right)\left(\frac{dT}{dx}\right), \tag{2.114}$$

where dT/dx is the *measured* temperature gradient.

Now for an ideal monatomic gas,

$$\bar{E} = \frac{1}{2}m\bar{v}^2 = \frac{3}{2}kT \quad \text{and} \quad \frac{d\bar{E}}{dT} = \frac{3}{2}k. \tag{2.115}$$

Now $c_A(d\bar{E}/dT)$ is the increase in energy when a unit volume is heated one degree, or the heat capacity per unit volume. The heat capacity per unit volume is the specific heat times the density, ϱC_V. Therefore

$$c_A\left(\frac{d\bar{E}}{dT}\right) = \varrho C_V, \tag{2.116}$$

therefore

$$H \doteq C_V\bar{v}L\left(\frac{dT}{dx}\right), \tag{2.117}$$

so

$$H \propto \frac{dT}{dx},$$

and the proportionality constant is known as the thermal conductivity $\varkappa$. Therefore

$$\varkappa = \varrho C_V\bar{v}L. \tag{2.118}$$

More exact analysis using kinetic theory yields the same form for $\varkappa$ except for a factor of $25\pi/64 = 1.23$.

Equation (2.118) relates the measurable macroscopic quantities $\varkappa$, ϱ, and C_V to the microscopic molecular properties $\bar{v}$ and L. Indeed, measurement of $\varkappa$ allows an accurate determination of the mean free path L and thereby of the molecular diameter.

Comparison of Eqs. (2.111) and (2.118) shows that

$$\varkappa \doteq C_V\eta. \tag{2.119a}$$

More accurate calculations show that

$$\varkappa \doteq \tfrac{5}{2}C_V\eta. \tag{2.119b}$$

The factor $\frac{5}{2}$ is exact for an intermolecular repulsive force which is an inverse fifth power of the intermolecular distance. It is 1% higher for the hard sphere model. The relation has been verified experimentally for the noble gases He, Ne, and Ar. For more complex molecules the error is appreciable. This is due to the internal molecular motions of vibration and rotation.

However, the theory has produced a remarkable result in Eq. (2.119b), by correlating a thermal property ($\varkappa$) and a mechanical property (η).

Again, comparing Eqs. (2.111) and (2.108), one obtains

$$\bar{\eta} \doteq \varrho D. \tag{2.120}$$

A more accurate calculation (hard spheres) shows that

$$\eta = \tfrac{5}{6} \varrho D. \tag{2.121}$$

2.10.5 Numerical Examples for a Typical Gas

Now it has been shown that

$$D = \bar{v}L, \qquad \eta = \varrho D, \qquad \text{and} \qquad \varkappa = C_{\text{v}}\eta,$$

where C_{v} is the specific heat (at constant volume).

The magnitudes of $\bar{v}$, L, η, and C_{v} will be estimated for argon (a gas of medium molecular weight and nearly ideal at room temperature).

Now,

$$\text{molecular weight} = 40 \quad \text{and} \quad \text{molecular diameter} = 4 \text{ Å}.$$

For 1 molecule of an ideal gas, the specific heat is

$$\frac{d\bar{E}}{dT} = \frac{3}{2} k.$$

Multiplying by N/M, the number of molecules in a gram, and since $Nk = \mathscr{R}$, the gas constant per mole, one has

$$C_{\text{v}} = \frac{3}{2} \frac{\mathscr{R}}{M} = \frac{3}{2} \left(\frac{2 \text{ cal/mole} \cdot \text{deg}}{40 \text{ gm/mole}} \right)$$
$$= \frac{3}{40} \text{ cal/gm} \cdot \text{deg} = 3 \times 10^{-7} \text{ erg/gm} \cdot \text{deg}.$$

The density ϱ is equal to the molecular weight divided by the volume of 1 mole, V. For a gas at standard conditions, $V = 22.4$ liters, and so

$$\varrho = \frac{M}{V} = \frac{40 \text{ gm/mole}}{22.4 \text{ liters/mole}} = 2 \times 10^{-3} \quad \text{gm/cm}^3.$$

The number of molecules per unit volume at standard conditions is

$$c_A = \frac{N}{V} = \frac{6.02 \times 10^{23} \text{ molecules/mole}}{22.4 \times 10^3 \text{ cm}^3\text{/mole}} \doteq 3 \times 10^{19} \quad \text{cm}^{-3}$$

and

$$L = (\sqrt{2} c_A \pi d^2)^{-1} = (\sqrt{2} \times 3 \times 10^{19} \text{ cm}^{-3}\ \pi \times 4^2 \times 10^{-6} \text{ cm}^2)^{-1}$$
$$\doteq 6 \times 10^{-6} \quad \text{cm}.$$

The mean velocity $\bar{v}$ can be approximated by the rms speed,

$$\bar{v} = v_{\text{rms}} = \frac{3\mathscr{R}T}{M} = \left(2 \frac{3}{2} \frac{\mathscr{R}}{M} T\right)^{1/2}$$
$$= (2 \times 0.3 \times 10^7 \text{ ergs/gm} \cdot \text{deg} \times 300 \text{ deg})^{1/2}$$
$$\doteq 4 \times 10^4 \quad \text{cm/sec}.$$

These values are all fairly representative for a gas at room temperature and atmospheric pressure.

The diffusion coefficient is then given by

$$D = \bar{v}L = 0.25 \quad \text{cm}^2\text{/sec}.$$

It is found that the time to diffuse a distance x is given by a relation of the form

$$t = \frac{x^2}{D}.$$

If $x = 1$ meter, then $t = 4 \times 10^4$ sec. This shows that diffusion is a slow process. Normally, mixing is accomplished much more quickly in fluids by convection or stirring. Slow as it is, diffusion in a gas is about 10^5 times faster than in a liquid, and it is even slower in a solid.

The viscosity is

$$\eta \doteq \varrho D \doteq 0.5 \times 10^{-3} \quad \text{gm/cm} \cdot \text{sec}.$$

The cgs unit of viscosity, gm/cm · sec, is called a *poise*. The viscosity of gases is usually given in micropoise (1 μP $= 10^{-6}$ poise); so, for argon it is 500 μP. The viscosity of liquids is usually given in centipoise (1 cP $= 10^{-2}$ poise). The viscosity of water at 20°C is 1.00 cP. So, a typical gas is perhaps 100 times less viscous than a typical liquid.

The thermal conductivity is

$$\varkappa \doteq C_V\eta = 1.5 \times 10^3 \quad \text{ergs/cm} \cdot \text{sec} \cdot \text{deg}$$
$$= 4 \times 10^{-5} \quad \text{cal/cm} \cdot \text{sec} \cdot \text{deg.}$$

So a gas has a relatively low thermal conductivity—a typical liquid has a value about 10 times larger (for water, $\varkappa = 1.4 \times 10^{-3}$ cal/cm · sec · deg). Solids are 10 to 100 times higher, and metallic solids can be as much as 10^4 higher (for copper, $\varkappa \sim 1$ cal/cm · sec · deg).

Gases, in spite of their low thermal conductivity, do not make good thermal insulators unless convection is prevented. Convection carries heat through a fluid much faster than conduction.

2.11 Fluid Flow Phenomena

Consider the flow of fluid induced by the uniform translatory motion of a plane spaced a distance Y above a stationary parallel plane (Fig. 2.18). If the fluid velocity increases linearly from zero (at the stationary plane) to V (at the moving plane), then

$$\text{rate of shear deformation} = \frac{du}{dy} = \frac{U}{Y}. \tag{2.122}$$

For many fluids it is found that the magnitude of the shearing stress is related to the rate of shear proportionally:

$$T = \eta \frac{du}{dy} = \eta \frac{U}{Y}. \tag{2.123}$$

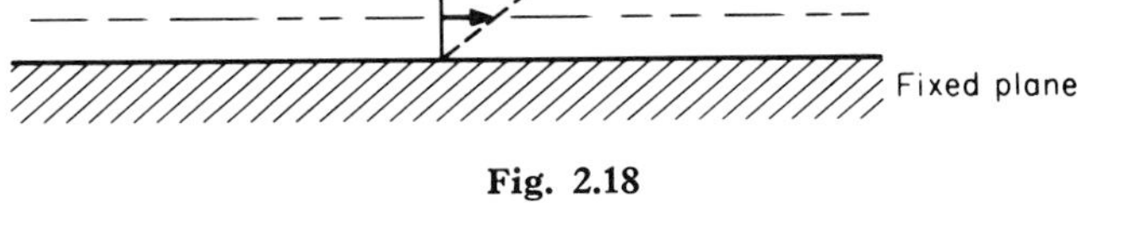

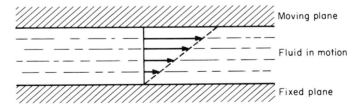

Fig. 2.18

Fluids which obey Eq. (2.123) in the above situation are known as *Newtonian fluids*. Generally in Newtonian fluids both normal and tangential components of stress are linearly dependent on the rates of deformation of the fluid. Gases under normal conditions, water, and low molecular weight solutions are typical Newtonian fluids.

Non-Newtonian fluids do not show the above linear relationship between stress and rate of shear deformation. In such fluids, the apparent coefficient of viscosity at constant temperature and pressure is not a constant but varies with the rate of shear and may even depend on the previous history of the fluid.

Only Newtonian fluids will be dealt with in this text. In fact, only fluids which have a very small coefficient of viscosity (such as air or water) will be of most interest. When such fluids flow at reasonable velocities it is found that viscous effects appear only in thin layers on the surface of objects or surfaces over which the fluid moves. The fluid velocity must be zero at the surface (no slip condition) and the falloff in velocity, and thus the velocity gradient, is large in this narrow boundary layer on the surface.

The boundary layer in a fluid flowing over a flat plate is sketched in Fig. 2.19. In a later chapter the concept of boundary layer flow will be defined in much more generality. The effects of boundary layers or fluid flow are very important, especially in regard to the resistance to motion (or drag) of bodies in contact with the fluid.

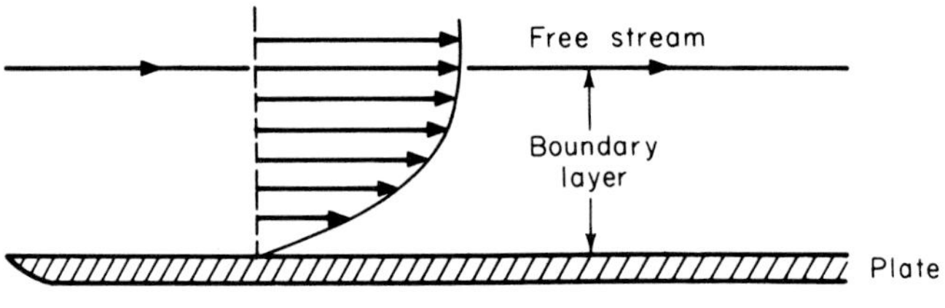

Fig. 2.19

2.11.1 Discontinuities in the Flow Field

Consider two streams of fluid, at different velocities, coming into contact (Fig. 2.20). The surface which separates the two streams at the point where they come into contact is a surface of discontinuity across which the velocity changes abruptly. Due to the action of viscous fric-

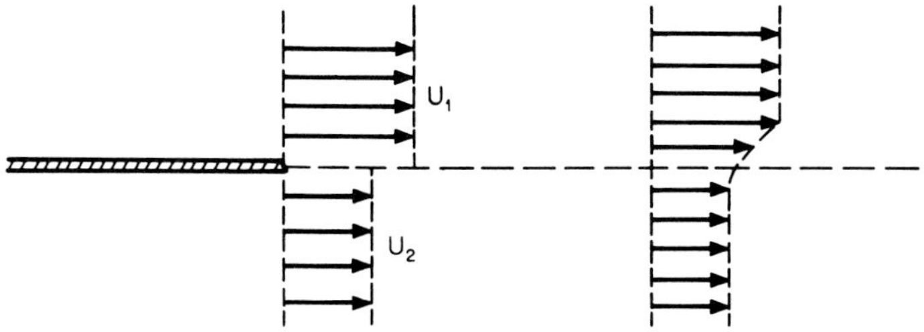

Fig. 2.20

tion, or the transport of momentum across the surface, the abrupt change in velocity is smoothed out and the discontinuity surface becomes a thin (boundary) separation layer. It is found in fact that the discontinuity surface is unstable and breaks into a large number of eddies (or vortices). Actually, these vortices, in effect, speed up the transverse transport of momentum leading to a rapid formation of the separation layer.

A discontinuity surface which breaks into vortices is also exhibited in flow over a sharp edge. This is illustrated in Fig. 2.21. The surface

Fig. 2.21

rolls up into a vortex, separates from the sharp edge, and then moves downstream. If the velocity is high enough, such separation and rollup into vortices occurs in flow over blunt bodies such as spheres and cylinders. In the latter case, a fairly stable succession of such vortices form alternately on either side of the cylinder, and this array is called a *Kármán vortex street* (Fig. 2.22).

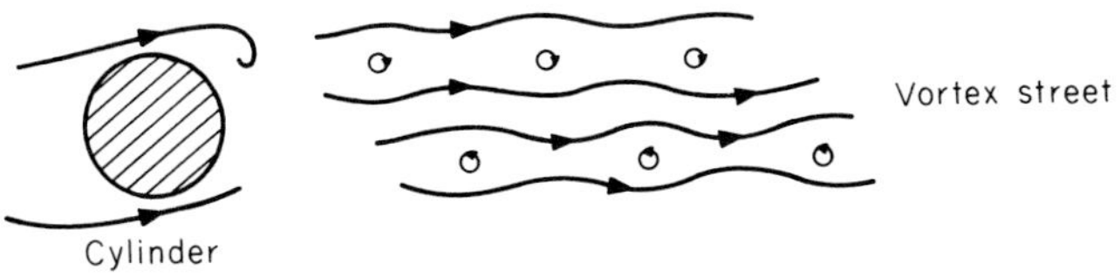

Fig. 2.22

Gases which flow at velocities approaching, or greater than, the speed of sound are compressible fluids. In fact, a gas flowing at the speed of sound is said to have a Mach number M of 1. So in subsonic flow,

$M < 1$ and in supersonic flow, $M > 1$. Obviously, M is defined as the ratio of the gas speed to the (local) speed of sound. In fluids flowing at supersonic speeds ($M > 1$), discontinuities in pressure, and velocity, can occur—these are known as *shock waves*. They are very thin regions across which the flow variables change very rapidly.

2.11.2 Turbulence

Generally speaking, for low-velocity fluid flows in pipes, or over plates or cylinders, the fluid travels in well-defined regular path- or streamlines. When the velocity, through the pipe say, is increased above a certain value, it is found that the fluid motion becomes very irregular with eddying motions of all sizes and intensity occurring randomly throughout the flow field. In the case of a pipe, when this transition occurs, the pressure drop per unit length increases greatly. Such fluid flow is called *turbulent flow*. The classification of flow into *laminar* or *turbulent flow* is perhaps the major classification in fluid dynamics. In 1883, Osborne Reynolds showed that in a pipe, the transition from laminar to turbulent flow depends on the value of the dimensionless expression

$$\frac{\varrho \bar{V} R}{\eta},$$

where $\bar{V}$ is the mean flow velocity, R the pipe radius, and η the coefficient of (shear) viscosity.

Boundary layer flows can also be turbulent or laminar. Moreover, turbulence can be present in flows unbounded by solid walls, for example, in jets or the wake region downstream of bodies. These are called *free turbulent flows*. The study of the stability of laminar flows and the transition to turbulence has formed a major topic of research in fluid dynamics for almost a century.

In laminar flow, transport of momentum by molecular diffusion is responsible for the property of (shear) viscosity of a flowing fluid. In turbulent flow the irregular motion of the fluid elements causes a diffusion process which dominates (except close to a surface) the molecular diffusion. This also results in additional stresses in the fluid, known as *Reynolds*, or *turbulent*, *stresses*. By analogy with the laminar case, an eddy coefficient of viscosity and eddy heat conductivity are defined.

As in laminar flow, there can be steady and unsteady turbulent flow. "Steady" turbulent flow implies that averages of flow variables over long enough times do not vary with time. This distribution of turbulent kinetic energy can be uniform, whence the name *isotropic turbulence*. The theory of isotropic "steady" turbulent flow is reasonably satisfactory today, but that for nonisotropic and nonsteady (as in transition) turbulence is still in an unsatisfactory state.

2.12 Physical Similarity and Dimensional Analysis

In mechanical and thermal transport processes (mass, heat, and momentum transfer), dimensionless numbers play a very important role. They can help in understanding the mechanisms behind transport processes. They can be used to permit the application of limited experimental results to a wide variety of processes and dynamic systems with different physical dimensions and physical properties.

Every physical property has dimensions. Dimensions can be the fundamental ones such as mass, length, and time or derived dimensions such as velocity and acceleration. However, all physical properties can be expressed in terms of a set of fundamental dimensions.

Combinations of physical quantities whose fundamental dimensions cancel produce dimensionless numbers.

The Reynolds number of a fluid flow field has already been referred to. It is defined as VL/ν, where V is the fluid velocity, L a characteristic length, and ν the fluid kinematic viscosity; V, L, and ν each have their own physical significance—their combination has a special significance. It represents the ratio of the inertial force per unit area to the viscous force per unit area. The combination supplies information on the state of the flow field (whether it is turbulent or not) not revealed by V, L, or ν.

Dimensional analysis involves the development of dimensionless numbers, and is based on the basic principle of dimensional homogeneity which states that dimensionally different quantities cannot be added together to represent a valid physical situation. Mathematical equations must be dimensionally homogeneous.

Dimensional analysis has been used in fluid dynamics for some time. Stokes, in 1850, showed that in geometrically similar flow systems the combination of fluid velocity V, characteristic length L, and kinematic viscosity ν, later called the Reynolds number, could be used as a criterion for dynamic similarity. Rayleigh, in 1899, first used the method in the problem of finding how temperature affects the viscosity of a gas.

Buckingham, in 1914, stated the Pi theorem—one of the foundations of modern dimensional analysis. Nusselt, in 1915, applied the method extensively to heat transfer. The Nusselt number is as important in heat transfer as the Reynolds number is in fluid flow.

Every dimensionally homogeneous equation can be changed to a form containing only dimensionless groups of variables.

Consider the equation

$$P_0 + P_1 + P_2 + \cdots + P_n = 0. \tag{2.124a}$$

By dividing across by one of the terms, say P_0, Eq. (2.124a) becomes

$$1 + \frac{P_1}{P_0} + \frac{P_2}{P_0} + \cdots + \frac{P_n}{P_0} = 0. \tag{2.124b}$$

This equation is dimensionless, and $P_1/P_0, P_2/P_0, \ldots, P_n/P_0$ are the dimensionless numbers.

Equation (2.124a) can now be written as

$$f(D_1, D_2, \ldots, D_n) = 0, \tag{2.125}$$

where $D_1 = P_1/P_0$, etc.

As a more specific case, consider the following equation (it is the famous Bernoulli equation for the flow of an incompressible nonviscous fluid in a pipe and will be derived later):

$$p + \frac{1}{2}\varrho V^2 + \varrho g y = 0 \qquad \text{or} \qquad \frac{p}{\varrho g} + \frac{V^2}{2g} + y = 0. \tag{2.126}$$

Dividing each term by $V^2/2g$, one obtains

$$\frac{2p}{\varrho V^2} + 1 + \frac{2gy}{V^2} = 0. \tag{2.127}$$

There are two dimensionless groups in this equation. So, instead of the five variables in Eq. (2.126), there are now only two dimensionless variables. The advantage of reducing the number of variables in a problem was stated by Langhaar as follows:

> A function of one independent variable may be plotted as a single curve. A function of two is represented by a family of curves, one curve for each value of the second variable. A function of three

> variables is represented by a set of charts, one chart for each value of the third variable. If, for example, five experimental points are required to plot a curve, twenty-five points are required for a chart of five curves, one hundred and twenty-five points to plot a set of 5 charts, etc. This situation gets quickly out of hand. . . .†

In transport phenomena, most problems require experimental results. The reduction of the number of variables amplifies the information obtainable from a few such results and is a practical first step in tackling the problem.

The similarity law implies that two processes are similar if the numerical value of the important dimensionless numbers of the processes are the same. The phenomena of mass, heat, and momentum transfer are often analogous and this sometimes allows the extension of concepts in one case to the other two.

The most useful dimensionless numbers are those indicating the relative importance of properties. For example, the Reynolds number can be considered to represent the ratio of the inertial forces to the viscous forces.

2.12.1 Derivation of Dimensionless Numbers

The most useful dimensionless numbers have been derived from dimensional analysis.

The dimensions of a physical quantity are concerned with the properties of that quantity and not with its magnitude. Moreover, while a physical quantity has only one set of dimensions, the dimensions can be measured in a variety of units (for example, MKS, cgs, and English).

The fundamental units are mass, length, and time (augmented by temperature or heat, and electric charge). Then there are derived dimensions, for example, volume, velocity, acceleration, energy, and force. The choice of dimensions is arbitrary, but a commonly used set is M (mass), L (length), T (time), Q (heat), and θ (temperature).

A typical derived dimension would be force F, where

$$F = MLT^{-2}. \tag{2.128}$$

† From H. L. Langhaar, "Dimensional Analysis and Theory of Models." Copyright 1951, Wiley, New York. Reproduced by permission of John Wiley & Sons, Inc.

TABLE I *Dimensions and Their Equivalents*

Dimensions	Symbol	$MLTQ,\theta$
Length	L	L
Time	T	T
Mass	M	M
Force	F	MLT^{-2}
Temperature	θ	θ
Heat	Q	Q
Area	A	L^2
Volume	V	L^3
Density	ϱ	ML^{-3}
Velocity	V	LT^{-1}
Acceleration	a	LT^{-2}
Mass flow rate	w	MT^{-1}
Thermal conductivity	k	$Q\theta^{-1}L^{-1}T^{-1}$
Diffusivity	D	L^2T^{-1}
Thermal diffusivity	$\varkappa=k/\varrho C_p$	L^2T^{-1}
Kinematic viscosity	ν	L^2T^{-1}
Dynamic viscosity	η	$ML^{-1}T^{-1}$
Pressure	p	$ML^{-1}T^{-2}$
Work	W	ML^2T^{-2}
Specific heat	C_p, C_v	$QM^{-1}\theta^{-1}$
Surface tension	σ	MT^{-2}
Molecular mean free path	λ	L

Dimensions and their equivalents are shown in Table I. When studying fluid motions, two systems are deemed to be similar or not, in the light of the following definitions.

Geometrical similarity In the two compared systems, if all linear dimensions are proportional to each other with an identical ratio, then they are geometrically similar.

Physical similarity If in addition to geometrical similarity, all other physical quantities in the two systems such as velocities, time, and temperature are respectively proportional to each other, then the systems are said to be physically similar.

When the mathematical equations describing the physics of a process are known, dimensional analysis can be used to obtain logical groupings of quantities for presentation of the results. On the other hand, when the mathematical equations are unknown, dimensional analysis can be used to reduce the number of variables and so reduce the number of experiments necessary (using the Pi theorem method).

2.12.2 Buckingham's Pi Theorem

This theorem states that "if an equation is dimensionally homogeneous, it can be reduced to a relation between a complete set of dimensionless products."

The number of dimensionless products required to form a complete set can be determined from a rule formulated by Van Driest in 1946 (and which can be rigorously proved). It states that "the number of dimensionless products in a complete set equals the total number of variables in the system minus the maximum number of these variables that will not form a dimensionless product." This leads to a more useful formulation of the Pi theorem. If there are R physical quantities and r fundamental dimensions, there exists a maximum number ($N_{\max}$) of these quantities which in themselves cannot form a dimensionless group. Then

$$N(\pi) = R - N_{\max}, \qquad (2.129)$$

where $N(\pi)$ is the maximum number of independent dimensionless numbers and $N_{\max}$ is usually equal to r.

The logical steps in applying the Pi theorem, are as follows:

(a) after careful consideration of the problem, list all the important variables in it and express them in their dimensional equivalents;

(b) the number of these variables minus the number of fundamental dimensions gives the number of dimensionless groups in the complete set;

(c) each variable must appear in at least one of the dimensionless groups;

(d) the dependent variable should appear only in one of the dimensionless groups.

An example will now be given to illustrate the use of the Pi theorem.

The problem will be that of determining the pressure drop for an incompressible gas or liquid flowing through a smooth pipe. The dependent variable is the pressure drop, and careful consideration of the physics of the problem shows that this is a function of the five (independent) variables,

η fluid dynamic viscosity,
ϱ fluid density,
V mean fluid velocity in the pipe,
L length of the pipe,
D diameter of the pipe.

Mathematically, then, one has

$$\Delta p = f(\eta, L, D, \varrho, V). \tag{2.130}$$

The variables have the dimensions

η $ML^{-1}T^{-1}$,
ϱ ML^{-3},
V LT^{-1},
L L,
D L.

There are six dimensions but only three fundamental dimensions, and so the number of independent dimensionless groups is $6 - 3 = 3$. The general expression for π is then

$$\pi = (\Delta p)^a(\eta)^b(L)^c(D)^d(\varrho)^e(V)^f.$$

Substituting the dimensions and grouping, one gets

$$\pi = M^{a+b+e}L^{-a-b+c+d-3e+f}\,T^{-2a-b-f}. \tag{2.131}$$

Since π is dimensionless, all the exponents of M, L, and T should be zero. Therefore,

$$a + b + e = 0, \quad -a - b + c + d - 3e + f = 0, \quad -2a - b - f = 0,$$

and so

$$d = -c - b, \qquad e = -a - b, \qquad f = -2a - b. \tag{2.132}$$

There are thus six unknowns and only three equations, and so three unknowns must be chosen arbitrarily. There are three independent dimensionless groups and these can be obtained by assigning three sets of values for a, b, and c.

(a) $a = 1$, $b = 0$, $c = 0$, so that $d = 0$, $e = -1$, $f = -2$. Therefore

$$\pi_1 = \frac{\Delta p}{\varrho V^2}.$$

(b) $a = 0$, $b = 1$, $c = 0$, so that $d = -1$, $e = -1$, $f = -1$. Therefore

$$\pi_2 = \frac{\eta}{D\varrho V}.$$

(c) $a = 0$, $b = 0$, $c = 1$, so that $d = -1$, $e = 0$, $f = 0$. Therefore

$$\pi_3 = \frac{L}{D}.$$

The answer is

$$f\left(\frac{\Delta \varrho}{\varrho V^2}, \frac{\eta}{D\varrho V}, \frac{L}{D}\right) = 0. \tag{2.133}$$

The center term is seen to be the inverse of the dimensionless Reynolds number.

2.12.3 Differential Method

As mentioned previously, when the mathematical equations describing a physical system are known, dimensional analysis can be used to get logical groupings of quantities.

Again, the flow of an incompressible liquid flowing through a smooth pipe will be considered—this time by the differential method. The differential equation of the fluid flow is

$$u\frac{\partial u}{\partial x} + v\frac{\partial u}{\partial r} = -\frac{1}{\varrho}\frac{\partial p}{\varrho x} + \frac{\nu}{r}\left(r\frac{\partial u}{\partial r}\right), \tag{2.134}$$

where u is the axial component of velocity and v is the radial component of velocity.

When two systems are geometrically and physically similar, all dimensions and physical properties have the same ratio:

$$K_L = \frac{L_1}{L_2} = \frac{r_1}{r_2} = \frac{D_1}{D_2}, \qquad K_\varrho = \frac{\varrho_1}{\varrho_2}, \qquad K_{(dp/dx)} = \frac{(dp/dx)_1}{(dp/dx)_2},$$

$$K_v = \frac{u_1}{u_2} = \frac{v_1}{v_2}, \qquad K_\nu = \frac{\nu_1}{\nu_2},$$

where $K_L, K_v, \ldots, K_{(dp/dx)}$ are constant ratios and u is the average velocity in the x direction. For system number one,

$$u_1 \frac{\partial u_1}{\partial x_1} + v_1 \frac{\partial u_1}{\partial r_1} = -\frac{1}{\varrho_1}\left(\frac{\partial p}{\partial x}\right)_1 + \frac{\nu_1}{r_1}\frac{\partial}{\partial r_1}\left(r_1 \frac{\partial u_1}{\partial r_1}\right), \tag{2.135}$$

and for system number two,

$$u_2 \frac{\partial u_2}{\partial x_2} + v_2 \frac{\partial u_2}{\partial r_2} = -\frac{1}{\varrho_2}\left(\frac{\partial p}{\partial x}\right)_2 + \frac{\nu_2}{r_2}\frac{\partial}{\partial r_2}\left(r_2 \frac{\partial u_2}{\partial r_2}\right). \tag{2.136}$$

Substituting for $u_2, x_2, r_2, (\partial p/\partial x)_2$, etc. in terms of the ratios K_1, etc. and u_1, x_1, r_1, etc. in Eq. (2.136), one gets the equation

$$\frac{K_v^2}{K_L} u_1 \frac{\partial u_1}{\partial x_1} + \frac{(K_v)^2}{K_L} v_1 \frac{\partial v_1}{\partial r_1}$$
$$= \frac{K_{(dp/dx)}}{K_\varrho}\left(\frac{1}{\varrho_1}\right)\left(\frac{\partial p}{\partial x}\right)_1 + \frac{K_\nu K_v}{K_L^2}\frac{\nu}{r_1}\frac{\partial}{\partial r_1}\left(r_1 \frac{\partial u_1}{\partial r_1}\right). \tag{2.137}$$

Comparing Eqs. (2.135) and (2.137), one has

$$\frac{K_{(dp/dx)}/K_\varrho}{K_v^2/K_L} = 1 \tag{2.138}$$

and

$$\frac{K_\nu K_v/K_L^2}{K_v^2/K_L} = 1. \tag{2.139}$$

Therefore

$$\frac{K_\nu}{K_v K_L} = 1. \tag{2.140}$$

Substituting for the ratios again, one gets

$$\frac{L_1}{r_1} = \frac{L_2}{r_2}, \qquad \frac{L_1}{D_1} = \frac{L_2}{D_2}, \qquad \frac{(dp/dx)_1 D_1}{\varrho_1 V_1^2} = \frac{(dp/dx)_2 D_2}{\varrho_2 V_2^2},$$

$$\frac{\nu_1}{D_1 V_1} = \frac{\nu_2}{D_2 V_2}.$$

So the three dimensionless numbers which represent similarity are

$$\left(\frac{L}{D}\right), \quad \left(\frac{\Delta p}{\varrho V^2}\right)\frac{D}{L}, \quad \left(\frac{\nu}{DV}\right) \qquad \text{or} \qquad \left(\frac{L}{D}\right), \quad \left(\frac{\Delta p}{\varrho V^2}\right), \quad \frac{1}{\mathrm{Re}}.$$

Thus, the results of the Pi theorem method and the differential method are the same. While the former method involves a certain skill in selecting the independent variables in the problem, the latter method requires that the differential equation of the system behavior be known. It must be noted that *neither* method can provide the precise equations which represent the physical phenomenon.

2.12.4 Dimensionless Quantities in Transport Processes

Each dimensionless number entering into transport phenomena has a certain physical significance, and it is interesting and useful to derive the physical significance for some of the more useful ones.

2.12.5 Fluid Dynamical Numbers

(i) *Reynolds number*, Re This is defined as

$$\mathrm{Re} = \frac{\varrho VL}{\eta} = \frac{VL}{\nu}.$$

Its physical interpretation can be deduced by writing it in the form

$$\mathrm{Re} = \frac{\varrho VL}{\eta} = \frac{\varrho V^2}{\eta V/L}.$$

The numerator of this expression can be regarded as the inertial force of the fluid per unit area. Consider an infinitesimal cylindrical element of a fluid in steady flow (Fig. 2.23). The volume of this element is

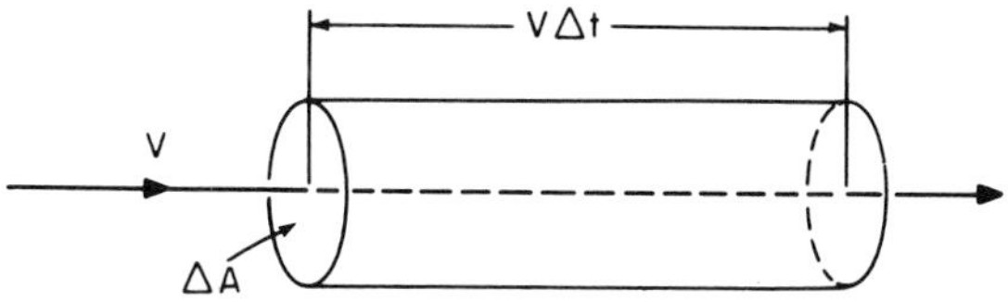

Fig. 2.23

$V \cdot \varDelta t \cdot \varDelta A = \varDelta v;$

$$\varrho V^2 = (\varrho V)(V) = \frac{(\varrho V)(\varDelta t)(\varDelta A)V}{(\varDelta t)(\varDelta A)}$$

$$= \frac{\varrho(\varDelta v)V}{(\varDelta t)(\varDelta A)} = \frac{\text{mass}}{(\varDelta t)(\varDelta A)} V$$

$$= \frac{\text{momentum}}{(\varDelta t)(\varDelta A)} = \frac{\text{force}}{(\varDelta A)}.$$

Thus, ϱV^2 is an inertial term. Near the solid wall,

$$\frac{\varDelta V}{\varDelta L} = \frac{V}{L},$$

so the denominator $\eta V/L$ can be interpreted as the shear stress or the viscous force per unit area:

$$\eta \frac{V}{L} = \eta\left(\frac{\varDelta V}{\varDelta L}\right) = \tau_{\mathrm{w}} = \text{shear stress.}$$

Thus,

$$\mathrm{Re} = \frac{\varrho V^2}{\eta V/L} = \frac{\text{inertial force of the fluid flow}}{\text{viscous force near the wall}}. \tag{2.141}$$

Another interpretation of the physical significance of the Reynolds number can be obtained. From the kinetic theory of gases,

$$\eta = 0.5\bar{c}\lambda,$$

where λ is the mean free path, $\bar{c}$ the molecular rms speed, and

$$\mathrm{Re} = \frac{\varrho VL}{\eta} = \frac{\varrho VL}{0.5\varrho\bar{c}\lambda} = 2\left(\frac{V}{\bar{c}}\right)\left(\frac{L}{\lambda}\right) \tag{2.142}$$

This is the product of a velocity ratio and a length ratio of gas in the macroscopic to gas in the microscopic.

In the first interpretation, then, equality of the value of the corresponding Reynolds numbers in two physically similar systems indicates that the ratio of the magnitudes of the inertial and viscous forces is the same at corresponding points in the systems. In the second interpretation, it would be the *product* of the velocity and length ratios that is the same.

Besides acting as a similarity parameter, the Reynolds number is intimately connected with the stability of laminar flow, as mentioned before. It must be noted that the Reynolds number has no significance in a study of flows of perfect fluids (nonviscous).

(ii) *Froude number*, Fr This is defined as

$$\mathrm{Fr} = \frac{V^2}{gL}. \tag{2.143}$$

Its physical interpretation may be derived as follows:

$$\mathrm{Fr} = \frac{V^2}{gL} = \frac{\varrho V^2 L^2}{\varrho g L^3} = \frac{(\varrho V^2)L^2}{(\varrho L^3)g} = \frac{(\varrho V^2)L^2}{mg}.$$

Now ϱV^2 is the inertial force of the fluid per unit area. So the numerator $(\varrho V^2)L^2$ is the inertial force of the fluid. The denominator is obviously the gravity force. Therefore,

$$\mathrm{Fr} = \frac{\text{inertial force}}{\text{gravity force}}. \tag{2.144}$$

The equality of the values of corresponding Froude numbers in physically similar systems, then, means that the ratio of the magnitudes of the inertial and gravity forces is the same at corresponding points in the systems.

In flow systems involving liquids with a free surface, Fr is an important similarity parameter, because the gravity surface waves involve an interaction between gravity and inertial forces. In flowing liquids with no free surface, the Froude number need not be considered when considering physical similarity of flow systems.

2.12.6 Mach Number

Classification of flow of a compressible fluid *usually* is based on the ratio of the flow speed to the local speed of sound in the fluid. The ratio is known as the Mach number.

If U is the flow velocity, and a the local speed of sound, then

$$\text{Mach number} = M = \frac{U}{a}. \tag{2.145}$$

For an ideal gas,

$$a = (g\gamma R T)^{1/2}, \tag{2.146}$$

where γ is the specific heat ratio for the gas, R the gas constant, g the gravitational constant, and T the temperature.

Two physical interpretations of M are possible.

(i) *For an ideal gas,*

$$M^2 = \frac{U^2}{a^2} = \frac{U^2}{g\gamma kT} = \frac{(2/w)(wU^2/2g)}{\gamma kT},$$

where w is the specific weight of the gas. Now, from kinetic theory,

$$T = \frac{1}{3}\left(\frac{m\bar{U}^2}{k}\right),$$

where m is the mass of a molecule, $\bar{U}^2$ is the mean square molecular velocity, and k is Boltzmann's constant; therefore

$$M^2 = \frac{(wU^2/2g)}{(wv\ R/3k)(m\bar{U}^2/2)}, \tag{2.147}$$

where $wU^2/2g$ is the macroscopic kinetic energy of a gas and $m\bar{U}^2/2$ is the microscopic kinetic energy of a gas. So

$$M^2 \propto \frac{\text{macroscopic kinetic energy of a gas}}{\text{microscopic kinetic energy of a gas}}.$$

(ii) *Also for an ideal gas,*

$$M = \frac{U}{a} = \frac{U}{(\gamma gkT)^{1/2}}.$$

From kinetic theory,

$$\bar{U} = \left(\frac{8\,R\,T}{\pi}\right)^{1/2}, \qquad a = \bar{U}\left(\frac{\gamma g\pi}{8}\right)^{1/2}.$$

Therefore

$$M = \frac{U}{\bar{U}(\gamma g\pi/8)^{1/2}} = \left(\frac{U}{\bar{U}}\right)\left(\frac{8}{\gamma g\pi}\right)^{1/2}. \tag{2.148}$$

So

$$\text{Mach number} \propto \frac{\text{macroscopic velocity } U}{\text{microscopic rms speed } \bar{U}}.$$

The equality of the values of corresponding Mach numbers in two physically similar systems implies that the ratio of the magnitudes of the macroscopic and microscopic kinetic energies is the same at corresponding points in the systems, according to the first interpretation.

The effect of the Mach number on a flow pattern is important, in practice, only for high speed flows of gases.

2.12.7 Weber Number, We

This similarity parameter arises when the physical system under consideration includes a liquid with a free surface. It arises in the study of capillary (surface tension) waves, and in the study of the breakup of liquid droplets and films.

The Weber number is defined as

$$\mathrm{We} = \frac{\varrho U^2 L}{\sigma}, \tag{2.149}$$

where σ is the surface tension, L a characteristic length (for example, the droplet diameter), and U the fluid velocity. To arrive at a physical interpretation, rearrange Eq. (2.149) as follows:

$$\mathrm{We} = \frac{\varrho U^2 L^3}{L(L\sigma)};$$

$\varrho U^2 L^3$ can be interpreted as the kinetic energy of the fluid, and $L(L\sigma)$ as the work done by the surface tension resistance. So,

$$\mathrm{We} = \frac{\text{kinetic energy of the fluid}}{\text{work done by surface tension resistance}}. \tag{2.150}$$

Equality of the corresponding We values in two physically similar systems implies that the ratios of the magnitudes of the inertial and surface tension forces is the same at corresponding points in the two systems.

As the velocity of a liquid droplet through a gas increases, the Weber number increases. There is a certain critical number at which the drop breaks up or atomizes. This is analogous to the role of the Reynolds number in the stability of laminar fluid flow.

The above parameters are the most important ones for systems involving fluid flow. With dynamical similarity accounted for, similarity in the resistance of objects moving through fluids can be expected. This resistance is also called the *drag force.*

Another similarity parameter can arise, however, if there is periodic fluid motion in the physical systems considered. Blunt bodies moving through a fluid shed vortices into the downstream wake. So what is the parameter which relates the vortex shedding frequencies in the two systems?

Now, the acceleration in a simple harmonic motion (which may be associated with vortex shedding) is proportional to the square of the frequency f and to the amplitude (with the dimensions of length). The ratio of this force to the inertial force gives the parameter

$$\frac{(\varrho L^3)(f^2L)}{\varrho U^2L} = \frac{f^2L^2}{U^2}. \tag{2.151}$$

It is the square root of this parameter which is normally used, and it is called the *Strouhal number* S. Thus, the similarity parameter in this case is

$$\mathrm{S} = \frac{fL}{U}. \tag{2.152}$$

In physically similar systems, then,

$$\left(\frac{f_1L_1}{U_1}\right) = \left(\frac{f_2L_2}{U_2}\right).$$

So f_2 can be determined if f_1 and the scale factors L_1/L_2 and U_1/U_2 are known.

2.12.8 Prandtl Number

Although heat transfer is not dealt with formally in this text, the conservation of energy equation is one of the equations of fluid motion and the specific heat and thermal conductivity of the fluid enter in this equation, as well as the viscosity. A similarity parameter known as the Prandtl number thus emerges, where

$$\text{Prandtl number} = \mathrm{Pr} = \frac{\eta C_\mathrm{p}}{k}. \tag{2.153}$$

This can be rearranged as

$$\mathrm{Pr} = \frac{\eta C_\mathrm{p}}{k} = \frac{\eta/\varrho}{k/\varrho C_\mathrm{p}} = \frac{\nu}{\varkappa} = \frac{\text{kinematic viscosity}}{\text{thermal diffusivity}}. \tag{2.154}$$

It expresses the relative speeds at which momentum and energy are propagated through the system.

2.12.9 Dimensionless Number in Mass Transfer

There are two useful dimensionless numbers that occur in systems in which diffusion of material occurs.

(i) *Lewis number*, Le By definition, this is

$$\mathrm{Le} = \frac{k}{\varrho C_{\mathrm{p}} D}, \tag{2.155}$$

where D is the mass diffusivity. Now,

$$\mathrm{Le} = \frac{k}{\varrho C_{\mathrm{p}} D} = \frac{\varkappa}{D} = \frac{\text{thermal diffusivity}}{\text{mass diffusivity}}. \tag{2.156}$$

So the Lewis number expresses the relative rates of propagation of energy and mass within a system.

(ii) *Schmidt number*, Sc This is defined as

$$\mathrm{Sc} = \frac{\eta}{\varrho D} = \frac{\nu}{D} = \frac{\text{kinetic viscosity}}{\text{mass diffusivity}}. \tag{2.157}$$

Therefore

$$\mathrm{Sc} = \frac{\text{momentum diffusivity}}{\text{mass diffusivity}}. \tag{2.158}$$

The Schmidt number therefore expresses the relative rates of propagation of momentum and mass in the system.

As a final example of the method of dimensional analysis, consider the flow of a fluid past an object such as a cylinder, which involves heat transfer. Suppose the fluid is a liquid with no free surface (no surface waves). Since liquids are incompressible, the Mach number will not be significant. The problem is to find how the heat-transfer coefficient by convection (that is, by fluid motion), h, depends on other physical quantities.

One would expect h to depend on

(a) diameter of the cylinder D,
(b) flow speed V (at a location far upstream of cylinder),
(c) fluid density ϱ,
(d) fluid viscosity η,
(e) coefficient of thermal conductivity k,
(f) specific heat C_{p}.

If the cylinder is effectively infinitely long, its length is not a factor. Then,

$$h = f(D, V, \varrho, \eta, C_{\mathrm{p}}, k). \tag{2.159}$$

The dimensions of the various physical properties are

$$h = \frac{L}{T^3\theta}, \qquad \text{where } \theta \text{ is the temperature,}$$

$$D = L, \qquad \eta = \frac{M}{LT},$$

$$V = \frac{L}{T}, \qquad C_p = \frac{L^2}{T^2\theta},$$

$$\varrho = \frac{M}{L^3}, \qquad \varkappa = \frac{ML}{T^3\theta},$$

and Q has been given the dimensions of energy.

The mass M is eliminated by dividing h, η, and k by ϱ, so that

$$\frac{h}{\varrho} = f\left(D, V, \frac{\eta}{\varrho}, C_p, \frac{\varkappa}{\varrho}\right). \tag{2.160}$$

The temperature can now be eliminated by multiplying h/ϱ and C_p by $\varrho/\varkappa$, and so

$$\frac{h}{\varkappa} = f\left(D, V, \frac{\eta}{\varrho}, \frac{C_p\varrho}{\varkappa}\right). \tag{2.161}$$

Now the dimensions are

$$\frac{h}{\varkappa} = \frac{1}{L}, \qquad \frac{\eta}{\varrho} = \frac{L^2}{T},$$

$$D = L, \qquad \frac{C_p\varrho}{\varkappa} = \frac{T}{L^2}.$$

$$V = \frac{L}{T},$$

Finally, L can be eliminated by multiplying $h/\varkappa$ by D, $C_p\varrho/\varkappa$ by D^2, and by dividing V by D and η/ϱ by D^2. The result is

$$\frac{hD}{\varkappa} = f\left(\frac{V}{D}, \frac{\eta}{\varrho D^2}, \frac{C_p\varrho D^2}{\varkappa}\right), \tag{2.162}$$

and

$$\frac{hD}{\varkappa} = 1, \qquad \frac{\eta}{\varrho D^2} = \frac{1}{T},$$

$$\frac{V}{D} = \frac{1}{T}, \qquad \frac{C_p\varrho D^2}{\varkappa} = T.$$

Now, time T is eliminated and the following dimensionless axpression is obtained:

$$\frac{hD}{\varkappa} = f\left(\frac{\varrho VD}{\eta}, \frac{C_p\eta}{\varkappa}\right) \tag{2.163}$$

or

$$h = \frac{\varkappa}{D} f\left(\frac{\varrho VD}{\eta}, \frac{C_p\eta}{\varkappa}\right). \tag{2.164}$$

The parameter

$$\mathrm{Nu} = \frac{hD}{\varkappa} \tag{2.165}$$

is very important in heat transfer and is called the Nusselt number;

$$\mathrm{Pr} = \frac{C_p\eta}{\varkappa} \tag{2.166}$$

is known as the Prandtl number, and the third parameter is the Reynolds number.

Thus, dimensionless analysis has shown that although there are seven variables in the flow problem, there are only three dimensionless parameters which are significant and uniquely related. It is these three parameters which would be used in the correlation of experimental data.

It is easy to show that the Nusselt number represents the ratio of the heat transfer by convection to that by pure conduction. The Prandtl number represents the relative speeds at which momentum and energy are propagated through the system. When the Prandtl number in a system is one, this means that the velocity and temperature distributions are the same.

2.12.10 The Knudsen Number

This number will be considered because it is significant in the concept of the continuum and also in the regime of rarefied gases. It is defined as the ratio of the molecular mean free path in a gas to some macroscopic characteristic length:

$$\mathrm{Kn} = \frac{\lambda}{L}. \tag{2.167}$$

For gases at ordinary pressure, the intermolecular distance is much larger than that for solids and liquids, but still much smaller than the macroscopic dimensions one usually encounters. So, the concept of a continuum gas is well fulfilled. At any arbitrary point (x, y, z) in the fluid field, as mentioned before, properties such as temperature, density, and pressure can be assigned with the assurance that there are enough particles in the neighborhood of the points so that collisions are frequent enough to produce local equilibrium.

At the solid–fluid interfaces, zero relative velocity can be assumed, and equality of the temperature of the solid and fluid. The reason is that for very small particle spacings, a particle undergoes many collisions while in a particular location and so comes to equilibrium with its local surroundings, which in this case is the wall.

If the fluid density is small, as at high atmospheric altitudes, or if the relevant physical scale of the system is small enough, such as in a hot-wire anemometer probe (a local velocity-measuring device), the length of the molecular mean free path is of the same or even greater magnitude than the macroscopic scale. Then the fluid can no longer be treated as a continuous fluid. The Knudsen number, the ratio of the mean free path to the macroscopic scale, has significance in connection with a rarefied gas field. Three regimes of the flow of gases may be distinguished.

(a) *Continuum flow*, in which the gas behaves as a continuous medium.

(b) *Slip flow*, the point at which the continuum model of flow first fails.

(c) *Free molecular flow*, in which the mean free path is so great that the molecular collisions occur primarily at the surface and at great distances from the undisturbed stream, so that the intermolecular collisions are rare and less important than collision with the wall.

The velocity and temperature distributions over a heated plate for these three kinds of flow are shown in Figs. 2.24A and 2.24B. The three regimes

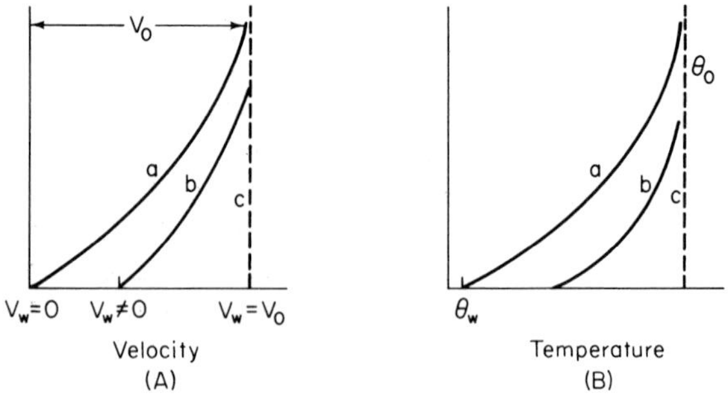

Fig. 2.24 (A) Velocity and (B) temperature distributions over a heated plate for (a) continuum flow, (b) slip flow, and (c) molecular flow.

may be delineated roughly as follows:

Flow type	Kn
Continuum	< 0.01
Slip	$0.01 < \text{Kn} < 10$
Free molecule	$\text{Kn} > 10$

The method of finding dimensionless numbers, how to define their physical significance, and how to use them has been dealt with in some detail. The method of dimensional analysis, of course, is not the only method of solving transport problems. However, it does provide sound predictions and is particularly useful for correlating experimental results.

In this chapter some of the more important basic physical properties of fluids have been expounded on. This should be enough to form a sound foundation for the study of the dynamics of fluids in the following chapters.

PROBLEMS

2.1 In a continuum in motion, suppose the normal and shear stresses σ_{zz}, τ_{xz}, τ_{yz}, and τ_{zy} in the z direction are all zero (this pertains to a two-dimensional flow). Suppose

$$\sigma_{xx} = -200 \text{ lb} \cdot \text{f/in}^2, \quad \sigma_{yy} = -250 \text{ lb} \cdot \text{f/in}^2, \quad \tau_{xy} = \tau_{yx} = 40 \text{ lb} \cdot \text{f/in}^2.$$

Determine the normal stress σ_{nn}.

2.2 If a stress field is given by the matrix

$$\begin{pmatrix} 2x + 4y^2 & -3x^2 & 0 \\ -2x^2 & 9xy - 5y^2 & 0 \\ 0 & 0 & 0 \end{pmatrix},$$

then determine the normal stress at the position (4, 2, 1) in the field.

2.3 In a perfect gas, at zero altitude the pressure and density are p_0 and ϱ_0, respectively. Show that at an altitude y, the pressure is given by

$$p = p_0 \exp\left(-\frac{\varrho_0 g y}{p_0}\right).$$

2.4 The surface tension σ of a liquid may be defined through the work term as

$$dW = \sigma\, dA,$$

where A is the area of the surface.

Show that if $\sigma = \sigma(T)$, then the internal energy u of the liquid may be expressed as

$$u = u_0 + \left(\sigma - T\frac{d\sigma}{dT}\right)A,$$

where u_0 is the value of the internal energy as defined for a perfect fluid.

2.5 If the distribution function of a gas is isotropic (it depends only on the speed v), show that

(a) the number flux is $n\langle v\rangle/4$ in the positive x direction;
(b) the momentum flux in the x direction is 2/3 of the kinetic energy per unit volume.

2.6 A flowing gas has the distribution function

$$f = n\left(\frac{m}{2\pi kT}\right)^{3/2} \exp\left(-\frac{m\{(v_x - u)^2 + v_y{}^2 + v_z{}^2\}}{2kT}\right),$$

where n, T, and u are constants.

(a) Show that the mean velocity $\langle \mathbf{v}\rangle$ is a constant, u, in the $+x$ direction.
(b) Find the number flux and the momentum flux.

2.7 A compromise equation between that for a perfect gas and the van der Waals equation is

$$p(v - b) = RT.$$

Calculate the bulk modulus and coefficient of thermal expansion resulting from this equation. Comment on the physical difference between this gas behavior and that of a perfect gas.

2.8 A gas has a coefficient of volume expansion α and an isothermal bulk modulus β which are, respectively,

$$\alpha = \frac{1}{T}\left(1 + \frac{3a}{vT^2}\right) \quad \text{and} \quad \beta = p\Big/\left(1 + \frac{a}{vT^2}\right).$$

(a) Show that these are compatible.
(b) Derive the equation of state for the gas.

2.9 Let D, ϱ, μ, l, and U be the drag force, fluid density, viscosity, reference length, and velocity, respectively, for a body moving through a fluid. Using dimensional analysis, derive an expression for D in terms of dimensionless groups and identify the Reynolds number.

2.10 Repeat Problem 2.9 using the π theorem.

2.11 Using dimensional analysis, establish an expression for the dynamic force (force exerted by a fluid stream on a body submerged and at rest in it) for the following cases:

(a) inviscid incompressible fluid,
(b) viscous incompressible fluid,
(c) inviscid compressible fluid,
(d) viscous compressible fluid.

Identify all the dimensionless groups such as Reynolds number, Mach number, specific heat ratio, and Prandtl number.

BIBLIOGRAPHY

G. O. Hirschfelder and C. F. Curtiss, "Molecular Theory of Gases and Liquids," 1st. ed. Wiley, New York, 1954.

O. K. Rice, "Statistical Mechanics, Thermodynamics and Kinetics." Freeman, San Francisco, 1967.

A. H. Cottrell, "The Mechanical Properties of Matter." Wiley, New York, 1964.

W. G. Vincenti and C. H. Kruger, Jr., "Physical Gas Dynamics." Wiley, New York, 1965.

G. K. Batchelor, "An Introduction to Fluid Dynamics," Chapter 1. Cambridge Univ. Press, London and New York, 1967.

L. Prandtl, "The Physics of Solids and Fluids." Blackie, Edinburgh, 1930.

W. J. Duncan, "Physical Similarly and Dimensional Analysis." Arnold, London, 1953.

C. M. Focken, "Dimensional Methods and Their Applications." Arnold, London, 1953.

L. J. Sedov, "Similarity and Dimensional Methods in Mechanics" (translated by M. Friedman). Academic Press, New York, 1959.

R. Bird, W. Stewart, and E. Lightfoot, "Transport Phenomena." Wiley, New York, 1960.

CHAPTER THREE *FLUID FLOW KINEMATICS*

Kinematics implies the study of motion. All such motion is subject to the conservation of mass. Two distinct methods of specifying the flow field have been developed. In the first—the Euler method—the flow quantities (velocity, pressure, and density) are defined as functions of position in space ($\mathbf{r}$) and time (t). The basic flow quantity is the vector velocity of the fluid, written $\mathbf{v}(\mathbf{r}, t)$. In the second method, the dynamical history of a selected fluid element is described. If $\mathbf{r}_0$ is the position of the center of mass of the fluid element at time t_0, then the basic flow quantity in the Lagrangian description is the velocity $\mathbf{v}(\mathbf{r}_0, t)$. While the Lagrange method is useful in certain special areas of fluid dynamics, it involves very cumbersome analysis and, unless stated otherwise, in this text the Euler method will be adopted. However, the concept of fluid elements, fluid surfaces, and fluid lines, which always consist of the same fluid particles and move with them, will be widely used in the Euler specification of the flow field.

3.1 The Eulerian Equations

Let u, v, w be the velocity components (in a Cartesian reference system) at the position (x, y, z) at time t. These components are functions of the independent variables x, y, z, t. For any given time t, they define the motion at all points in the space occupied by the fluid. For a given point (x, y, z), they describe how the velocity is changing with time at that point.

The motion considered is continuous in that u, v, w are finite and continuous functions of position and their spatial derivatives are finite everywhere (but not necessarily continuous).

Consider any function $f(x, y, z, t)$ which varies with the fluid motion (it could be velocity, density, or some such quantity of interest). To determine the rate at which f is changing, one notes that at the time $t + \delta t$ the fluid element, which was at the position (x, y, z) originally, is in the position $(x + u\,\delta t, y + v\,\delta t, z + w\,\delta t)$, and so the corresponding value of f is

$$\begin{aligned} &f(x + u\,\delta t, y + v\,\delta t, z + w\,\delta t, t + \delta t) \\ &\quad = f + u\,\delta t \frac{\partial f}{\partial x} + v\,\delta t \frac{\partial f}{\partial y} + w\,\delta t \frac{\partial f}{\partial z} + \delta t \frac{\partial f}{\partial t}. \end{aligned} \tag{3.1}$$

Following Stokes, suppose the symbol D/Dt denotes differentiation following the fluid motion; then the new value of f is $f + (Df/Dt)\,\delta t$ and so

$$\frac{Df}{Dt} = \frac{\partial f}{\partial t} + u \frac{\partial f}{\partial x} + v \frac{\partial f}{\partial y} + w \frac{\partial f}{\partial z}. \tag{3.2}$$

Vectorially, this can be written as

$$\frac{Df}{Dt} = \frac{\partial f}{\partial t} + (\mathbf{v} \cdot \operatorname{grad})f. \tag{3.3}$$

The first term on the right-hand side of Eq. (3.3) represents a temporal rate of change of f, and the second term the convective rate of change.

If $\mathbf{v}$ is introduced for f in Eq. (3.3), then the total derivative of the velocity with respect to time is

$$\mathbf{a} = \frac{D\mathbf{v}}{Dt} = \frac{\partial \mathbf{v}}{\partial t} + (\mathbf{v} \cdot \operatorname{grad})\mathbf{v}. \tag{3.4}$$

In indicial notation this can be written as

$$a_i = \frac{dv_i}{dt} = \frac{\partial v_i}{\partial t} + \frac{\partial v_i}{\partial x_j}\,\frac{\partial x_j}{\partial t}. \tag{3.5}$$

In Cartesian rectangular coordinates, the operator $(\mathbf{v} \cdot \text{grad})$ is defined as

$$(\mathbf{v} \cdot \text{grad}) = u\,\frac{\partial}{\partial x} + v\,\frac{\partial}{\partial y} + w\,\frac{\partial}{\partial z}. \tag{3.6}$$

It is this term in the Euler relation for acceleration which is responsible for the nonlinearities in the equations of motion of fluid dynamics.

3.2 The Equation of Continuity (Conservation of Mass)

The concept of a continuous medium has been introduced in Chapter 2. From it the continuity equation is derived. Apart from heterogenous and noncontinuous fluids, the equation simply expresses the law of conservation of mass. The quantity of fluid entering a certain volume in space must be balanced by that quantity leaving, unless compression occurs.

Let V be an arbitrary volume fixed in space, bounded by a surface S, and containing a fluid of density ϱ (Fig. 3.1). The volume element δV is small enough so that ϱ can be regarded as constant through it.

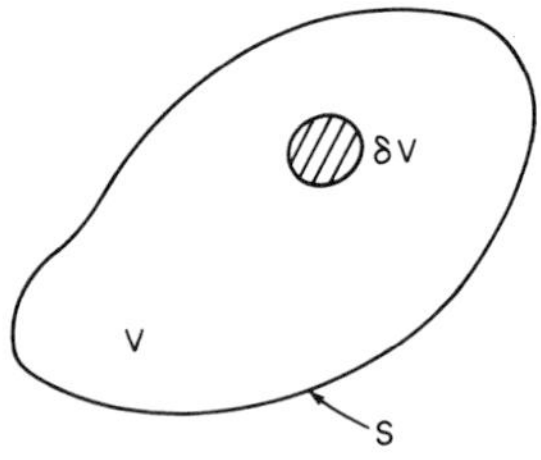

Fig. 3.1

The time rate of increase of mass in V is

$$\frac{\partial}{\partial t}\int_V \varrho\,dV.$$

If the volume lies in a flow field, then fluid enters V through part of its bounding surface S and leaves through another part. For an element of surface dS, the outward mass flux is $(\varrho\mathbf{v}) \cdot \mathbf{n}\, dS$, where $\mathbf{n}$ is the outward directed normal. The total outward flux is

$$\int_S (\varrho\mathbf{v}) \cdot \mathbf{n}\, dS.$$

The sum of the net outward convection of mass plus the time rate of increase of mass in volume must be zero:

$$\int_S (\varrho\,\mathbf{v}) \cdot \mathbf{n}\, dS + \frac{\partial}{\partial t}\int_V \varrho\, dV = 0. \tag{3.7}$$

Using Gauss's theorem on the surface integral, one has

$$\int_V \left[\frac{\partial \varrho}{\partial t} + \boldsymbol{\nabla} \cdot (\varrho\,\mathbf{v})\right] dV = 0. \tag{3.8}$$

Since this is true for arbitrary elementary volumes,

$$\frac{\partial \varrho}{\partial t} + \boldsymbol{\nabla} \cdot (\varrho\,\mathbf{v}) = 0 \tag{3.9}$$

or

$$\frac{\partial \varrho}{\partial t} + \mathbf{v} \cdot \boldsymbol{\nabla}\varrho + \varrho\boldsymbol{\nabla} \cdot \mathbf{v} = 0.$$

Therefore

$$\frac{D\varrho}{Dt} + \varrho\boldsymbol{\nabla} \cdot \mathbf{v} = 0, \tag{3.10}$$

where $\boldsymbol{\nabla} \cdot \mathbf{v}$ is called the dilation of the fluid at a point in the field.

This is the general continuity equation for nonhomogeneous or compressible fluids.

In rectangular Cartesian coordinates, Eq. (3.10) has the form

$$\frac{D\varrho}{Dt} + \varrho\left(\frac{\partial u}{\partial x} + \frac{\partial v}{\partial y} + \frac{\partial w}{\partial z}\right) = 0. \tag{3.11}$$

For steady motion, $\partial\varrho/\partial t = 0$ and the continuity relation [from Eq. (3.9)] is

$$\boldsymbol{\nabla} \cdot (\varrho\mathbf{v}) = 0. \tag{3.12}$$

In the case of homogeneous and incompressible fluids, the continuity equation becomes simply

$$\boldsymbol{\nabla} \cdot \mathbf{v} = 0. \tag{3.13}$$

This covers the case of the ideal fluid, which is defined as being inviscid and incompressible. All real gases are compressible and liquids are slightly so. It is found, however, that as long as the Mach number does not exceed about 0.3, the fluid can be regarded as incompressible to a first approximation.

3.3 Streamline

A line in the fluid whose tangent is everywhere parallel to $\mathbf{v}$ at every instant is a *streamline*. Since the velocity must be tangent or parallel to the line element, and since the cross product of parallel vectors is zero, the equation for the streamline is

$$d\boldsymbol{l} \times \mathbf{v} = 0. \tag{3.14}$$

In rectangular Cartesian coordinates, this leads directly to the relations

$$\frac{dx}{u(\mathbf{r}, t)} = \frac{dy}{v(\mathbf{r}, t)} = \frac{dz}{w(\mathbf{r}, t)}. \tag{3.15}$$

When the flow is steady, the streamlines have the same shape at all times. A streamtube is formed by all the streamlines that pass through a given closed curve in the fluid.

3.3.1 Stream Function

It has been shown that for the flow of an incompressible fluid, or a steady flow of a compressible fluid [Eqs. (3.12) and (3.13)], the continuity, or mass conservation, equation reduces to the statement that the divergence of a vector is zero (either $\varrho\mathbf{v}$ or $\mathbf{v}$). If the flow field is restricted to be either two-dimensional (rectangular Cartesian) or axially symmetrical, the divergence is the sum of only two derivatives. For example, for the two-dimensional planar case, with $v = (u, v, 0)$ and u, v not dependent on z, the continuity equation has the form

$$\frac{\partial u}{\partial x} + \frac{\partial v}{\partial y} = 0. \tag{3.16}$$

Thus $u\,\delta y - v\,\delta x$ is an exact differential, call it $\delta\psi$, and

$$u = \frac{\partial\psi}{\partial y}, \qquad v = -\frac{\partial\psi}{\partial x}, \tag{3.17}$$

and the unknown (scalar) function $\psi(x, y, t)$ is defined by

$$\psi - \psi_0 = \int (u\,dy - v\,dx), \tag{3.18}$$

where ψ_0 is a constant and the line integral is taken along an arbitrary curve joining a reference point O to some point P (Fig. 3.2). The function ψ is called the *stream function* and for this two-dimensional flow field is associated with Lagrange.

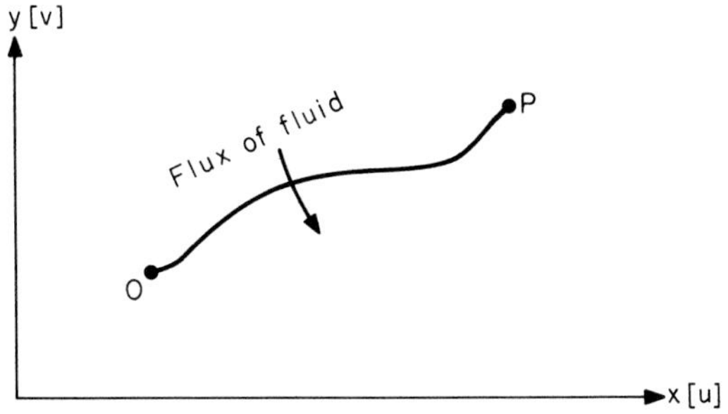

Fig. 3.2

The right-hand side of Eq. (3.18) gives the flux of fluid across a line joining O and P. The flux is regarded as positive if it moves in an anticlockwise manner relative to P. It is independent of the path taken, provided all circuits formed by any two paths are reducible, that is, any circuit can be shrunk to a point without passing out of the region occupied by the incompressible fluid. This also imposes the condition that such regions are simply connected.

It is immediately apparent from the definition of $\psi - \psi_0$ and of a streamline that ψ must be constant along any streamline. Moreover, as there can be no flux across the surface of a solid body, such surfaces must form streamlines (Fig. 3.3).

For an axially symmetric flow, the continuity equation becomes

$$\nabla \cdot \mathbf{v} = \frac{\partial u}{\partial z} + \frac{1}{r}\,\frac{\partial(rv)}{\partial r} = 0. \tag{3.19}$$

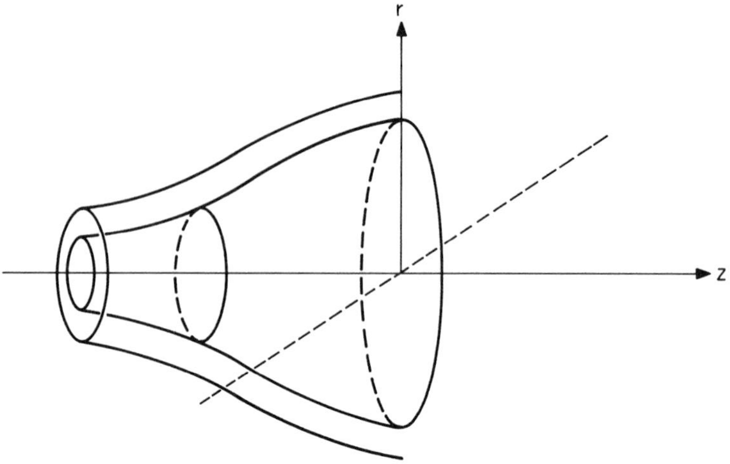

Fig. 3.3

In this case

$$\delta\psi = ru\,\delta r - rv\,\delta z. \tag{3.20}$$

Thus

$$\psi - \psi_0 = \int r(u\,dr - v\,dz) \tag{3.21}$$

and

$$u = \frac{1}{r}\frac{\partial\psi}{\partial r}, \qquad v = -\frac{1}{r}\frac{\partial\psi}{\partial z}. \tag{3.22}$$

In this case, $\psi - \psi_0$ is the flux across a surface generated by rotating an arbitrary curve joining O to P in an axial plane about the axis of symmetry (z).

The ψ in this geometry is called the Stokesian stream function.

3.4 Kinematics of Deformation

G. G. Stokes (1845) showed for viscous fluids and H. von Helmholtz (1958) for ideal fluids that the motions of deformable bodies, such as fluids, can be broken down into a sum of a translation, a rotation, and a strain.

The strain can be subdivided further into a pure linear strain and a pure shearing (angular) strain.

In fluid media the parameter of prime interest is the rate of deformation, or the velocity, of the fluid. It is expedient to consider carefully the relative motion near a point in the fluid. Let the velocity at some point defined by the vector $\mathbf{r}_0$ be $\mathbf{v}_0$, so that in the neighborhood of this point the velocity is given by a Taylor series expansion:

$$\mathbf{v} = \mathbf{v}(\mathbf{x}) = \mathbf{v}_0 + \frac{\mathbf{r} - \mathbf{r}_0}{1!}[\nabla \mathbf{v}]_0 + \frac{(\mathbf{r} - \mathbf{r}_0)^2}{2!}[\nabla\nabla \mathbf{v}]_0 + \cdots. \tag{3.23}$$

In a small neighborhood ($\mathbf{r} - \mathbf{r}_0$ small) the higher-order terms may be neglected and a linear relation for the velocity is obtained. This will be shown to involve the four components listed above. In rectangular Cartesian coordinates there are three component equations:

$$\begin{aligned} u &= u_0 + (x - x_0)\frac{\partial u}{\partial x} + (y - y_0)\frac{\partial u}{\partial y} + (z - z_0)\frac{\partial u}{\partial z}, \\ v &= v_0 + (x - x_0)\frac{\partial v}{\partial x} + (y - y_0)\frac{\partial v}{\partial y} + (z - z_0)\frac{\partial v}{\partial z}, \\ w &= w_0 + (x - x_0)\frac{\partial w}{\partial x} + (y - y_0)\frac{\partial w}{\partial y} + (z - z_0)\frac{\partial w}{\partial z}. \end{aligned} \tag{3.24}$$

The difference between the velocity at the reference point and the velocity at some point near it is characterized by nine velocity derivatives. Indicially this can be written

$$\delta v_i = r_j \frac{\partial v_i}{\partial r_j}, \tag{3.25}$$

correct to first order in the small distance r between the two points.

The $\partial v_i/\partial r_j$ is a second-order tensor and can be decomposed into parts which are symmetric and antisymmetric, respectively:

$$\frac{\partial v_i}{\partial r_j} = \frac{1}{2}\left(\frac{\partial v_i}{\partial r_j} + \frac{\partial v_j}{\partial r_i}\right) + \frac{1}{2}\left(\frac{\partial v_i}{\partial r_j} - \frac{\partial v_j}{\partial r_i}\right). \tag{3.26}$$

Let

$$e_{ij} = \frac{1}{2}\left(\frac{\partial v_i}{\partial r_j} + \frac{\partial v_j}{\partial r_i}\right), \tag{3.27}$$

$$\xi_{ij} = \frac{1}{2}\left(\frac{\partial v_i}{\partial r_j} - \frac{\partial v_j}{\partial r_i}\right). \tag{3.28}$$

Now, v_1, v_2, and v_3 correspond to u, v, and w in the rectangular Cartesian system, and r_1, r_2, and r_3 correspond to x, y, and z. Using (3.26) in

Eq. (3.24) and gathering terms, one obtains

$$u = u_0 + \frac{1}{2}\left[\left(\frac{\partial u}{\partial x} - \frac{\partial u}{\partial x}\right)(x - x_0) + \left(\frac{\partial u}{\partial y} - \frac{\partial v}{\partial x}\right)(y - y_0) + \left(\frac{\partial u}{\partial z} - \frac{\partial w}{\partial x}\right)(z - z_0)\right] + \frac{1}{2}\left[\left(\frac{\partial u}{\partial x} + \frac{\partial u}{\partial x}\right)(x - x_0) + \left(\frac{\partial u}{\partial y} + \frac{\partial v}{\partial x}\right)(y - y_0) + \left(\frac{\partial u}{\partial z} + \frac{\partial w}{\partial x}\right)(z - z_0)\right], \tag{3.29a}$$

$$v = v_0 + \frac{1}{2}\left[\left(\frac{\partial v}{\partial x} - \frac{\partial u}{\partial y}\right)(x - x_0) + \left(\frac{\partial v}{\partial y} - \frac{\partial v}{\partial y}\right)(y - y_0) + \left(\frac{\partial v}{\partial z} - \frac{\partial w}{\partial y}\right)(z - z_0)\right] + \frac{1}{2}\left[\left(\frac{\partial v}{\partial x} + \frac{\partial u}{\partial y}\right)(x - x_0) + \left(\frac{\partial v}{\partial y} + \frac{\partial v}{\partial y}\right)(y - y_0) + \left(\frac{\partial v}{\partial z} + \frac{\partial w}{\partial y}\right)(z - z_0)\right], \tag{3.29b}$$

$$w = w_0 + \frac{1}{2}\left[\left(\frac{\partial w}{\partial x} - \frac{\partial u}{\partial z}\right)(x - x_0) + \left(\frac{\partial w}{\partial y} - \frac{\partial v}{\partial z}\right)(y - y_0) + \left(\frac{\partial w}{\partial z} - \frac{\partial w}{\partial z}\right)(z - z_0)\right] + \frac{1}{2}\left[\left(\frac{\partial w}{\partial z} + \frac{\partial u}{\partial z}\right)(x - x_0) + \left(\frac{\partial w}{\partial y} + \frac{\partial v}{\partial z}\right)(y - y_0) + \left(\frac{\partial w}{\partial z} + \frac{\partial w}{\partial z}\right)(z - z_0)\right]. \tag{3.29c}$$

In vector notation these equations may be expressed as

$$\mathbf{v} = \mathbf{v}_0 + (\mathbf{r} - \mathbf{r}_0) \times \boldsymbol{\omega} + (\mathbf{r} - \mathbf{r}_0)\dot{T}. \tag{3.30}$$

The first term of Eq. (3.30), u_0, v_0, w_0, is a translatory contribution, or a displacement rate.

The second term, which can be written indicially as

$$\delta v_i^{(a)} = -\tfrac{1}{2}\varepsilon_{ijk} r_j \omega_k, \tag{3.31}$$

is the antisymmetric contribution to the change in velocity. In fact,

$$\xi_{ij} = \frac{1}{2}\left(\frac{\partial v_i}{\partial r_j} - \frac{\partial v_j}{\partial r_i}\right) = \omega_k,$$

and

$$2\omega_1 = \frac{\partial v_3}{\partial r_2} - \frac{\partial v_2}{\partial r_3}, \qquad 2\omega_2 = \frac{\partial v_1}{\partial r_3} - \frac{\partial v_3}{\partial r_1}, \qquad 2\omega_3 = \frac{\partial v_2}{\partial r_1} - \frac{\partial v_1}{\partial r_2}. \tag{3.32}$$

In rectangular Cartesian coordinates,

$$\boldsymbol{\omega} = \frac{1}{2}\left[\left(\frac{\partial w}{\partial y} - \frac{\partial v}{\partial z}\right)\mathbf{i} + \left(\frac{\partial u}{\partial z} - \frac{\partial w}{\partial x}\right)\mathbf{j} + \left(\frac{\partial v}{\partial x} - \frac{\partial u}{\partial y}\right)\mathbf{k}\right]; \quad (3.33)$$

$\delta\mathbf{v}^{(a)}$ is the velocity produced at position **r** relative to a point about which there is a rigid body rotation with angular velocity $\boldsymbol{\omega}$ where

$$\boldsymbol{\zeta} = 2\boldsymbol{\omega} = \nabla \times \mathbf{v}. \quad (3.34)$$

The vector $\boldsymbol{\zeta}$ is called the *local vorticity* of the fluid. Thus, if in a certain region occupied by a flowing fluid, the curl of the velocity vector is zero throughout, then the flow field is said to be *irrotational.*

The third component of Eq. (3.30) involves the deformation, or strain, of the fluid, both linear (extensional) and shear. The strain is described in terms of the strain rate tensor $\dot{T} = e_{ij}$ defined by the square array

$$\dot{T} = \begin{bmatrix} \dot{e}_{xx} & \dot{e}_{xy} & \dot{e}_{xz} \\ \dot{e}_{yx} & \dot{e}_{yy} & \dot{e}_{yz} \\ \dot{e}_{zx} & \dot{e}_{zy} & \dot{e}_{zz} \end{bmatrix}$$

$$= \begin{bmatrix} \dfrac{\partial u}{\partial x} & \dfrac{1}{2}\left(\dfrac{\partial u}{\partial y} + \dfrac{\partial v}{\partial x}\right) & \dfrac{1}{2}\left(\dfrac{\partial u}{\partial z} + \dfrac{\partial w}{\partial x}\right) \\ \dfrac{1}{2}\left(\dfrac{\partial v}{\partial x} + \dfrac{\partial u}{\partial y}\right) & \dfrac{\partial v}{\partial y} & \dfrac{1}{2}\left(\dfrac{\partial v}{\partial z} + \dfrac{\partial w}{\partial y}\right) \\ \dfrac{1}{2}\left(\dfrac{\partial w}{\partial x} + \dfrac{\partial u}{\partial z}\right) & \dfrac{1}{2}\left(\dfrac{\partial w}{\partial y} + \dfrac{\partial v}{\partial z}\right) & \dfrac{\partial w}{\partial z} \end{bmatrix}. \quad (3.35)$$

This is obviously a symmetric tensor. The terms on the principal diagonal indicate the linear, or extensional, strain rates. The sum of these terms, or the trace of the tensor, is just the divergence of the velocity and represents the fluid dilation ϑ:

$$\vartheta = \dot{e}_1 + \dot{e}_2 + \dot{e}_3 = \frac{\partial u}{\partial x} + \frac{\partial v}{\partial y} + \frac{\partial w}{\partial z} = \nabla \cdot \mathbf{v}. \quad (3.36)$$

For two-dimensional flow the two components of linear strain are shown diagramatically in Fig. 3.4.

In indicial notation, the dilation is given by

$$e_{ii} = \frac{\partial v_i}{\partial r_i}. \quad (3.37)$$

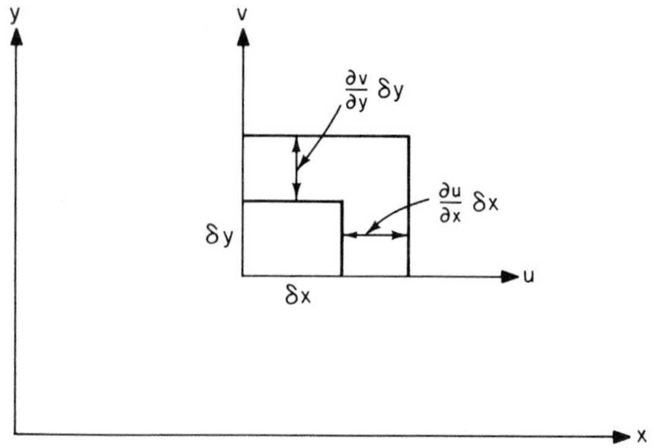

Fig. 3.4

The off-diagonal terms of the rate-of-strain tensor lead to the appearance of shearing stresses and thus energy dissipation. They are therefore of no interest in ideal fluid motions where viscosity is ignored, but they are of prime importance in viscous fluid, or boundary layer, flows.

Each pair of symmetric terms may be shown to represent a rate of angular distortion of a fluid element. Again, consider the two-dimensional rectangular coordinate system case (Fig. 3.5).

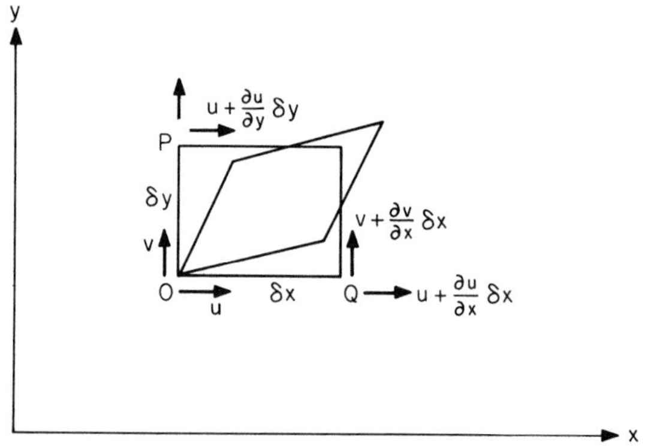

Fig. 3.5

The angular velocity of side OQ is

$$\left(\frac{v_Q - v_O}{\delta x}\right) = \frac{\partial v/\partial x \, \delta x}{\delta x} = \frac{\partial v}{\partial x},$$

and that of side OP is

$$\frac{u_P - u_O}{\delta y} = -\frac{\partial u}{\partial y}.$$

The rate of angular distortion at O is then

$$\dot{\nu} = \frac{\partial \nu}{\partial t} = \omega_{OQ} - \omega_{OP} = \frac{\partial v}{\partial x} - \left(-\frac{\partial u}{\partial y}\right) = \frac{\partial v}{\partial x} + \frac{\partial u}{\partial y}.$$

This is one of the off-diagonal terms in the rate-of-strain tensor. The other may be interpreted similarly. To summarize then, it has been seen that to first order in linear dimensions of a small neighborhood surrounding the point **r**, the velocity field in this region is the superposition of

(a) a uniform translation with velocity $\mathbf{v}(\mathbf{r})$,

(b) a pure straining motion characterized by the rate-of-strain tensor $\dot{T} = e_{ij}$, which can itself be broken down into an isotropic expansion and a pure shear strain, and

(c) a rigid body rotation with angular velocity $\boldsymbol{\omega}$. Indicially, this is summarized by the equation for the velocity at the position **r**.

$$v_i(\mathbf{r}) = v_i(\mathbf{r}_0) + \frac{\partial}{\partial S_i}\left(\tfrac{1}{2}S_i S_k e_{jk}\right) + \tfrac{1}{2}\varepsilon_{ijk}\omega_j S_k, \tag{3.38}$$

where $\mathbf{S} = \mathbf{r} - \mathbf{r}_0$ and e_{ij} and ω_j are evaluated at the point $\mathbf{r}_0$.

3.5 Circulation and Vorticity

Kelvin in 1869 introduced the concept of circulation. The *circulation* around any closed curve (circuit) in the fluid is defined by the integral

$$\Gamma = \oint_C \mathbf{v} \cdot d\mathbf{l} = \oint_C v_i \, dl_i. \tag{3.39}$$

Consider now that the circuit moves with the fluid. The rate of change of the circulation around this moving circuit is calculated as

$$\frac{D\Gamma}{Dt} = \frac{D}{Dt}\oint_C \mathbf{v} \cdot d\mathbf{l} = \oint_C \frac{D\mathbf{v}}{Dt} \cdot d\mathbf{l} + \oint_C \mathbf{v} \cdot \frac{D}{Dt}(\mathbf{l}).$$

It is obvious from Fig. 3.6 that

$$\mathbf{v}\, dt + d\mathbf{l}' = d\mathbf{l} + (\mathbf{v} + d\mathbf{v})\, dt$$

or

$$\frac{d\mathbf{l}' - d\mathbf{l}}{dt} = \frac{D}{Dt}(d\mathbf{l}) = d\mathbf{v}.$$

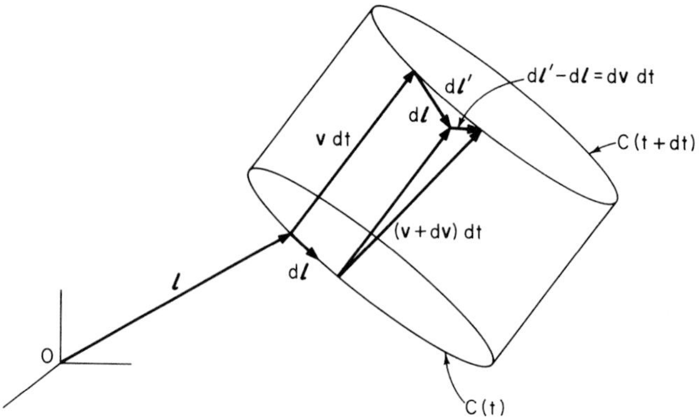

Fig. 3.6

Thus

$$\frac{D\Gamma}{Dt} = \oint_C \frac{D\mathbf{v}}{Dt} \cdot d\boldsymbol{l} + \oint_C \mathbf{v} \cdot d\mathbf{v} = \oint_C \frac{D\mathbf{v}}{Dt} \cdot d\boldsymbol{l} + \oint_C d\left(\frac{v^2}{2}\right). \tag{3.40}$$

As the flow speed v is a single-valued function of spatial position, the second integral on the right-hand side of Eq. (3.40) is zero. Hence

$$\frac{D\Gamma}{Dt} = \oint_C \frac{D\mathbf{v}}{Dt} \cdot d\boldsymbol{l}; \tag{3.41}$$

this is known as *Kelvin's equation*.

Only for flows in which

$$\oint_C \frac{D\mathbf{v}}{Dt} \cdot d\boldsymbol{l} = 0$$

will $D\Gamma/Dt = 0$. Such fluid flows are termed *circulation preserving*.

It has already been pointed out that the vorticity of the fluid $\boldsymbol{\zeta}$ is defined by

$$\boldsymbol{\zeta} = \boldsymbol{\nabla} \times \mathbf{v}; \tag{3.42}$$

$\boldsymbol{\zeta}$ will be a function of position in the fluid and represents, at each point, twice the angular velocity of a fluid element.

If ζ_n is the component of vorticity normal to the surface S bounded by the circuit C (at any point in S), then

$$\Gamma = \int_S \zeta_n \, dS. \tag{3.43}$$

Now

$$\zeta_n = \mathbf{n} \cdot (\nabla \times \mathbf{v}),$$

where **n** is the unit normal vector. The relation between the circulation and the vorticity is thus given by the equation

$$\Gamma = \int_S (\mathbf{n} \cdot \boldsymbol{\zeta})\, dS = 2 \int_S (\mathbf{n} \cdot \boldsymbol{\omega})\, dS, \tag{3.44}$$

and

$$\mathbf{n} \cdot \boldsymbol{\omega} = \frac{1}{2} \frac{d\Gamma}{dS}.$$

When the fluid flow is such that the fluid elements do not rotate (that is when $\boldsymbol{\zeta} = 0$ throughout the fluid), the circulation is zero for any closed circuit in the flow region.

The vorticity in a plane incompressible fluid flow is connected to the Lagrangian stream function as follows (the only component is the z component):

$$\zeta_z = 2\omega_z = \frac{\partial v}{\partial x} - \frac{\partial u}{\partial y} = \frac{\partial}{\partial x}\left(-\frac{\partial \psi}{\partial x}\right) - \frac{\partial}{\partial y}\left(\frac{\partial \psi}{\partial y}\right)$$

$$= -\frac{\partial^2 \psi}{\partial x^2} - \frac{\partial^2 \psi}{\partial y^2} = -\nabla^2 \psi. \tag{3.45}$$

In the case of axisymmetric motion and the Stokesian stream function ψ_S, the vorticity is given by the relation

$$\zeta_\varphi = -\frac{1}{r} \frac{\partial^2 \psi_S}{\partial z^2} + \frac{1}{r^2} \frac{\partial \psi_S}{\partial r} - \frac{1}{r} \frac{\partial^2 \psi_S}{\partial r^2} = -\frac{1}{r} D^2 \psi_S, \tag{3.46}$$

where

$$D^2 \equiv \frac{\partial^2}{\partial z^2} - \frac{1}{r} \frac{\partial}{\partial r} + \frac{\partial^2}{\partial r^2}.$$

3.6 Irrotational Motion and the Velocity Potential

An ideal fluid is one in which no energy dissipation occurs—it has no viscosity. Vorticity and rotation are directly attributable to the action of viscous forces, and so ideal fluid flow is characterized by the absence of vorticity or rotation. In general, such fluid motion is said to be *ir-*

rotational and mathematically this is expressed by the relation

$$\boldsymbol{\zeta} = 2\boldsymbol{\omega} = \mathbf{0}. \tag{3.47}$$

It is desirable also to have a physical conception for this condition. One cannot classify all motions in a curved path as rotational nor all rectilinear motions as irrotational. As a test, one can imagine a small rotatable cruciform placed in the flow region under consideration. This device actually averages the angular velocity of two orthogonal lines and thus gives a measure of the rotation. If it does not rotate in moving along with the flow, then the motion is irrotational—whether the flow lines are curved or straight.

Flow between two parallel plates, one of which moves with a velocity higher than the other, is rotational even though the streamlines are rectilinear (Fig. 3.7). The cruciform placed in such a flow would rotate.

On the other hand, the curved streamlines in the bend of a pipe do not indicate rotation (Fig. 3.8). Finally, the solid body rotation in the core of a vortex is rotational and the cruciform placed in such a vortex core would rotate (Fig. 3.9).

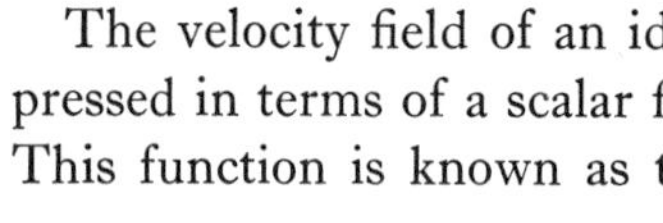

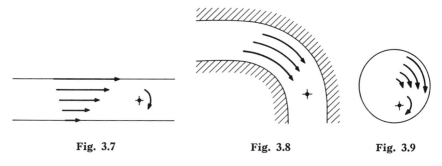

Fig. 3.7

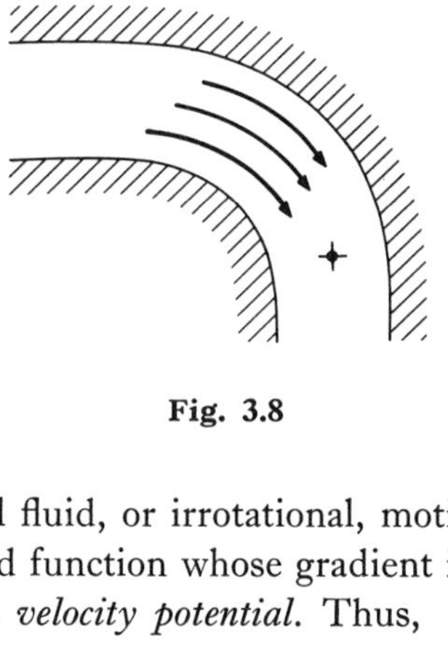

Fig. 3.8

Fig. 3.9

The velocity field of an ideal fluid, or irrotational, motion can be expressed in terms of a scalar field function whose gradient is the velocity. This function is known as the *velocity potential.* Thus,

$$\mathbf{v} = \nabla\varphi. \tag{3.48}$$

Sometimes the negative gradient is used, but which sign is chosen does not matter, as long as it is adhered to strictly. The existence of such a potential for a flow depends on the irrotational nature of that flow.

Consider a two-dimensional motion, for which a potential exists. Thus,

$$u = \frac{\partial\varphi}{\partial x}, \qquad v = \frac{\partial\varphi}{\partial y}. \tag{3.49}$$

In rectangular Cartesian coordinates, the irrotational condition is

$$\zeta_z = \frac{\partial v}{\partial x} - \frac{\partial u}{\partial y} = 0. \tag{3.50}$$

Using Eq. (3.49), one obtains

$$\frac{\partial v}{\partial x} = \frac{\partial}{\partial x}\left(\frac{\partial \varphi}{\partial y}\right), \qquad \frac{\partial u}{\partial y} = \frac{\partial}{\partial y}\left(\frac{\partial \varphi}{\partial x}\right),$$

and obviously

$$\frac{\partial v}{\partial x} - \frac{\partial u}{\partial y} = 0.$$

Thus, the irrotational condition and the existence of a potential are irrevocably connected.

For irrotational fluid flow in a simply connected region, the potential difference between two points A and B is given by

$$\varphi_A - \varphi_B = \int_A^B \mathbf{v} \cdot dl, \tag{3.51}$$

the result being independent of the path between A and B. The potential φ can therefore be written

$$\varphi = \int (u\,dx + v\,dy + w\,dz). \tag{3.52}$$

This implies that the integrand is a perfect differential, and

$$d\varphi = u\,dx + v\,dy + w\,dz.$$

This leads to the mathematical definition of the potential. For the above expression to be a perfect differential, the following relations must be satisfied:

$$\frac{\partial u}{\partial y} = \frac{\partial v}{\partial x}, \qquad \frac{\partial u}{\partial z} = \frac{\partial w}{\partial x}, \qquad \frac{\partial v}{\partial z} = \frac{\partial w}{\partial y}.$$

Then

$$\zeta_z = \left(\frac{\partial v}{\partial x} - \frac{\partial u}{\partial y}\right) = 0 = \zeta_y = \zeta_x$$

and, moreover,

$$d\varphi = \frac{\partial \varphi}{\partial x}\,dx + \frac{\partial \varphi}{\partial y}\,dy + \frac{\partial \varphi}{\partial z}\,dz,$$

whence

$$u = \frac{\partial \varphi}{\partial x}, \qquad v = \frac{\partial \varphi}{\partial y}, \qquad w = \frac{\partial \varphi}{\partial z}. \tag{3.53}$$

The ideal fluid continuity equation is

$$\nabla \cdot \mathbf{v} = 0;$$

thus

$$\nabla \cdot \nabla \varphi = \nabla^2 \varphi = 0. \tag{3.54}$$

The potential, then, for all possible irrotational motions of an incompressible fluid must satisfy the Laplace equation. The solution of flow problems is thus reduced to finding solutions to Laplace's equation. These solutions are known as *harmonic functions.*

3.6.1 Connection between the Potential and Stream Functions

It is obvious that these functions must be closely related. In the case of plane flow,

$$\frac{\partial \varphi}{\partial x} = \frac{\partial \psi}{\partial y}, \qquad \frac{\partial \varphi}{\partial y} = -\frac{\partial \psi}{\partial x}. \tag{3.55}$$

These are the Cauchy–Riemann equations. In the case of axisymmetric flow,

$$u = \frac{1}{r}\frac{\partial \psi_S}{\partial r}, \qquad v = -\frac{1}{r}\frac{\partial \psi_S}{\partial z}, \tag{3.56}$$

and these are not the Cauchy–Riemann equations. It has already been pointed out that the Stokesian stream function does not obey the Laplace equation.

However, in both the plane and axisymmetric cases, the potential and stream functions are orthogonal functions. Lines of constant ψ (streamlines) are orthogonal to lines of constant φ (equipotential lines). This follows from the facts that

(a) $\mathbf{v}$ is normal to lines, or surfaces, of constant φ, as $\mathbf{v} = \nabla \varphi$, and
(b) $\mathbf{v}$ is tangent to ψ lines.

Both these functions may be included in a single function—the *complex potential*:

$$\Omega = \varphi + i\psi = f(z). \tag{3.57}$$

As both φ and ψ are functions of x and y, the complex potential Ω is a function of the complex coordinate $z = x + iy$.

Since

$$u = \frac{\partial \varphi}{\partial x} = \frac{\partial \psi}{\partial y} \quad \text{and} \quad v = \frac{\partial \varphi}{\partial y} = -\frac{\partial \psi}{\partial x},$$

the differential of Ω with respect to z must be closely related to the velocity. It is called the *complex velocity*,

$$w = u - iv = \frac{d\Omega}{dz} = qe^{-i\vartheta}, \tag{3.58}$$

where q is the speed of the flow.

3.6.2 Kinetic Energy Theorem

Motions in which the velocity potential is single-valued are called *acyclic*. For acyclic irrotational flow, a very useful energy theorem can be simply deduced. If the fluid volume is V and the enclosed surface is S, the kinetic energy of motion is

$$E = \int_V \frac{\varrho v^2}{2}\, dV = \frac{\varrho}{2} \int_V (\nabla\varphi) \cdot (\nabla\varphi)\, dV$$

for incompressible flow. Using Green's theorem, one has

$$\int_V (\nabla\varphi) \cdot (\nabla\varphi)\, dV = -\int_V \varphi\, \nabla^2\varphi\, dV + \int_S \varphi \frac{\partial \varphi}{\partial n}\, dS,$$

where the normal points outward from S. The first integral on the right-hand side is zero (by Laplace's equation) and so

$$E = \frac{\varrho}{2} \int_S \varphi \frac{\partial \varphi}{\partial n}\, dS, \tag{3.59}$$

which is a unique relation between the kinetic energy in a simply connected region and the potential at its surface. In the case of flows with *localized* regions of vorticity (for example, in circulation around an airfoil), this theorem can also be very useful.

3.7 Boundary Conditions for Ideal Fluids

Gauss's mean value theorem states that the mean value of the potential φ over any spherical surface throughout whose interior $\nabla^2\varphi = 0$ equals the value of φ at the center of the sphere. Thus, φ cannot be a maximum or a minimum in the interior of such a region, and so the maximum (or minimum) velocities will occur only at the boundaries of potential fluid motions. Stagnation points (or points of zero velocity) will occur only at the boundaries. When singularities (for example, point sources) are used to develop flow patterns, no real boundary exists. But the $\psi = 0$ streamline, which separates the internal and external flows, is taken to represent the boundary. The extreme values of the velocity will then occur on this streamline.

At boundaries, the fluid cannot penetrate and so the relative normal velocity must be zero. So the kinematic boundary condition is

$$\left.\frac{\partial\varphi}{\partial n}\right|_S = \left.\frac{\partial\psi}{\partial S}\right|_S = 0 \qquad \text{or} \qquad \mathbf{n}\cdot\mathbf{v}\,|_S = 0. \tag{3.60}$$

The stream function must, therefore, be constant along the boundary surface. As walls or surfaces will form streamlines, the value of ψ is usually taken as zero along them (for steady motion). If the surface is moving with a velocity U through the fluid and at an angle ϑ to the outward normal to S, then the boundary condition assumes the form

$$v_n = \left.\frac{\partial\varphi}{\partial n}\right|_S = U\cos\vartheta. \tag{3.61}$$

In the case of free surfaces, such as an air–water interface, an interesting situation arises due to the constant pressure condition at the interface. It is worth examining this in some detail, even at the risk of being somewhat out of place in the general development of the text.

The general form of the Bernoulli equation (the integral form of Euler's equation) will be dealt with in the next chapter. It is as follows:

$$\int\frac{dp}{\varrho} + \frac{v^2}{2} + gh + \frac{\partial\varphi}{\partial t} = \Pi(t), \tag{3.62}$$

where g is the gravitational constant, Π the Bernoulli constant, and h the elevation above some reference point. If the reference axis is taken as the x axis, laying in the plane of the undisturbed fluid surface, then the elevation of the surface is

$$h = h(x, t) \qquad \text{and} \qquad y = h.$$

The surface motion will be assumed small so that the kinetic energy term $(v^2/2)$ can be neglected.

Equation (3.62) now becomes

$$gh = -\frac{p}{\varrho} - \frac{\partial \varphi}{\partial t} + \Pi.$$

The pressure p is constant at the surface and so can be included in Π. Therefore

$$h = \frac{\Pi}{g} - \frac{1}{g}\left(\frac{\partial \varphi}{\partial t}\right)_{y=h}. \tag{3.63}$$

The vertical velocity component at the free surface is

$$v_{\text{surf}} = \frac{\partial h}{\partial t} = \left.\frac{\partial \varphi}{\partial y}\right|_{y=h}.$$

Using this in Eq. (3.63), one therefore obtains

$$\frac{\partial^2 \varphi}{\partial t^2} + g\frac{\partial \varphi}{\partial y} = 0, \tag{3.64}$$

which is the velocity potential relation for the free surface.

To convert Eq. (3.64) into a dimensionless form, let L and V be a characteristic length and velocity, respectively. Then

$$y = Ly', \qquad v = Vv', \qquad \varphi = (VL)\varphi', \qquad t = \left(\frac{L}{V}\right)t'.$$

Substitution of these terms into Eq. (3.64) yields the dimensionless equation

$$\frac{\partial^2 \varphi'}{\partial t'^2} + \frac{gL}{V^2}\frac{\partial \varphi'}{\partial y'} = 0. \tag{3.65}$$

Now

$$\frac{gL}{V^2} = \frac{1}{\text{Fr}},$$

where Fr is the Froude number characterizing the motion.

For low values of Fr, the second term predominates and $\partial\varphi/\partial y = 0$ at the surface. The flow behaves as if the interface was rigid, and φ satisfies the Neuman, or second, boundary value problem (see below). For large values of Fr, the first term predominates and φ is a constant on the surface, putting this into the category of a Dirichlet, or first, boundary value problem of potential theory.

3.8 Types of Potential Problems and Methods of Obtaining Solutions

The solutions of the Laplace equation are quite general, but the boundary conditions specify the solutions to specific flow situations. Potential problems are grouped into three major types, depending on how the boundary conditions are given.

(a) The potential φ is defined over the boundary of the region. Flows satisfying this condition are called *Dirichlet flows* and fall into the First Boundary Value Problem of potential theory. In the area of ideal fluid dynamics, the boundaries are defined by lines of constant ψ. Since φ and ψ are interchangeable, this is not significant.

(b) The normal derivative of φ is defined on the boundary. This is the Neuman, or Second, Boundary Value Problem of potential theory. Since the derivative of φ is a velocity component, this type really involves specification of the velocity on the boundary.

(c) When *both* φ and its normal derivative are specified on the boundary, one has the Cauchy, or Third, Boundary Value Problem. Over specification of the problem is a frequent difficulty in this case.

Apart from direct and analytical solutions of the Laplace equations, the method of singularities (or combinations of elementary flow solutions) and the method of conformal transformation (or change of variable) are particularly powerful methods of solution. These will be dealt with in subsequent chapters.

3.9 Example of an Irrotational Flow System

The existence of a potential implies that the velocity field must satisfy the following two equations:

$$\nabla \cdot \mathbf{v} = 0, \qquad \nabla \times \mathbf{v} = 0. \tag{3.66}$$

As an example of such a velocity field (or distribution as it is sometimes called), consider the neighborhood of a point where $\mathbf{v} = 0$. Such a point is called a *stagnation point*, and it can arise either in the body of a fluid or at the boundary. Excluding geometrical singularities on a boundary, φ is well behaved near O (its derivatives are finite and continuous), and so φ can be expanded in the neighborhood of O as a Taylor series in Cartesian coordinates r_i with origin at O:

$$\varphi = \varphi_0 + a_i r_i + \tfrac{1}{2} a_{ij} r_i r_j + O(h^3),$$

where $h^2 = r_i r_i$ and the tensor a_{ij} is symmetric.

Now $\nabla \varphi = 0$ at the origin and thus the coefficients are zero. Moreover, the Laplace equation $\nabla^2 \varphi = 0$ holds everywhere and so $a_{ii} = 0$. Only the off-diagonal terms of the rate-of-strain tensor a_{ij} remain, and the motion is one of pure strain (no change in volume). The velocity is $\partial\varphi/\partial r_i$ and thus is a linear distribution given by

$$v_i = a_{ij} r_j$$

to the order $O(h^2)$.

Using axes parallel to the principal axes of the tensor a_{ij} with reference coordinates (x, y, z), the velocity components are

$$u = ax, \qquad v = by, \qquad w = -(a + b)z, \tag{3.67}$$

where a and b are constants of the flow field.

As has already been pointed out, in the case of two-dimensional or axisymmetric flow a stream function can be written down. For two-dimensional flow, and with axes parallel to the principal axis of the tensor a_{ij} at the origin,

$$\varphi = \tfrac{1}{2} k(x^2 - y^2), \qquad \psi = kxy, \tag{3.68}$$

where k is a constant.

The streamlines near O are therefore rectangular hyperbolas, asymptotic to the two orthogonal branches of the streamlines through O.

The equipotential lines cross these lines at right angles and have asymptotes at 45° to the axes (Fig. 3.10). The flow system represented is one in which fluid strikes a flat surface perpendicularly.

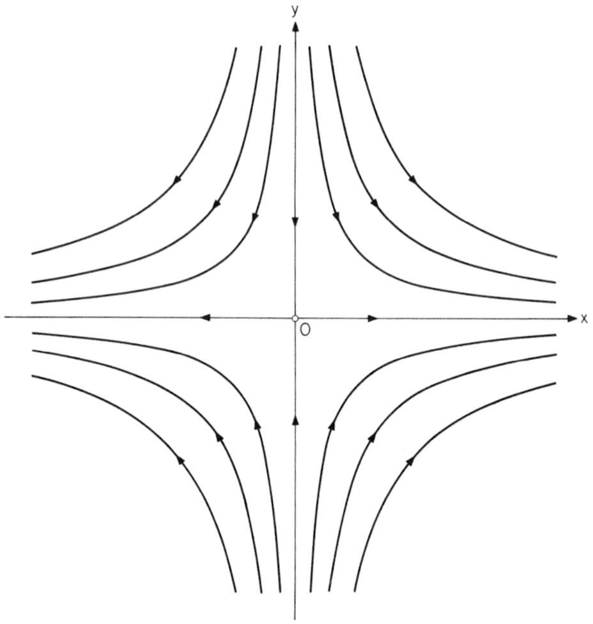

Fig. 3.10 Streamlines near a stagnation point.

3.10 Velocity Field with Nonzero Dilation and Vorticity

In this case no potential can be defined for the field and

$$\nabla \cdot \mathbf{v} = g, \qquad \nabla \times \mathbf{v} = \boldsymbol{\omega}. \tag{3.69}$$

Now it has been seen that the instantaneous motion near a point in the flow field comprises

(a) an isotropic dilation such that the rate of volume increase of a fluid element, per unit volume, is g,
(b) a pure strain motion without change of volume, and
(c) a rigid body rotation with an angular velocity $\boldsymbol{\omega}$.

The distributions (or fields) of g and $\boldsymbol{\omega}$ will largely determine the velocity field.

The case in which the vorticity field is zero everywhere will be considered first:

$$\nabla \cdot \mathbf{v}_\omega = g, \qquad \nabla \times \mathbf{v}_\omega = 0. \tag{3.70}$$

These equations can be satisfied if

$$\mathbf{v}_\omega = \nabla \varphi_\omega, \qquad \nabla^2 \varphi_\omega = g. \tag{3.71}$$

A known solution to this Poisson-type equation is

$$\varphi_\omega(\mathbf{r}) = -\frac{1}{4\pi} \int \frac{g'}{S}\, dV(\mathbf{r}'), \tag{3.72}$$

where the prime denotes evaluation at the point $\mathbf{r}'$, $\mathbf{S} = \mathbf{r} - \mathbf{r}'$, and the integration is over the volume occupied by the fluid.

Then, by Eq. (3.71),

$$\mathbf{v}_\omega(\mathbf{r}) = -\frac{1}{4\pi} \int g' \nabla_r\left(\frac{1}{S}\right) dV(\mathbf{r}') = \frac{1}{4\pi} \int \frac{\mathbf{S}}{S^3}\, g'\, dV(\mathbf{r}'). \tag{3.73}$$

The velocity $\mathbf{v}_\omega$ at the point $\mathbf{r}$ may be regarded as the sum of contributions from the different volume elements of the fluid. The contribution from an element at $\mathbf{r}'$ is

$$\delta \mathbf{v}_\omega(\mathbf{r}) = \frac{\mathbf{S}}{S}\, g' \frac{\delta V(\mathbf{r}')}{4\pi S^2}; \tag{3.74}$$

$g'\, \delta V(\mathbf{r}')$ is just the volume flux across any closed surface enclosing the point $\mathbf{r}'$. The velocity field given by Eq. (3.74) has zero rate of dilation everywhere except in the volume element $\delta V(\mathbf{r}')$ containing the point $\mathbf{r}'$. The rate of expansion in this element is g'. The velocity field in Eq. (3.73) thus has the specified rate of dilation everywhere. Each element $\delta V(\mathbf{r}')$ performs like a source of volume in an otherwise dilation-free fluid; the rate of emission of volume (or *source strength*) is $g(\mathbf{r}')\, \delta V(\mathbf{r}')$.

3.10.1 Zero Dilation and Nonzero Vorticity

The next case to be considered is when the dilation is zero everywhere and the vorticity field is specified:

$$\nabla \times \mathbf{v}_\mathrm{D} = \boldsymbol{\omega}, \qquad \nabla \cdot \mathbf{v}_\mathrm{D} = 0 \tag{3.75}$$

(and $\nabla \cdot \boldsymbol{\omega} = 0$ everywhere).

This equation can be satisfied by choosing

$$\mathbf{v}_\mathrm{D} = \nabla \times \mathbf{A}_\mathrm{D}, \tag{3.76}$$

where $\mathbf{A}_D$ is a vector potential. Then

$$\nabla \times (\nabla \times \mathbf{A}_D) = \nabla(\nabla \cdot \mathbf{A}_D) - \nabla^2 \mathbf{A}_D = \boldsymbol{\omega}. \tag{3.77}$$

If $\nabla \cdot \mathbf{A}_D = 0$ everywhere, then

$$\nabla^2 \mathbf{A}_D = -\boldsymbol{\omega}. \tag{3.78}$$

A solution of this equation is

$$\mathbf{A}_D(\mathbf{r}) = \frac{1}{4\pi} \int \frac{\boldsymbol{\omega}'}{S}\, dV(\mathbf{r}'). \tag{3.79}$$

This solution must now be tested to see if $\nabla \cdot \mathbf{A}_D = 0$:

$$\begin{aligned} \nabla \cdot \left\{ \frac{1}{4\pi} \int \frac{\boldsymbol{\omega}'}{S}\, dV(\mathbf{r}') \right\} &= \frac{1}{4\pi} \int \boldsymbol{\omega}' \cdot \nabla_r \left(\frac{1}{S} \right) dV(\mathbf{r}') \\ &= -\frac{1}{4\pi} \int \nabla_{r'} \left(\frac{\boldsymbol{\omega}'}{S} \right) dV(\mathbf{r}') \\ &= -\frac{1}{4\pi} \int \frac{\boldsymbol{\omega}' \cdot \mathbf{n}}{S}\, dA(\mathbf{r}'), \end{aligned}$$

where the integral is over the total boundary of the fluid. This integral vanishes when the specified vorticity has zero normal component at each point of the boundary. This would be true for a fluid extending to infinity in all directions, the fluid velocity being zero there.

Within the restriction of $\boldsymbol{\omega} \cdot \mathbf{n}$ on the boundary then, from Eqs. (3.76) and (3.77),

$$\mathbf{v}_D = \frac{1}{4\pi} \int \nabla_r \times \left(\frac{\boldsymbol{\omega}'}{S} \right) dV(\mathbf{r}') = -\frac{1}{4\pi} \int \frac{\mathbf{S} \times \boldsymbol{\omega}'}{S^3}\, dV(\mathbf{r}'), \tag{3.80}$$

and $\mathbf{v}_D$ can be regarded as being comprised of contributions from the various volume elements of the fluid,

$$\delta \mathbf{v}_D = -\frac{\mathbf{S} \times \boldsymbol{\omega}'\, \delta V(\mathbf{r}')}{4\pi S^3} \tag{3.81}$$

$\boldsymbol{\omega}$ cannot be nonzero and uniform inside a volume element and zero outside, as such a field would have a nonzero divergence. Thus, Eq. (3.81) is a velocity distribution which cannot exist by itself.

Combining both cases, one concludes that if $\mathbf{v}$ is a velocity field which is consistent with specified fields of dilation (g) and vorticity ($\boldsymbol{\zeta}$), then

$\mathbf{v} - \mathbf{v}_\omega - \mathbf{v}_D$ is both solenoidal and rotational:

$$\mathbf{v} = \mathbf{v}_\omega + \mathbf{v}_\mathrm{D} + \mathbf{v}_\mathrm{I}, \tag{3.82}$$

where $\mathbf{v}_\mathrm{I}$ is a vector satisfying the equations

$$\nabla \cdot \mathbf{v}_\mathrm{I} = \nabla \times \mathbf{v}_\mathrm{I} = 0 \tag{3.83}$$

at all points in the fluid. This is the case of irrotational motion of an incompressible fluid referred to above, and it involves pure strain. The $\mathbf{v}_\mathrm{I}$ satisfies the Laplace equation and is determined by the boundary conditions on the fluid.

3.11 Sources and Sinks—Singularities in the Rate of Dilation

A nonzero dilational velocity flow field will be considered, in which the dilation $\nabla \cdot \mathbf{v}$ has a singularity at a specific point in the fluid. The field will also be regarded as vorticity-free and so is irrotational.

Consider that g has a large value in the neighborhood (an infinitesimal sphere ε) of a point $\mathbf{r}'$ and that it is zero at all other points occupied by the fluid. Since the nonzero values of g are in the neighborhood of $\mathbf{r}'$, Eq. (3.73) becomes

$$\mathbf{v}_\omega(\mathbf{r}) = \frac{1}{4\pi} \frac{\mathbf{S}}{S} \int_\varepsilon g' \, dV(\mathbf{r}'), \tag{3.84}$$

where $\mathbf{S} = \mathbf{r} - \mathbf{r}'$.

Suppose ε now contracts to a point but g' increases so that $\int_\varepsilon g' \, dV(\mathbf{r}')$ remains constant, equal to a value μ, say. This is, of course, the normal concept of a point source of fluid, and the associated flow field is given exactly by the equation

$$\mathbf{v}_\omega(\mathbf{r}) = \frac{\mu}{4\pi} \cdot \frac{\mathbf{S}}{S^3}, \qquad \varphi_\omega = -\frac{\mu}{4\pi S}; \tag{3.85}$$

μ is termed the source strength. When μ is negative, the net flow is into the point, and the singularity represents a sink. The μ physically represents the total outward flux of fluid volume across any spherical surface enclosing the point $\mathbf{r}'$.

Another type of singularity may be obtained by considering a source and a sink, with equal strength $+\mu$ and $-\mu$, respectively, placed close to one another at positions $\mathbf{r}' + \frac{1}{2}\,\delta\mathbf{r}'$ and $\mathbf{r}' - \frac{1}{2}\,\delta\mathbf{r}'$, respectively. The distance between them, $\delta\mathbf{r}'$, is now allowed to go to zero and the strength μ is allowed to go to infinity, in such a way that

$$\varkappa = \lim_{\delta\mathbf{r}'=0} \mu\,\delta\mathbf{r}'.$$

This singularity is known as a doublet of strength $\varkappa$ at the position $\mathbf{r}'$. The irrotational flow field of the doublet (or dipole) is obtained by superposing the fields due to the source and sink and then allowing the separation to go to zero as μ goes to infinity. Therefore

$$\begin{aligned}\varphi_\omega(\mathbf{r}) &= \lim_{\substack{\delta\mathbf{r}'\to 0\\ \mu\to\infty}} \frac{\mu}{4\pi}\left\{-\frac{1}{(\mathbf{r}-\mathbf{r}'-\frac{1}{2}\,\delta\mathbf{r}')}+\frac{1}{(\mathbf{r}-\mathbf{r}'+\frac{1}{2}\,\delta\mathbf{r}')}\right\}\\ &= -\frac{1}{4\pi}\boldsymbol{\varkappa}\left(\nabla_{\mathbf{r}}\cdot\frac{1}{S}\right)=\frac{1}{4\pi}\boldsymbol{\varkappa}\left(\nabla_{\mathbf{r}}\frac{1}{S}\right)\end{aligned} \tag{3.86}$$

and

$$\begin{aligned}\mathbf{v}_\omega(\mathbf{r}) = \nabla\varphi_\omega(\mathbf{r}) &= \frac{1}{4\pi}\boldsymbol{\varkappa}\cdot\nabla_{\mathbf{r}}\left(\nabla_{\mathbf{r}}\frac{1}{S}\right)=\frac{1}{4\pi}\boldsymbol{\varkappa}\cdot\nabla_{\mathbf{r}}\left(-\frac{\mathbf{S}}{S^3}\right)\\ &= \frac{1}{4\pi}\left\{-\frac{\boldsymbol{\varkappa}}{S^3}+3\,\frac{\boldsymbol{\varkappa}\cdot\mathbf{S}}{S^5}\,\mathbf{S}\right\}.\end{aligned} \tag{3.87}$$

This is a three-dimensional field and so a stream function cannot be defined for it in general. However, it is axially symmetric (about the $\boldsymbol{\varkappa}$ axis) and so a Stokesian stream function can be derived.

Setting the origin of a spherical polar coordinate system at the point $\mathbf{r}'$ and taking S as the radial distance, the radial component of $\mathbf{v}_\omega$ is

$$\frac{1}{S^2\sin\vartheta}\,\frac{\partial\psi}{\partial\vartheta}=\frac{\mathbf{S}\cdot\mathbf{v}_\omega}{S}.$$

Now

$$\begin{aligned}\mathbf{S}\cdot\mathbf{v}_\omega &= \frac{1}{4\pi}\left\{-\frac{\mathbf{S}\cdot\boldsymbol{\varkappa}}{S^3}+3\,\frac{\boldsymbol{\varkappa}\cdot\mathbf{S}}{S^5}\,(\mathbf{S}\cdot\boldsymbol{\varkappa})\right\}\\ &= \frac{1}{4\pi}\left\{-\frac{\mathbf{S}\cdot\boldsymbol{\varkappa}}{S^3}+\frac{3(\mathbf{S}\cdot\boldsymbol{\varkappa})}{S^3}\right\}=\frac{1}{2\pi}\,\frac{\mathbf{S}\cdot\boldsymbol{\varkappa}}{S^3}.\end{aligned}$$

Thus

$$\frac{1}{S^2\sin\vartheta}\,\frac{\partial\psi}{\partial\vartheta}=\frac{1}{2\pi}\,\frac{\mathbf{S}\cdot\boldsymbol{\varkappa}}{S^4}=\frac{\varkappa}{2\pi}\,\frac{\cos\vartheta}{S^3},$$

where $\varkappa = |\boldsymbol{\varkappa}|$ and

$$\psi = \frac{\varkappa}{4\pi} \frac{\sin^2 \vartheta}{S}. \tag{3.88}$$

In an axial plane the streamline pattern close to a source doublet is as shown in Fig. 3.11.

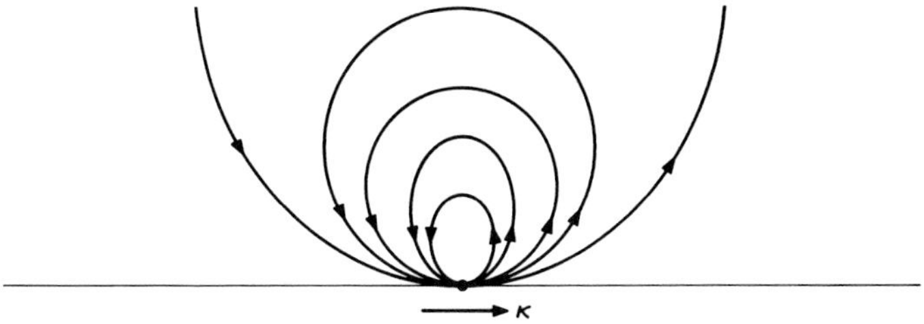

Fig. 3.11

3.11.1 Line Singularities

It is also useful to define line singularities. In an irrotational field this implies two-dimensional sources and sinks. Fluid flows out radially at a uniform rate from such a source. The flux (volume flow per second per unit area) per unit length of a line source measures the line density of source strength. If the line density is the value μ at all points of a line parallel to the z axis, each element of the line $\delta z'$ acts as a point source of strength $\mu\,\delta z'$ and the irrotational velocity field $(u_\omega, v_\omega, 0)$ produced by the whole line is

$$\begin{aligned} u_\omega(x, y) &= \frac{\mu}{4\pi} \int_{-\infty}^{+\infty} \frac{x - x'}{S^3}\, dz' = \frac{\mu}{4\pi} \frac{x - x'}{l^2}, \\ v_\omega(x, y) &= \frac{\mu}{4\pi} \int_{-\infty}^{\infty} \frac{y - y'}{S^3}\, dz' = \frac{\mu}{4\pi} - \frac{y - y'}{l^2}, \end{aligned} \tag{3.89}$$

where $l^2 = (x - x')^2 + (y - y')^2$. The corresponding potential function is

$$\varphi_\omega(x, y) = \frac{\mu}{2\pi} \log l. \tag{3.90}$$

A diagram of two-dimensional incompressible source flow is shown in Fig. 3.12. As expected from Eq. (3.90), the equipotential lines are circles with the origin at the source.

Note The constant of integration has been taken as zero.

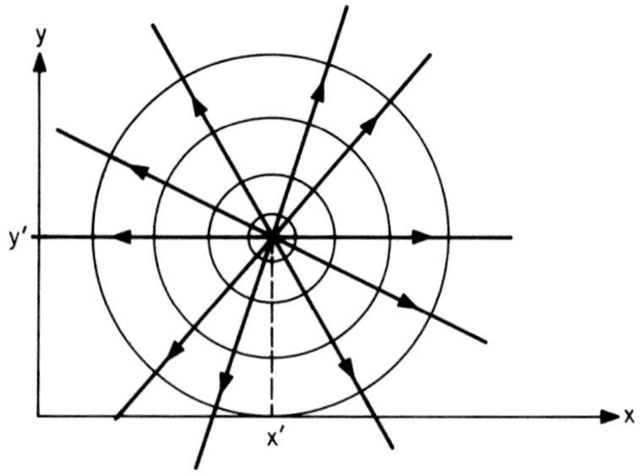

Fig. 3.12

In terms of polar coordinates σ and ϑ, the radial velocity is given by

$$\frac{\partial \varphi_\omega}{\partial \sigma} = |\,\mathbf{v}_r\,| = \frac{\mu}{2\pi\sigma} \qquad \text{and} \qquad \frac{\partial \psi}{\partial \vartheta} = \sigma\,|\,\mathbf{v}_r\,|$$

so that

$$\begin{aligned} \psi &= \sigma\,|\,\mathbf{v}_r\,|\,\vartheta + \text{constant} \\ &= \frac{\mu}{2\pi}\,\vartheta + \text{constant} \end{aligned} \tag{3.91}$$

for $0 \leq \vartheta \leq 2\pi$, where ψ is the stream function of the irrotational velocity field associated with this line source.

Another kind of line singularity is that formed by considering a vortex tube (in which $\boldsymbol{\zeta} \neq 0$) to contract onto a curve with the vortex strength remaining constant, and equal to $\varkappa$, say. This then becomes a line vortex of strength $\varkappa$. It must not be confused with the vortex line, which is merely a line in a fluid whose tangent at all points is parallel to the local vorticity vector $\boldsymbol{\zeta}$. The vorticity distribution is that associated with the line vortex, with zero vorticity in the surrounding fluid. The velocity field associated with this vorticity distribution is determinable from Eq. (3.80). Now, if $\delta \boldsymbol{l}$ is an elemental vector length of the line vortex lying in the volume δV, then

$$\int_{\delta V} \boldsymbol{\zeta}\, dV = \varkappa\, \delta \boldsymbol{l},$$

$$\mathbf{v}_\mathrm{D} = -\frac{\varkappa}{4\pi} \int \frac{\mathbf{S} \times d\boldsymbol{l}(\mathbf{r}')}{S^3}, \tag{3.92}$$

where $\mathbf{S} = \mathbf{r} - \mathbf{r}'$.

There is a direct analogy between this equation and the Biot–Savart law in electromagnetic theory. This law gives the magnetic field produced by a steady current passing around a closed conducting loop:

$$\mathbf{H} = \frac{\mathbf{B}}{\mu} = I \int \frac{\mathbf{S} \times dl(\mathbf{r}')}{S^3}.$$

The field strength **H** is seen to be analogous to the velocity **v**, and the current I analogous to the circulation $\varkappa$ (apart from the constant 4π). Helmholtz was the first to point out this analogy between hydrodynamics and electrodynamics. Sources and sinks correspond to magnetic poles. The analogy has been widely used in evaluating complex wing and propeller problems. In experimental work the vortex lines are represented by current-carrying wires.

Consider the case of a straight line vortex of infinite length. The velocity $\mathbf{v}_D$ will be everywhere in the azimuthal direction about the line vortex and with a magnitude

$$|\,\mathbf{v}_\mathrm{D}\,| = \frac{\varkappa r_0}{4\pi} \int_{-\infty}^{+\infty} \frac{dl}{(r_0{}^2 + l^2)^{1/2}} = \frac{\varkappa}{2\pi r_0} \tag{3.93}$$

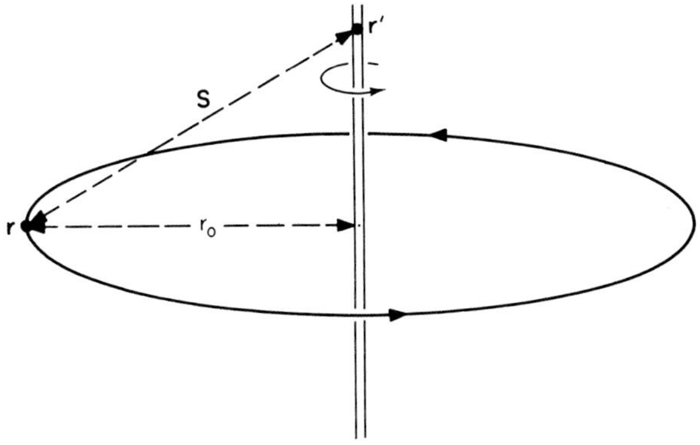

Fig. 3.13

(see Fig. 3.13). As a line vortex cannot terminate anywhere, the ends may be regarded as being joined by a line vortex in the form of a large semicircle of radius R. The contribution to the velocity field due to this semicircle is of the order of $1/R$ and so is negligible. The stream function for this flow field can also be determined with ease:

$$|\,\mathbf{v}_\mathrm{D}\,|_{\mathrm{azi}} = -\frac{\partial\psi}{\partial r} \qquad \text{[from Eq. (3.93)]},$$

$$\psi = -\frac{\varkappa}{2\pi} \log r_0. \tag{3.94}$$

In a planar flow field, of course, the line vortex singularity becomes a point vortex singularity.

3.11.2 Vortex Sheets

Vorticity can also be localized to surfaces in a fluid. This occurs in practice in the case of flow over cylinders and other blunt bodies, and also in the flow fields associated with airfoils. Mathematically, they can be characterized as follows. If $\boldsymbol{\zeta}$ is the vorticity, then consider the integral

$$\boldsymbol{\Gamma} = \int \boldsymbol{\zeta}\, dn, \tag{3.95}$$

where n is a distance normal to the sheet and the integral is over some small range ε containing the surface. Now let $\varepsilon \to 0$, with $\boldsymbol{\Gamma}$ remaining constant. This leads to the concept of the sheet vortex of strength per unit width, $|\boldsymbol{\Gamma}| = \Gamma$.

With this vorticity distribution, Eq. (3.80) gives

$$\mathbf{v}_{\mathrm{D}}(\mathbf{r}) = -\frac{1}{4\pi}\int \frac{\mathbf{S} \times \boldsymbol{\Gamma}'}{S}\, dA(\mathbf{r}'), \tag{3.96}$$

and the integration is over the area of the sheet.

With a single plane sheet of uniform vorticity,

$$\begin{aligned}
\mathbf{v}_{\mathrm{D}}(\mathbf{r}) &= \frac{1}{4\pi}\boldsymbol{\Gamma} \times \int \frac{\mathbf{S}}{S^3}\, dA(\mathbf{r}') \\
&= \frac{1}{4\pi}\boldsymbol{\Gamma} \times \int \frac{\mathbf{n}\cdot\mathbf{S}}{S^3}\, \mathbf{n}\, dA(\mathbf{r}') \\
&= \frac{1}{4\pi}\boldsymbol{\Gamma} \times \mathbf{n}\left(\int \frac{\mathbf{n}\cdot\mathbf{S}\, dA(\mathbf{r}')}{S^3}\right).
\end{aligned}$$

Thus,

$$\mathbf{v}_{\mathrm{D}}(\mathbf{r}) = \tfrac{1}{2}\boldsymbol{\Gamma} \times \mathbf{n}; \tag{3.97}$$

$\mathbf{n}$ is the unit normal to the sheet directed to the side on which the point $\mathbf{r}$ lies.

It is seen that the velocity is in a direction parallel to the sheet, with magnitude $\frac{1}{2}\Gamma$. But the direction of the velocity is in opposite directions on either side of the sheet. The strength of the sheet vortex is thus the velocity difference across it. It has been shown that sheet vortices are unstable and tend to roll up into a series of larger vortices (this will be referred to in Chapter 5 on Vortex Dynamics).

As mentioned previously, a powerful method of solving the Laplace equations and boundary conditions arising for various ideal flow systems is to use the superposition of the flow fields of the above singularities in various appropriate combinations.

PROBLEMS

3.1 Ascertain which of the following sets of velocity components u and v satisfy the equation of continuity for a two-dimensional flow of an incompressible fluid.

(i) $u = C \sin xy$, $v = -C \sin xy$ (ii) $u = Cx$, $v = -Cy$ (iii) $u = -C(x/y)$, $v = C \ln xy$.

3.2 In a planar flow of incompressible fluid, the x component of velocity is given by

$$u = \tfrac{1}{2}x^2 + 2x - 4y.$$

Use the equation of continuity to derive the expression for v.

3.3 Derive the equation of the streamlines for the flow

$$v = -i(3y^2) - j(9x).$$

Draw a graph of the streamline passing the point (2, 1).

3.4 Derive the equation of the streamlines for the planar flow of an incompressible fluid as given by

$$u = e^x \cosh(y), \qquad v = -e^x \sinh(y).$$

3.5 Derive the rates of strain for the following velocity fields and comment on the nature of the results:

(i) $u = c$, $v = w = 0$ (ii) $u = cx$, $v = cy$, $w = -4cz$ (iii) $u = 4cy$, $v = w = 0$

where c is a constant.

3.6 Derive the stress components for the results obtained in Problem 3.5.

3.7 Show that the following equations describe a rigid body motion:

$$u = a_1 + a_2 y - a_3 z, \qquad v = a_4 - a_2 y + a_5 z, \qquad w = a_6 + a_3 x - a_5 z,$$

where a_i, $i = 1, \ldots, 5$, are arbitrary constants.

3.8 For the flow fields

(i) $\mathbf{v} = (3x^2 - xy)\mathbf{i} + (y^2 - 6xy + 2yz^2)\mathbf{j} - (z^2 + xy^2)\mathbf{k}$,
(ii) $\mathbf{v} = (5t^2 + 6t)\mathbf{i} + (y^2 - z^2 + 2)\mathbf{j} - (2yz + 3y)\mathbf{k}$,

determine the vorticity and the mean rates of rotation of a fluid element

(a) for (i), at the point (3, 3, 1),
(b) for (ii), at the point (5, 2, 5) and at time $t = 6$.

In the case of (ii), determine the surface on which this flow is always irrotational, and determine the velocity and acceleration of a fluid element at the point (5, 2, 5) and at $t = 6$.

3.9 (a) The velocity potential of a flow is $\varphi = -xy$. Determine the velocity components at the point (1. 2, 1, 0). Construct the equipotential lines for $\varphi = -0.8$ and $\varphi = -1.2$ for the interval $0.7 \leq x \leq 1.4$.

(b) Find the velocity potential of a uniform stream parallel to the xy plane moving with a speed U inclined at an angle α with the x axis.

3.10 Show that the velocity potential of the flow produced by a line source of uniform strength is given by

$$\varphi = -c \ln r,$$

where r is the radial distance from any position along the line and c is a constant.

3.11 Show that the velocity potential of a free vortex is given by

$$\varphi = -\frac{\varkappa}{2\pi}\,\vartheta,$$

where $\varkappa$ is the circulation round the axis and ϑ is the angle of the radial plane to the horizontal axis.

3.12 Show that the stream functions ψ and velocity potential φ of the following flows satisfy the Laplace equations $\nabla^2\psi = 0$ and $\nabla^2\varphi = 0$, respectively:

(a) two-dimensional rectilinear flow,
(b) source flow,
(c) doublet flow.

3.13 If the stream functions ψ_1 and ψ_2 satisfy Laplace's equation, show that the function $\psi = \psi_1 + \psi_2$ also satisfies it. Use $\psi_1 = U_\infty y$ and $\psi_2 = \varkappa\vartheta/2\pi$ as examples.

3.14 Determine the circulation around the closed path $x^2 + y^2 = 1$ when $\mathbf{v} = -6y\mathbf{i} + 8v\mathbf{j}$.

3.15 A sphere of radius r_0 moves uniformly at a velocity v_0 in a fluid initially at rest. In spherical coordinates the velocity potential is

$$\varphi = \tfrac{1}{2}v_0 \frac{r^3}{r_0^2} \cos\vartheta.$$

Determine the total kinetic energy of this flow field.

BIBLIOGRAPHY

J. Serrin, Mathematical Principles of Classical Fluid Mechanics, *in* "Encyclopedia of Physics" (edited by S. Flügge), Vol. 8. Springer, Berlin, 1959.

G. K. Batchelor, "An Introduction to Fluid Dynamics," Chapter 2. Cambridge Univ. Press, New York and London, 1967.

A. Sommerfeld, "Mechanics of Deformable Bodies" (translated by G. Kuerti). Academic Press, New York, 1950.

J. Robertson, "Hydrodynamics in Theory and Application," Chapter 2. Prentice-Hall, Englewood Cliffs, New Jersey, 1965.

E. Kochin, J. Kibel, and N. Roze, "Theoretical Hydromechanics" (translated by D. Boyanovitch), Chapter 1. Wiley (Interscience), New York, 1964.

S. Yuan, "Foundations of Fluid Mechanics," Chapters 3 and 4. Prentice-Hall, Englewood Cliffs, New Jersey, 1970.

CHAPTER FOUR *THE EQUATIONS OF FLUID MOTION*

4.1 Conservation of Momentum

The conservation of mass and its consequence, the equation of continuity, have already been examined. In continuum mechanics another quantity that is conserved is the linear momentum of a material body of fluid. This leads to the equation of motion for the fluid. In its most elementary form this equates the rate of change of momentum of a body of the fluid with the sum of all forces acting on that body of fluid.

Consider a fluid mass in motion at time t occupying a volume V and bounded by a surface A. Let δV be a small element of volume (Fig. 4.1). If the fluid is isotropic and homogeneous and of density ϱ, the mass of the element is $\varrho\,\delta V$ and it moves with velocity $\mathbf{v}(\mathbf{r}, t)$.

The inertial force on this element is $-\varrho\,\delta V(D\mathbf{v}/Dt)$. This force equals the rate of change of linear momentum of the element and for the whole fluid mass is given by

$$F_{\mathrm{I}} = -\iiint \frac{D\mathbf{v}}{Dt}\,\varrho\,\delta V. \tag{4.1}$$

The force on the body is the sum of body and surface forces. If electromagnetic forces are omitted, the body

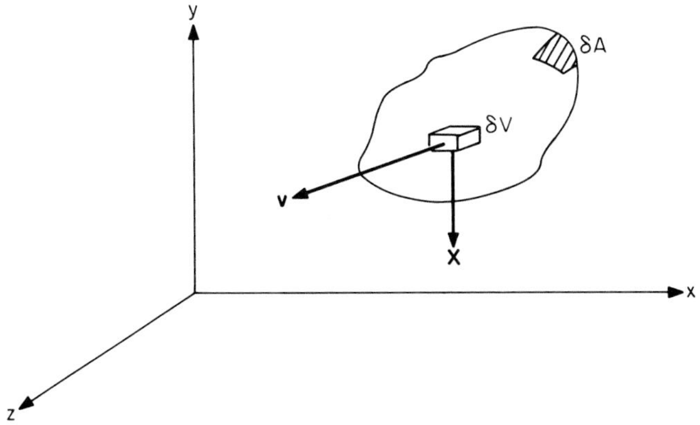

Fig. 4.1

force is usually gravitational and is often so weak as to be negligible. If the body force per unit mass acting on the fluid element is $\mathbf{X}$, then $\mathbf{X}\varrho\, \delta V$ is the body force on the element. The total body force on the fluid mass is then

$$\mathbf{F}_{\mathrm{B}} = \iiint_V \mathbf{X}\varrho\, dV. \tag{4.2}$$

Consider a surface element $\delta\mathbf{A}$. The three stresses acting at a point in a fluid are given in terms of the normal and shear stresses acting on the fluid element [see Eq. (2.11)] by

$$\begin{aligned} \mathbf{F}_x &= \mathbf{i}\sigma_{xx} + \mathbf{j}\sigma_{xy} + \mathbf{k}\sigma_{xz}, \\ \mathbf{F}_y &= \mathbf{i}\sigma_{yx} + \mathbf{j}\sigma_{yy} + \mathbf{k}\sigma_{yz}, \\ \mathbf{F}_z &= \mathbf{i}\sigma_{zx} + \mathbf{j}\sigma_{zy} + \mathbf{k}\sigma_{zz}. \end{aligned} \tag{4.3}$$

Then the surface force acting on a surface element $\delta\mathbf{A}$ is given by

$$\mathbf{s} = \mathbf{i}S_x + \mathbf{j}S_y + \mathbf{k}S_z = \mathbf{i}(\mathbf{F}_x \cdot \delta\mathbf{A}) + \mathbf{j}(\mathbf{F}_y \cdot \delta\mathbf{A}) + \mathbf{k}(\mathbf{F}_z \cdot \delta\mathbf{A}). \tag{4.4}$$

The resultant surface force on the fluid mass is then

$$\mathbf{F}_{\mathrm{S}} = \mathbf{i}\iint \mathbf{F}_x \cdot d\mathbf{A} + \mathbf{j}\iint \mathbf{F}_y \cdot d\mathbf{A} + \mathbf{k}\iint \mathbf{F}_z \cdot d\mathbf{A}. \tag{4.5}$$

Using Gauss's theorem, one obtains

$$\mathbf{F}_{\mathrm{S}} = \mathbf{i}\iiint \nabla \cdot \mathbf{F}_x\, dV + \mathbf{j}\iiint \nabla \cdot \mathbf{F}_y\, dV + \mathbf{k}\iiint \nabla \cdot \mathbf{F}_z\, dV. \tag{4.6}$$

Now $\mathbf{F}_I = \mathbf{F}_B + \mathbf{F}_S$, so that

$$\iiint_V \left\{-\varrho \frac{D\mathbf{v}}{Dt} + \varrho\mathbf{X} + [\mathbf{i}(\nabla \cdot \mathbf{F}_x) + \mathbf{j}(\nabla \cdot \mathbf{F}_y) + \mathbf{k}(\nabla \cdot \mathbf{F}_z)]\right\} dV = 0. \tag{4.7}$$

Equation (4.7) is true for arbitrary elementary volumes and so

$$\varrho \frac{D\mathbf{v}}{Dt} = \varrho\mathbf{X} + \mathbf{i}(\nabla \cdot \mathbf{F}_x) + \mathbf{j}(\nabla \cdot \mathbf{F}_y) + \mathbf{k}(\nabla \cdot \mathbf{F}_z). \tag{4.8}$$

This is the general equation of fluid motion.

Substituting the stress components for $\mathbf{F}_x$, $\mathbf{F}_y$, and $\mathbf{F}_z$ into Eq. (4.8), one obtains the following equations of motion (in Cartesian coordinates):

$$\begin{aligned} \varrho \frac{Du}{Dt} &= \varrho X_x + \frac{\partial \sigma_{xx}}{\partial x} + \frac{\partial \sigma_{xy}}{\partial y} + \frac{\partial \sigma_{xz}}{\partial z}, \\ \varrho \frac{Dv}{Dt} &= \varrho X_y + \frac{\partial \sigma_{xy}}{\partial x} + \frac{\partial \sigma_{yy}}{\partial y} + \frac{\partial \sigma_{yz}}{\partial z}, \\ \varrho \frac{Dw}{Dt} &= \varrho X_z + \frac{\partial \sigma_{zx}}{\partial x} + \frac{\partial \sigma_{zy}}{\partial y} + \frac{\partial \sigma_{zz}}{\partial z}, \end{aligned} \tag{4.9}$$

X_x, X_y, and X_z are the components of the body force $\mathbf{X}$ in the x, y, and z directions, respectively, and the Eulerian derivative is

$$\frac{D}{Dt} \equiv \frac{\partial}{\partial t} + \mathbf{v} \cdot \nabla = \frac{\partial}{\partial t} + u\frac{\partial}{\partial x} + v\frac{\partial}{\partial y} + w\frac{\partial}{\partial z}. \tag{4.10}$$

As developed in Chapter 2, the stress components are connected to the rates of strain by the following relations:

$$\begin{aligned} \sigma_{xx} &= 2\mu\varepsilon_{xx} + \gamma\nabla \cdot \mathbf{v} - p, & \sigma_{xy} &= \sigma_{yx} = 2\mu\varepsilon_{xy}, \\ \sigma_{yy} &= 2\mu\varepsilon_{yy} + \gamma\nabla \cdot \mathbf{v} - p, & \sigma_{yz} &= \sigma_{zy} = 2\mu\varepsilon_{yz}, \\ \sigma_{zz} &= 2\mu\varepsilon_{zz} + \gamma\nabla \cdot \mathbf{v} - p, & \sigma_{zx} &= \sigma_{xz} = 2\mu\varepsilon_{zx}, \end{aligned} \tag{4.11}$$

where

$$\gamma = -\tfrac{2}{3}(\mu - \mu_B), \tag{4.12}$$

μ is the shear viscosity, and μ_B is the bulk viscosity (negligible for a perfect gas).

Using these expansions for the stress components in Eq. (4.9), one gets the final form of the equations of motion:

$$
\begin{aligned}
\varrho \frac{Du}{Dt} = \varrho\left(\frac{\partial u}{\partial t} + \mathbf{v} \cdot \nabla u\right) &= \varrho X_x - \frac{\partial p}{\partial x} + \frac{\partial}{\partial x}\left\{\mu\left[2\frac{\partial u}{\partial x} - \tfrac{2}{3}(\nabla \cdot \mathbf{v})\right]\right\} \\
&\quad + \frac{\partial}{\partial y}\left[\mu\left(\frac{\partial u}{\partial y} + \frac{\partial v}{\partial x}\right)\right] \\
&\quad + \frac{\partial}{\partial z}\left[\mu\left(\frac{\partial w}{\partial x} + \frac{\partial u}{\partial z}\right)\right], \\
\varrho \frac{Dv}{Dt} = \varrho\left(\frac{\partial v}{\partial t} + \mathbf{v} \cdot \nabla v\right) &= \varrho X_y - \frac{\partial p}{\partial y} + \frac{\partial}{\partial y}\left\{\mu\left[2\frac{\partial v}{\partial y} - \tfrac{2}{3}(\nabla \cdot \mathbf{v})\right]\right\} \\
&\quad + \frac{\partial}{\partial z}\left[\mu\left(\frac{\partial v}{\partial z} + \frac{\partial w}{\partial y}\right)\right] \\
&\quad + \frac{\partial}{\partial x}\left[\mu\left(\frac{\partial u}{\partial y} - \frac{\partial v}{\partial x}\right)\right], \qquad (4.13) \\
\varrho \frac{Dw}{Dt} = \varrho\left(\frac{\partial w}{\partial t} + \mathbf{v} \cdot \nabla w\right) &= \varrho X_z - \frac{\partial p}{\partial z} + \frac{\partial}{\partial z}\left\{\mu\left[2\frac{\partial w}{\partial z} - \tfrac{2}{3}(\nabla \cdot \mathbf{v})\right]\right\} \\
&\quad + \frac{\partial}{\partial x}\left[\mu\left(\frac{\partial w}{\partial x} + \frac{\partial u}{\partial z}\right)\right] \\
&\quad + \frac{\partial}{\partial y}\left[\mu\left(\frac{\partial v}{\partial z} + \frac{\partial w}{\partial y}\right)\right].
\end{aligned}
$$

These equations are better known as the Navier–Stokes equations of motion for a viscous compressible fluid. It must be noted that:

(a) The equations as they stand apply to laminar motion only, in which diffusion of momentum is governed purely by molecular processes and not by convective motions.

(b) The equations apply to Newtonian (or near) fluids, that is, to fluids in which the stress is linearly related to the rate of strain. Visco-elastic fluids are therefore not governed by these equations. The motion of fluids containing a high concentration of particulate matter would likewise not be governed by them.

(c) The bulk viscosity has been ignored in the above form of the equations.

There are three commonly occurring cases in which great simplification of Eqs. (4.13) occur.

(i) *Fluids with constant viscosity* As μ is now a constant, the equations become, in vector notation,

$$\varrho\left[\frac{\partial \mathbf{v}}{\partial t} + (\mathbf{v} \cdot \nabla)\mathbf{v}\right] = \varrho\mathbf{X} - \nabla p + \mu \nabla^2\mathbf{v} + \frac{\mu}{3}\nabla(\nabla \cdot \mathbf{v}). \quad (4.14)$$

(ii) *Fluids with constant viscosity and density* These fluids are called incompressible isotropic fluids. The equations become

$$\frac{\partial \mathbf{v}}{\partial t} + (\mathbf{v} \cdot \nabla)\mathbf{v} = \mathbf{X} - \frac{\nabla p}{\varrho} + \nu \nabla^2\mathbf{v}, \quad (4.15)$$

where $\nu = \mu/\varrho$ is the kinematic viscosity of the fluid and $\nabla \cdot \mathbf{v} = 0$ for an incompressible fluid.

(iii) *Fluids with zero viscosity, or inviscid fluids* Equation (4.14) now reduces to the simple form

$$\frac{\partial \mathbf{v}}{\partial t} + (\mathbf{v} \cdot \nabla)\mathbf{v} = \mathbf{X} - \frac{\nabla p}{\varrho}. \quad (4.16)$$

This equation is known as Euler's equation for inviscid flow and applies to both incompressible and compressible fluids. For liquids, such as water, and gases at low subsonic velocities, the former property applies. For gases at transonic and supersonic velocity, the latter case applies, the density being a function of both pressure and temperature.

4.2 Conservation of Energy—the Energy Equation

As discussed in Chapter 2, transport of thermal energy can also occur in a fluid when temperature gradients exist. The principle of conservation of energy leads to another equation of fluid flow—the energy equation.

The rate-of-strain tensor components in Cartesian coordinates are

$$\begin{aligned}
\varepsilon_{xx} &= \frac{\partial u}{\partial x}, & \varepsilon_{xy} &= \frac{\partial u}{\partial y} + \frac{\partial v}{\partial x} = \varepsilon_{yx}, \\
\varepsilon_{yy} &= \frac{\partial v}{\partial y}, & \varepsilon_{yz} &= \frac{\partial v}{\partial z} + \frac{\partial w}{\partial y} = \varepsilon_{zy}, \\
\varepsilon_{zz} &= \frac{\partial w}{\partial z}, & \varepsilon_{zx} &= \frac{\partial w}{\partial x} + \frac{\partial u}{\partial z} = \varepsilon_{xz},
\end{aligned} \quad (4.17)$$

where u, v, and w are the three velocity components in the x, y, and z directions, respectively.

The rate of work done by the stresses on a unit volume of the fluid is

$$\sigma_{xx}\varepsilon_{xx} + \sigma_{yy}\varepsilon_{yy} + \sigma_{zz}\varepsilon_{zz} + \sigma_{xy}\varepsilon_{xy} + \sigma_{yz}\varepsilon_{yz} + \sigma_{zx}\varepsilon_{zx}. \tag{4.18}$$

The First Law of Thermodynamics states that if a small quantity of heat is added to a simple system, it is used up in changing the kinetic energy of the molecules and in the external mechanical work done by the system. It is expressed as

$$dQ = dE + dW, \tag{4.19}$$

where dQ is the infinitesimal quantity of heat added to a unit mass of the system, dW the mechanical work done per unit mass by the system on its surroundings, and dE the increase in internal energy per unit mass of the system.

In terms of the time variation of energy, the above law can be written as

$$\varrho\frac{dQ}{dt} + (\sigma_{xx}\varepsilon_{xx} + \sigma_{yy}\varepsilon_{yy} + \sigma_{zz}\varepsilon_{zz} + \sigma_{xy}\varepsilon_{xy} + \sigma_{yz}\varepsilon_{yz} + \sigma_{zx}\varepsilon_{zx}) = \varrho\frac{DE}{Dt}. \tag{4.20}$$

Substituting in Eq. (4.18) for the stress components and the rates of strain, one gets the following expression for the rate of work done:

$$\begin{aligned} -p\left(\frac{\partial u}{\partial x} + \frac{\partial v}{\partial y} + \frac{\partial w}{\partial z}\right) &- \tfrac{2}{3}\mu(\nabla \cdot \mathbf{v})^2 + 2\mu\left[\left(\frac{\partial u}{\partial x}\right)^2 + \left(\frac{\partial v}{\partial y}\right)^2 + \left(\frac{\partial w}{\partial z}\right)^2\right] \\ &+ \mu\left[\left(\frac{\partial v}{\partial x} + \frac{\partial u}{\partial y}\right)^2 + \left(\frac{\partial w}{\partial y} + \frac{\partial v}{\partial z}\right)^2 + \left(\frac{\partial u}{\partial z} + \frac{\partial w}{\partial x}\right)^2\right] \\ &= -p\left(\frac{\partial u}{\partial x} + \frac{\partial v}{\partial y} + \frac{\partial w}{\partial z}\right) + \Phi, \end{aligned} \tag{4.21}$$

where Φ is called the *dissipation function*. It is the time rate of energy dissipation per unit volume of the fluid through the action of viscosity.

The form of this dissipation function might suggest that dissipation and vorticity are interdependent. That they are, in fact, independent of each other, can be shown as follows. Consider the incompressible case, for simplicity.

A vorticity function (a measure of vorticity) can be defined by taking the scalar product of the vorticity vector with itself:

$$\Omega = \zeta \cdot \zeta = \left(\frac{\partial w}{\partial y} - \frac{\partial v}{\partial z}\right)^2 + \left(\frac{\partial w}{\partial z} - \frac{\partial w}{\partial x}\right)^2 + \left(\frac{\partial v}{\partial x} - \frac{\partial u}{\partial y}\right)^2, \tag{4.22}$$

where $\boldsymbol{\zeta}$ is the vorticity vector, and Φ and Ω are functions of nine variables —the nine derivatives of the three velocity components. Eight of these variables are independent; because the fluid is assumed to be incompressible, one gradient may be expressed in terms of two others. Let the eight independent ones be denoted by a, b, c, d, e, f, g, and h, and let the dependent gradient be denoted by k. Then

$$\Phi = \mu[2a^2 + 2b^2 + 2k^2 + (d+e)^2 + (f+g)^2 + (h+c)^2],$$
$$\Omega = [(d-e)^2 + (f-g)^2 + (h-c)^2].$$

Since the matrix

$$\begin{vmatrix} \dfrac{\partial\Phi}{\partial a} & \dfrac{\partial\Phi}{\partial b} & \dfrac{\partial\Phi}{\partial d} & \dfrac{\partial\Phi}{\partial e} & \dfrac{\partial\Phi}{\partial f} & \dfrac{\partial\Phi}{\partial g} & \dfrac{\partial\Phi}{\partial h} & \dfrac{\partial\Phi}{\partial c} \\ \dfrac{\varrho\Omega}{\partial a} & \dfrac{\partial\Omega}{\partial b} & \dfrac{\partial\Omega}{\partial d} & \dfrac{\partial\Omega}{\partial e} & \dfrac{\partial\Omega}{\partial f} & \dfrac{\partial\Omega}{\partial g} & \dfrac{\partial\Omega}{\partial h} & \dfrac{\partial\Omega}{\partial c} \end{vmatrix}$$

can be shown to be of rank two, Φ and Ω are mathematically independent, that is to say, dissipation is independent of vorticity. An example of this is the rigidly rotating core of a vortex, which has vorticity but no dissipation.

The development of the energy equation will be presented next.

The First Law of Thermodynamics, written in Eq. (4.20), can now be expressed as

$$\varrho \frac{dQ}{dt} + \Phi = \varrho \frac{DE}{Dt} + p\nabla \cdot \mathbf{v}. \tag{4.23}$$

The equation of continuity for a compressible fluid is

$$\frac{D\varrho}{Dt} + \varrho \nabla \cdot \mathbf{v} = 0; \tag{4.24}$$

Therefore

$$\frac{p}{\varrho} \nabla \cdot \mathbf{v} = -\frac{p}{\varrho^2} \frac{D\varrho}{Dt} = \frac{D}{Dt}\left(\frac{p}{\varrho}\right) - \frac{1}{\varrho} \frac{Dp}{Dt}. \tag{4.25}$$

Using this equation in Eq. (4.23), one obtains

$$\varrho \frac{dQ}{dt} + \Phi = \varrho \frac{Dh}{Dt} - \frac{Dp}{Dt}, \tag{4.26}$$

where $h = E + (p/\varrho)$ is the enthalpy of the fluid per unit mass. It remains to specify Q, the heat added per unit mass.

The Fourier heat conduction law implies that the heat flow across an element of fluid surface in δt seconds is proportional to the gradient in the temperature,

$$\delta Q \propto -k\,\delta A\,\delta t\,\frac{\partial T}{\partial n}, \tag{4.27}$$

where k is the coefficient of thermal conductivity. The negative sign implies that the heat flows from points of higher temperatures to points of lower temperatures.

Consider a closed surface A enclosing an arbitrary volume V. The heat flow out in δt seconds is

$$-\delta t \iint k\,\frac{\partial T}{\partial n}\,dA = -\delta t \iint k(\nabla T)\cdot d\mathbf{A} = -\delta t \iiint \nabla\cdot[k\,\nabla T]\,dV. \tag{4.28}$$

The heat added to the system per unit volume is then

$$\delta t\,\nabla\cdot[k\,\nabla T] = \varrho\,\delta Q. \tag{4.29}$$

In the limit as $\delta t \to 0$, therefore,

$$\varrho\,\frac{dQ}{dt} = \nabla\cdot[k\,\nabla T]. \tag{4.30}$$

Expanding this in Cartesian coordinates, one has

$$\varrho\,\frac{dQ}{dT} = \frac{\partial}{\partial x}\left(k\,\frac{\partial T}{\partial x}\right) + \frac{\partial}{\partial y}\left(k\,\frac{\partial T}{\partial y}\right) + \frac{\partial}{\partial z}\left(k\,\frac{\partial T}{\partial z}\right). \tag{4.31}$$

Substituting this value for $\varrho\,dQ/dt$ in Eq. (4.26), one obtains

$$\nabla\cdot(k\,\nabla T) + \Phi = \varrho\,\frac{Dh}{Dt} - \frac{Dp}{Dt}. \tag{4.32}$$

This is the thermal energy equation for the fluid when there are no heat sources or sinks in it (thus, chemical reaction is excluded), and when there is no heating by radiant energy.

In Cartesian coordinates the energy equation assumes the form

$$\begin{aligned}
&\frac{\partial}{\partial x}\left(k\,\frac{\partial T}{\partial x}\right) + \frac{\partial}{\partial y}\left(k\,\frac{\partial T}{\partial y}\right) + \frac{\partial}{\partial z}\left(k\,\frac{\partial T}{\partial z}\right) + \Phi \\
&\quad = \varrho\,\frac{\partial(C_{\mathrm{p}}T)}{\partial t} + \varrho u\,\frac{\partial(C_{\mathrm{p}}T)}{\partial x} + \varrho v\,\frac{\partial(C_{\mathrm{p}}T)}{\partial y} + \varrho w\,\frac{\partial(C_{\mathrm{p}}T)}{\partial z} \\
&\quad\quad - \left(\frac{\partial p}{\partial t} + u\,\frac{\partial p}{\partial x} + v\,\frac{\partial p}{\partial y} + w\,\frac{\partial p}{\partial z}\right),
\end{aligned} \tag{4.33}$$

where $C_{\mathrm{p}}T = h$ is the enthalpy per unit mass of fluid.

Again there are certain specific conditions of interest under which the energy equation simplifies considerably.

(i) *Constant viscosity and thermal conductivity of the viscous incompressible fluid* The equation in this case becomes

$$k \nabla^2 T + \Phi_{k\mu} = \varrho C_v \frac{\partial T}{\partial t} + \varrho C_v \mathbf{v} \cdot (\nabla T), \tag{4.34}$$

where C_v is the specific heat at constant volume,

$$\Phi_{k\mu} = 2\mu\left[\left(\frac{\partial u}{\partial x}\right)^2 + \left(\frac{\partial v}{\partial y}\right)^2 + \left(\frac{\partial w}{\partial z}\right)^2\right] + \mu\left[\left(\frac{\partial v}{\partial x} + \frac{\partial u}{\partial y}\right)^2 + \left(\frac{\partial w}{\partial y} + \frac{\partial v}{\partial z}\right)^2 + \left(\frac{\partial u}{\partial z} + \frac{\partial w}{\partial x}\right)^2\right], \tag{4.35}$$

and

$$C_v = C_p - R,$$

where C_p is the specific heat at constant pressure and R is the gas constant.

(ii) *Inviscid fluid (no dissipation)* In this case, Eq. (4.32) becomes

$$\nabla \cdot (k \nabla T) = \varrho \frac{Dh}{Dt} - \frac{Dp}{Dt}. \tag{4.36}$$

If the flow is adiabatic, then $dQ = 0$ and the thermal energy equation reduces to

$$\varrho \frac{Dh}{Dt} - \frac{Dp}{Dt} = 0. \tag{4.37}$$

4.3 Equation of State

For compressible fluids, a relation is also required between the pressure, specific volume, and the temperature. This is the equation of state. For a perfect gas (no intermolecular interactions), the equation is

$$\frac{p}{\varrho} = RT. \tag{4.38}$$

For real gases, the intermolecular interactions must be accounted for and equations such as that of van der Waals used instead of the perfect gas relation given in Eq. (4.38).

4.4 Diffusion Equation

There is one more equation which can arise in fluid dynamics. This corresponds to the transport, or diffusion, of mass. It arises when there is more than one component in the fluid and concentration gradients exist.

In a two-dimensional system, and for two components only, the equation has the form

$$\frac{\partial}{\partial x}(cu) + \frac{\partial}{\partial y}(cv) = \frac{\partial}{\partial y}\left(D_{12}\frac{\partial c}{\partial y}\right), \tag{4.39}$$

where c is the molecular concentration of component 1 in the mixture, and D_{12} is the coefficient of binary diffusion for the two components.

As this text deals almost exclusively with the dynamics of single-component fluids, this equation will not be dealt with in any detail.

The three momentum equations [Eqs. (4.13)]—the Navier–Stokes equations—are nonlinear partial differential equations. Few exact solutions are possible.

4.5 Some Exact Solutions of the Navier–Stokes Equations

In the case of parallel flows in which there is only one velocity component different from zero, exact solutions of the Navier–Stokes equations can be derived for the case of incompressible fluids of constant viscosity.

As the component v and w are zero, the equation of continuity demands that $\partial u/\partial x = 0$ and so u is independent of x. Thus there is only one component equation and it has the form

$$\varrho\frac{\partial u}{\partial t} = -\frac{dp}{dx} + \mu\left(\frac{\partial^2 u}{\partial y^2} + \frac{\partial^2 u}{\partial z^2}\right). \tag{4.40}$$

One notes that the convective terms are not present and the equation is a linear one.

4.5.1 Steady Parallel Flow through a Straight Channel

Consider a channel with two parallel flat walls, and let the distance between the walls be $2b$.

Equation (4.40) now can be written

$$\frac{dp}{dx} = \mu \frac{d^2u}{dy^2}. \tag{4.41}$$

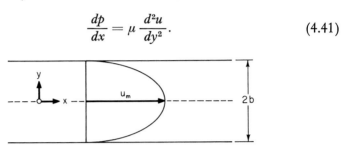

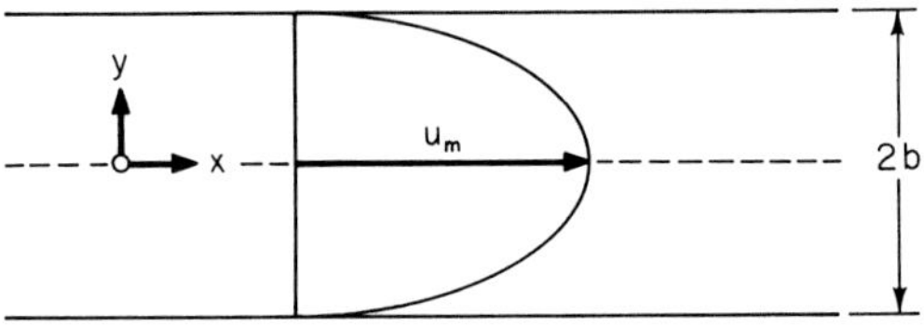

Fig. 4.2

The boundary condition on the velocity is $u = 0$ for $y = \pm b$ (Fig. 4.2). Since $\partial p/\partial y = 0$, p is independent of y, and so from Eq. (4.41), dp/dn is a constant. The solution is thus

$$u(y) = -\frac{1}{2\mu}\frac{dp}{dx}(b^2 - y^2). \tag{4.42}$$

The velocity profile is parabolic. The solution can also be written as

$$u(y) = u_{\mathrm{m}}\left(1 - \frac{y^2}{b^2}\right), \tag{4.43}$$

where

$$u_{\mathrm{m}} = -\frac{b^2}{2\mu}\frac{dp}{dx}$$

is the maximum velocity in the flow occurring at $y = 0$.

4.5.2 Steady Couette Flow

In this case one of the walls moves in its own plane with a velocity U and the other is at rest.

The boundary conditions are now

$$u = 0 \quad \text{at} \quad y = 0, \qquad u = U \quad \text{at} \quad y = h.$$

The solution is

$$u = \frac{y}{h} U - \frac{h^2}{2\mu} \frac{dp}{dx} \frac{y}{h} \left(1 - \frac{y}{h}\right). \tag{4.44}$$

For zero pressure gradient,

$$u = \frac{y}{h} U. \tag{4.45}$$

This is the case of simple Couette flow, or simple shear flow.

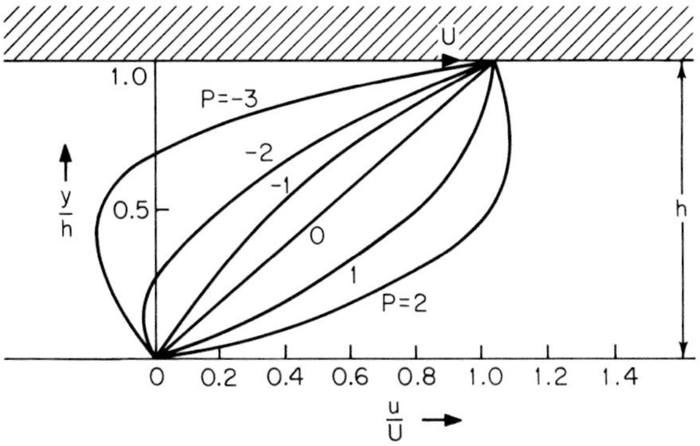

Fig. 4.3

The shape of the velocity profile is determined by the dimensionless pressure gradient (Fig. 4.3),

$$P = \frac{h^2}{2\mu U}\left(-\frac{dp}{dx}\right).$$

When $P > 0$, or for a pressure decreasing in the positive x direction, the velocity is positive over the whole channel. When $P < 0$, the velocity over part of the channel can be negative. Backflow can occur near the wall which is at rest. Of course, for $P = 0$, the case is that of simple shear flow.

4.5.3 Hagen–Poiseuille Flow in a Circular Pipe

The steady laminar flow through a long straight pipe of circular cross section is considered. To maintain this flow, the shearing force on the surface of any cylindrical shape of fluid must be balanced by the difference of pressure between the ends.

The only component of velocity which is nonzero is the axial one u_z, and this is a function of the radius r only, since there is axial symmetry. The equation of motion in cylindrical coordinates is

$$\mu\left(\frac{d^2u_z}{dr^2} + \frac{1}{r}\frac{du_z}{dr}\right) = \frac{dp}{dz}, \tag{4.46}$$

and $\partial p/\partial r = 0$. Thus dp/dz is a constant. The boundary condition on u_z is $u_z = 0$ at $r = r_0$, where r_0 is the radius of the tube.

The solution of Eq. (4.46) with this boundary condition is

$$u_z = -\frac{1}{4\mu}\frac{dp}{dz}(r_0^2 - r^2) = u_{zm}\left(1 - \frac{r^2}{r_0^2}\right), \tag{4.47}$$

where

$$\frac{dp}{dz} = \frac{p_2 - p_1}{L}$$

and L is the length of the pipe section with a pressure drop of $(p_2 - p_1)$ across it (Fig. 4.4).

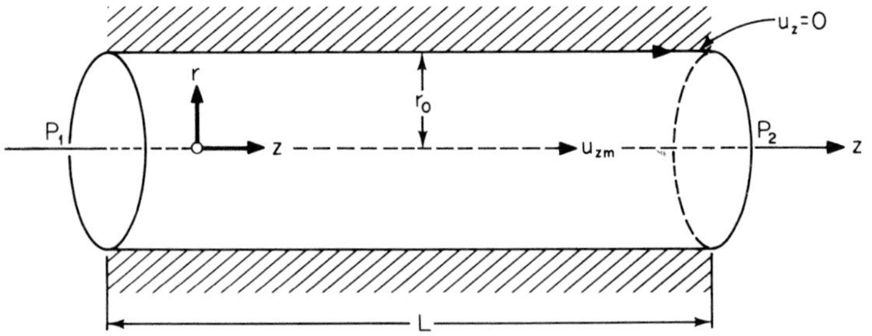

Fig. 4.4

The velocity profile is again seen to be parabolic. The maximum velocity occurs on the axis of the pipe and is

$$u_{zm} = \frac{p_1 - p_2}{L}\frac{r_0^2}{4\mu}. \tag{4.48}$$

The volume of fluid flowing through the pipe is

$$V = \tfrac{1}{2}u_{zm}\pi r_0^2 = \frac{\pi r_0^4}{8\mu}\frac{p_1 - p_2}{L}. \tag{4.49}$$

This relation was originally obtained from experiment by Hagen in 1839 and by Poiseuille in 1840. It is often used to determine the viscosity of a fluid.

It must be noted that the parabolic profile is attained only at a sufficient distance from the entrance of the pipe.

4.6 Unsteady Flow

Some nonsteady parallel flows will now be considered. The unsteady motion of a flat plate in its own plane also can be solved exactly. The only component of velocity different from zero will be that parallel to the plate. The velocity and pressure in the fluid become functions of time and the distance normal to the plate. The system is shown in Fig.

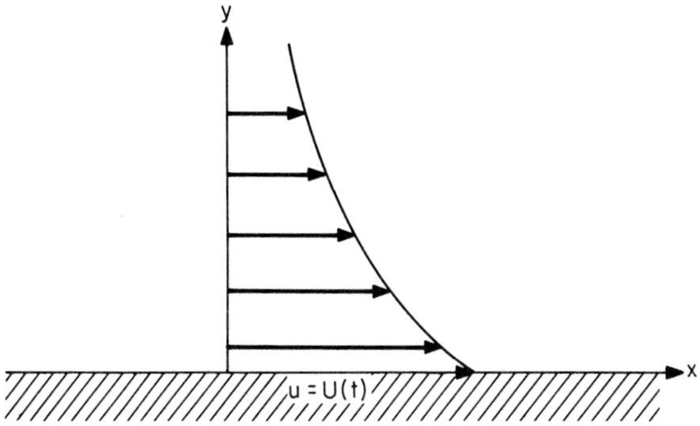

Fig. 4.5

4.5. The Navier–Stokes equations for a constant viscosity incompressible fluid reduce to the form

$$\frac{\partial u}{\partial t} = \gamma \frac{\partial^2 u}{\partial y^2}. \tag{4.50}$$

The convective acceleration terms have vanished identically. The boundary conditions depend on how the plate moves. Stokes's first and second problems are the two most significant cases.

4.6.1 Stokes's First Problem

In this case the plate is suddenly accelerated from rest to a velocity U_0. The boundary conditions are

$$\begin{aligned} t \leq 0: &\quad u = 0 &&\text{for all } y, \\ t > 0: &\quad u = U_0 &&\text{for } y = 0, \\ &\quad u = 0 &&\text{for } y = \infty. \end{aligned} \tag{4.51}$$

The differential equation is identical with the equation of heat conduction. It describes the propagation of heat in the region $y > 0$ when at $t = 0$ the wall is suddenly raised to a temperature above that of its surroundings.

The partial differential equation (4.50) can be reduced to an ordinary differential equation by the substitution of the dimensionless variable

$$\eta = \frac{y}{2(\nu t)^{1/2}}. \tag{4.52}$$

If one assumes that the solution has the form

$$u = U_0 f(\eta), \tag{4.53}$$

then the differential equation obtained for $f(\eta)$ is

$$f'' + 2\eta f' = 0, \tag{4.54}$$

where

$$f' = \frac{\partial f}{\partial \eta}, \qquad f'' = \frac{\partial^2 f}{\partial \eta^2}.$$

The boundary conditions transform to

$$f = 1 \quad \text{at} \quad \eta = 0, \qquad f = 0 \quad \text{at} \quad \eta = \infty.$$

The solution is thus

$$u = U_0 \operatorname{erfc} \eta, \tag{4.55}$$

where

$$\operatorname{erfc} \eta = \frac{2}{\sqrt{\pi}} \int_\eta^\infty \exp(-\eta^2)\, d\eta = 1 - \operatorname{erf} \eta = 1 - \frac{2}{\sqrt{\pi}} \int_0^\eta \exp(-\eta^2)\, d\eta.$$

This is known as the complementary error function and, like the error function itself, is well tabulated. The error function is also known as the probability integral.

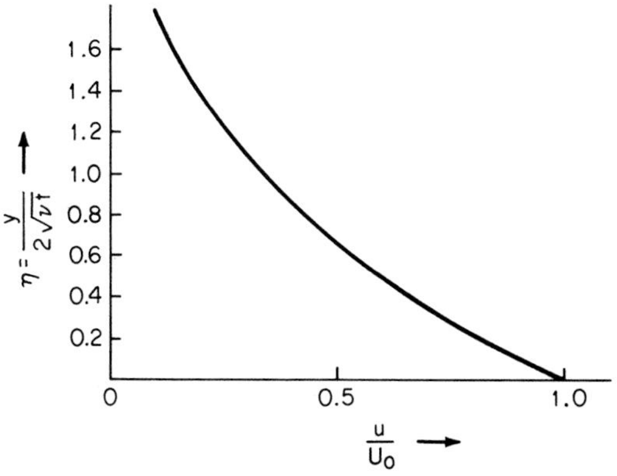

Fig. 4.6

A universal velocity distribution may be obtained by plotting η versus u/U_0 (Fig. 4.6). The velocity profiles for various times are said to be "similar"—they can be superposed on each other merely by changing the scale for η.

4.6.2 Stokes's Second Problem

This involves the flow near a plate which is oscillating harmonically with constant magnitude and frequency. Due to the no-slip condition at the wall, the fluid velocity there must equal that of the wall.

The boundary condition now is given by

$$y = 0: \qquad u(0, t) = U_0 \cos \omega t, \tag{4.56}$$

where U_0 is now the maximum velocity of the plate.

As is well known from the theory of heat conduction, the fluid velocity $u(y, t)$ is the solution of Eq. (4.50) with the boundary condition (4.56). It is

$$u(y, t) = U_0 e^{-ky} \cos(\omega t - ky), \tag{4.57}$$

where $k = (\omega/2\nu)^{1/2}$.

The velocity of the fluid decreases exponentially as the distance from the plate increases. The smaller the viscosity and the higher the frequency, the faster the rate of decrease of u with increase in y, for a given U_0. If one defines a variable

$$\eta = ky = y(\omega/2\nu)^{1/2},$$

then

$$u(y, t) = U_0 e^{-\eta} \cos(\omega t - \eta). \tag{4.58}$$

The velocity distribution near the oscillating wall is shown in Fig. 4.7. A fluid layer at a distance y from the wall has a phase lag $y(\omega/2\nu)^{1/2}$ with respect to the wall motion. Two fluid layers a distance $2\pi/k = 2\pi(2\omega/\nu)^{1/2}$ apart will oscillate in phase. This distance can be regarded as a wavelength of the motion and is often called the depth of penetration of the viscous wave.

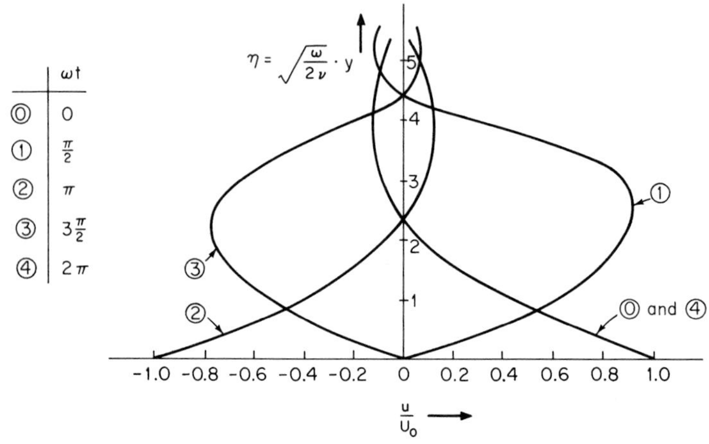

Fig. 4.7

4.7 Laminar Flow near a Rotating Disk

The motion of the viscous fluid on one side of an infinite plane lamina which is rotating at constant angular velocity about the axis $r = 0$ is considered.

The cylindrical coordinate system shown in Fig. 4.8 is used, and the fluid is considered to be infinite in extent, being bounded only by the plane $z = 0$. This problem was solved exactly by von Kármán.† The boundary conditions are

$$\begin{aligned} v_z = v_r &= 0, \quad v_\vartheta = \omega r \qquad \text{at} \quad z = 0, \\ v_r = v_\vartheta &= 0 \qquad\qquad\qquad\;\; \text{at} \quad z = \infty. \end{aligned} \tag{4.59}$$

† T. von Kármán, *Z. Angew. Math. Mech.* **1**, 244–247 (1921).

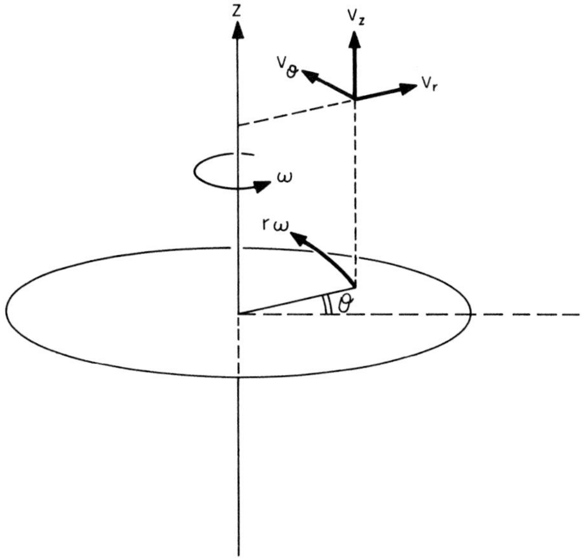

Fig. 4.8

The axial velocity does not vanish at $z = \infty$ but goes to some finite negative limit. A steady axial flow toward the rotating lamina occurs as the lamina throws fluid out radially from its surface.

Taking account of the rotational symmetry and considering steady, incompressible flow with no body forces one gets the equations of motion:

$$v_r \frac{\partial v_r}{\partial r} - \frac{v_\vartheta^2}{r} + v_z \frac{\partial v_r}{\partial z} = -\frac{1}{\varrho} \frac{\partial p}{\partial r} + \nu\left\{\frac{\partial^2 v_r}{\partial r^2} + \frac{\partial}{\partial r}\left(\frac{v_r}{r}\right) + \frac{\partial^2 v_r}{\partial z^2}\right\}, \quad (4.60)$$

$$v_r \frac{\partial v_\vartheta}{\partial r} + \frac{v_r v_\vartheta}{r} + v_z \frac{\partial v_\vartheta}{\partial z} = \nu\left\{\frac{\partial^2 v_\vartheta}{\partial r^2} + \frac{\partial}{\partial r}\left(\frac{v_\vartheta}{r}\right) + \frac{\partial^2 v_\vartheta}{\partial z^2}\right\}, \quad (4.61)$$

$$v_r \frac{\partial v_z}{\partial r} + v_z \frac{\partial v_z}{\partial z} = -\frac{1}{\varrho} \frac{\partial p}{\partial z} + \nu\left\{\frac{\partial^2 v_z}{\partial r^2} + \frac{1}{r} \frac{\partial v_z}{\partial r} + \frac{\partial^2 v_z}{\partial z^2}\right\}. \quad (4.62)$$

The equation of continuity is

$$\frac{\partial v_r}{\partial r} + \frac{v_r}{r} + \frac{\partial v_z}{\partial z} = 0. \tag{4.63}$$

To integrate this system of equations, one introduces a dimensionless distance from the wall,

$$z_1 = z\left(\frac{\omega}{\nu}\right)^{1/2}. \tag{4.64}$$

One also assumes the following forms for the velocity and pressure components:

$$v_r = r\omega F(z_1), \qquad v_\vartheta = r\omega G(z_1), \qquad v_z = (\nu\omega)^{1/2} H(z_1). \tag{4.65}$$

Substituting these in the equations of motion, one gets

$$F^2 - G^2 + F'H = F'', \quad 2FG + G'H = G'', \quad HH' = P' + H'', \tag{4.66}$$

where F', for example, implies $\partial F/\partial z_1$. The equation of continuity yields

$$2F + H' = 0. \tag{4.67}$$

The boundary conditions in Eq. (4.59) transform to

$$\begin{aligned} z_1 &= 0 \qquad F = 0, \quad G = 1, \quad H = 0, \quad P = 0, \\ z_1 &= \infty \qquad F = G = 0. \end{aligned} \tag{4.68}$$

These equations have been solved by W. C. Cochran and the values of F, G, H, and P versus z_1 are tabulated in Schlichting's textbook on boundary layer theory. The graphs of F, G, and H are shown in Fig. 4.9. It is seen

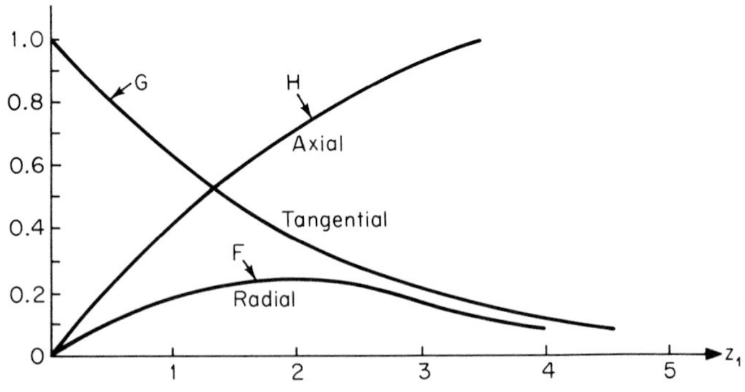

Fig. 4.9

that F and G tend to zero exponentially and become almost zero for some finite value of z_1. The peripheral velocity has dropped to half the disk velocity at a distance from the plate of the order of $(\nu/\omega)^{1/2}$. When ν/ω is small, this will be a very thin layer.

The calculation is strictly applicable only to an infinitely large disk or lamina. But the results may be used for a finite disk as long as its radius R is much larger than the depth of the layer carried with it, that is;

$$R \gg \left(\frac{\nu}{\omega}\right)^{1/2}.$$

Due to the viscous drag of the fluid, a turning moment must be applied to the disk continuously.

The circumferential component of the shear stress is

$$\tau_{z\varphi} = \mu\left(\frac{\partial v}{\partial z}\right)_{z=0}. \tag{4.69}$$

Then the incremental turning moment of an annular disk element of width dr at radius r is

$$dM = -2\pi r\, dr\, \tau_{z\varphi},$$

and so the moment for a complete disk wetted on one side is

$$M = -2\pi \int_0^R r^2 \tau_{z\varphi}\, dr. \tag{4.70}$$

From Eq. (4.65),

$$\tau_{z\varphi} = \varrho r \nu^{1/2} \omega^{3/2} G'(0). \tag{4.71}$$

Thus the total moment for a disk of radius R wetted on both sides by a fluid of viscosity ν is

$$2M = -\pi\varrho R^4 (\nu\omega^3)^{1/2} G'(0) = 0.616\pi\varrho R^4 (\nu\omega^3)^{1/2}, \tag{4.72}$$

where $-G'(0)$ has been tabulated as 0.616.

Defining a dimensionless moment coefficient as

$$C_M = \frac{2M}{\frac{1}{2}\varrho\omega^2 R^5},$$

one gets

$$C_M = \frac{2\pi(0.616)\sqrt{\nu}}{R\sqrt{\omega}}. \tag{4.73}$$

If a Reynolds number based on disk radius and edge velocity is defined,

$$\mathrm{Re} = \frac{R^2\omega}{\nu},$$

then

$$C_M = \frac{3.87}{\sqrt{\mathrm{Re}}}. \tag{4.74}$$

It is found that experimental results confirm Eq. (4.74) up to a Reynolds number of about 3×10^5. The flow then transitions to turbulence and a turbulent analysis of rotating disk flow must be used.

The quantity of liquid pumped outward by the action of one side of the disk is

$$Q = 2\pi R \int_{z=0}^{\infty} v_r\, dz = 0.886\pi R^2(\nu\omega)^{1/2}. \tag{4.75}$$

The same quantity of fluid must flow axially toward the disk.

Finally, it is to be noted that the pressure difference across the layer carried by the disk is about $\varrho\nu\omega$, which is very small for small viscosity fluids. Moreover, there is no radial pressure gradient.

There are several other cases for which exact solutions of the Navier–Stokes equations have been obtained—the above are possibly the most interesting and instructive.

Other solutions of these equations of viscous fluid motion are possible and are based on two approximations:

(i) *Treat the fluid as nondissipative* When this is done, the Navier–Stokes equations reduce to a set of equations which can be solved by the methods of potential theory. Regions in which this approximation is valid are called regions of potential flow. These methods will be dealt with subsequently in this chapter.

(ii) *Consider the case in which the velocity gradients are large* This occurs in the thin (boundary) layer of fluid around a solid body in a moving fluid and in the mixing region of jets and wakes. This approximation was first introduced by Prandtl in 1904, and the form of the equations is known as the boundary layer equations.

The subjects of vorticity, boundary layers, and methods of solution will be presented in later chapters.

4.8 Inviscid Incompressible Flows

Euler's equation (4.16) holds for both compressible and incompressible flow. It is convenient to record it again at this stage:

$$\frac{\partial \mathbf{v}}{\partial t} + (\mathbf{v} \cdot \nabla)\mathbf{v} = \mathbf{X} - \frac{\nabla p}{\varrho}.$$

For a conservative system the body force $\mathbf{X}$ can be derived from a potential χ, such that

$$X_x = -\frac{\partial \chi}{\partial x}, \qquad X_y = -\frac{\partial \chi}{\partial y}, \qquad X_z = -\frac{\partial \chi}{\partial z}. \tag{4.76}$$

Now

$$\nabla(\mathbf{v} \cdot \mathbf{v}) = 2(\mathbf{v} \cdot \nabla)\mathbf{v} + 2\mathbf{v} \times (\nabla \times \mathbf{v}). \tag{4.77}$$

Using this relation in Euler's equation, one obtains

$$\frac{\partial \mathbf{v}}{\partial t} - \mathbf{v} \times \boldsymbol{\zeta} = \mathbf{X} - \frac{\nabla p}{\varrho} - \nabla\left(\frac{v^2}{2}\right) = -\nabla\left(\chi + \frac{p}{\varrho} + \frac{v^2}{2}\right), \tag{4.78}$$

where $\boldsymbol{\zeta}$ is the vorticity vector. For steady flow, this equation becomes

$$\mathbf{v} \times \boldsymbol{\zeta} = \nabla\left(\chi + \frac{p}{\varrho} + \frac{v^2}{2}\right)$$

or

$$(\mathbf{v} \times \boldsymbol{\zeta}) \cdot d\mathbf{r} = d\left(\chi + \frac{p}{\varrho} + \frac{v^2}{2}\right). \tag{4.79}$$

Along a streamline, the left-hand side of this equation is zero ($\mathbf{v} \times \boldsymbol{\zeta}$ is orthogonal to a streamline) and so

$$\chi + \frac{p}{\varrho} + \frac{v^2}{2} = C, \tag{4.80}$$

where C is a constant along a specific streamline but varies from line to line.

For irrotational flow, $\boldsymbol{\zeta} = 0$ and

$$\frac{\partial \mathbf{v}}{\partial t} = \frac{\partial(\nabla \varphi)}{\partial t} = \nabla \frac{\partial \varphi}{\partial t}.$$

Equation (4.78) now becomes

$$\nabla\left(\frac{\partial \varphi}{\partial t} + \chi + \frac{p}{\varrho} + \frac{v^2}{2}\right) = 0. \tag{4.81}$$

Integration of this equation, after taking the dot product with $d\mathbf{r}$, yields

$$\frac{\partial \varphi}{\partial t} + \chi + \frac{v^2}{2} + \frac{p}{\varrho} = G(t), \tag{4.82}$$

where $G(t)$ is an arbitrary function of time.

If the flow field is both steady and irrotational, then this equation reduced to Bernoulli's equation for incompressible fluids:

$$\frac{v^2}{2} + \frac{p}{\varrho} + \chi = C. \tag{4.83}$$

The constant now has the *same* value all through the flow field. In many cases the force field is gravitational and the equation assumes the more common form

$$\frac{v^2}{2} + \frac{p}{\varrho} + gh = C, \tag{4.84}$$

where h is the height above some reference level in the field.

Two examples illustrating the application of the general time-dependent Bernoulli equation (4.82) and the steady form as given in Eq. (4.84) will now be presented.

(i) *An infinite mass of nonviscous liquid of constant density is initially at rest and has a spherical cavity of radius r_0 embedded in it* A pressure is applied uniformly over the surface of the cavity and the liquid is forced to move out radially. The motion is thus irrotational. There is no pressure at infinity and no body forces are considered.

Suppose the radius of the cavity is r_0 at time t and that the applied pressure is Kr_0^{-3}, where K is a constant. It is required that the pressure at a distance r from the cavity center be determined. The potential of motion is

$$\varphi = \frac{A}{r},$$

and equating velocities at the bubble boundary, one gets

$$\frac{dr_0}{dt} = \frac{A}{r_0^2};$$

therefore

$$A = r_0^2 \frac{dr_0}{dt},$$

so that

$$\varphi = \frac{r_0^2}{r}\left(\frac{dr_0}{dt}\right). \tag{i}$$

Using Bernoulli's equation for unsteady motion, Eq. (4.82), one obtains

$$\frac{\partial \varphi}{\partial t} + \chi + \frac{v^2}{2} + \frac{p}{\varrho} = G(t). \qquad \text{(ii)}$$

In this case $\chi = 0$, $v = A/r^2$, and

$$\frac{\partial \varphi}{\partial t} = \frac{1}{r}\frac{dA}{dt}.$$

Now, since $r \to \infty$ when $p = 0$, hence $G(t) = 0$ for all t. Thus, after substitution for A, Eq. (ii) becomes

$$\frac{p}{\varrho} = \frac{\{r_0^2(d^2r_0/dt^2) + 2r_0(dr_0/dt)^2\}}{r} + \frac{1}{2}\frac{r_0^4(dr_0/dt)^2}{r^4} = 0. \qquad \text{(iii)}$$

Since $p = Kr_0^{-3}$ when $r = r_0$, Eq. (iii) becomes

$$\frac{K}{\varrho r_0^3} = \frac{3}{2}\left(\frac{dr_0}{dt}\right)^2 + r_0\frac{d^2r_0}{dt^2}. \qquad \text{(iv)}$$

Since

$$\frac{d^2r_0}{dt^2} = \frac{1}{2}\frac{d}{dr_0}\left(\frac{dr_0}{dt}\right)^2 \quad \text{and} \quad \frac{dr_0}{dt} = 0$$

when $r_0 = r$, Eq. (iv) integrates to

$$r_0^3\left(\frac{dr_0}{dt}\right)^2 = \left(\frac{2K}{\varrho}\right)\log\frac{r_0}{r_1}.$$

Substituting in Eq. (iii), one has

$$p = \left(\frac{K}{r_0^2 r}\right)\left[1 + \left\{1 - \frac{r_0^3}{r^3}\right\}\log\frac{r_0}{r_1}\right],$$

which is the required result.

(ii) *Consider the closed conduit of rectangular cross section, shown in Fig. 4.10* The center axis of the conduit is a circle of radius R. The flow is postulated as being steady and nondissipative, and the streamlines are all circular arcs centered on the axis of the bend.

The pressure difference between points 1 and 2 on the inside and outside edges of the bend, respectively, is required.

Consider a streamline of radius r. The equation of motion for any fluid particle on this streamline is then Euler's equation. The equation

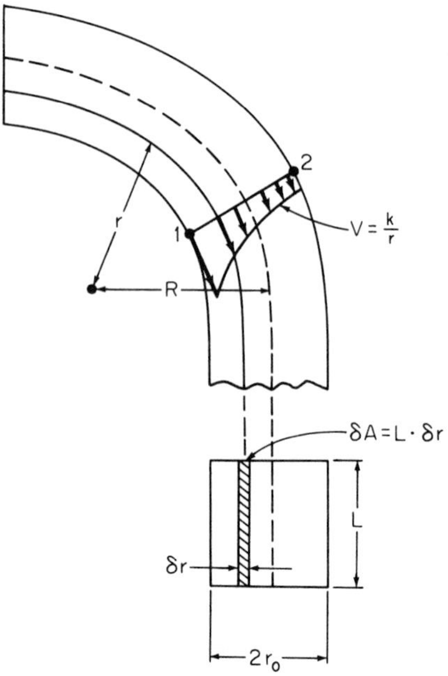

Fig. 4.10

for the radial direction component, and for steady incompressible flow, is

$$\frac{1}{\varrho}\frac{\partial p}{\partial r} + g\frac{\partial z}{\partial r} = \frac{V^2}{r} \tag{i}$$

or

$$\frac{\partial}{\partial r}\left(\frac{p}{\varrho} + gz\right) = \frac{V^2}{r}. \tag{ii}$$

The Bernoulli equation is

$$\frac{p}{\varrho} + gz + \frac{V^2}{2} = \text{constant}$$

for any streamline. It yields

$$\frac{\partial}{\partial r}\left(\frac{p}{\varrho} + gz\right) + V\frac{\partial V}{\partial r} = 0, \tag{iii}$$

Substituting this in Eq. (ii), one therefore obtains

$$\frac{V^2}{r} + V\frac{\partial V}{\partial r} = 0$$

or

$$\frac{\partial r}{r} + \frac{\partial V}{V} = 0. \tag{iv}$$

Integration gives

$$\ln r + \ln V = \ln(rV) = \text{constant};$$

therefore

$$V = \frac{K}{r}, \tag{v}$$

where K is a constant of integration.

The velocity profile is thus the arc of a hyperbola, as shown in Fig. 4.10. To determine K, one uses the continuity principle. If Q is the flow rate in the conduit, then

$$Q = \iint_A V\,dA = K \iint_A \frac{dA}{r};$$

therefore

$$K = \frac{Q}{\displaystyle\iint_A \frac{dA}{r}} = \frac{Q}{L \displaystyle\int_{R-r_0}^{R+r_0} \frac{dr}{r}} = \frac{Q}{L\left[\ln\left(\dfrac{R+r_0}{R-r_0}\right)\right]}. \tag{vi}$$

Replacing V by K/r in Eq. (i), one can write this equation as

$$\partial\left(\frac{p}{\varrho} + gz\right) = \frac{K^2}{r^3}\,\partial r.$$

Integrating between points 1 and 2 and using the above expression for K, one obtains

$$\left(\frac{p}{\varrho} + gz\right)_2 - \left(\frac{p}{\varrho} + gz\right)_1 = -\frac{K^2}{2r^2}\Bigg|_{R-r_0}^{R+r_0}$$

$$= \frac{1}{2}\left\{\frac{Q}{L\ln\left(\dfrac{R+r_0}{R-r_0}\right)}\right\}^2\left[\frac{1}{(R-r_0)^2} - \frac{1}{(R+r_0)^2}\right];$$

therefore

$$p_2 - p_1 = \varrho g(z_2 - z_1) + 2\varrho R r_0\left\{\frac{Q}{(R^2 - r_0^2)\,L\ln\left(\dfrac{R+r_0}{R-r_0}\right)}\right\}^2.$$

This is the pressure difference which was to be determined.

4.9 Method of Singularities

As mentioned in Chapter 3, velocity potentials and stream functions for the flow around solid bodies can be constructed by introducing singularities into a field, which represents the undisturbed flow pattern.

The basic singularities are sources, sinks, doublets, and vortices, and these are combined in various ways along with a uniform streamflow.

The stream function and velocity potential for uniform flow in straight lines will be derived first.

Suppose the flow is parallel to the positive x axis. The velocity components are then

$$u = U_\infty = \text{constant}, \qquad v = 0. \tag{4.85}$$

Now

$$u = \frac{\partial \varphi}{\partial x} = \frac{\partial \psi}{\partial y} = U_\infty, \qquad v = \frac{\partial \varphi}{\partial y} = -\frac{\partial \psi}{\partial x} = 0.$$

Integrating the first equation and taking the constants as zero, one obtains

$$\varphi = U_\infty x = U_\infty r \cos \vartheta, \tag{4.86}$$

$$\psi = U_\infty y = U_\infty r \sin \vartheta. \tag{4.87}$$

Obviously the streamlines are straight lines parallel to the x axis and the potential lines are straight lines parallel to the y axis (Fig. 4.11). It can be easily shown that both φ and ψ satisfy the Laplace equation and are harmonic functions.

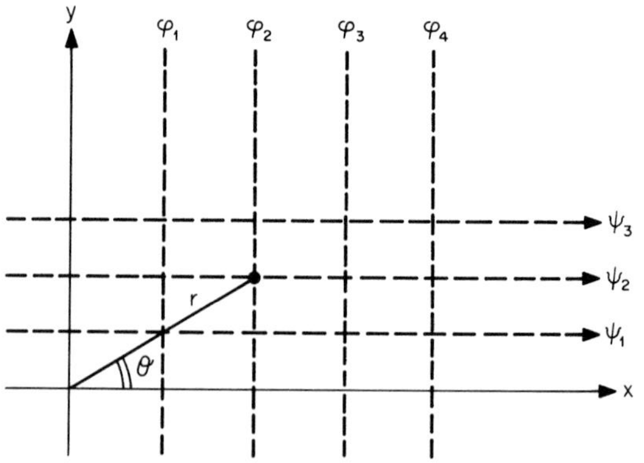

Fig. 4.11

4.9.1 Two-Dimensional Source Flow

This comes under the heading of a line singularity, as dealt with in Chapter 3.

Equation (3.90) gave for the velocity potential of this radial flow

$$\varphi(x, y) = \frac{\mu}{2\pi} \ln r = \frac{\mu}{2\pi} \ln(x^2 + y^2)^{1/2} \tag{4.88}$$

when the origin is placd at x', y', and μ is the strength of the line source, or the total volumetric rate of flow from it.

Equation (3.91) gave the stream function as

$$\psi = \frac{\mu}{2\pi} \vartheta, \tag{4.89}$$

where the constant has been taken as zero and $0 \leq \vartheta \leq 2\pi$. By taking the constant as zero, the positive x axis is taken as the zero streamline. From Eq. (4.89) it is seen that the streamlines are radial lines emanating from the center, as shown in Fig. 4.12. The equation of the equipotential lines is obtained by setting the right-hand side of Eq. (4.88) to a constant, whence $r =$ constant. This is the equation of a family of concentric circles. Clearly these two families of lines are orthogonal to each other.

The velocity components are

$$u = v_r \cos \vartheta = \frac{\mu}{2\pi r} \frac{x}{r} = \frac{\mu}{2\pi} \frac{x}{(x^2 + y^2)^{1/2}}, \tag{4.90}$$

$$v = v_r \sin \vartheta = \frac{\mu}{2\pi r} \frac{y}{r} = \frac{\mu}{2\pi} \frac{y}{(x^2 + y^2)^{1/2}}. \tag{4.91}$$

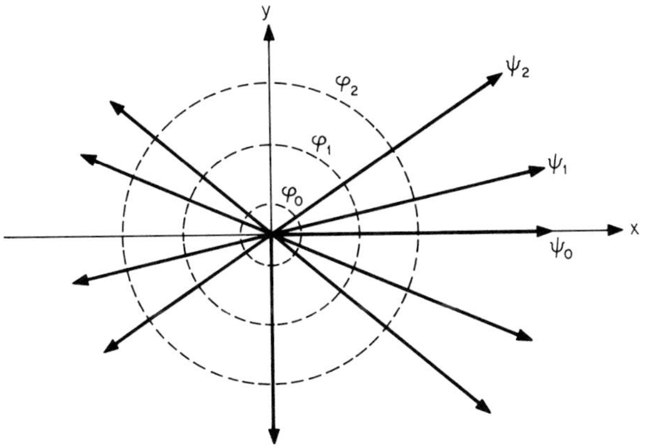

Fig. 4.12

The circulation about the singularity can be found by drawing a small circle of radius r about it. Then

$$\Gamma = \oint \mathbf{V} \cdot dl = \int_0^{2\pi} v_\vartheta r \, d\vartheta = 0,$$

since v_ϑ is zero.

When the value of μ is negative, this represents a sink singularity and the flow is directed radially inward.

4.9.2 Line Vortex Flow (or a Two-Dimensional Free Vortex)

It was shown in Chapter 3, Eq. (3.94), that the stream function for the flow field around this singularity is

$$\psi = -\frac{\varkappa}{2\pi} \ln r, \tag{4.92}$$

where r is the perpendicular distance from the point in the field to the line, and $\varkappa$ is the vortex strength. From Eq. (3.93), the tangential velocity is

$$v_\vartheta = \frac{1}{r} \frac{\partial \varphi}{\partial \vartheta} = \frac{\varkappa}{2\pi r},$$

and thus

$$\varphi = \frac{\varkappa}{2\pi} \vartheta. \tag{4.93}$$

In this case the *radial* lines are now equipotential and the *concentric* circles become streamlines (Fig. 4.13). The counterclockwise direction of the free vortex motion is considered positive. Now,

$$\nabla^2 \psi = \frac{\partial^2 \psi}{\partial r^2} + \frac{1}{r} \frac{\partial \psi}{\partial r} + \frac{1}{r^2} \frac{\partial^2 \psi}{\partial \vartheta^2} = \frac{\varkappa}{2\pi r^2} - \frac{\varkappa}{2\pi r^2} + 0 = 0$$

and

$$\nabla^2 \varphi = \frac{\partial^2 \varphi}{\partial r^2} + \frac{1}{r} \frac{\partial \varphi}{\partial r} + \frac{1}{r^2} \frac{\partial^2 \varphi}{\partial \vartheta^2} = 0 + 0 + 0 = 0,$$

so that both φ and ψ are again harmonic functions, as required.

Since the tangential velocity in this flow field is inversely proportional to the radius, the circulation, or strength, of the vortex, $\varkappa$, is the same

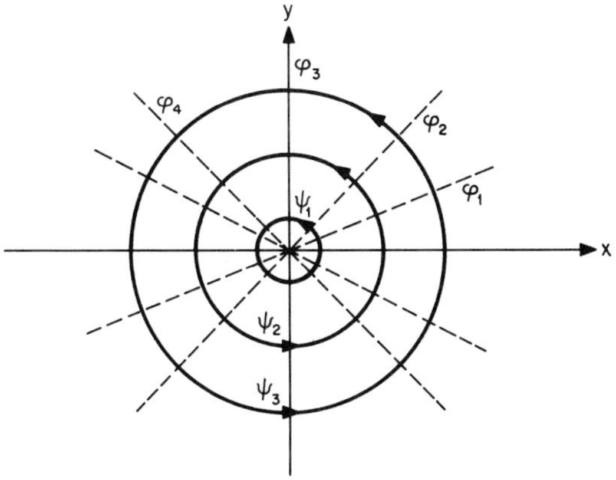

Fig. 4.13

on all streamlines:

$$\varkappa = \oint \mathbf{V} \cdot d\mathbf{l} = \int_0^{2\pi} v_\vartheta r \, d\vartheta = \int_0^{2\pi} C \, d\vartheta = 2\pi C.$$

The circulation is the same on any other simple closed path enclosing the line vortex, as can be seen by applying Stokes's theorem to the annular region formed by this path and a small path round the origin.

Since the circulation around the vortex is finite, the basic characteristic of irrotational flow is violated. It is due to the fact that the streamlines *enclose* a singular point. At $r = 0$ the velocity is infinite, whereas all derivatives of a harmonic potential must be finite. Such points must therefore be excluded from the region in which the Laplace equation holds.

4.9.3 Two-Dimensional Doublet

The three-dimensional doublet was discussed in Chapter 3. The two-dimensional doublet is formed similarly—by placing a source and sink a distance a apart and letting a go to zero as the strength μ goes to infinity—in such a way that $2a\mu$ approaches the constant value m. The expressions for the stream function and potential function in rectangular Cartesian coordinates are

$$\psi = -\frac{m}{2\pi} \frac{y}{x^2 + y^2}, \tag{4.94}$$

$$\varphi = \frac{m}{2\pi} \frac{x}{x^2 + y^2}. \tag{4.95}$$

The equation of streamlines for a doublet is

$$-\frac{m}{2\pi}\frac{y}{x^2+y^2}=C \qquad \text{(a constant).}$$

Rearranging this equation, one gets

$$x^2+\left(y+\frac{m}{2\pi C}\right)^2=\left(\frac{m}{4\pi C}\right)^2.$$

The streamlines are thus a family of circles with radii equal to $m/4\pi C$ and with centers at $(0, -m/4\pi C)$ on the y axis. Also, when $y=0$, $x=0$ for all values of C. The circular streamlines must all pass through the center of the doublet. The equation of the equipotential lines is

$$\frac{m}{2\pi}\frac{x}{x^2+y^2}=B \qquad \text{(a constant)}$$

or

$$y^2+x^2-\frac{m}{2\pi B}x=0.$$

Thus these are also circles touching the origin, but with centers on the x axis, as shown in Fig. 4.14.

Since the doublet was formed from a source on the negative x axis and a sink on the positive x axis, the flow must be clockwise in the upper half-plane and anticlockwise in the lower half-plane.

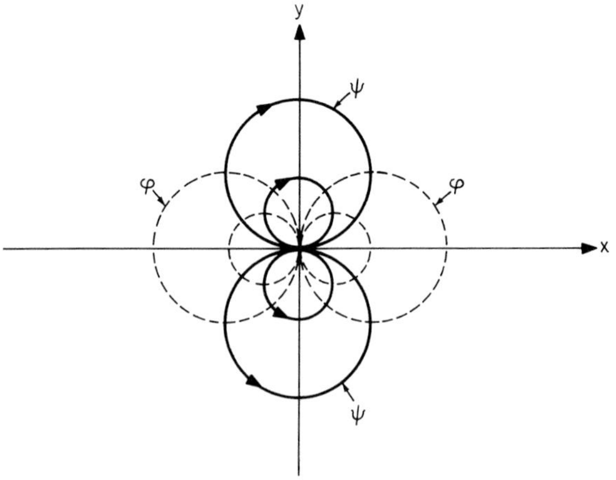

Fig. 4.14

4.10 Superposition of Flows

4.10.1 Uniform Flow and a Doublet

Uniform flow and a doublet will be seen to represent the flow past a cylinder with no circulation around the cylinder.

The uniform flow has a velocity U_∞ in the direction of the positive x axis and the doublet is placed at the origin of the coordinate system

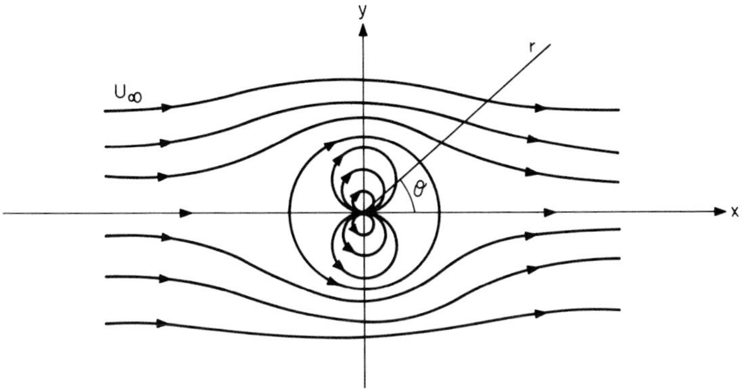

Fig. 4.15

(Fig. 4.15). In polar coordinates, the stream function for the combined system is

$$\psi = U_\infty r \sin \vartheta - \frac{m \sin \vartheta}{2\pi r}, \tag{4.96}$$

which is merely the sum of Eqs. (4.87) and (4.94).

The equation of the streamlines for the combined flow is then

$$U_\infty r \sin \vartheta - \frac{m \sin \vartheta}{2\pi r} = C \qquad \text{(a constant).} \tag{4.97}$$

The pattern of streamlines shown in Fig. 4.15 is obtained by choosing various values for C.

The zero streamline ($C = 0$) gives the equation

$$\left(U_\infty - \frac{m}{2\pi r^2}\right) \sin \vartheta = 0. \tag{4.98}$$

This equation is satisfied if

$$\begin{aligned} &\text{(i)} \quad \vartheta = 0 \quad \text{or} \quad \pi, \\ &\text{(ii)} \quad r = \left(\frac{m}{2\pi U_\infty}\right)^{1/2} = R. \end{aligned} \tag{4.99}$$

The ψ_0 streamline is thus a circle of radius R, along with the x axis. As there can be no flow *across* a streamline, this circle can be considered as the circular cylinder.

Substituting for the value of m from Eq. (4.99) in Eq. (4.96), one obtains

$$\psi = U_\infty\left(r - \frac{R^2}{r}\right) \sin \vartheta. \tag{4.100}$$

Similarly, it can be shown that the velocity potential for the combined flow is

$$\varphi = U_\infty\left(r + \frac{R^2}{r}\right) \cos \vartheta. \tag{4.101}$$

It can easily be shown that φ and ψ obey Laplace's equation.

The components of fluid velocity are

$$\begin{aligned} v_\vartheta &= -\frac{\partial \psi}{\partial r} = -U_\infty\left(1 + \frac{R^2}{r^2}\right) \sin \vartheta, \\ v_r &= \frac{1}{r} \frac{\partial \psi}{\partial \vartheta} = U_\infty\left(1 - \frac{R^2}{r^2}\right) \cos \vartheta. \end{aligned} \tag{4.102}$$

The boundary conditions on the velocity are (Fig. 4.16)

$$\begin{aligned} &v_r = \frac{\partial \varphi}{\partial r} = 0 && \text{at} \quad r = R, \\ &u = U_\infty \quad \text{and} \quad v = 0 && \text{as} \quad r \to \infty. \end{aligned} \tag{4.103}$$

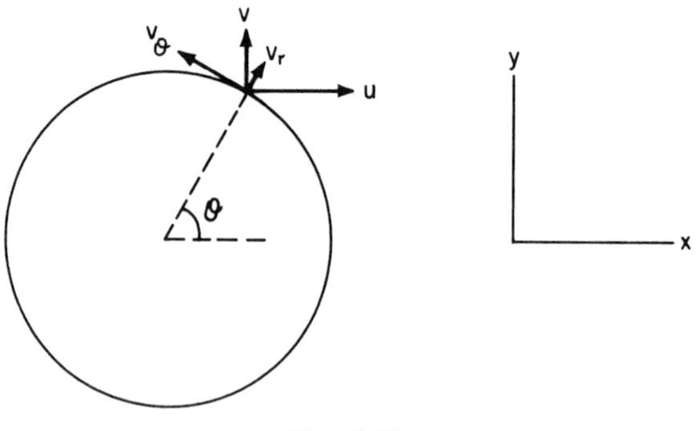

Fig. 4.16

At $r = R$,

$$v_\vartheta \,|_R = -2U_\infty \sin \vartheta. \tag{4.104}$$

and is in the clockwise direction. The values of v_ϑ at $\vartheta = 0$ and π are zero and these are called ***stagnation points***. The positions of maximum velocity are at $\vartheta = \pm\pi/2$ and the value is

$$|\, v_\vartheta \,|_{\max} = 2U_\infty. \tag{4.105}$$

The pressure at any point on the cylinder surface is obtained by use of the Bernoulli equation for steady flow:

$$p + \frac{\varrho}{2} v_\vartheta^2 = p_\infty + \frac{\varrho}{2} U_\infty^2. \tag{4.106}$$

Using Eq. (4.104) for v_ϑ, one has

$$p - p_\infty = \frac{\varrho U_\infty^2}{2} (1 + 2 \sin^2 \vartheta). \tag{4.107}$$

The position of maximum pressure is thus for $\vartheta = 0$ and π and is given by

$$p - p_\infty \,|_{\max} = \tfrac{1}{2}\varrho U_\infty^2. \tag{4.108}$$

The minimum pressure points are for $\vartheta = \pi/2$ and $3\pi/2$ and the value is

$$p - p_\infty \,|_{\min} = -\tfrac{3}{2}\varrho U_\infty^2. \tag{4.109}$$

The positions of maximum pressure are obviously the stagnation points.

4.10.2 Lift and Drag on the Cylinder

The *lift* and *drag* will be defined as forces per unit length on the cylinder in directions perpendicular and parallel, respectively, to the uniform flow. The lift force per unit length is given by

$$L = \int_0^{2\pi} -(p - p_\infty) R \, d\vartheta \sin \vartheta. \tag{4.110}$$

Using Eq. (4.107), one sees that the lift is zero. Likewise, the drag force per unit length is given by

$$D = \int_0^{2\pi} (p - p_\infty) R \, d\vartheta \cos\vartheta$$

$$= \int_0^{2\pi} - \left[p_\infty + \varrho \frac{U_\infty{}^2}{2} - \frac{\varrho}{2} (2U_\infty \sin\vartheta)^2 \right] R \cos\vartheta \, d\vartheta$$

$$= 0. \tag{4.111}$$

This result for drag on the cylinder is in complete contradiction to experimental observation. This was first noted by d'Alembert and is known as the *d'Alembert paradox*. The reason lies in the assumption of irrotational flow in the whole flow field. It is known that even in fluids of low viscosity (such as water and air), the frictional effect of the flowing fluid in the region close to a solid surface cannot be neglected. The viscous action in the boundary layer becomes important.

Despite this weakness of irrotational flow theory, it has been useful in some areas of aerodynamics.

4.10.3 Uniform Flow, Doublet, and a Vortex

This combination is found to represent the flow past a cylinder with circulation. The stream function and velocity potential are, respectively,

$$\psi = U_\infty \left(r - \frac{R^2}{r^2} \right) \sin\vartheta + \frac{\varkappa}{2\pi} \ln r, \tag{4.112}$$

$$\varphi = U_\infty \left(r + \frac{R^2}{r^2} \right) \cos\vartheta - \frac{\varkappa}{2\pi} \vartheta. \tag{4.113}$$

The velocity components are therefore,

$$\begin{aligned} v_\vartheta &= -\frac{\partial\psi}{\partial r} = -U_\infty\left(1 + \frac{R^2}{r^2}\right) \sin\vartheta - \frac{\varkappa}{2\pi r}, \\ v_r &= \frac{1}{r}\frac{\partial\psi}{\partial\vartheta} = U_\infty\left(1 - \frac{R^2}{r^2}\right)\cos\vartheta, \end{aligned} \tag{4.114}$$

where $\varkappa$ is the strength (circulation) of the vortex.

It is seen that the radial velocity component is not changed by the circulation.

At the surface of the cylinder ($r = R$),

$$v_\vartheta = -2U_\infty \sin\vartheta - \frac{\varkappa}{2\pi R}. \tag{4.115}$$

At the stagnation points, $v_\vartheta = 0$, and so

$$\sin(-\vartheta_S) = \frac{\varkappa}{4\pi R U_\infty} = \sin(\pi + \vartheta_S), \tag{4.116}$$

where $\varkappa/4\pi R U_\infty$ is less than or equal to zero.

When $\varkappa = 4\pi R U_\infty$, the two stagnation points coincide at $r = R$ and $\vartheta = -\pi/2$. (Fig. 4.17). When $\varkappa$ is greater than this, the stagnation point moves away from the cylinder surface. This can be shown as follows.

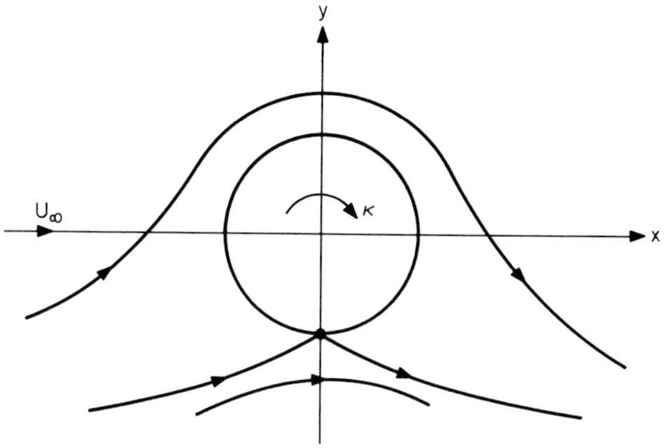

Fig. 4.17

The condition for a stagnation point away from the cylinder is that *both* velocity components be zero. Therefore

$$\begin{aligned} 0 &= -U_\infty\left(1 + \frac{R^2}{r^2}\right)\sin\vartheta - \frac{\varkappa}{2\pi r}, \\ 0 &= U_\infty\left(1 - \frac{R^2}{r^2}\right)\cos\vartheta. \end{aligned} \tag{4.117}$$

The second equation implies that $\vartheta = \pi/2$ or $3\pi/2$. But the first equation can be satisfied only if $\vartheta = 3\pi/2$, since $\varkappa$ cannot be negative. Using this value,

$$U_\infty\left(1 + \frac{R^2}{r^2}\right) = \frac{\varkappa}{2\pi r}.$$

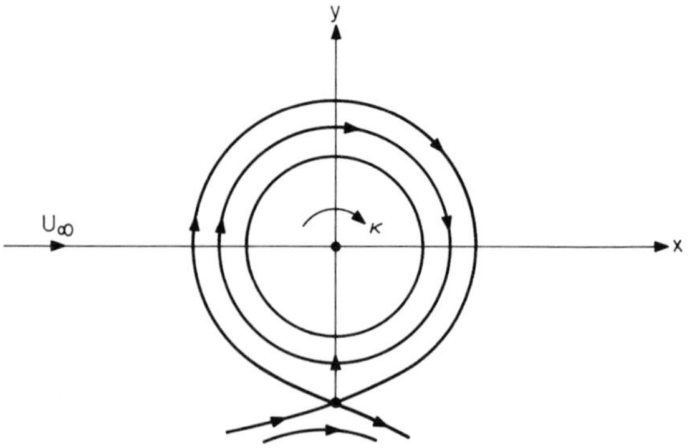

Fig. 4.18

The values of r at which the stagnation points occur are thus given by

$$r = \frac{\varkappa}{4\pi U_\infty} + \frac{1}{2}\left[\left(\frac{\varkappa}{2\pi U_\infty}\right)^2 - 4R^2\right]^{1/2}. \tag{4.118}$$

Also, the points must lie along the negative y axis (Fig. 4.18).

For smaller values of $\varkappa$ the stagnation points lie on the cylinder, and the position for any R, $\varkappa$, and U_∞ is obtained by solving Eq. (4.116) for ϑ_S (Fig. 4.19).

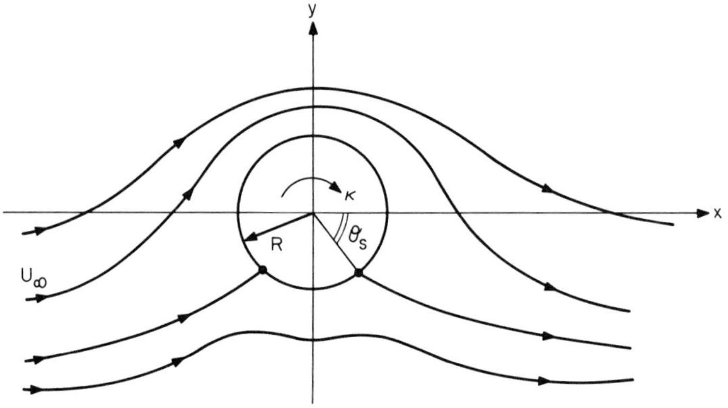

Fig. 4.19

4.11 Drag and Lift for the Cylinder with Circulation

At large distances from the cylinder, the velocity and the pressure p_∞ in the approaching stream are uniform. The equation for the pressure p_S at any point on the surface of the cylinder is obtained from Bernoulli's equation:

$$p_S = p_\infty + \tfrac{1}{2}\varrho U_\infty^2 - \tfrac{1}{2}\varrho U_S^2. \tag{4.119}$$

From Eq. (4.114), the tangential velocity at the cylinder surface, which is U_S, is

$$(v_\vartheta)_S = U_S = -2U_\infty \sin\vartheta - \frac{\varkappa}{2\pi R}.$$

Using this in Eq. (4.119), one obtains

$$p_S = p_\infty + \frac{1}{2}\varrho\left[U_\infty^2 - \left(2U_\infty \sin\vartheta + \frac{\varkappa}{2\pi R}\right)^2\right]. \tag{4.120}$$

This pressure is normal to the surface of the cylinder and produces an elemental force $p_S R\, d\vartheta$ on an elementary area $R d\vartheta$ per unit length of the cylinder. The total lift and drag forces are thus

$$L = \int_0^{2\pi} - p_S R \sin\vartheta\, d\vartheta, \tag{4.121}$$

$$D = \int_0^{2\pi} - p_S R \cos\vartheta\, d\vartheta. \tag{4.122}$$

Because of the symmetry of the flow pattern about a vertical axis through the cylinder center, the integral for D vanishes, and so irrotational flow theory predicts zero drag again in this case.

By substituting for p_S from Eq. (4.120), the lift L, per unit length, can be determined:

$$L = \int_0^{2\pi}\left[\frac{\varrho U_\infty^2}{2} - \frac{\varrho}{2}\left(2U_\infty \sin\vartheta + \left(\frac{\varkappa}{2\pi R}\right)^2\right]\sin\vartheta\, R\, d\vartheta.$$

Now

$$\int_0^{2\pi} \sin\vartheta\, d\vartheta = \int_0^{2\pi} \sin^3\vartheta\, d\vartheta = 0.$$

Then

$$L = \frac{\varrho U_\infty \varkappa}{\pi}\int_0^{2\pi} \sin^2\vartheta\, d\vartheta = \varrho U_\infty \varkappa. \tag{4.123}$$

This is known as the Kutta–Joukowski theorem.

The lift is thus directly proportional to

(a) the fluid density ϱ,
(b) the stream velocity U_∞,
(c) the circulation $\varkappa$.

By the use of the mathematics of complex variables and conformal mapping, it is possible to transform a two-dimensional irrotational flow about a circular cylinder to a flow about airfoil-type bodies. In this way, the Kutta–Joukowski theorem is used also to compute the lift on airfoils.

The lift force produced by circulation of fluid about the cylinder can also be produced by rotation of the cylinder in a fluid stream. In fact, this can be demonstrated experimentally. This phenomenon was first observed by the German scientist Magnus in 1852. It is usually referred to as the Magnus effect. The curved trajectory of a spinning baseball is a typical consequence of the effect.

4.12 Solution by Change in Variable—Conformal Mapping

A powerful method in the solution of partial differential equations involves the linear combination of the original variable into new variables which are more suitable for solution. In the special applications of fluid dynamics its use leads to the conformal mapping technique.

The complex potential for a plane flow has been defined as

$$\Omega = \varphi + i\psi = f(x + iy) = f(z). \tag{4.124}$$

Suppose that a solution to the Laplace equation is known for certain boundary conditions. Now consider a new plane defined by the variables (r, s), where $t = r + is$, in which a solution to the Laplace equation is to be determined. Suppose F is the mapping function between the z and t planes, so that

$$t = F(z) \tag{4.125}$$

or, conversely,

$$z = G(t). \tag{4.126}$$

The transformation must satisfy the Cauchy–Riemann relations, and thus if the original function $f(z)$ is analytic in the z plane, the new function $f_1(t)$ must be analytic in the t plane.

The transformation ensures that $f_1(t)$ satisfies the boundary conditions in the t plane, just as $f(z)$ did in the t plane. Because of the uniqueness theorem, there can be only one analytic function which satisfies the boundary conditions in the t plane. The solution must be the required harmonic function.

As the vanishing of the Laplacian is invariant under the transformation, angles are preserved from one plane to the other and thus the title, conformal transformation.

The transformation also conserves the character of hydrodynamic singularities, that is, a source will remain a source, a sink a sink, and so on. That other singularities transform into similar singularities follows from the fact that they are all derivable from the source or sink. In some transformations, mathematical singularities are produced by the transformation. In hydrodynamics, however, these represent stagnation points and are not hydrodynamic singularities in the accepted sense.

Finally, it can be easily demonstrated that the kinetic energy and vorticity of a flow system are invariant to a conformal transformation.

Consider an area A_w in the w plane, which is transformed to an area A_z in the z plane by the conformal transformation $z = f(w)$. Let q be the fluid velocity and ϱ its density; then the kinetic energy per unit length in the direction normal to the plane is

$$T = \int_A \frac{\varrho q^2}{2}\, dA = \frac{\varrho}{2} \int_A q^2\, dA,$$

and ϱ is assumed constant; or

$$\frac{2T}{\varrho}\bigg|_w = \int_{A_w} q_w^{\,2}\, dA_w \qquad \text{and} \qquad \frac{2T}{\varrho}\bigg|_z = \int_{A_z} q_z^{\,2}\, dA_z.$$

Now

$$|\, q_w \,| = \left| \frac{d\Omega}{dw} \right|, \qquad |\, q_z \,| = \left| \frac{d\Omega}{dz} \right|,$$

where Ω is the complex potential. Since

$$\frac{d\Omega}{dz} = \frac{d\Omega}{dw}\frac{dw}{dz},$$

hence

$$q_w^{\,2} = q_z^{\,2}\left(\frac{dz}{dw}\right)^2.$$

Now the differential areas dA_w and dA_z must be similar, and their corresponding elements are changed in the linear ratio $|\, dw/dz \,|$. Hence

$$dA_w = \left(\frac{dw}{dz}\right)^2 dA_z,$$

and so

$$q_w^2\, dA_w = q_z^2\left(\frac{dz}{dw}\right)^2\left(\frac{dw}{dz}\right)^2 dA_z = q_z^2\, dA_z.$$

Since the domains in z and w are similar, the kinetic energy of the flow pattern remains unchanged on transformation.

It can also easily be shown that

$$\int\left(\frac{\partial^2\psi}{\partial x^2} + \frac{\partial^2\psi}{\partial y^2}\right) dS = \text{constant},$$

and since the integrand is the vorticity and this is true for arbitrary area elements, thus vorticity is also invariant to conformal transformation.

4.12.1 Examples of Simple Transformations

(1) $z = \cos t$ The transformation of a rectangular network in the t plane into the corresponding set of orthogonal lines in the z plane is sketched in Fig. 4.20. Now $z = x + iy$ and $t = r + is$. Therefore

$$\begin{aligned} z = x + iy &= \cos(r + is) \\ &= \cos r \cos is - \sin r \sin is \\ &= \cos r \cosh s - \sin r \sinh s \end{aligned}$$

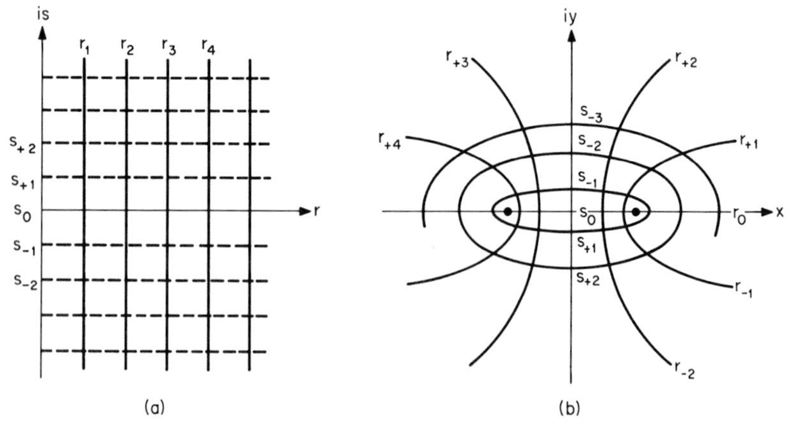

Fig. 4.20 (a) t plane; (b) z plane.

or

$$x = \cosh s \cos r, \qquad y = \sinh s \sin r.$$

Thus, lines of constant s become the ellipses

$$\frac{x^2}{\cosh^2 s} + \frac{y^2}{\sinh^2 s} = 1,$$

and lines of constant r become the hyperbolas

$$\frac{x^2}{\cos^2 r} - \frac{y^2}{\sin^2 r} = 1,$$

both with foci at $z = \pm 1$ (corresponding to $t = 0, \pi$). These are singular points, since $dz/dt = -\sin t = 0$.

The strip from $r = 0$ to $r = \pi$ in the t plane is mapped into the entire z plane. Every similar strip in the t plane requires an entire plane for mapping.

Hydrodynamically, the pattern in the z plane can be regarded as representing the circulatory flow about an ellipse of length greater than 2, or about a flat plate of length 2. Then the t plane would be an Ω plane with the orthogonal lines representing those of constant φ and ψ. The s lines in Figs. 4.20a and b are then streamlines.

Another hydrodynamic interpretation is possible if the t plane was considered to be an $i\Omega$ plane. The r lines are then streamlines and the pattern in the z plane then represents the flow through a plane orifice (or slot) of width 2. It could also represent the flow from a two-dimensional line source located on the x axis between $x = \pm 1$.

(2) *Uniform flow* The complex potential for uniform flow is

$$\Omega = U_\infty z$$

since

$$\varphi = U_\infty x \qquad \text{and} \qquad \psi = U_\infty y.$$

Actually U_∞ is constant, and it can be complex, real, or pure imaginary. Suppose U_∞ is the conjugate of a complex number $a + ib$; that is, $U_\infty = a - ib$. Then

$$\frac{d\Omega}{dz} = U_\infty = a - ib = Re^{-i\alpha},$$

where

$$R = (a^2 + b^2)^{1/2} \qquad \text{and} \qquad \alpha = \tan^{-1}\frac{b}{a}.$$

Therefore

$$\delta z = \frac{dz}{d\Omega}\,\delta\Omega = \frac{1}{R}\,e^{i\alpha}\,\delta\Omega.$$

This indicates that elements in the Ω plane are multiplied by a constant factor $1/R$ and rotated through an angle α when transformed into the z plane. So the flow network in the z plane is that of a uniform flow at an angle α to the positive x axis (Fig. 4.21). Now

$$\frac{d\Omega}{dz} = U_\infty = a - ib = u - iv.$$

Therefore

$$a = u \qquad \text{and} \qquad b = v.$$

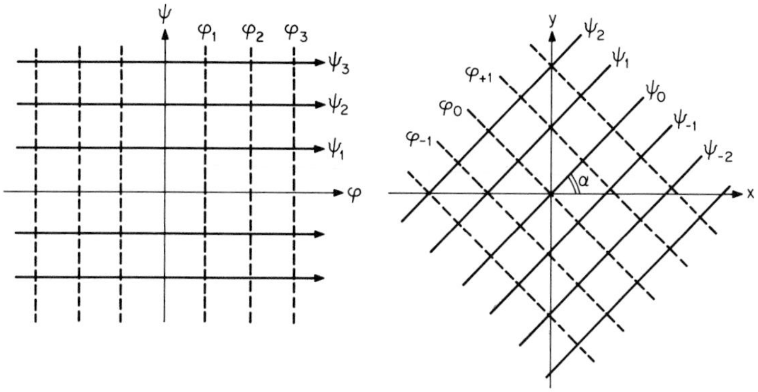

Fig. 4.21

When U_∞ is real, $\alpha = 0$. The flow network in the z plane is that of the Ω plane scaled up by the factor $1/R$.

When U_∞ is pure imaginary, the flow network in the Ω plane is rotated through 90° when transformed onto the z plane.

As $d\Omega/dz = U_\infty$ (a constant), there are no singular points in the plane.

(3) *Elementary foil theory* A very important class of problems treated with the help of conformal transformation is that of plane two-dimensional flow about airfoils, or hydrofoils, in an unconfined fluid. Many useful devices, such as wings, propellors, turbines, and fans, use airfoils as their principal element.

N. E. Joukowsky in 1910 developed a special transformation for airfoil shapes. It can be written as

$$z = t + \frac{a^2}{t}. \tag{4.127}$$

This transforms a circle of radius a into a foil. There are two analytic singularities in the inverse transform at $t = \pm a$, since

$$\frac{dt}{dz} = \left(1 - \frac{a^2}{t^2}\right)^{-1}$$

becomes infinite there. The singularities obviously occur either in or on the perimeter of the circle in t plane. A foil with a sharp trailing edge is obtained by passing the circle being transformed through one of these singular points, which point thus becomes the trailing edge.

Foil theory will not be developed here and the reader is referred to other textbooks for such material. However, the case of a degenerate foil, or flat plate, will be considered.

(i) *Plate parallel to the stream* Direct application of the Joukowsky transformation to the unit circle [$a = 1$ in Eq. (4.127)],

$$z = t + \frac{1}{t}, \tag{4.128}$$

produces the flow around a flat plate of length $4a = 4$ lying on the x axis at the origin (Fig. 4.22).

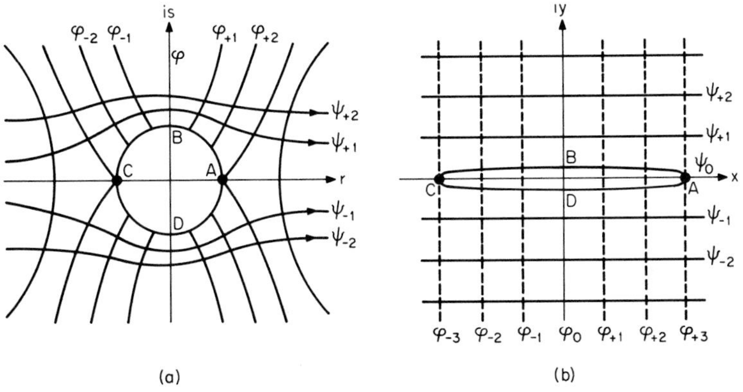

Fig. 4.22 (a) t plane; (b) z plane.

The flow pattern is seen to consist of straight streamlines parallel to the x axis. The plate is infinitely thin, in fact, and is itself a streamline.

One can verify the transformation by considering the points A, B, and C in the t plane. The coordinates are $(+1, 0)$, $(0, +1)$ and $(-1, 0)$, respectively. These transform to points $(+2, 0)$, $(0, 0)$, and $(-2, 0)$, or A, B, and C, respectively, in the z plane. In polar coordinates, the contour in the t plane is given by

$$t = e^{i\vartheta}.$$

It becomes $2 \cos \vartheta$ in the z plane. Now,

$$\Omega_t = U\left(t + \frac{1}{t}\right)$$

and so $\Omega_z = Uz$, the complex potential in the z plane.

(ii) *Plate normal to the stream* Multiplication of the flow system above by i will result in a rotation by $\pi/2$. Thus, use of the Joukowsky transformation and i produces the flow about a plate normal to the uniform stream.

It is basically a succession of three transformations (Fig. 4.23). The first transformation, $\zeta = it$, rotates the known flow pattern about the circle in a stream in the positive x direction to that for a stream in the positive y direction. Application of the Joukowsky transformation leads to a plate along the x axis with flow in the positive y direction. The transformation $z = -iw$ rotates the whole pattern by $\pi/2$ into the desired position.

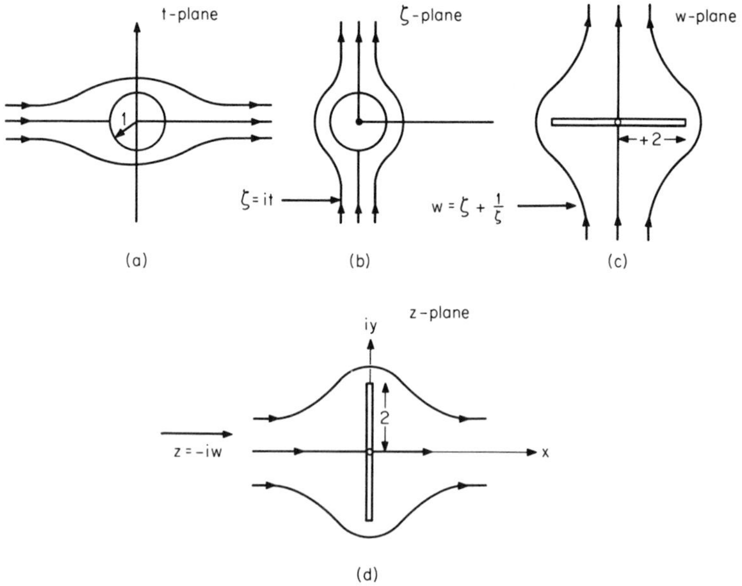

Fig. 4.23 (a) t plane; (b) ζ plane; (c) w plane; (d) z plane.

Combining these successive transformations gives the overall result:

$$z = t - \frac{1}{t}.$$

This is merely the Joukowsky transformation with $a = i$.

The result is of little or no practical value, since separation of the flow will occur at the corners of the plate. The concept of free streamlines must then be introduced.

In the next chapter on Vortex Dynamics, the restrictive properties of incompressibility and zero viscosity will again be resorted to in the interests of simplification. However, as study of ideal fluids is of limited interest today, no further consideration will be given in this text to the solution of ideal fluid flow equations.

Having developed the general equations of real fluid flow in this chapter, the student will now be in a position to proceed to study the problems of fluid flows with vorticity and viscosity.

PROBLEMS

4.1 Show that the relation between the components e_{ij} of the rate-of-strain tensor and p_{ij} of the stress tensor in a viscous fluid is of the form

$$p_{ij} = (-p' + \tfrac{1}{2}\lambda e_k e_k)\,\delta_{ij} + \mu e_{ij},$$

referred to any set of Cartesian axes. Then write down the momentum equations for an incompressible viscous fluid.

Show that for an incompressible flow the vorticity $\boldsymbol{\zeta}$ satisfies the equation

$$\frac{\partial \boldsymbol{\zeta}}{\partial t} - (\nabla \times \mathbf{v}) \times \boldsymbol{\zeta} = \frac{\mu}{\varrho} \nabla^2 \boldsymbol{\zeta},$$

where μ is the viscosity and ϱ is the fluid density.

4.2 The velocity components for an incompressible steady flow with constant viscosity are

$$u(y) = y\frac{U_\infty}{h} + \frac{h^2}{2\mu}\left(-\frac{dp}{dx}\right)\frac{y}{h}\left(1 - \frac{y}{h}\right), \qquad v = w = 0.$$

If there is no body force, does $u(y)$ satisfy the equation of motion, given that h, U_∞, and dp/dx are constants and $p = p(x)$?

4.3 In a simple Couette flow, the temperatures of the stationary and moving plates are T_w and T_∞, respectively (both constant). The velocity and temperature distributions in the field are

$$u = \frac{U_\infty y}{h}, \qquad v = 0, \qquad p = \text{constant},$$

$$\frac{T - T_w}{T_\infty - T_w} = \frac{y}{h} + \frac{\mu U^2}{2k(T_\infty - T_w)}\left(\frac{y}{h}\right)\left(1 - \frac{y}{h}\right),$$

where μ, h, and k are constants. Show that these distributions are solutions of the energy equation.

4.4 Show that the equation of motion,

$$\frac{\partial u}{\partial t} = \nu \frac{\partial^2 u}{\partial y^2},$$

where ν is a constant, is satisfied by the velocity field

$$u(\eta) = U_\infty\left[1 - \frac{2}{\sqrt{\pi}} \int_0^\eta \exp(-\eta^2)\, d\eta\right],$$

where $\eta = y/2(\nu t)^{1/2}$ (dimensionless) and U_∞ is a constant.

4.5 Plane Poiseuille flow satisfies the boundary conditions $T = T_0$ for $y = \pm h/2$ and has the following velocity and temperature distributions:

$$u(y) = U_{\max}\left[1 - \left(\frac{2y}{h}\right)^2\right], \qquad T(y) = T_0 + \mu \frac{U_{\max}^2}{3k}\left[1 - \left(\frac{2y}{h}\right)^4\right].$$

Show that the energy equation is satisfied.

4.6 Using the Navier–Stokes equations, show that for a planar incompressible laminar viscous flow the stream function $\psi = \psi(x, y, t)$ satisfies the following equation:

$$\nabla^2 \frac{\partial \psi}{\partial t} + \frac{\partial \psi}{\partial y} \nabla^2 \frac{\partial \psi}{\partial x} - \frac{\partial \psi}{\partial x} \nabla^2 \frac{\partial \psi}{\partial y} = \frac{\mu}{\varrho} \nabla^4 \psi,$$

where

$$\nabla^4 \equiv \nabla^2(\nabla^2) \equiv \frac{\partial^4}{\partial x^4} + \frac{\partial^4}{2\, \partial x^2\, \partial y^2} + \frac{\partial^4}{\partial y^4}.$$

4.7 The velocity field of a viscous incompressible flow is given by

$$\mathbf{v} = 8x^2 z\mathbf{i} - 10y^2 z^2\mathbf{j} + (4yz^3 - 10xz^2)\mathbf{k}.$$

If the fluid density is 1.1 gm/cm³, the fluid viscosity 2.1 centipoise, and the negative z axis the direction in which gravity acts, calculate the pressure gradient at the point (0.3, 0.6, 1).

4.8 Oil of kinematic viscosity 0.07 kg/m s flows through a 25-mm-diameter pipe with a mean velocity of 1.0 m/s. Determine the pressure drop over a 60-m-length of pipe.

4.9 A flat plate is moved suddenly with a velocity U_∞ in oil, which has a density $\varrho = 900$ kg/m³ and a viscosity $\mu = 0.08$ kg/m s. Determine the velocity ratio u/U_∞ as a function of vertical distance from the plate at times $t = 0.01$, 0.05, and 0.15 second.

4.10 For plane Couette flow, determine the vorticity distribution across the space if the moving plate has a velocity U and the plates are spaced h apart. Determine the variation in the Bernoulli constant and in the pressure p across this flow.

4.11 A two-dimensional irrotational flow is to be represented by the function

$$\varphi = \frac{k}{2}(x^2 - y^2).$$

If the pressure is zero at the point $r = 2\mathbf{i} + 4\mathbf{j}$, determine the maximum pressure (point of zero velocity) in the flow field.

4.12 The velocity potential for a three-dimensional axially symmetric flow is

$$\varphi = \frac{c}{r^2}\cos\vartheta,$$

where c is constant. Derive the Stokes stream function for such a flow.

4.13 Sketch the streamlines and equipotential lines in the z plane for the following complex potentials:

(i) $w(z) = z^2 + 3z$,
(ii) $w(z) = 2\ln z + 3i\ln z$.

4.14 Determine the velocity at $z = 6 + 8i$ in the z plane for the following potentials:

(i) $w(z) = z + \ln z$,
(ii) $w(z) = 2(i/z) + 3z^2$,
(iii) $w(z) = iz^2 + 4z$.

4.15 A uniform flow past a half-body can be represented by the combination of a uniform flow and a source flow. Find the complex potential for a uniform flow with $\mathbf{v} = \mathbf{v}_0\mathbf{j}$ and a two-dimensional source of strength q located at the origin.

Determine an expression for the velocity on the surface of the body and locate the stagnation point in the field.

4.16 A hemisphere of radius R and density ϱ_c rests on a flat plate over which a rectilinear flow with velocity U_∞ moves. The gas density is ϱ.

Note By Archimedes' principle the buoyant force equals the weight of displaced fluid. What must the density of the hemisphere be if it is to remain on the plate?

4.17 Write the successive conformal transformations necessary to obtain a flow in the z plane which represents a uniform flow with speed U past a flat plate with circulation $\varkappa$ around the plate. The plate length is $6a$ and the flow approaches it at an angle of attack α.

BIBLIOGRAPHY

S. W. YUAN, "Foundations of Fluid Mechanics," Chapters 5–8. Prentice-Hall, Englewood Cliffs, New Jersey, 1970.

J. OWCZAREK, "Introduction to Fluid Mechanics," Chapters 5–7. International Textbook Co., Scranton, Pennsylvania, 1968.

G. K. BATCHELOR, "Fluid Dynamics," Chapter 3. Cambridge Univ. Press, London and New York, 1967.

R. S. BRODKEY, "The Phenomena of Fluid Motions," Chapter 4. Addison-Wesley, Reading, Massachusetts, 1967.

CHAPTER FIVE *VORTEX DYNAMICS*

It is appropriate at this stage to study vortex motions. These can often be treated by the same analytic methods used for the potential and stream functions as long as the singular point, line, or surface in which the vorticity is concentrated is excluded from the analysis. To avoid the singularity, it is frequently postulated that the vortex core is finite and characterized by solid body rotation.

The effect of viscous action is to cause an increase in the core size and an attenuation of the vortex strength as measured by the circulation it induces.

The analytic treatment of vortex motions started with Hermann von Helmholtz's classic paper "On Integrals of the Hydrodynamic Equations Corresponding to Vortex Motions" in 1858. Later Kelvin, in 1869, demonstrated the necessity of Helmholtz's theorems for the existence of vortex motions. The material presented here is drawn from the presentation of authors such as Lamb and Basset.

A fundamental theorem due to Stokes in 1849 showed that a velocity field can be decomposed into the sum of an irrotational and a rotational field, as long as the vectors and their derivatives vanish at infinity. Thus,

$$\mathbf{v} = \mathbf{v}_i + \mathbf{v}_r = \nabla\varphi + \nabla \times \mathbf{A}, \tag{5.1}$$

where φ is a scalar potential and $\mathbf{A}$ is a vector potential.

It is stipulated that $\nabla \cdot \mathbf{A} = 0$ so that the following interesting relation results:

$$\nabla \times \mathbf{v} = \nabla \times (\nabla \times \mathbf{A}) = \nabla(\nabla \cdot \mathbf{A}) - \nabla^2\mathbf{A} = -\nabla^2\mathbf{A} = \boldsymbol{\zeta}, \quad (5.2)$$

where $\boldsymbol{\zeta}$ is the vorticity. Moreover,

$$\vartheta = \nabla \cdot \mathbf{v} = \nabla^2\varphi + \nabla \cdot (\nabla \times \mathbf{A}) = \nabla^2\varphi = 0 \quad (5.3)$$

as the rotational field obeys the continuity equation and so its divergence is zero. The ϑ is the dilation in the field, or divergence.

The correspondence between Eqs. (5.2) and (5.3) is obvious and both are forms of the Poisson equation.

5.1 Vortices and Circulation

Rotation and vorticity have already been defined in Chapter 4.

$$\boldsymbol{\zeta} = 2\boldsymbol{\omega} = \nabla \times \mathbf{v}. \quad (5.4)$$

For a planar flow with rotation about the z axis,

$$\zeta_z = 2\omega_z = \left(\frac{\partial v_y}{\partial x} - \frac{\partial v_x}{\partial y}\right). \quad (5.5)$$

The concept of circulation was introduced by Thompson (Lord Kelvin) in 1869. It is defined as the line integral of the tangential component of velocity around a closed contour C. Thus,

$$\text{circulation} = \Gamma = \int_C \mathbf{v} \cdot dl = \oint \mathbf{v} \cdot dl. \quad (5.6)$$

The circulation is a scalar, but has directional properties in so far that its sign depends on the sense of the circulation. By application of Stokes's theorem,

$$\Gamma = \int_C \mathbf{v} \cdot dl = \int_S \mathbf{n} \cdot \nabla \times \mathbf{v}\, dS = \int_S \zeta_n\, dS, \quad (5.7)$$

where $\mathbf{n}$ is the unit normal vector to a surface S bounded by the contour C, and ζ_n is the component of vorticity normal to the surface at any point.

Equation (5.7) is called Kelvin's relation. Its differential form is

$$\zeta_n = \frac{d\Gamma}{dS}. \tag{5.8}$$

Thus, the circulation of the velocity about any closed curve is equal to the surface integral of the normal component of the vorticity over any surface which is bounded by that curve. The relationship between the vorticity and the stream function was established in Eqs. (3.45) and (3.46). For a plane flow in the xy plane,

$$\zeta_z = -\frac{\partial^2 \psi}{\partial x^2} - \frac{\partial^2 \psi}{\partial y^2} = -\nabla^2 \psi. \tag{5.9}$$

In the case of axisymmetric motion,

$$-r\zeta_\varphi = \frac{\partial^2 \psi_S}{\partial r^2} - \frac{1}{r}\frac{\partial \psi_S}{\partial r} + \frac{\partial^2 \psi_S}{\partial z^2}, \tag{5.10}$$

where ψ_S is the Stokesian stream function.

5.2 Kelvin's Circulation Theorem

This states that the circulation around a closed path containing a given set of fluid particles remains constant in time as long as the external force field acting is conservative. In other words, in ideal fluids, vortex strength is constant in time, and it remains so for certain fluid particles. It is a fluid-bound property. Fluid particles in irrotational flow cannot acquire vorticity (unless nonconservative forces act) and vice versa. Thus, rotational and irrotational flow regions in ideal-fluid motion are mutually exclusive.

Proof of the theorem The substantive derivative of the circulation is

$$\frac{D}{Dt}\Gamma = \frac{D}{Dt}\oint \mathbf{v}\cdot dl = \oint \frac{D\mathbf{v}}{Dt}\cdot dl + \oint \mathbf{v}\cdot\frac{D(dl)}{Dt}.$$

Now, using the Euler equation,

$$\frac{D\mathbf{v}}{Dt} = -\frac{\nabla}{\varrho}(p + V),$$

where p is the pressure of the fluid and V is the force potential. Therefore

$$\oint \frac{D\mathbf{v}}{Dt} \cdot dl = -\oint \frac{1}{\varrho} (\nabla p + \nabla V) \cdot dl = -\frac{1}{\varrho} (\delta p + \delta V)_C = 0$$

since the change in $p + V$ going around a closed path C is zero. Moreover,

$$\oint \mathbf{v} \cdot \frac{D(dl)}{Dt} = \oint \mathbf{v} \cdot d\left(\frac{Dl}{Dt}\right) = \oint \mathbf{v} \cdot d\mathbf{v} = \oint d\left(\frac{v^2}{2}\right).$$

Since the velocity is single-valued, this integral is also zero:

$$\frac{D\Gamma}{Dt} = 0 \qquad \text{or} \qquad \Gamma = \text{constant.}$$

Since it is the substantive derivative that is involved, this implies that the constancy of circulation is a fluid-bound property. The contour C moves with the particles. In the Euler sense, this would not be true.

As mentioned before, in the regions of high velocity gradients, the action of viscosity will invalidate this theorem. However, if such regions are localized in the flow field—for example, the boundary layer long a solid surface—the remaining large region of flow can be regarded as obeying the Kelvin theorem. Even in regions in which viscous effects are significant, over short enough time intervals the flow can be considered to obey the Kelvin theorem.

5.3 Helmholtz Vortex Theorems

These are closely linked to the Kelvin theorem. For irrotational fluid motion the vortex theorems state that:

(a) A vortex line or filament can neither start nor end in the fluid. Thus, it must appear either as a closed loop or it must terminate, or start, on a boundary.

(b) Vortex tubes (composed of vortex lines) always contain the same fluid particles and vice versa—the vorticity is "frozen" into the material.

(c) In a nonviscous fluid the strength of any vortex filament is constant in time.

The proofs of these theorems will not be given, because they follow directly from Kelvin's theorem.

It is interesting to note that because the simplest closed loop is a circular ring, this explains why vortex rings are so commonly observed.

5.4 Origins of Circulation

Consider again the substantive derivative of the vortex strength, or circulation:

$$\frac{D\Gamma}{Dt} = \oint \frac{D\mathbf{v}}{Dt} \cdot dl + \oint \mathbf{v} \cdot \frac{D(dl)}{Dt}.$$

The second term on the right-hand side is zero as before. Using the Navier–Stokes equation for viscous fluid motion, one obtains

$$\frac{D\mathbf{v}}{Dt} = \frac{\mathbf{X}}{\varrho} - \frac{\nabla p}{\varrho} + \nu \nabla^2 \mathbf{v} = \frac{\mathbf{X}}{\varrho} - \frac{\nabla p}{\varrho} + \nu(\nabla\vartheta - \nabla \times \boldsymbol{\zeta}),$$

where ϑ is the dilation of the fluid, $\boldsymbol{\zeta}$ the vorticity of the fluid, and

$$\nabla^2\mathbf{v} = \nabla(\nabla \cdot \mathbf{v}) - \nabla \times (\nabla \times \mathbf{v}) = \nabla\vartheta - \nabla \times \boldsymbol{\zeta}.$$

Then,

$$\frac{D\Gamma}{Dt} = \oint \frac{\mathbf{X}}{\varrho} \cdot dl - \oint \frac{\nabla p}{\varrho} \cdot dl + \nu \oint \nabla\vartheta \cdot dl - \nu \oint (\nabla \times \boldsymbol{\zeta}) \cdot dl.$$

Now, by Stokes's theorem,

$$\oint \nabla\vartheta \cdot dl = \int_S \nabla \times \nabla\vartheta \, dS = 0.$$

Therefore

$$\frac{D\Gamma}{Dt} = \oint \frac{\mathbf{X}}{\varrho} \cdot dl - \oint \frac{\nabla p}{\varrho} \cdot dl - \nu \oint (\nabla \times \boldsymbol{\zeta}) \cdot dl. \qquad (5.11)$$

This equation indicates that the circulation can result from three types of action:

(a) from nonconservative body forces, through the first term in Eq. (5.11), namely,

$$\oint \frac{\mathbf{X}}{\varrho} \cdot dl;$$

(b) from pressure–density forces, through the second term,

$$\oint \frac{\nabla p}{\varrho} \cdot dl;$$

(c) from vorticity diffusion via viscosity, as per the last term.

Coriolis forces and electromagnetic forces are typical nonconservative body forces. Coriolis forces are those responsible for large scale meteorological phenomena such as tornados.

A medium for which the density is a function of pressure only is called *barotropic*. One in which the density is not a function of pressure alone, in other words,

$$\varrho \neq f(p),$$

is called a *baroclinic* medium.

It is interesting to note that the equation of continuity and the three Euler equations of motion are not sufficient in this case to solve a flow problem. For the motion of baroclinic fluids, one has to take an extra factor into consideration—the flux of energy.

The atmosphere is normally baroclinic, and surfaces of equal pressure and equal density intersect at finite angles. Circulation transforms the fluid into a barotropic state. The theorem of V. Bjerknes (1900) must then be considered.

The second term on the right-hand side of Eq. (5.11) can be transformed as follows:

$$-\oint \frac{\nabla p}{\varrho} \cdot dl = \int_S \nabla \times \left(\frac{\nabla p}{\varrho}\right) \cdot d\mathbf{S}$$
$$= \int_S \left(\nabla\left(\frac{1}{\varrho}\right) \times \nabla p + \frac{1}{\varrho} \nabla \times \nabla p\right) \cdot d\mathbf{S}.$$

Since curl grad is zero, in the absence of viscous effects and body forces, Eq. (5.11) becomes

$$\frac{D\Gamma}{Dt} = -\int_S \nabla \frac{1}{\varrho} \times \nabla p \cdot d\mathbf{S}. \tag{5.12}$$

This is Bjerknes' theorem. When the pressure and density are uniquely related—as in a barotropic fluid—then grad $1/\varrho$ is parallel to ∇p, and so Kelvin's theorem is the result. Thus, circulation results from the action of pressure–density forces.

Equation (5.12) may also be written

$$\frac{d\Gamma}{dt} = -\oint_C \frac{1}{\varrho}\,dp. \tag{5.13}$$

This determines the variation of the circulation along a contour C in the fluid.

Let $w = 1/\varrho$ be the specific volume of the fluid. Surfaces for which w is constant are called *isosteric* surfaces. Those for which p is constant are called *isobaric* surfaces. When $\varrho = f(p)$—that is, a barotropic fluid—ϱ would have a constant value on an isobaric surface. The isobaric and isosteric surfaces coincide. If they do not, then these surfaces intersect each other. If isobaric surfaces at p values differing by unity are drawn, and similarly isosteric surfaces, then the entire fluid space will be broken into a number of tubes formed by two consecutive isobaric surfaces and two consecutive isosteric surfaces. These may be called isobaro–isosteric unit tubes.

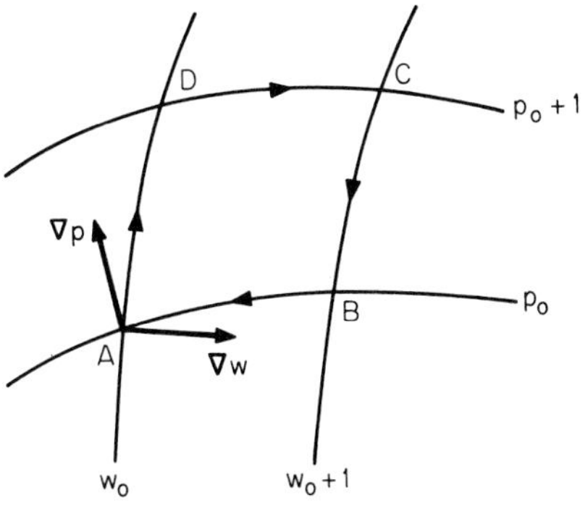

Fig. 5.1

Now consider the contour C and the number of unit isobaro–isosteric tubes it surrounds. Let one of these tubes be $ADCB$ as in Fig. 5.1. The value of the integral $-\oint_C w\,dp$ around $ADCB$ is computed as follows. On sides BA and DC, p is constant and so $dp = 0$. On side CB, $w = w_0 + 1$, while p varies from $p_0 + 1$ to p_0, and so

$$-\int_{CB} w\,dp = -(w_0+1)\int_{p_0+1}^{p_0} dp = w_0 + 1.$$

On side AD, $w = w_0$ while p varies from p_0 to $p_0 + 1$, and so

$$-\int_{AD} w\,dp = -w_0\int_{p_0}^{p_0+1} dp = -w_0.$$

Therefore

$$-\oint_C w\,dp = 1.$$

Note If the contour was followed in the opposite sense ($ABCD$), then

$$-\oint_C w\,dp = -1.$$

Thus, the integral $-\oint_C w\,dp$ around a single unit tube is ± 1, and if the contour surrounds N_+ positive tubes, or N_- negative tubes,

$$-\oint_C w\,dp = N_+, \qquad -\oint_C w\,dp = N_-, \tag{5.14}$$

respectively. If it surrounds a mixture of N_+ positive and N_- negative tubes, then

$$-\oint_C w\,dp = N_+ - N_-. \tag{5.15}$$

Thus, the integral $-\oint_C w\,dp$ represents the difference between the number of positive and negative isobaro–isosteric unit tubes crossing the contour C.

Equation (5.13) may be written

$$\frac{d\Gamma}{dt} = N_+ - N_-. \tag{5.16}$$

The intersection of isobaric and isosteric surfaces is, then, a reason for the formation of vortices.

Vorticity from the action of pressure/density forces can also be produced through the discontinuity surfaces of curved shock and flame fronts. A change in entropy occurs across a flame front, and this is accompanied by the appearance of vorticity. It is related to the ratio of stagnation temperatures and the radius of curvature by the following relation which was derived in 1951 by H. S. Tsien:

$$\zeta = \frac{q}{R}\left(1 - \frac{T_{02}}{T_{01}}\right), \tag{5.17}$$

where T_{01} is the stagnation temperature upstream, T_{02} the stagnation temperature downstream, and R the curvature of the flame front. Hadamard in 1903 showed that curved shock waves also produce vorticity.

Finally, vorticity occurs in the tangential flow along solid surfaces and in the discontinuity surfaces which appear in separated flows. It results from the action of viscous shearing stresses.

Now $\Gamma = \int_S \zeta_n \, dS$, and transforming Eq. (5.11) via Stokes's theorem to a surface integral, one gets

$$\frac{D\boldsymbol{\zeta}}{Dt} = -\nu \nabla \times (\nabla \times \boldsymbol{\zeta}) = -\nu \nabla(\overset{0}{\cancel{\nabla} \cdot \boldsymbol{\zeta}}) + \nu \nabla^2 \boldsymbol{\zeta};$$

Therefore

$$\frac{D\boldsymbol{\zeta}}{Dt} = \nu \nabla^2 \boldsymbol{\zeta}, \tag{5.18}$$

and this is a diffusion equation for vorticity.

5.5 Generation of a Vortex Ring by an Impulsive Pressure Acting over a Circular Area

Consider first the motion produced by impulsive forces. These are forces which act for very short times and so the acceleration terms $\partial u/\partial t$, etc. are much larger than the inertial terms $u \, \partial u/\partial x$, etc.

The equations of motion for a perfect fluid may then be written:

$$\frac{\partial u}{\partial t} = -\frac{1}{\varrho}\frac{\partial p}{\partial x} + \frac{f_x}{\varrho}, \qquad \frac{\partial v}{\partial t} = -\frac{1}{\varrho}\frac{\partial p}{\partial y} + \frac{f_y}{\varrho},$$
$$\frac{\partial w}{\partial t} = -\frac{1}{\varrho}\frac{\partial p}{\partial z} + \frac{f_z}{\varrho}, \tag{5.19}$$

where f_x, f_y, and f_z are the components of the impulsive force.

Along with the equation of continuity, these are then four linear equations in the four unknowns u, v, x, and p. The pressure can be eliminated from the equations by cross differentiation. Thus, the third equation of (5.19) is differentiated with respect to y, the second with respect to z, and they are then subtracted. The resulting equations are

$$\frac{\partial \zeta_x}{\partial t} = \frac{1}{\varrho}\left(\frac{\partial f_z}{\partial y} - \frac{\partial f_y}{\partial z}\right), \qquad \frac{\partial \zeta_y}{\partial t} = \frac{1}{\varrho}\left(\frac{\partial f_x}{\partial z} - \frac{\partial f_z}{\partial x}\right),$$
$$\frac{\partial \zeta_z}{\partial t} = \frac{1}{\varrho}\left(\frac{\partial f_y}{\partial x} - \frac{\partial f_x}{\partial y}\right). \tag{5.20}$$

Now consider the space lying between the planes $z = 0$ and $z = h$, bounded by a cylindrical surface of radius a, with its axis parallel to the z axis (Fig. 5.2). Outside this region the external forces are taken to be zero; inside it they are directed in the negative z direction. The force is constant across the region but falls rapidly to zero at the cylindrical surface.

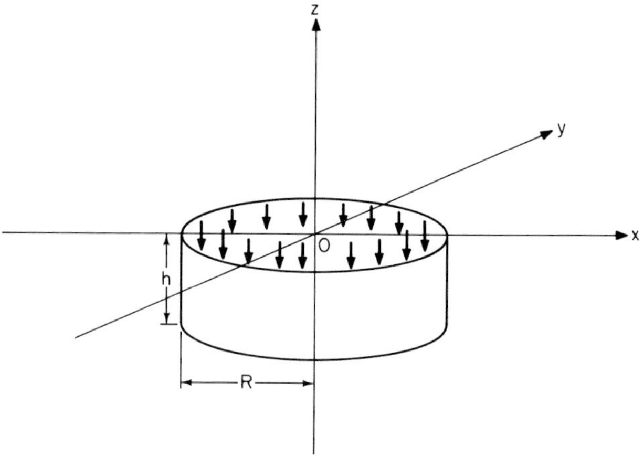

Fig. 5.2

As f_z is the only force component acting in this case, Eq. (5.20) reduces to

$$\frac{\partial \zeta_x}{\partial t} = \frac{1}{\varrho} \frac{\partial f_z}{\partial y}, \qquad \frac{\partial \zeta_y}{\partial t} = -\frac{1}{\varrho} \frac{\partial f_z}{\partial x}, \qquad \frac{\partial \zeta_z}{\partial t} = 0. \qquad (5.21)$$

Since f_z varies only in the neighborhood of the cylindrical surface, it is only in this region that vorticity will be produced. Supposing that the motion starts from rest, and integrating Eq. (5.21) with respect to time, one obtains

$$\zeta_x = \frac{1}{\varrho} \frac{\partial i_z}{\partial y}, \qquad \zeta_y = -\frac{1}{\varrho} \frac{\partial i_z}{\partial x}, \qquad \zeta_z = 0, \qquad (5.22)$$

where i_z is the impulse (time integral) per unit volume of the force f_z.

It is seen that the system of vortex lines generated in this way will consist of circles, with planes parallel to the xy plane and axes coinciding with z. If the height h is much smaller than the circle radius R, then it is sufficient to combine these vortex lines into one single circular vortex lying in the xy plane and of radius R.

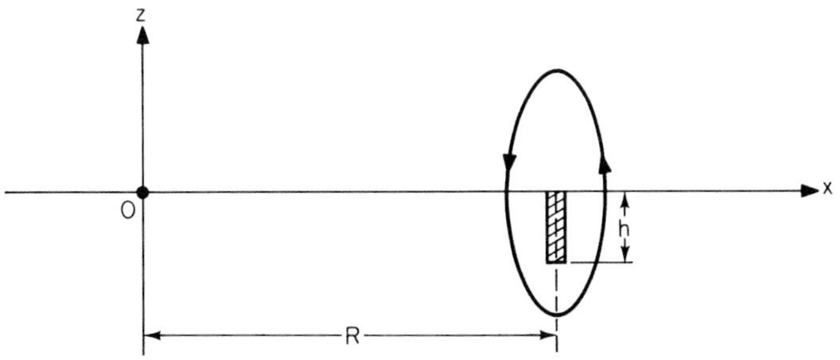

Fig. 5.3

Consider the section of the vortex in the xz plane, with circulation of fluid about it (Fig. 5.3). The hatched region is that where f_z falls to zero.

The total strength Γ of the vortex is obtained by integrating ζ_y over this region. Therefore

$$\Gamma = \iint dx\, dy\, \zeta_y = -\frac{1}{\varrho}\int dz \int dx \frac{\partial i_z}{\partial x} = \frac{1}{\varrho}\int i_z\, dz. \quad (5.23)$$

Since i_z is independent of z between the planes $z = 0$ and $z = h$,

$$\Gamma = \frac{i_z h}{\varrho}. \quad (5.24)$$

When h is very small, $i_z h$ is the intensity per unit area of the impulse of the external forces (or the impulse of the pressure),

$$\Gamma = \frac{I}{\varrho}. \quad (5.25)$$

A continuous force can be regarded as a sequence of impulses—each generating a vortex ring. These impulses move off from the region of formation, with the fluid. In the limit of an infinite number of pulses, the rings merge into a continuous sheet of vorticity. This is illustrated in Fig. 5.4. The cylindrical surface becomes a vortex sheet. As shown previously, the strength Γ of a vortex sheet is determined by the difference of velocity on either side of the sheet and equals the circulation around a strip of the sheet having unit length in the direction of the cylinder axis. Consider the case in which the fluid is moving under a constant pressure P at a velocity V. The circulation generated in unit time is $V\Gamma$ and so

$$V\Gamma = \frac{P}{\varrho}. \quad (5.26)$$

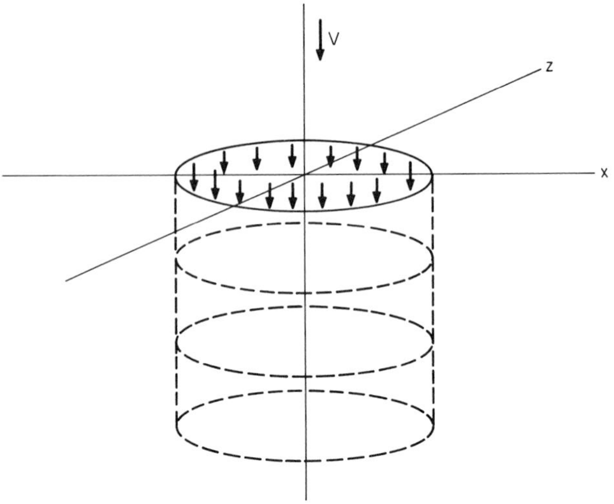

Fig. 5.4

Vortex sheets are unstable and tend to roll up into a sequence of discrete vortices, thus forming vortex streets. This phenomenon will be considered later in the chapter.

5.6 Determination of the Velocity Field from a Given Vortex Field

It will be considered that the fluid fills the whole space and is at rest at infinity, also that the vorticity $\boldsymbol{\zeta}$ and the divergence ϑ of the velocity are given at every point of the space. It is required to determine the velocity vector **v**.

In Chapter 3, the velocity field **v** was shown to be related to the vorticity field $\boldsymbol{\zeta}$, by the equation

$$\mathbf{v}(\mathbf{x}) = -\frac{1}{4\pi}\int \frac{\mathbf{S} \times \boldsymbol{\zeta}'}{S^3}\, dV(\mathbf{x}'),$$

where $\mathbf{S} = \mathbf{x} - \mathbf{x}'$ and the prime denotes a point in the field other than **x**.

For the case of a straight-line vortex of infinite length and strength Γ, where

$$\int_{\delta V} \boldsymbol{\zeta}\, dV = \Gamma\, \delta \boldsymbol{l},$$

and so

$$|\mathbf{v}| = \frac{\Gamma\varrho}{4\pi}\int_{-\infty}^{\infty}\frac{dl}{(\varrho^2 + l^2)^{3/2}} = \frac{\Gamma}{2\pi\varrho}, \tag{5.27}$$

where ϱ is the perpendicular distance from the point in the field to the line vortex. The velocity $\mathbf{v}$ is everywhere in the azimuthal direction about the line vortex, with direction corresponding to positive circulation about it.

The projection of the line vortex in a perpendicular plane is a point and the streamlines are concentric circles about this point vortex.

If this is the xy plane, then, with the line vortex intersecting the plane at the point (ξ, η), the component velocities are easily shown to be

$$v_x = -\frac{\Gamma}{2\pi}\frac{y-\eta}{\varrho^2}, \qquad v_y = -\frac{\Gamma}{2\pi}\frac{x-\xi}{\varrho^2}, \tag{5.28}$$

where $\varrho^2 = (x - \xi)^2 + (y - \eta)^2$.

Thus, under the influence of a single point vortex, the fluid particles move along circles with center at the vortex and with velocities inversely proportional to the distance of the moving point from the vortex:

$$|\mathbf{v}| = v = \frac{\Gamma}{2\pi}\frac{1}{\varrho}.$$

This is the velocity relation for the potential, or free vortex, since the fluid motion is irrotational except at the line vortex itself (the origin of the field). When ϱ is zero, v is infinite, and so the origin is a singularity.

As a consequence of the symmetry of the motion around a point (or line) vortex, the vortex will not translate. Self-induced motion results when the vortex line is curved, or when two (or more) straight-line vortices are near one another. The latter case will be considered first.

5.7 Two Straight-Line Vortices and Their Motion

Consider two parallel rectilinear vortex filaments. From Eq. (5.28), the complex velocity can be constructed:

$$v_x - iv_y = -\frac{\Gamma}{2\pi}\frac{y-\eta+i(x-\xi)}{\varrho^2} = \frac{\Gamma}{2\pi i}\frac{z^* - z_0^*}{(z - z_0)(z^* - z_0^*)}, \tag{5.29}$$

where

$$z_0 = \xi + i\eta, \qquad z^* = x - iy, \qquad z_0{}^* = \xi - i\eta.$$

The complex velocity equals the derivative of the complex potential w with respect to z,

$$v_x - iv_y = \frac{dw}{dz}, \tag{5.30}$$

so that

$$w = \left(\frac{\Gamma}{2\pi i}\right) \log(z - z_0). \tag{5.31}$$

The fluid motion takes place in a plane perpendicular to the two parallel filaments, and this will be taken as the z plane. Suppose the vortex strengths are Γ_1 and Γ_2, respectively. Then, the complex potential for the pair is

$$w(z) = \frac{\Gamma_1}{2\pi i} \log(z - z_0) + \frac{\Gamma_2}{2\pi i} \log(z - z_2), \tag{5.32}$$

and the complex velocity is

$$v_x - iv_y = \frac{dw}{dz} = \frac{\Gamma_1}{2\pi i} \frac{1}{z - z_1} + \frac{\Gamma_2}{2\pi i} \frac{1}{z - z_2}. \tag{5.33}$$

Now $v_x - iv_y = dz^*/dt$, and so one gets the differential equation

$$\frac{dz^*}{dt} = \frac{\Gamma_1}{2\pi i}(z - z_1) + \frac{\Gamma_2}{2\pi i}(z - z_2). \tag{5.34}$$

The vortex at the point z_1 moves under the influence of the other vortex at z_2 (there is no self-inductance effect in a rectilinear vortex). Thus, the first vortex will move in a circle about the second one, and the second in a circle about the first. The intervortex distance remains fixed during the motion. This motion will now be proven.

To obtain the velocity of the first vortex, in Eq. (5.34) the first term on the right-hand side is omitted and the z is replaced by z_1. Therefore

$$\frac{dz_1{}^*}{dt} = \frac{\Gamma_2}{2\pi i}(z_1 - z_2). \tag{5.35}$$

Similarly,

$$\frac{dz_2{}^*}{dt} = \frac{\Gamma_1}{2\pi i}(z_2 - z_1). \tag{5.36}$$

Separation of these equations into real and imaginary parts results in the following system of differential equations:

$$\frac{dx_1}{dt} = -\frac{\Gamma_2}{2\pi}\frac{y_1 - y_2}{r^2}, \tag{5.37a}$$

$$\frac{dx_2}{dt} = \frac{\Gamma_1}{2\pi}\frac{y_1 - y_2}{r^2}, \tag{5.37b}$$

$$\frac{dy_1}{dt} = \frac{\Gamma_2}{2\pi}\frac{x_1 - x_2}{r^2}, \tag{5.37c}$$

$$\frac{dy_2}{dt} = -\frac{\Gamma_1}{2\pi}\frac{x_1 - x_2}{r^2}, \tag{5.37d}$$

where $r^2 = (x_1 - x_2)^2 + (y_1 - y_2)^2$. Multiplying (5.37a) by Γ_1 and (5.37c) by Γ_2 and adding them, one gets

$$\Gamma_1 \frac{dx_1}{dt} + \Gamma_2 \frac{dx_2}{dt} = 0;$$

therefore

$$\Gamma_1 x_1 + \Gamma_2 x_2 = \text{constant}. \tag{5.38a}$$

Also, multiplying (5.37b) by Γ_1 and (5.37d) by Γ_2 and adding, one has

$$\Gamma_1 \frac{dy_1}{dt} + \Gamma_2 \frac{dy_2}{dt} = 0;$$

therefore

$$\Gamma_1 y_1 + \Gamma_2 y_2 = \text{constant}. \tag{5.38b}$$

From (5.38a) and (5.38b), it follows that

$$\frac{\Gamma_1 x_1 + \Gamma_2 x_2}{\Gamma_1 + \Gamma_2} = \text{constant}, \tag{5.39a}$$

$$\frac{\Gamma_1 y_1 + \Gamma_2 y_2}{\Gamma_1 + \Gamma_2} = \text{constant}. \tag{5.39b}$$

These are called "integrals of motion of the centroid" of the system of two vortices. They imply that the point

$$x_c = \frac{\Gamma_1 x_1 + \Gamma_2 x_2}{\Gamma_1 + \Gamma_2}, \tag{5.40a}$$

$$y_c = \frac{\Gamma_1 y_1 + \Gamma_2 x_2}{\Gamma_1 + \Gamma_2}, \tag{5.40b}$$

which is the centroid of the two vortices, remains fixed at all times during the motion.

Equations (5.37) are now transformed as follows: Eq. (5.37c) is subtracted from (5.37a), and (5.37d) from (5.37b). This gives

$$\frac{d(x_1 - x_2)}{dt} = -\frac{\Gamma_1 + \Gamma_2}{2\pi} \frac{y_1 - y_2}{r^2}, \tag{5.41a}$$

$$\frac{d(y_1 - y_2)}{dt} = \frac{\Gamma_1 + \Gamma_2}{2\pi} \frac{x_1 - x_2}{r^2}. \tag{5.41b}$$

Multiplying (5.41a) by $(x_1 - x_2)$ and (5.41b) by $(y_1 - y_2)$ and adding, one gets

$$(x_1 - x_2)\frac{d(x_1 - x_2)}{dt} + (y_1 - y_2)\frac{d(y_1 - y_2)}{dt} = 0.$$

Integration of this equation gives

$$(x_1 - x_2)^2 + (y_1 - y_2)^2 = \text{constant} \tag{5.42}$$

or $r =$ constant.

The distance between the two vortices remains constant. Thus the two vortices must rotate about the centroid with constant distance between them—which was to be shown.

When $\Gamma_1 = -\Gamma_2$, that is, the vortices have equal and opposite rotation, the centroid will lie at infinity, since $\Gamma_1 + \Gamma_2 = 0$. It will now be shown that the vortices move in translation with constant velocity, perpendicular to the straight line joining them (Fig. 5.5). Suppose that

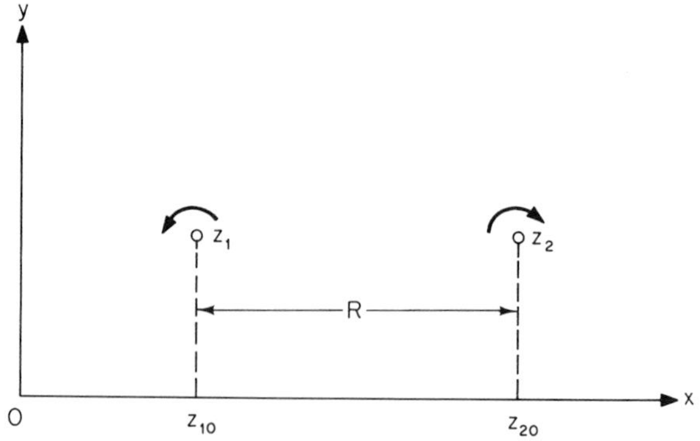

Fig. 5.5

initially the two vortices are on the Ox axis separated by a distance R. From the equations of motion,

$$\frac{dz_1^*}{dt} = \frac{dz_2^*}{dt} = -\frac{\Gamma_1}{2\pi i}(z_1 - z_2),$$

$$z_2^* - z_1^* = \text{constant} = R,$$

$$z_2 - z_1 = \frac{1}{R};$$

therefore

$$\frac{dz_1^*}{dt} = \frac{dz_2^*}{dt} = \frac{\Gamma_1}{2\pi i R} = -\frac{\Gamma_1 i}{2\pi R}.$$

Separating the real and imaginary parts, one obtains

$$v_{1x} = v_{2x} = 0, \qquad v_{1y} = v_{2y} = \frac{\Gamma_1}{2\pi R}. \tag{5.43}$$

Thus, the vortices move parallel to the Oy axis.

This case is relevant to the situation in which a vortex lies close to a wall. There is an image vortex of equal but opposite rotation in the wall, and so the vortex will move along the wall.

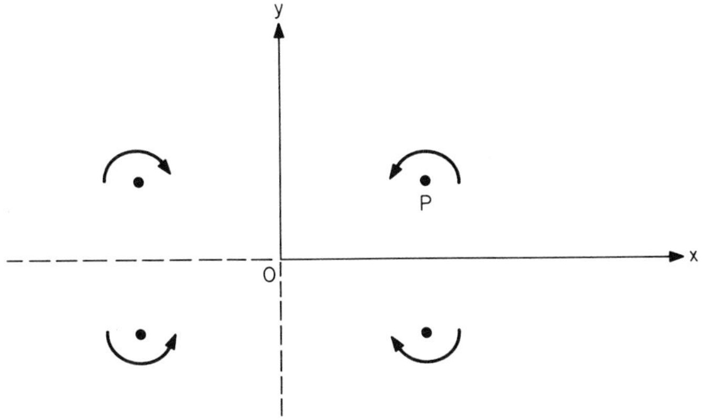

Fig. 5.6

An interesting situation arises if the vortex is in a corner. Then there are three image vortices, as shown in Fig. 5.6. The images consist of two negative vortices at the points $(-x, y)$, $(x, -y)$ and a positive vortex at the point $(-x, -y)$. Since the vortex at $P(x, y)$ cannot produce any translation of itself, its motion is due solely to that produced by the combined effect of its images. Therefore

$$\dot{x} = \frac{\Gamma}{2\pi y} - \frac{\Gamma y}{2\pi(x^2 + y^2)} = \frac{\Gamma x^2}{2\pi y(x^2 + y^2)},$$
$$\dot{y} = -\frac{\Gamma}{2\pi x} + \frac{\Gamma x}{2\pi(x^2 + y^2)} = -\frac{\Gamma y^2}{2\pi x(x^2 + y^2)}. \tag{5.44}$$

Thus

$$\frac{\dot{x}}{x^3} + \frac{\dot{y}}{y^3} = 0 \quad \text{or} \quad \frac{1}{x^2} + \frac{1}{y^2} = \frac{1}{a^2},$$

where a is a constant. Therefore

$$\frac{x^2 + y^2}{x^2 y^2} = \frac{1}{a^2} \quad \text{or} \quad \frac{r^2}{x^2 y^2} = \frac{1}{a^2},$$

whence

$$\frac{r}{xy} = \frac{1}{a} = \frac{r}{r^2 \cos\vartheta \sin\vartheta},$$

and so

$$r \sin 2\vartheta = 2a. \tag{5.45}$$

This is the equation of the curve described by the vortex. Also, since

$$x\dot{y} - y\dot{x} = -\tfrac{1}{2}\Gamma,$$

the vortex executes the curve a particle would describe if repelled from the origin with a force $3\Gamma^2/16\pi^2 r^3$.

In this connection, it can be shown that if two vortices of strength Γ exist in a fluid, as proved above, each will describe a circle whose center is the middle of the line joining them, with velocity $\Gamma/4\pi R$, where R is the distance between them, and they move as if there existed a stress (or tension) between them, of magnitude† $\Gamma^2/16\pi^2 R^3$.

5.8 The Kármán Vortex Streets

Bénard in 1908 was the first to investigate the appearance of vortices behind a body moving in a fluid. The body he used was a cylinder.

† G. Greenhill, Plane vortex motion, *Quart. J.* **25**, 20 (1877).

He observed that at a high enough fluid velocity (or Reynolds number based on the cylinder diameter), which depends on the viscosity and width of the body, vortices start to shed behind the cylinder, alternately from the top and the bottom of the cylinder. Initially, these vortices have the velocity of the fluid relative to the body (**v**), but then their velocity decreases and they tend to disperse laterally.

At some distance behind the body, the vortices are arranged at a definite distance l apart and with a definite separation h between the two rows. The senses of the rotation in the two rows are opposite (Fig. 5.7). In 1912 von Kármán expounded a theory of such vortex streets and the drag which a cylinder would experience due to their formation.

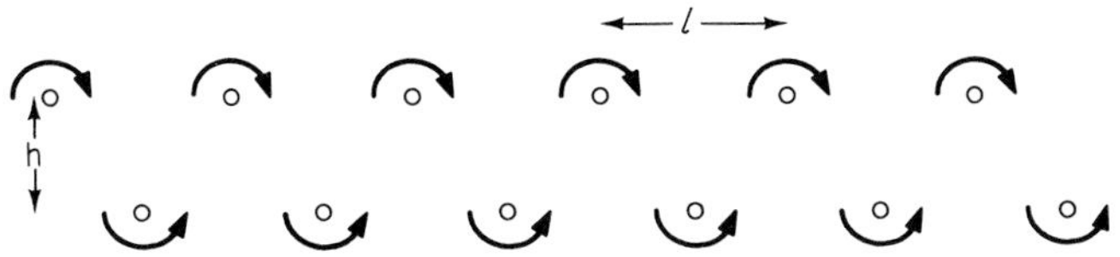

Fig. 5.7

5.8.1 Von Kármán's Theory of a Single Vortex Street

The vortex sheet, or single row of rectilinear vortices, has already been referred to as characterizing a surface of discontinuity. To avoid the question of vortex generation, consider an infinite row of point vortices, distributed along a straight line at equal distances from each other and with the same strength Γ. Let their positions be denoted by z_0, z_{+1},

Fig. 5.8

$z_{-1}, z_{+2}, z_{-2}, \ldots$ (Fig. 5.8). Now, for a single vortex of positive sense, the complex potential is

$$\Omega_1 = \varphi_1 + i\psi_1 = \frac{\Gamma}{2\pi}(\vartheta - i \ln r) = -i\frac{\Gamma}{2\pi} \ln (z - z_0), \quad (5.46)$$

where r and ϑ are the polar coordinates from the location of the vortex at z_0.

For a row of vortices, as described above, the complex potential is

$$\Omega = -\frac{i\Gamma}{2\pi}\,[\ln(z - z_0) + \ln(z - z_0 - l) + \ln(z - z_0 + l) + \ln(z - z_0 - 2l) + \ln(z - z_0 + 2l) + \cdots + \ln(z - z_0 \pm nl)]\,. \tag{5.47}$$

Therefore

$$\Omega = \frac{\Gamma}{2\pi i}\ln\left[\frac{z - z_0}{l}\left(1 - \frac{(z - z_0)^2}{l^2}\right)\left(1 - \frac{(z - z_0)^2}{4l^2}\right)\cdots\left(1 - \frac{(z - z_0)^2}{n^2l^2}\right)\right] + \text{constant},$$

and since

$$\sin \pi x = \pi x \prod_{k=1}^{\infty}\left(1 - \frac{x^2}{k^2}\right),$$

therefore

$$\Omega = \frac{\Gamma}{2\pi i}\log \sin \frac{\pi}{l}(z - z_0). \tag{5.48}$$

A few streamlines (lines of constant $\varphi = \mathscr{I}\,\Omega$) are drawn in Fig. 5.9. The complex velocity at the point z in the flow field is the sum of the velocities caused by each vortex:

$$v_x - iv_y = \frac{\Gamma}{2\pi i}\left\{\frac{1}{z - z_0} + \sum_{k=1}^{\infty}\left(\frac{1}{z - z_0 - kl} + \frac{1}{z - z_0 + kl}\right)\right\}. \tag{5.49}$$

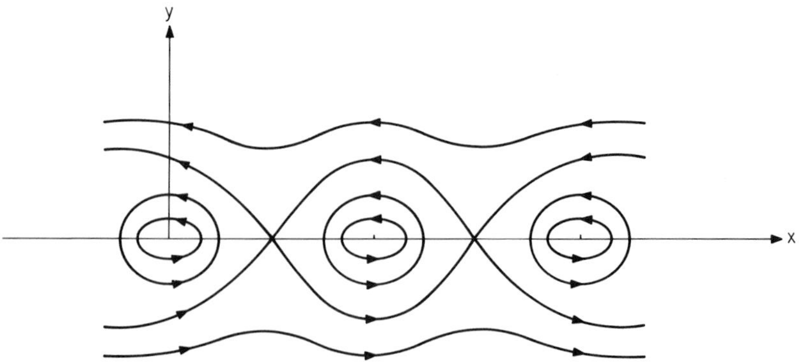

Fig. 5.9

On summation it is found that

$$v_x - iv_y = \frac{\Gamma}{2li} \cot \frac{\pi}{l} (z - z_0). \tag{5.50}$$

This can be directly obtained from Eq. (5.48) by using the relation

$$v_x - iv_y = \frac{d\Omega}{dz}. \tag{5.51}$$

To find the velocity of the vortex street itself—that is, the velocity of the vortices which constitute it—it is enough to consider the motion of the point z_0, as obviously all vortices must move with the same velocity. The velocity at z_0 is due to the influence of all the vortices, except that at $z = z_0$. Substituting $z = z_0$ in all terms and omitting the first term in Eq. (5.49), it is seen that all the terms cancel in pairs, and so the velocity of the vortex (v_0) is zero. A single vortex street will remain stationary, as one might expect from symmetry considerations.

Helmholtz, Rayleigh, Kelvin, and others have studied the stability of the single row of like vortices. More recent work† has used the power of the digital computer and automatic plotters to demonstrate the behavior graphically. Briefly, the vortex sheet is observed to become undulatory and then to curl up into a sequence of large vortices.

5.8.2 Two Vortex Streets

In the wake behind a body such as a cylinder, there are bound to be two vortex streets formed—one from each side of the body.

Let the intervortex distance in each street be l and the distance between streets be h. Also, suppose that the strength in the upper street is Γ_1 and in the lower Γ_2. Suppose one of the upper row vortices is at z_1 and the one closest to it in the lower row is at z_2. The complex potential then is

$$\Omega = \frac{\Gamma_1}{2\pi i} \log \sin \frac{\pi}{l} (z - z_1) + \frac{\Gamma_2}{2\pi i} \log \sin \frac{\pi}{l} (z - z_2),$$

and the complex velocity at the point z is

$$v_x - iv_y = \frac{d\Omega}{dz} = \frac{\Gamma_1}{2li} \cot \frac{\pi}{l} (z - z_1) + \frac{\Gamma_2}{2li} \cot \frac{\pi}{l} (z - z_2).$$

† L. Rosenhead, *Proc. Roy. Soc.* **A 175**, 436 (1940); F. H. Abernathy, and R. E. Kronauer, *J. Fluid Mech.* **13**, 1–20 (1962).

Each street can be considered as a single unit with regard to its motion, and so it is enough to study the velocities of two vortices; for example those at z_1 and z_2. The vortex at z_1 will move under the influence of the second street, since it has been established that a single street does not move. So the velocity $v_{1x} - iv_{1y}$ of the vortex at z_1 is obtained by omitting the first term in the expression $v_x - iv_y$ and by setting $z = z_1$ in the second term:

$$v_{1x} - iv_{1y} = \frac{\Gamma_2}{2li} \cot \frac{\pi}{l} (z_1 - z_2). \tag{5.52}$$

Similarly, for the vortex at z_2,

$$v_{2x} - iv_{2y} = -\frac{\Gamma_1}{2li} \cot \frac{\pi}{l} (z_1 - z_2). \tag{5.53}$$

These are the velocities of the streets. Only the case of "rigid" streets will be considered here; that is, l and h remain constant during the motion. So

$$v_{1x} - iv_{1y} = v_{2x} - iv_{2y},$$

and therefore

$$\Gamma_2 = -\Gamma_1.$$

Thus, the strengths, or circulations, of the streets must be the same in magnitude and opposite in sign.

It will be assumed, for simplicity, that the streets move parallel to the x axis. Therefore

$$v_{1y} = v_{2y} = 0.$$

Now,

$$z_1 - z_2 = b + hi, \tag{5.54}$$

where $b = x_1 - x_2$, and

$$\cot \frac{\pi}{l} (b + hi) = \frac{\sin(2\pi b/l)}{\cosh(2\pi h/l) - \cos(2\pi b/l)} - \frac{i \sinh(2\pi h/l)}{\cosh(2\pi h/l) - \cos(2\pi b/l)},$$

where

$$\sinh x = \frac{e^x - e^{-x}}{2i}, \qquad \cosh x = \frac{e^x + e^{-x}}{2}.$$

Therefore,

$$v_{1x} = v_{2x} = \frac{\Gamma}{2l} \frac{\sinh(2\pi h/l)}{\cosh(2\pi h/l) - \cos(2\pi b/l)}, \tag{5.55}$$

$$v_{1y} = v_{2y} = -\frac{\Gamma}{2l} \frac{\sin(2\pi b/l)}{\cosh(2\pi h/l) - \cos(2\pi b/l)}. \tag{5.56}$$

It has been assumed that $v_{1y} = v_{2y} = 0$. Therefore

$$\sin \frac{2\pi b}{l} = 0, \tag{5.57}$$

and so either $b = 0$ or $b = l/2$. In the first case the vortices in both streets line up with each other and a symmetric distribution results. Von Kármán showed that this is an unstable situation, and that the only stable situation possible is when $b = l/2$, which gives a staggered arrangement of vortices (Fig. 5.10). The velocity of the street for this staggered arrangement is

$$v_{1x} = \frac{\Gamma}{2l} \tanh \frac{\pi h}{l}. \tag{5.58}$$

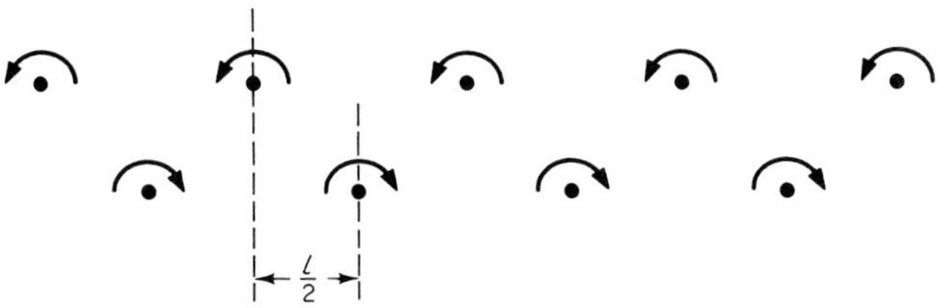

Fig. 5.10

5.9 Circular Vortex Filament—Vortex Rings

Consider an indefinitely thin circular vortex filament, every element of which is rotating with angular velocity about the tangent to the circle of which the element forms a part.

The z axis will be considered to pass through the center of the circle and in a direction perpendicular to the plane of the circle. When the flow pattern is identical in any of the planes passing through the axis z, the system is axisymmetric and any point in the field is described by the coordinates (r, z). The equation of the streamlines is then

$$r(w\,dr - u\,dz) = 0, \tag{5.59}$$

where u is the axial component of fluid velocity and w is the radial component.

The equation of continuity is

$$\frac{d(ru)}{dr} + r\frac{dw}{dz} = 0. \tag{5.60}$$

Thus, the left-hand side of Eq. (5.59) is a perfect differential, and so

$$w = \frac{1}{r}\frac{d\psi}{dr}, \qquad u = -\frac{1}{r}\frac{d\psi}{dz}, \tag{5.61}$$

where ψ is the Stokes stream function, and

$$2\omega = \frac{\partial u}{\partial z} - \frac{\partial w}{\partial r}. \tag{5.62}$$

Substituting for u and w from Eq. (5.61), one obtains equation for ψ and ω:

$$\frac{\partial^2 \psi}{\partial z^2} + \frac{\partial^2 \psi}{\partial r^2} - \frac{1}{r}\frac{\partial \psi}{\partial r} + 2r\omega = 0. \tag{5.63}$$

A circular vortex ring is supposed to comprise a large number of circular vortex filaments. At all points in the core of the ring, Eq. (5.63) holds, while at exterior points, ψ satisfies the equation

$$\frac{\partial^2 \psi'}{\partial z^2} + \frac{\partial^2 \psi'}{\partial r^2} - \frac{1}{r}\frac{\partial \psi'}{\partial r} = 0. \tag{5.64}$$

Let $\psi = \chi r$, and then Eqs. (5.63) and (5.64) become

$$\frac{\partial^2 \chi}{\partial z^2} + \frac{\partial^2 \chi}{\partial r^2} + \frac{1}{r}\frac{\partial \chi}{\partial r} - \frac{\chi}{r^2} + 2\omega = 0 \tag{5.65}$$

and

$$\frac{\partial^2 \chi'}{\partial z^2} + \frac{\partial^2 \chi'}{\partial r^2} + \frac{1}{r}\frac{\partial \chi'}{\partial r} - \frac{\chi'}{r^2} = 0. \tag{5.66}$$

These equations imply that $\chi \cos \vartheta$ is the potential of a distribution of matter of density $\omega \cos \vartheta/2\pi$, which occupies the same region of space as the vortex ring. To determine this potential one proceeds as follows.

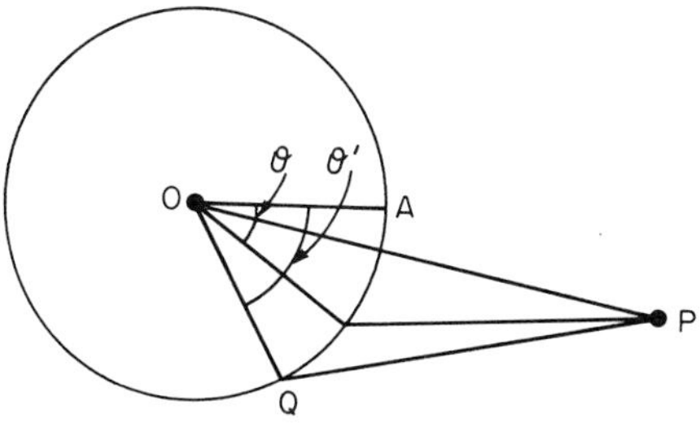

Fig. 5.11

Suppose O is the center of the ring (Fig. 5.11), and OA a reference line from which ϑ is measured. P is any point whose coordinates are z, r, ϑ, and Q is a point on the center line of the ring whose coordinates are z', r_0, ϑ'. Now, the distance from P to Q is given by

$$[(z - z')^2 + r^2 + r_0^2 - 2rr_0 \cos(\vartheta' - \vartheta)]^{1/2},$$

and if σ is the cross section of the core of the ring ($=\pi a^2$), then

$$\chi \cos \vartheta = \frac{r_0 \omega \sigma}{2\pi} \int_{\vartheta}^{2\pi+\vartheta} \frac{\cos \vartheta' \, d\vartheta'}{[(z - z')^2 + r^2 + r_0^2 - 2rr_0 \cos(\vartheta' - \vartheta)]^{1/2}}. \tag{5.67}$$

Let $\vartheta' - \vartheta = \varepsilon$, and so

$$\chi \cos \vartheta = \frac{\sigma \omega r_0}{2\pi} \int_0^{2\pi} \frac{(\cos \vartheta \cos \varepsilon - \sin \vartheta \sin \varepsilon) \, d\varepsilon}{[(z - z')^2 + r^2 + r_0^2 - 2rr_0 \cos \varepsilon]^{1/2}}. \tag{5.68}$$

The second integral vanishes, and so

$$\psi = \chi r = \frac{\sigma \omega r r_0}{\pi} \int_0^{\pi} \frac{\cos \varepsilon \, d\varepsilon}{[(z - z')^2 + r^2 + r_0^2 - 2rr_0 \cos \varepsilon]^{1/2}}. \tag{5.69}$$

This determines ψ at any point outside the vortex core.

The integral in Eq. (5.69) can be conveniently expressed in terms of complete elliptic integrals. By defining the modulus as

$$k = \frac{2(rr_0)^{1/2}}{[z^2 + (r + r_0)^2]^{1/2}} \tag{5.70}$$

and letting $2\eta = \varepsilon$, Eq. (5.69) becomes

$$\begin{aligned} \psi &= \sigma \omega \pi^{-1} (rr_0)^{1/2} \int_0^{\pi/2} \frac{2 \cos^2 \eta - 1}{(1 - k^2 \cos^2 \eta)^{1/2}} \, d\eta \\ &= \sigma \omega \pi^{-1} (rr_0)^{1/2} \int_0^{\pi/2} \frac{(1 - 2 \sin^2 \eta)}{(1 - k^2 \cos^2 \eta)^{1/2}} \, d\eta. \end{aligned} \tag{5.71}$$

Now the elliptic integrals of the first and second kind are defined as

$$K(k) = \int_0^{\pi/2} \frac{d\eta}{(1 - k^2 \sin^2\eta)^{1/2}}, \qquad E(k) = \int_0^{\pi/2} (1 - k^2 \sin^2\eta)^{1/2}\, d\eta,$$

and the integral in Eq. (5.71) can be expressed in the form

$$I = \left\{\left(\frac{2}{k} - k\right)K(k) - \frac{2}{k}\, E(k)\right\}. \tag{5.72}$$

Then

$$\psi = \sigma\omega\pi^{-1}(rr_0)^{1/2}\left\{\left(\frac{2}{k} - k\right)K(k) - \frac{2}{k}\, E(k)\right\}. \tag{5.73}$$

Let

$$U = \left(\frac{2}{k} - k\right)K(k) - \frac{2}{k}\, E(k),$$

and since $\Gamma = \pi a^2\omega = \sigma\omega$, Eq. (5.73) may be written

$$\psi = \Gamma(rr_0)^{1/2}\,\frac{U}{\pi}. \tag{5.74}$$

At the surface of the vortex ring, z and r are very nearly equal to z' and r_0, respectively. Then $k \sim 1$, and so if†

$$L = \log\frac{4}{k'}, \tag{5.75}$$

where $k' = (1 - k^2)^{1/2}$, then approximately

$$K(k) = L + \tfrac{1}{2}k'(L - 1), \tag{5.76}$$

$$E(k) = 1 + \tfrac{1}{2}k'^2(L - \tfrac{1}{2}), \tag{5.77}$$

and so

$$U = L - 2 + \tfrac{3}{4}k'^2(L - 1). \tag{5.78}$$

Now

$$\begin{aligned} w &= \frac{1}{r}\,\frac{\partial\psi}{\partial r} = \frac{\Gamma}{\pi}\left(\frac{r_0}{r}\right)^{1/2}\left(\frac{\partial U}{\partial r} + \frac{U}{2r}\right), \\ u &= -\frac{1}{r}\,\frac{\partial\psi}{\partial z} = -\frac{\Gamma}{\pi}\left(\frac{a}{r}\right)^{1/2}\frac{\partial U}{\partial z} \end{aligned} \tag{5.79}$$

† A. Cayley, "Elliptic Functions," 2nd ed., Dover, New York, 1895.

and

$$k'^2 = \frac{(z - z')^2 + (r - r_0)^2}{(z - z')^2 + (r + r_0)^2}.$$

Since a is the core radius, at the surface of the ring

$$k' \doteqdot \frac{a}{2r_0}, \tag{5.80}$$

and

$$\frac{dk'}{dr} = \frac{2r_0(r - r_0) - a^2}{4r_0^2 a}, \tag{5.81}$$

$$\frac{dk'}{dz} = \frac{z - z'}{2r_0 a}. \tag{5.82}$$

Also

$$\frac{dU}{dk'} = -\frac{1}{k'} + \frac{3}{2} k'\left(L - \frac{3}{2}\right) = -\frac{2r_0}{a} + \frac{3a}{4r_0}\left(L - \frac{3}{2}\right). \tag{5.83}$$

Therefore,

$$u = \frac{\Gamma}{\pi}\left\{\frac{2r_0}{a} - \frac{3a}{4r_0}\left(L - \frac{3}{2}\right)\right\}\frac{z - z'}{2r_0 a}. \tag{5.84}$$

When $z = z'$, then $u = 0$ and so the radius of the ring remains unchanged. Also,

$$w = -\frac{\Gamma}{\pi}\left\{\frac{2r_0}{a} - \frac{3a}{4r_0}\left(L - \frac{3}{2}\right)\right\}\frac{2r_0(r - r_0) - a^2}{4r_0^2 a} + \frac{\Gamma}{2\pi r_0}(L - 2). \tag{5.85}$$

To obtain the velocity of the ring through the surrounding fluid, we must put $r = r_0$, and then

$$w = \frac{\Gamma}{2\pi r_0}(L - 1) = \frac{\Gamma}{2\pi r_0}\left(\log\frac{8r_0}{a} - 1\right). \tag{5.86}$$

Thus, the ring will move forward in the direction of the cyclic motion through its aperture, with the constant velocity as given by Eq. (5.86).

A simple calculation shows that the velocity produced by the vortex at the center of the ring is

$$\frac{\Gamma}{r_0} = \frac{\pi a^2 \omega}{r_0}.$$

Hence, an isolated vortex ring in an unbounded ideal fluid will move without sensible change of size in a direction perpendicular to its plane, with a constant velocity. This velocity is small compared to that of the liquid adjacent to the central line of the core, but large with respect to the velocity of the fluid at the center of the ring.

5.9.1 Finite Core

For a rectilinear vortex, the tangential velocity in the irrotational region outside the core has been shown to be given by

$$q_\vartheta = \frac{\Gamma}{2\pi r}, \tag{5.87}$$

where $q_\vartheta = (u^2 + w^2)^{1/2}$, Γ is the vortex strength, and r is the radial distance from the center of the core. To avoid an infinite velocity, obviously the core must be of finite size—as indeed experimental observation of a vortex ring quickly shows.

If it is assumed that (a) the core is circular, and (b) the core has a uniform distribution of vorticity across it, then

$$\zeta = \frac{1}{r}\frac{\partial}{\partial r}(rq_\vartheta) = \text{constant} \qquad \text{or} \qquad \zeta r\, dr = d(rq_\vartheta).$$

Therefore

$$\frac{\zeta r^2}{2} = rq_\vartheta + C. \tag{5.88}$$

Thus, the tangential velocity is now proportional to the radius—a characteristic of solid body rotation.

The constant of integration C can be found as follows. If a is the core radius, then

$$\Gamma = \int_0^{2\pi}\left[-\frac{C}{a} + \frac{\zeta a}{2}\right]a\, d\vartheta = -2\pi C + \zeta a^2\pi.$$

Since ζ is constant, therefore $\Gamma = \zeta a^2\pi$, and so $C = 0$. Thus,

$$q_\vartheta = \frac{r\zeta}{2} = r\omega \tag{5.89}$$

or

$$q_\vartheta = \frac{r\zeta}{2} = \frac{\Gamma}{2\pi}\frac{r}{a^2}. \tag{5.90}$$

The fluid in the core moves with constant angular velocity ($\omega = \Gamma/2\pi a^2$). Thus, in the case of both rectilinear and ring vortices, it is appropriate to consider the velocity field in the core as being that corresponding to solid body rotation. At the edge of the core ($r = a$), the velocities in the two regions must match.

The combination of the rotational core and surrounding irrotational region is known as the Rankine combined vortex, and the overall velocity distribution is sketched in Fig. 5.12. The action of viscosity is to cause the vorticity to diffuse outward and the core to increase in size. The discontinuity in slope of the velocity at $r = a$ is also smoothed out.

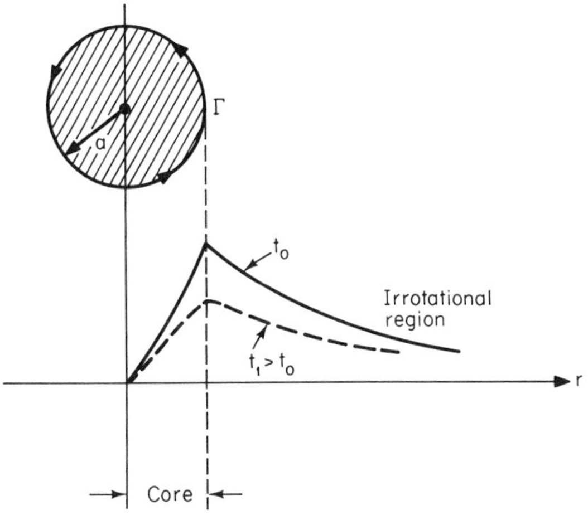

Fig. 5.12

5.9.2 Pressure Distribution in the Combined Vortex

Again a rectilinear vortex will be considered for simplicity. For the irrotational region,

$$q_\vartheta = \frac{\Gamma}{2\pi r}.$$

Using this in Bernoulli's equation, one obtains

$$p + \frac{\varrho}{2} q^2 + \varrho gh = p + \varrho gh + \frac{\varrho}{2} \frac{\Gamma^2}{4\pi^2 r^2} = \Pi.$$

Taking conditions far from the core as a reference ($r = \infty$, and using the subscript 0), the pressure variation in this region can be represented as

$$(p + \varrho gh) - (p + \varrho gh)_0 = -\frac{\varrho}{8\pi^2}\frac{\Gamma^2}{r^2}. \tag{5.91}$$

The pressure varies as the inverse radius squared. If the core was not finite, the pressure at the center of the vortex ($r = 0$) would become negatively infinite.

In the rotational core, since vorticity exists, then the Bernoulli constant will vary with r. The more basic Euler equation must now be resorted to.

From the Euler equations, the acceleration component normal to the streamlines is given by

$$a_n = \frac{\partial q}{\partial t} + \frac{q^2}{n} = -\frac{1}{\varrho}\frac{\partial}{\partial n}(p + \varrho gh). \tag{5.92}$$

For current streamlines (as in a vortex core), $n = r$, and for steady flow, therefore,

$$\frac{\partial(p + \varrho gh)}{\partial r} = -\frac{\varrho q_\vartheta{}^2}{r}. \tag{5.93}$$

Now,

$$q_\vartheta = \frac{\Gamma}{2\pi a}\left(\frac{r}{a}\right),$$

and so

$$\frac{\partial(p + \varrho gh)}{\partial r} = \frac{\varrho q_\vartheta{}^2}{r} = \frac{\varrho r \Gamma^2}{4\pi^2 a^4}.$$

Integration of this equation gives the pressure relation

$$p + \varrho gh = \frac{\varrho}{2}\frac{\Gamma^2}{4\pi^2 a^2}\left(\frac{r}{a}\right)^2 + C. \tag{5.94}$$

The pressure increases as the square of the radius. The C is evaluated at $r = a$, so that

$$C = (p + \varrho gh)_0 - \frac{\varrho \Gamma^2}{4\pi^2 a^2}.$$

Therefore

$$(p + \varrho gh) - (p + \varrho gh)_0 = \frac{\varrho}{2}\frac{\Gamma^2}{4\pi^2 a^2}\left[\left(\frac{r}{a}\right)^2 - 2\right]. \tag{5.95}$$

This is the pressure variation and it is sketched qualitatively in Fig. 5.13. Due to the pressure in the core being negative, it is possible that vaporization of the liquid (if the liquid state is involved) could occur unless the pressure in the surrounding fluid is high enough.

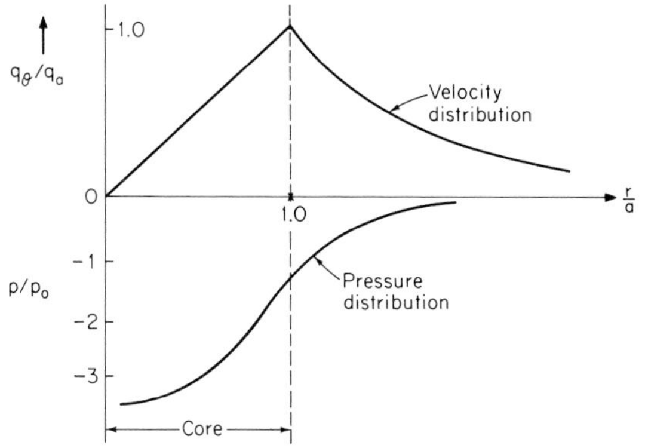

Fig. 5.13

Basset has determined the velocity distribution in the core of a vortex ring proper, assuming that the vorticity is constant in the core, but not postulating that the core is exactly circular in steady motion. In view of the fact that so little is known about the structure of the vortex core—it may not be valid to treat it as a continuum at all—the solid body rotation approximation would appear to be as satisfactory as any. At least experimental velocity profiles obtained for large cores suggest that the actual profile closely resembles that associated with solid body rotation.

The size of the core, in an ideal fluid, is another imponderable. Observations on gaseous vortex cores formed by impulses at orifices of 2 or 3 cm diameter suggest sizes of at least one-tenth of this diameter. Vortices in liquid helium, on the other hand, are estimated to have cores of about 1 micron in diameter.

In any case, as mentioned previously, and to be dealt with later, the action of viscosity will cause the core size to increase indefinitely from its instant of formation.

Finally, with respect to vortex rings in ideal fluids, the early work on their stability, vibrations, and interactions with one another, carried out in the last century by Thompson, Hicks, Basset, Stokes, and others, must be referred to by any serious student of this phenomenon. Such work was stimulated by the then popular vortex theory of the atom.

5.9.3 Confined Vortex Ring

In the model of a vortex used so far, the vorticity has been distributed uniformly through a core whose cross section is circular and of radius a, where a is much smaller than the mean radius r_0, of the ring. This is Lamb's model in effect, and the energy and impulse (momentum) are given by the relations

$$E = \frac{1}{2}\,\varrho\Gamma^2 r_0\left[\ln\left(\frac{8r_0}{a}\right) - \frac{7}{4}\right], \tag{5.96}$$

$$I = \pi\varrho\Gamma r_0^{\,2}. \tag{5.97}$$

The velocity of the ring was deduced to be

$$v_R = \frac{\Gamma}{2\pi r_0}\left(\ln\left(\frac{8r_0}{a}\right) - \frac{7}{4}\right). \tag{5.98}$$

These formulas were derived for an unbounded fluid.

It is certain that the presence of walls will affect the energy and impulse of a vortex of given strength, ring diameter, and core radius. This will become significant when the ring radius becomes close to the channel radius. Since most of the kinetic energy in the velocity field of a vortex ring is contained in the fluid close to the core, it is to be expected that the energy of a ring whose distance from the wall is small compared to the ring radius will be asymptotically equal to that of the same length of a straight-line vortex at the same distance from a plane wall. It is known that the energy of this configuration approaches zero as the core is positioned closer to the wall.

As the impulse I does not depend[†] on the size or shape of the channel—as long as it is simply connected—only the energy need be examined in this context.

Several attempts to deal with this problem have been made in recent years, but that of Fineman and Chase[‡] will be considered here exclusively. In must be pointed out, however, that an unsatisfactory aspect is that their treatment predicts a zero critical velocity for vortex rings of liquid helium. This is the velocity at which superfluidity would be destroyed, so that superfluidity could not exist in liquid helium. Further reference to this will be made in Chapter 10.

† C. C. Lin, *in* Enrico Fermi International School of Physics, Italy, Chapter 1. 1961.
‡ J. C. Fineman and C. E. Chase, *Phys. Rev.* **129**, No. 1 (1963).

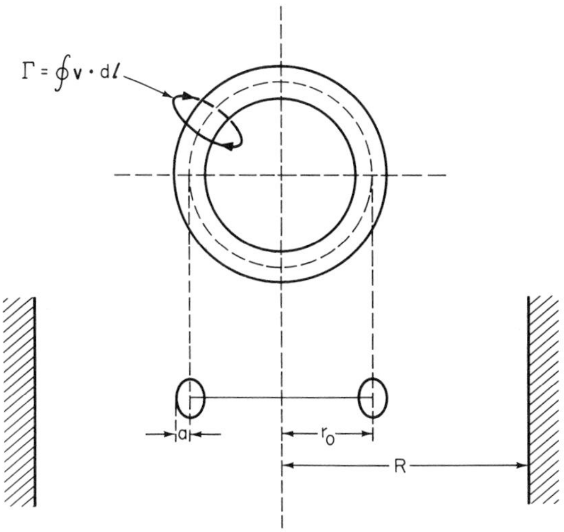

Fig. 5.14

Consider a circular vortex ring of strength Γ in an incompressible inviscid fluid, confined coaxially in a long tube of radius R (Fig. 5.14).

The velocity field outside the core is assumed to be that of a ring, with infinitesimal core, of radius r_0. This circle will be termed the *source circle.*

The vortex core is considered to be empty and to be bounded by streamlines. This model is virtually a hollow vortex tube with all the vorticity concentrated in a central vortex filament. The core radius is then the radial distance from the central vortex filament to the bounding streamlines. The vortex core is taken to be circular in cross section and the radius is much smaller than the ring radius r_0. This is, of course, purely a mathematical model for the core.

Since the core is empty, the fluid kinetic energy inside the core is zero and the energy is all surface energy.

For a numerical solution of the problem, a cylindrical coordinate system is set up, with the source circle in the $z = 0$ plane (Fig. 5.15).

Since the fluid is incompressible, $\nabla \cdot \mathbf{v} = 0$ and $\nabla \times \mathbf{v} = 0$ except for the singularity at the source circle. The velocity field in this irrotational region is the gradient of a potential φ that is a solution of Laplace's equation, $\nabla^2\varphi = 0$, in any simply connected region. This region is the whole inside of the channel, apart from a barrier consisting of a disk bounded by the source circle. Across this barrier, φ changes discontinuously because of the circulation Γ through the ring and is constant over

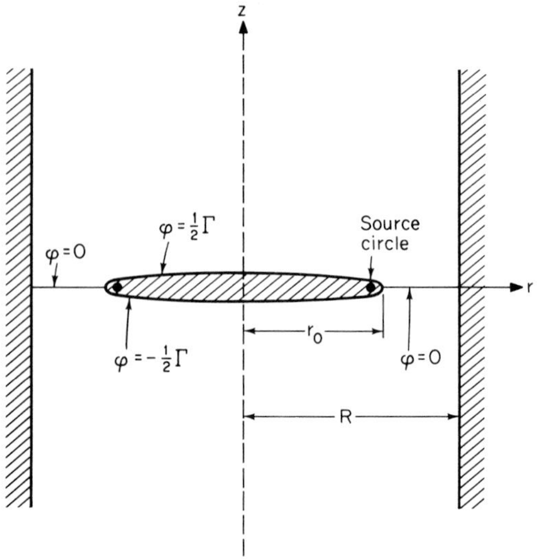

Fig. 5.15

the barrier. The boundary conditions at $z = 0$ must then be as follows:

$$\begin{aligned} \varphi &= \pm\tfrac{1}{2}\Gamma, \qquad 0 \le r < r_0, \\ &= 0, \qquad r_0 < r \le R, \end{aligned} \qquad z \to \pm 0. \tag{5.99}$$

The velocity at the wall must be tangential and zero, so that the boundary condition at $r = R$ is

$$\frac{\partial \varphi}{\partial r} = 0. \tag{5.100}$$

Since the problem is cylindrically symmetrical, the solution has the form

$$\varphi = \sum_{n=0}^{\infty} A_n J_0(k_n r) e^{-k_n |z|}, \tag{5.101}$$

where the A_n's and k_n's are constants.

From Eq. (5.100),

$$k_n = \frac{x_n}{R}, \tag{5.102}$$

where x_n is the nth root of $J_1(x) = 0$. Expanding Eq. (5.99) in the orthogonal functions $J_0(x_n r/R)$ and equating term by term to Eq. (5.101) with $z = 0$, one obtains

$$A_n = \frac{\Gamma r_0 J_1(x_n r_0/R)}{x_n R[J_0(x_n)]^2}. \tag{5.103}$$

The kinetic energy E equals the integral of $\frac{1}{2}\varrho(\nabla\varphi)^2$ over the volume of the channel outside the core. Integrating by parts and applying the divergence theorem to $\varphi\,\nabla\varphi$, one has

$$E = \tfrac{1}{2}\varrho \iint \varphi\,\nabla\varphi \cdot d\mathbf{S}, \tag{5.104}$$

where the integral is over the surface of the simply connected region. As the velocity is tangential at the walls and at the core boundaries, the only contribution to E is from the flux of fluid through the barrier, and so

$$\begin{aligned} E &= -\tfrac{1}{2}\Gamma\varrho \int_0^{r_0-a} \left(\frac{\partial\varphi}{\partial z}\right)_{z=0} 2\pi r\,dr \\ &= \pi\varrho\Gamma^2 \frac{r_0^{\,2}}{R} \sum_{n=0}^{\infty} \frac{J_1(x_n r_0/R)J_1[x_n(r_0-a)/R]}{x_n[J_0(x_n)]^2}. \end{aligned} \tag{5.105}$$

Fineman and Chase have computed this energy versus r_0/R for the series of a/R values from 10^{-2} to 10^{-7}.

It must be noted that for $r_0/R = 1$ the energy is predicted to be zero for all a/R values (Fig. 5.16).

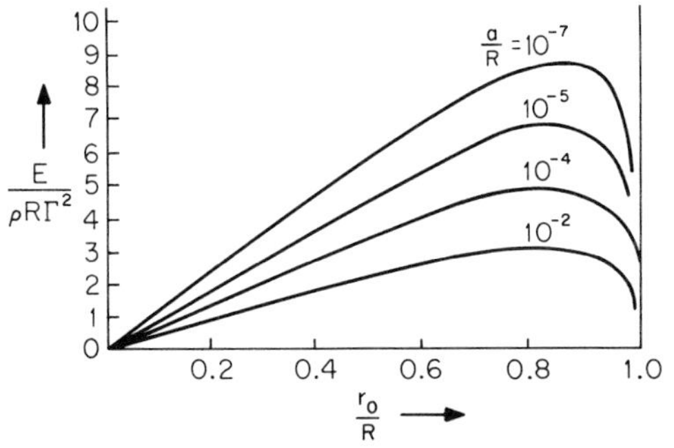

Fig. 5.16

When $R - r_0 \ll r_0$, that is, when the ring is almost as large as the tube, as mentioned previously, E should be the energy of a straight-line vortex of length $2\pi r_0$ at a distance $s = R - r_0$ from the wall. The velocity field of this configuration can be derived from that of a vortex line in an unbounded fluid by superposing the field of an opposite image line at a distance s behind the wall (see Fig. 5.17). The fluid velocity in the plane

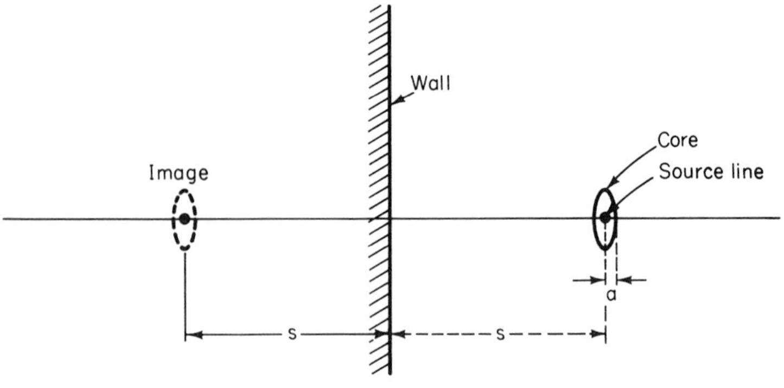

Fig. 5.17

of the source lines is

$$v = \left(\frac{\Gamma}{2\pi}\right)\left[\frac{1}{x - s} - \frac{1}{x + s}\right], \tag{5.106}$$

where x is the distance from the wall. In this case,

$$\begin{aligned} E' &= \left(\frac{\varrho\Gamma^2}{4\pi}\right)\int_{s+a}^{\infty}[(x - s)^{-1} - (x + s)^{-1}]\,dx\,2\pi r_0 \\ &= \frac{\varrho\Gamma^2 r_0}{2}\ln\left[2\left(\frac{R - r_0}{a}\right) + 1\right]. \end{aligned} \tag{5.107}$$

This gives a graph of E versus r/R_0 almost identical to that from Eq. (5.105) for a/R values of 10^{-7} or less, and $r_0 > 0.9R$.

The treatment of confined vortices is mathematically a very difficult problem, compounded by the presence of singularities and difficult boundary conditions.

5.10 Viscous Compressible Vortex

A tractable analysis is possible in this case only for a rectilinear vortex. The Rankine combined vortex solution, in which the vortex is regarded as consisting of a rotational central core and an irrotational surrounding region, will be dealt with.

The geometry is shown in Fig. 5.18.

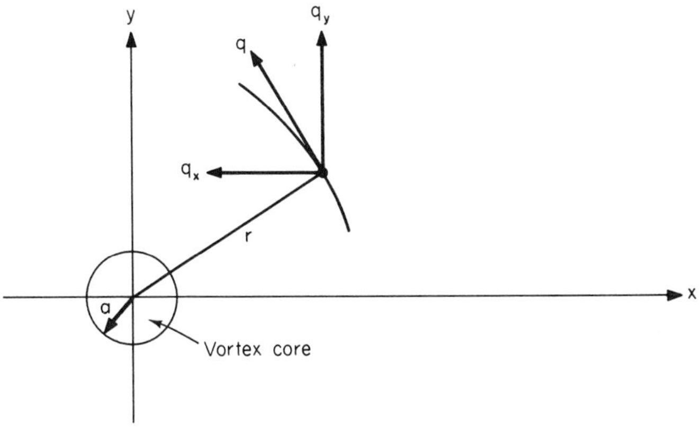

Fig. 5.18

Using the subscripts i and o for inside the core, and outside the core, the velocity components are

$$q_{ox} = -\frac{\Gamma y}{2\pi r^2}, \qquad q_{oy} = \frac{\Gamma x}{2\pi r^2},$$
$$q_{ix} = -\frac{\Gamma y}{2\pi a^2}, \qquad q_{iy} = \frac{\Gamma x}{2\pi a^2}, \tag{5.108}$$

where $q = (q_x{}^2 + q_y{}^2)^{1/2}$ is the tangential velocity and a is the core radius.

For circular streamlines, it has already been pointed out that the radial pressure variation is given by

$$\frac{1}{\varrho}\frac{\partial p}{\partial r} = \frac{q^2}{r}, \tag{5.109}$$

where ϱ is the fluid density. This is often called the radial momentum equation and is the starting point for a viscous compressible analysis.

Due to the action of viscosity, the tangential velocity q, for a given radial value, is not a constant in time. Following Lamb's analysis of incompressible viscous circular flow, one obtains

$$q = \frac{\Gamma_0}{2\pi r}\left[1 - \exp\left(-\frac{r^2}{4\nu t}\right)\right], \tag{5.110}$$

where Γ_0 is the initial vortex strength,

$$\Gamma(t) = \Gamma_0\left[1 - \exp\left(-\frac{r^2}{4\nu t}\right)\right], \tag{5.111}$$

and ν is the kinematic viscosity of the fluid; t is the time measured from the instant of formation of the vortex. Use of this value for q in a compressible analysis is justified only if the ratio of the maximum tangential vortex velocity to the local speed of sound (the vortex Mach number) is small.

The vortex core boundary velocity is obtained by maximizing Eq. (5.111) and is

$$q_{\max} = q_m = \frac{0.715\Gamma_0}{4\pi(1.256\nu t)^{1/2}} \tag{5.112}$$

and

$$a = r_m = 2(1.256\nu t)^{1/2}. \tag{5.113}$$

It is seen that for $t = 0$ this implies a zero core size and thus an infinite tangential velocity. One must therefore conclude that there is a virtual origin and that the time (t) is measured from a point in time subsequent to this (the time of formation of a real vortex is finite).

On defining a new variable $\eta = r/a$, Eq. (5.111) becomes

$$\frac{q}{q_m} = \frac{1.4}{\eta}[1 - \exp(-1.256\eta^2)], \tag{5.114}$$

and Eq. (5.109) becomes

$$\frac{1}{\varrho q_m^2}\frac{\partial p}{\partial \eta} = \frac{1}{\eta}\left(\frac{q}{q_m}\right)^2. \tag{5.115}$$

Now, $p = p(\eta)$ only, and so combining Eqs. (5.114) and (5.115), one obtains

$$\frac{1}{\varrho q_m^2}\,dp = \frac{1.96}{\eta}\,[1 - \exp(-1.256\eta^2)]^2\,d\eta. \tag{5.116}$$

Once again, for a low enough vortex speed,† it is apparently reasonable to assume that this radial thermodynamic process is isentropic. It can be shown by the use of Crocco's theorem (which deals with the variation of entropy normal to a streamline) that the entropy production throughout the entire vortex is approximately proportional to the square of the maximum vortex Mach number. Thus,

$$p = Ce^{\gamma}, \tag{5.117}$$

† H. Liepmann and A. Roshko, "Elements of Gas Dynamics." Wiley, New York, 1957.

where γ is the specific heat ratio and C is a constant. Then

$$\frac{dp}{\varrho} = \gamma C \varrho^{\gamma-2}\, d\varrho$$

and Eq. (5.116) becomes

$$\varrho^{\gamma-2} q_m^{-2}\, dp = \frac{1.96}{\gamma C} \frac{1}{\eta^3} [1 - \exp(-1.256\eta^2)]\, d\eta. \tag{5.118}$$

Integrating from $\eta \to \infty$, one obtains

$$\left.\frac{\varrho^{\gamma-1}}{(\gamma-1)q_m{}^2}\right|_{\varrho}^{\varrho_a} = \frac{1.96}{\gamma C} \int_{\eta}^{\infty} \frac{1}{\eta^3} [1 - \exp(-1.256\eta^2)]\, d\eta, \tag{5.119}$$

where ϱ_a is the ambient fluid density. The integral on the right-hand side has been evaluated and the result is

$$\frac{\varrho^{\gamma-1} - \varrho_a^{\gamma-1}}{(\gamma-1)q_m{}^2} = \frac{1.96}{\gamma C}\left\{A(\eta) + 1.256\left[\sum_{l=1}^{\infty} \frac{(-1.256\eta^2)^l}{l l!} - \sum_{m=1}^{\infty} \frac{(-2.512\eta^2)^m}{m m!}\right]\Bigg|_{\eta}^{\infty}\right\}, \tag{5.120}$$

where

$$A(\eta) = \frac{1}{2\eta^2} [1 - \exp(-1.256\eta^2]^2. \tag{5.121}$$

Let

$$B(\eta) = 1.256\left[\sum_{l=1}^{\infty} \frac{(-1.256\eta^2)^l}{l l!} - \sum_{m=1}^{\infty} \frac{(-2.512\eta^2)^m}{m m!}\right]. \tag{5.122}$$

Therefore

$$\varrho^{\gamma-1} - \varrho_a^{\gamma-1} = \frac{1.96(\gamma-1)q_m{}^2}{\gamma C} \{A(\eta) + B(\eta_\infty) - B(\eta)\}. \tag{5.123}$$

Rearrangement gives

$$\frac{\varrho}{\varrho_a} = [1 - \beta H(\eta)]^{1/(\gamma-1)}$$

or

$$\frac{\Delta\varrho}{\varrho_a} \equiv \frac{\varrho - \varrho_a}{\varrho} = [1 - \beta H(\eta)]^{1/(\gamma-1)} - 1, \tag{5.124}$$

where

$$\beta = \frac{1.96(\gamma-1)q_m{}^2}{\gamma C \varrho_a^{\gamma-1}} = \frac{\Gamma_0{}^2(\gamma-1)}{16\pi^2(1.256\nu)\gamma C\varrho_a^{\gamma-1}} \frac{1}{t}, \tag{5.125}$$

$$H(\eta) = A(\eta) + B(\eta_\infty) - B(\eta). \tag{5.126}$$

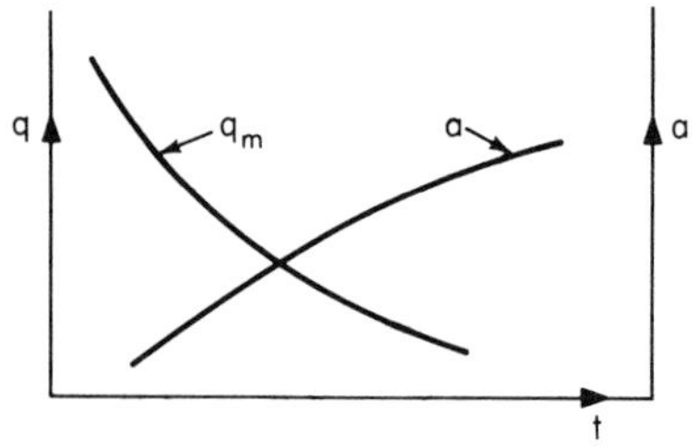

Fig. 5.19

The value of the constant C is obtained from the relation

$$v_s^2 = \gamma C \varrho_a^{\gamma-1}, \tag{5.127}$$

where v_s is the speed of sound in the ambient fluid.

The variation of q_m and a with time is shown in Fig. 5.19. The a increases as $t^{1/2}$ and q falls off exponentially.

Graphs for $\Delta\varrho/\varrho_a$ versus η for a series of t values are shown in Fig. 5.20. The density near the core center is a minimum, but as the core increases in size and its strength (induced circulation) diminishes, this minimum becomes less and less pronounced.

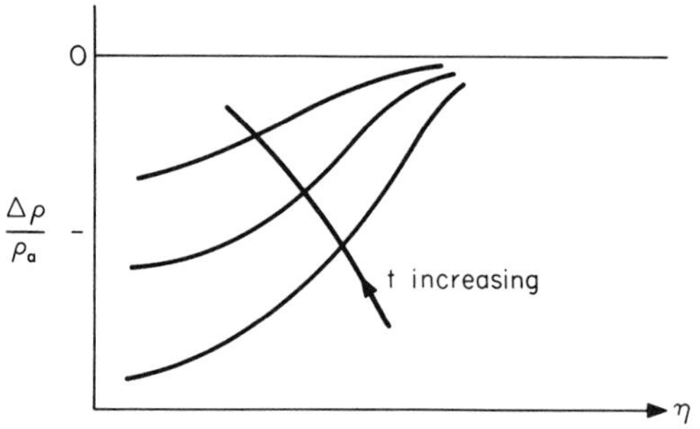

Fig. 5.20

The phenomenon of a vortex in a viscous compressible fluid is, of course, much more realistic than a vortex in a perfect fluid. But meaningful analysis is considerably more difficult. For example, replacement of the incompressible approximation for q and of the assumption of isentropicity would lead to failure to obtain a solution in closed form.

Sound generation by vortices in fluids and their interaction with shock and flame fronts are some of the interesting phenomenon still to be dealt with satisfactorily.

5.11 Hydrodynamic–Electromagnetic Analogy

As has been mentioned earlier, there is a close analogy† between calculations involving vortices (in a nonviscous fluid, or superfluid) and calculations involving current filaments in a magnetic field.

In electromagnetism (mks units),

$$\nabla \cdot \mathbf{B} = 0, \qquad \oint \mathbf{B} \cdot dl = \mu I, \qquad W_B = \frac{B^2}{2\mu}, \tag{5.128}$$

where **B** is the magnetic field intensity, δl an element of a filament, I the current in the filament, μ the permeability of the surrounding medium, and W_B is the magnetic field energy density. Now the hydrodynamic equations for a vortex field are as follows:

$$\nabla \cdot \mathbf{v} = 0, \qquad \oint \mathbf{v} \cdot dl = \varkappa, \qquad W = \frac{\varrho v^2}{2}, \tag{5.129}$$

where **v** is the velocity vector, $\varkappa$ the vortex strength (circulation), ϱ the fluid density, and W the velocity field energy density. Equations (5.128) and (5.129) are equivalent with the substitutions

$$\frac{\mathbf{B}}{\mu} \to \mathbf{v}, \qquad I \to \varkappa, \qquad \mu \to \varrho. \tag{5.130}$$

The analogy will now be used to derive the energy of a rectilinear vortex per unit length and the energy of a vortex ring.

Now the magnetic energy of a current carrying circuit is given by

$$W_B = \tfrac{1}{2} L_{\parallel} I^2$$

where $L_{\parallel}$ is the self-inductance of the circuit. Using the substitutions given in Eq. (5.130), one gets the hydrodynamic energy of a circuit of vortices:

$$W = \frac{\varkappa^2}{2} L_{\parallel}, \tag{5.131}$$

where $L_{\parallel}$ is the magnetic formula with μ replaced by ϱ.

To apply Eq. (5.131), consider a wire of cross-sectional radius a and permeability μ' in a medium of permeability μ. Assume that the current

† See A. Fetter and R. Donnelly, *Phys. Fluids* **9**, 619 (1966).

in the wire is uniformly distributed across the wire. For a long straight piece of wire with its axis at $r = 0$, and if $\mu = 0$ for $r > R$, the inductance per unit length of the wire† is

$$L_{\parallel} = \frac{\mu}{2\pi}\left[\ln\frac{8R}{a} + \frac{\mu'}{4\mu}\right]. \tag{5.132}$$

A circular loop of wire of radius R in an infinite medium has a self-inductance

$$L_{\parallel} = R\mu\left[\ln\frac{8R}{a} - 2 + \frac{\mu'}{\mu}\right]. \tag{5.133}$$

In the special case for which $\mu' = \mu$, the self-inductance of two parallel wires a distance $2b$ apart ($b > a$), carrying equal currents in opposite directions, is given by

$$L_{\parallel} = \frac{\mu}{\pi}\left[\ln\frac{2b}{a} + \frac{1}{4}\right]. \tag{5.134}$$

The analogy implies that the medium of permeability μ outside the wire represents a fluid of density ϱ outside. The wire carrying a total current I represents a vortex strength $\varkappa$. The fact that the wire itself has a permeability μ' implies that the core of the analogous vortex is filled with fluid of density ϱ'. Since it has been postulated that the current is uniformly distributed across the section of the wire, the fluid in the vortex core is undergoing solid body rotation at a frequency ω, equal to the angular frequency of the fluid at the edges of the core.

Using Eqs. (5.130) and (5.132), one obtains the energy per unit length of a rectilinear vortex in a cylindrical container of radius R:

$$E = \frac{\varrho\varkappa^2}{4\pi}\left[\ln\frac{R}{a} + \frac{\varrho'}{4\varrho}\right] \tag{5.135}$$

The energy of a vortex ring of radius R in an infinite fluid is

$$E = \frac{\varrho\varkappa^2 R}{2}\left[\ln\frac{8R}{a} - 2 + \frac{\varrho'}{4\varrho}\right]. \tag{5.136}$$

For the special case in which $\varrho' = \varrho$, the energy of the vortex ring is

$$E = \frac{\varrho\varkappa^2 R}{2}\left[\ln\frac{8R}{a} - \frac{3}{4}\right]. \tag{5.137}$$

† W. R. Smythe, "Static and Dynamic Electricity," 2nd ed., Chapter 8. McGraw-Hill, New York, 1950.

The strict analogy between hydrodynamics and electromagnetism breaks down (see Fetter and Donnelly) when one considers the dynamic behavior of vortices. But there is still a strong similarity. The Magnus force on a section of vortex of length δl moving at a velocity $\mathbf{v}_v$ in a fluid whose local velocity at Δl is $\mathbf{v}_l$ is given by

$$\delta\mathbf{F}_M = \delta l\, \varrho(\mathbf{v}_l - \mathbf{v}_M) \times \boldsymbol{\varkappa}. \tag{5.138}$$

Using the substitutions given in Eq. (5.130) in reverse, one finds that this would be analogous to the force on a current-carrying wire of length δl in a local magnetic field $\mathbf{B}_l$, if the Lorentz force law was modified to read

$$\delta\mathbf{F}_M = -\ \delta l[\mathbf{i} \times (\mathbf{B}_l - \mathbf{B}_v)], \tag{5.139}$$

where $\mathbf{v}_v \rightarrow \mathbf{B}_v/\mu$. Now, the correct Lorentz relation is

$$\delta\mathbf{F}_M = \delta l\, (\mathbf{i} \times \mathbf{B}). \tag{5.140}$$

Thus the vortex filament behaves dynamically like a current filament but with two apparent modifications. The first is that the sign of the Lorentz force is changed. Secondly, the analogous current filament is capable of inducing, by its own motion at a velocity $\mathbf{v}_v$, a compensating field $\mathbf{B}_v = \mu\mathbf{v}_v$, which must be subtracted from the local field. The modified force law is in itself an interesting problem and may have implications in electromagnetism.

PROBLEMS

5.1 Consider a planar flow in concentric circles about an axis with the velocity magnitude proportional to the radius. Derive expressions for

(a) the velocity components as functions of the coordinates,
(b) the dilation,
(c) the rotation,
(d) the acceleration,
(e) the shearing strain rate.

5.2 Repeat Problem (5.1) for the case in which the velocity magnitude is inversely proportional to the radius.

5.3 Show that if the forces acting on a fluid have a potential V and the density is a function of the pressure, then the Bernoulli integral is

$$H = P + \tfrac{1}{2}v^2 + V = f(t),$$

where

$$P = \int \frac{dp}{\varrho},$$

and the vortex lines coincide with the streamlines.

5.4 A fluid rotates about the Oz axis as a rigid body with angular velocity ω. Determine the velocity field.

5.5 Fluid elements rotate about the Oz axis with velocities inversely proportional to the distances of the particles from this axis such that

$$u = \frac{c}{(x^2 + y^2)^{1/2}} \cos(u, x) = -\frac{cy}{x^2 + y^2},$$

$$v = \frac{c}{(x^2 + y^2)^{1/2}} = \frac{cx}{x^2 + y^2},$$

$$w = 0.$$

Determine the velocity field.

5.6 Determine expressions for the pressure distribution as a function of radial position

(a) in the irrotational region of a two-dimensional vortex (use the Bernoulli equation),
(b) in the rotational core of the vortex (use the Euler equation and explain why).

5.7 Two vortexes of strength 20 cm^2/sec but opposite sense are located a distance 4 cm apart. How fast will they move and in what direction?

5.8 Find the equations of the streamlines of

(a) two vortexes of equal strength,
(b) two vortexes of equal, but opposite, strengths.

In each case sketch the streamline pattern.

5.9 Show that the equations of the streamlines of two Kármán vortex streets, distributed in a staggered pattern, is

$$\frac{\sin x + \sinh y}{\cosh y} = C,$$

where the constant C can be varied within the limits $-\sqrt{2} \rightarrow +\sqrt{2}$ (the limits of stability).

5.10 Prove that three infinitely long rectilinear cylindrical vortexes of equal strengths will be in stable steady motion when situated at the vertices of an equilateral triangle of sides much larger than the vortex radii. Also, that if slightly displaced, the time of oscillation equals the time of revolution of the system in its undisturbed state.

5.11 Prove that the velocity field due to a circular vortex at a large distance from itself is the same, approximately, as that of a doublet of strength $\frac{1}{2}\Gamma a^2$, where Γ is the vortex strength and a its radius.

5.12 Show that if the barotropic condition ($\varrho = \varrho(p)$) is satisfied,

$$\frac{\partial}{\partial z}\left(\frac{1}{\varrho}\frac{\partial p}{\partial x}\right) - \frac{\partial}{\partial x}\left(\frac{1}{\varrho}\frac{\partial p}{\partial z}\right) = 0.$$

BIBLIOGRAPHY

J. W. Robertson, "Hydrodynamics in Theory and Application," Chapter 4. Prentice-Hall, Englewood Cliffs, New Jersey, 1965.

H. Lamb, "Hydrodynamics," 6th ed. Dover, New York, 1945.

A. B. Basset, "A Treatise on Hydrodynamics," Volumes I and II. Dover Reprints, New York, 1961.

G. Greenhill, On plane vortex motion, *Quart. J. Pure Appl. Math.* **15**, 10–29 (1877).

C. Truesdell, "The Kinematics of Vorticity." Indiana Univ. Press, Bloomington, 1954.

"Advances in Applied Mechanics" (H. Dryden and T. von Kármán, eds.), Volume 6, pp. 273–287. Academic Press, New York, 1960.

N. Kochin, J. Kibel, and N. Roze, "Theoretical Hydromechanics," Chapter 5. Wiley (Interscience), New York, 1964.

CHAPTER SIX *VORTICITY AND THE LAMINAR BOUNDARY LAYER*

6.1 Introduction

In earlier chapters vorticity has been shown to be equal to half the angular velocity of the fluid particles. The pattern of flow in many problems in hydrodynamics is determined by the vorticity distribution. Indeed, the solution of hydrodynamical problems is facilitated by finding those regions, in which vorticity is concentrated and in which it is weak. Thus, where vorticity can sensibly be neglected the velocity distribution can be obtained from a potential function which satisfies Laplace's equation. In this case the problem is essentially linear. When vorticity is present, the problem of finding the flow is more difficult, because the equations become nonlinear due to the presence of the term $\mathbf{u} \times \boldsymbol{\zeta}$ in the equations of motion. One further advantage of considering the vorticity first is that pressure does not enter the equations of motion, thus making the variables vorticity and velocity rather than pressure and velocity.

6.2 The Vorticity Equation

The momentum equation may be written [see Eq. (4.78) and Section 5.4]

$$\frac{\partial \mathbf{u}}{\partial t} + \operatorname{grad}\left(\frac{p}{\varrho} + \frac{1}{2}\mathbf{u}^2\right) + \mathbf{u} \times \boldsymbol{\zeta} = \nu \operatorname{curl} \boldsymbol{\zeta}. \tag{6.1}$$

Taking the curl of this equation eliminates the pressure and one gets

$$\frac{\partial \boldsymbol{\zeta}}{\partial t} + \operatorname{curl}(\mathbf{u} \times \boldsymbol{\zeta}) = \nu \operatorname{curl} \operatorname{curl} \boldsymbol{\zeta}. \tag{6.2}$$

In Cartesian coordinates this equation becomes

$$\frac{D\boldsymbol{\zeta}}{Dt} = \nu \nabla^2 \boldsymbol{\zeta}. \tag{6.3}$$

The left side of this equation gives the rate at which vorticity is transported through the fluid by virtue of the motion of the fluid particles; the right side gives the molecular diffusion of the vorticity. It will be noted that in those regions in which the right side is negligible, $D\boldsymbol{\zeta}/Dt = \mathbf{0}$. This means that in inviscid flows, for instance, the vorticity is frozen into the particles of the fluid. Physically this is because in an inviscid fluid shear stresses are zero, so that there is no mechanism by which vorticity can be transferred from one fluid particle to another.

An important case is that of two-dimensional flow in the xy plane. Then the vorticity vector is always parallel to the z axis,

$$\boldsymbol{\zeta} = \zeta(x, y)\mathbf{k}.$$

In this case the vorticity equation reduces to the scalar equation

$$\frac{D\zeta}{Dt} = \nu \nabla^2 \zeta. \tag{6.4}$$

This equation may be compared with the energy equation which is, in incompressible flows,

$$\frac{DT}{Dt} = \frac{\nu}{Pr} \nabla^2 T.$$

(In this case viscous heating is neglected.)

There is therefore a direct analogy in the two-dimensional case between the transport of heat and vorticity. This is important because it enables the general distribution of vorticity to be found from a comparable problem in the flow of heat.

6.3 The Creation of Vorticity

The definition $\boldsymbol{\zeta} = \text{curl}\,\mathbf{u}$ implies that div $\boldsymbol{\zeta} \equiv$ div curl $\mathbf{u} \equiv 0$. When Gauss's theorem is used, it is clear that this means that there are no sources or sinks of vorticity *in the fluid*. Vorticity is created by the frictional shear stress at the solid boundaries of the fluid. A simple example is the two-dimensional flow past a solid wall coincident with the x axis. Then

$$\frac{\partial \zeta}{\partial y} = \frac{\partial}{\partial y}\left(\frac{\partial v}{\partial x} - \frac{\partial u}{\partial y}\right) = -\left(\frac{\partial^2 u}{\partial x^2} + \frac{\partial^2 u}{\partial y^2}\right), \tag{6.5}$$

using the two-dimensional incompressible continuity equation. On the wall, $\partial^2 u/\partial x^2 = 0$, so

$$\left(\frac{\partial \zeta}{\partial y}\right)_{\text{w}} = -\left(\frac{\partial^2 u}{\partial y^2}\right)_{\text{w}}. \tag{6.6}$$

The wall therefore acts as a source of vorticity of strength, $-\partial^2 u/\partial y^2$ per unit length, where y is the direction normal to the wall.

In the three-dimensional flow, $\boldsymbol{\zeta}$ is a vector with three components. However, here too a similar result holds.

6.4 A Point Source of Vorticity

A simple example of how vorticity is transported in a fluid is provided by the two-dimensional flow given by a uniform flow U_0 parallel to the x axis which is slightly perturbed by a unit line source of vorticity at the origin.

The equation of vorticity in this case is

$$\frac{\partial \zeta}{\partial x} \doteq \frac{\nu}{U_0} \nabla^2 \zeta, \tag{6.7}$$

where $v\,\partial\zeta/\partial y$ has been neglected in comparison with $U_0\,\partial\zeta/\partial x$ and where $u\,\partial\zeta/\partial x$ has been approximated by $U_0\,\partial\zeta/\partial x$. The above equation may be simplified by putting $\zeta = e^{kx} f(x, y)$, where $k = U_0/2\nu$. Then

$$\nabla^2 f = k^2 f. \tag{6.8}$$

A solution of this equation which behaves like a line source at the origin is obtained by separation of variables and is

$$f = K_0(kr), \tag{6.9}$$

where K_0 is the Bessel function of the second kind of order zero and $r = (x^2 + y^2)^{1/2}$. Thus

$$\zeta \doteq e^{kr\cos\vartheta} K_0(kr). \tag{6.10}$$

The physical orders of magnitude may be evaluated in terms of the dimensionless length

$$kr = \frac{U_0}{2\nu}\, r,$$

which gives the ratio of r to a quantity which has the physical dimensions of length $2\nu/U_0$. (This quantity is called the viscous length.) Now at large kr, that is, distances large compared with the viscous length (say five or six times larger),

$$K_0(kr) \sim \left(\frac{\pi}{2kr}\right)^{1/2} e^{-kr}. \tag{6.11}$$

Thus at large radial distances in general, vorticity decays exponentially as the ratio of r to the viscous length. An exception to this is where ϑ is small, that is, close to the positive x axis, where

$$\zeta \sim \left(\frac{\pi}{2kx}\right)^{1/2} \exp[-kr(1 - \cos\vartheta)] \sim \left(\frac{\pi}{2kx}\right)^{1/2} \exp\left[-kx\left(\frac{\vartheta^2}{2}\right)\right] \tag{6.12}$$

and where the decay is much smaller and is of order $1/(kx)^{1/2}$. This region is called the wake region of the flow. It is a parabolic region whose boundaries are defined by $ky^2/x \sim 1$ (Fig. 6.1). The shaded areas of Fig. 6.1 are regions where vorticity is appreciable. Outside these regions, vorticity $\sim e^{-kr}/(kr)^{1/2}$.

This vorticity is sensibly confined to two regions, namely within a circle of radius equal to the viscous length $2\nu/U_0$ and within a wake region defined by ky^2/x small.

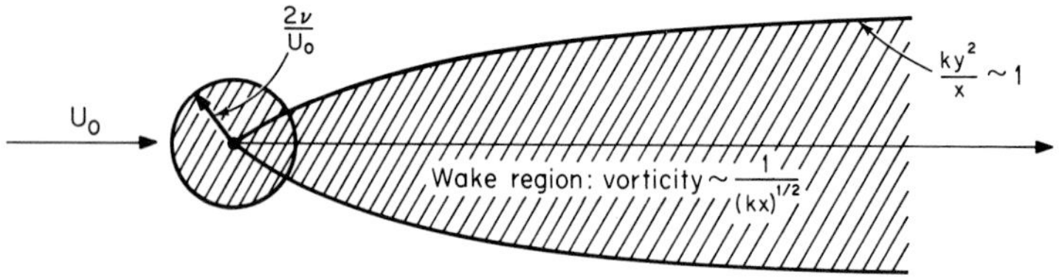

Fig. 6.1

6.5 Flow past a Solid Body from Infinity

Consider now the uniform two-dimensional flow past a solid body in the (xy) plane. At large distances upstream from the body, $\mathbf{u} \doteq U_0\mathbf{i}$ and the vorticity is zero. The vorticity in the flow issues from the solid body which acts as a distributed source of vorticity in the fluid.

An intuitive picture may be obtained from the thermal analogy described in the previous section. Thus, if the body acts as a source of heat in the fluid, then this heat is diffused molecularly around the body and is convected downstream with approximate speed U_0.

Two simple limiting cases are of interest: first, when the length L of the body is much smaller than the viscous length $2\nu/U_0$. In this case the body may at distances comparable to $2\nu/U_0$ be regarded as a point source of vorticity, and the distribution is that given in the previous section.

The second instance is that where L is much greater than $2\nu/U_0$. Here the body acts as a distributed source of vorticity. From a single source, vorticity propagates through a radial distance roughly equal to $2\nu/U_0$ by diffusion.

Thus, the vorticity from the surface of the body diffuses into a layer of thickness $2\nu/U_0$ around the body. Outside this layer the vorticity is exponentially small. This vorticity is washed downstream with speed U_0. At large radial distances the body may be regarded as a point source of vorticity, and the wake therefore occupies the parabolic region $ky^2/x \sim 1$. Closer to the body it is difficult to predict the precise distribution of vorticity, for the reason that the flows induced by the distributed vortex sources interact nonlinearly in this region. In this instance a steady flow has been assumed—often, however, the flow becomes unsteady.

6.6 The Reynolds Number Effect

In the examples treated in the previous section there existed two fundamental lengths: namely, the length L of the body and the viscous length $2\nu/U_0$. The ratio of these lengths $L/(2\nu/U_0)$ is usually defined to be one half the Reynolds number of the flow. Thus in the case where L is small compared with $2\nu/U_0$, the Reynolds number is small and the body is effectively immersed in a sea of vorticity. In the second example the Reynolds number is large and the vorticity is effectively confined to a thin skin around the body; that is to say, thin relative to the length of the body.

It has been shown earlier that when a viscous fluid flows over a body, the vorticity is confined to a thin layer on the surface of the body when the Reynolds number is large. This holds provided the vorticity does not escape from near the surface of the body. It will be assumed in this chapter that the conditions necessary for this proviso are satisfied.

6.7 Equations of Motion

For simplicity in analysis, only two-dimensional flows will be considered. Take as a concrete case the flow in the xy plane of speed U_0 over a flat plate coincident with the positive x axis. Then the vorticity ζ, which is in the z direction, satisfies the equation

$$u\frac{\partial \zeta}{\partial x} + v\frac{\partial \zeta}{\partial y} = \nu\, \nabla^2 \zeta. \tag{6.13}$$

It will be assumed that significant changes take place in ζ over a length L in the x direction and over a length ε in the y direction. (ε gives a measure of the thickness of the boundary layer region.) Then

$$\frac{\partial}{\partial x} \sim \frac{1}{L} \quad \text{and} \quad \frac{\partial}{\partial y} \sim \frac{1}{\varepsilon}.$$

The orders of magnitude of the terms in (6.13) are then given by

$$\begin{array}{ccccccc} u\dfrac{\partial \zeta}{\partial x} & + & v\dfrac{\partial \zeta}{\partial y} & = & \nu\Big(\dfrac{\partial^2 \zeta}{\partial x^2} & + & \dfrac{\partial^2 \zeta}{\partial y^2}\Big). \\[2ex] \dfrac{U_0\zeta}{L} & & \dfrac{v\zeta}{\varepsilon} & & \dfrac{\nu\zeta}{L^2} & & \dfrac{\nu\zeta}{\varepsilon^2} \end{array} \tag{6.14}$$

The dominant term on the right side of (6.13) has an order of magnitude $\nu\zeta/\varepsilon^2$, while the terms on the left side both have the same order of magnitude $U_0\zeta/L$; this follows from the fact that the continuity equation implies that

$$\frac{U_0}{L} \sim \frac{v}{\varepsilon} \tag{6.15}$$

inside the boundary layer. Now ζ satisfies two boundary conditions in the y direction: it is zero at the edge of the boundary layer and it has an unknown value on the x axis. Thus a second-order differential equation in y is required. Hence, however small ν may be, it is necessary that the right-side terms of (6.14) should remain in the equation and exert an influence comparable with the terms on the left-hand side. This implies that

$$\frac{U_0\zeta}{L} \sim \frac{\nu\zeta}{\varepsilon^2}, \qquad \text{that is,} \quad \frac{\varepsilon^2}{L^2} \sim \frac{\nu}{U_0 L} \equiv \frac{1}{\mathrm{Re}_L}, \tag{6.16}$$

where Re_L is the Reynolds number of the flow. This means that at the edge of the boundary layer the streamlines are inclined at an angle $1/(\mathrm{Re}_L)^{1/2}$ to the x axis; that is, the plate pushes aside the streamlines of the potential flow close to it.

Consider now the momentum equations

$$u\frac{\partial u}{\partial x} + v\frac{\partial u}{\partial y} = -\frac{1}{\varrho}\frac{\partial p}{\partial x} + \nu\left(\frac{\partial^2 u}{\partial x^2} + \frac{\partial^2 u}{\partial y^2}\right), \tag{6.17}$$

$$u\frac{\partial v}{\partial x} + v\frac{\partial v}{\partial y} = -\frac{1}{\varrho}\frac{\partial p}{\partial y} + \nu\left(\frac{\partial^2 v}{\partial x^2} + \frac{\partial^2 v}{\partial y^2}\right). \tag{6.18}$$

An estimate of the behavior of p in the boundary layer may be found from (6.18).

The orders of magnitude of the terms in (6.18) are as follows:

$$\begin{array}{ccccc} u\dfrac{\partial v}{\partial x} + & v\dfrac{\partial v}{\partial y} = & -\dfrac{1}{\varrho}\dfrac{\partial p}{\partial y} + & \nu\left(\dfrac{\partial^2 v}{\partial x^2} + \right. & \left.\dfrac{\partial^2 v}{\partial y^2}\right). \\ \dfrac{U_0 v}{L} & \dfrac{v^2}{\varepsilon} & & \dfrac{\nu v}{L^2} & \dfrac{\nu v}{\varepsilon^2} \end{array} \tag{6.19}$$

Hence

$$-\frac{1}{\varrho}\frac{\partial p}{\partial y} \sim \frac{\varepsilon U_0^2}{L^2},$$

so that

$$-\frac{p}{\varrho U_0^2} \sim \frac{\varepsilon^2}{L^2} + f(x),$$

where $f(x)$ is an arbitrary function of x. Thus

$$-\frac{p}{\varrho U_0^2} \sim \frac{1}{\mathrm{Re}_L} + f(x). \tag{6.20}$$

It follows that the pressure is sensibly constant over a cross section of the boundary layer; and since pressure is continuous at the edge of the boundary layer, $f(x)$ is equal to the value of $-p/\varrho U_0^2$ in the potential region close to the boundary layer. Pressure may then be said to be imprinted on the boundary layer by the outside flow.

Turn now to Eq. (6.17). This equation may be approximated by

$$u\frac{\partial u}{\partial x} + v\frac{\partial u}{\partial y} = -\frac{1}{\varrho}\frac{\partial p_0}{\partial x} + \nu\frac{\partial^2 u}{\partial y^2}, \tag{6.21}$$

where p_0 is the pressure in the potential region just outside the boundary layer. In obtaining (6.21), terms of order Re_L^{-1} have been neglected; thus $\partial^2 u/\partial x^2$ is neglected in comparison with $\partial^2 u/\partial y^2$, and the slight variation in pressure across the boundary layer, which is of order Re_L^{-1}, has been ignored.

In summary, the equations of motion inside the boundary layer on the surface of a body are

$$\frac{\partial u}{\partial x} + \frac{\partial v}{\partial y} = 0, \tag{6.22}$$

$$u\frac{\partial u}{\partial x} + v\frac{\partial u}{\partial y} = -\frac{1}{\varrho}\frac{\partial p_0}{\partial x} + \nu\frac{\partial^2 u}{\partial y^2}, \tag{6.23}$$

$$p = p_0(x), \tag{6.24}$$

where terms of relative order Re_L^{-1} have been ignored. Equations (6.22)–(6.24) are known as the boundary layer equations.

The chief simplification effected by replacing the Navier–Stokes equations by (6.22)–(6.24) is that pressure is no longer an unknown; however, (6.23) is still nonlinear so that the superposition of solutions remains impossible. Thus, exact general methods of solution of boundary layer problems do not exist.

6.8 Exact Solutions of the Boundary Layer Equations

A few exact solutions of the boundary layer equations exist for simple flow patterns. A number of these will be given in this section.

6.8.1 Flow over a Flat Plate (Blasius Flow)

One of the most important problems is that of uniform flow with speed U_0 over a flat plate (Fig. 6.2). Since the speed in the potential region outside the boundary layer is constant, it follows from Bernoulli's theorem

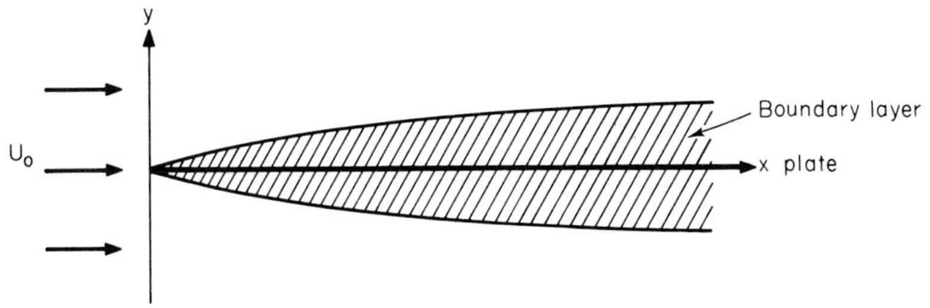

Fig. 6.2

that the pressure is constant. Thus, the pressure term drops out of (6.23), and the equations of motion inside the boundary layer are

$$u \frac{\partial u}{\partial x} + v \frac{\partial u}{\partial y} = \nu \frac{\partial^2 u}{\partial y^2}, \tag{6.25}$$

$$\frac{\partial u}{\partial x} + \frac{\partial v}{\partial y} = 0. \tag{6.26}$$

Equation (6.26) may be eliminated by introducing a stream function ψ such that $u = \partial\psi/\partial y$, and $v = -\partial\psi/\partial x$. Then (6.25) becomes

$$\frac{\partial \psi}{\partial y} \frac{\partial^2 \psi}{\partial x\, \partial y} - \frac{\partial \psi}{\partial x} \frac{\partial^2 \psi}{\partial y^2} = \nu \frac{\partial^3 \psi}{\partial y^3}. \tag{6.27}$$

The boundary conditions on ψ are, at $y = 0$,

$$\psi = 0, \qquad \frac{\partial \psi}{\partial y} = 0,$$

and at the edge of the boundary layer,

$$\frac{\partial \psi}{\partial y} \to U_0.$$

The simple geometry of this problem implies that the velocity at a point in the boundary layer depends only on its relative position in the boundary layer; for example, the velocity at half the boundary layer thickness is constant for all x sufficiently large. In mathematical terms, if the boundary layer thickness is proportional to x^q, then

$$u = f\left(\frac{y}{x^q}\right),$$

that is,

$$\psi = x^q f\left(\frac{y}{x^q}\right). \tag{6.28}$$

The value of q is found to be $\frac{1}{2}$ when (6.28) is inserted in (6.27). The equation for f may be made nondimensional by putting

$$\psi = (2\nu U_0 x)^{1/2}\, F(\eta),$$

where

$$\eta = \left(\frac{U_0 x}{2\nu}\right)^{1/2} \frac{y}{x}.$$

Then F satisfies the ordinary differential equation

$$F''' + FF'' = 0, \tag{6.29}$$

where the primes denote differentiation with respect to η. The boundary conditions on F are

$$\eta = 0: \quad F = F' = 0,$$

$$\eta \to \infty: \quad F' \to 1.$$

A solution of this equation has been determined by numerical methods and is shown graphically in Fig. 6.3. The behavior of the solution at large values of η is of interest. At large η, $F' \to 1$, so that $F \sim a + \eta$, where a is a constant. Inserting this value for F in (6.29), one gets

$$F''' + (a + \eta)F'' \sim 0.$$

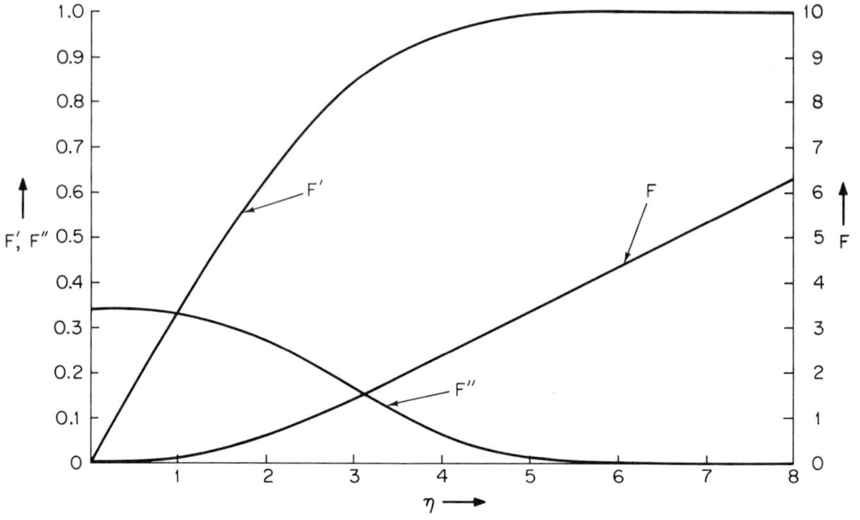

Fig. 6.3

Thus

$$F'' \sim a_1 \exp\left(-a\eta - \frac{\eta^2}{2}\right) \qquad \text{and} \qquad F' \sim 1 + b_1 \exp\left(-a\eta - \frac{\eta^2}{2}\right),$$

(where a_1 and b_1 are constants) at large values of η. These expressions show clearly that the final velocity is attained exponentially and that the vorticity which is proportional to F'' tends exponentially to zero at large η. A rough order of magnitude for the thickness of the boundary layer is obtained by observing that u/U_0 is equal to 0.95 at $\eta \doteq 2.8$, which gives a boundary layer thickness of order $2.8\sqrt{2}\, x/(\mathrm{Re}_x)^{1/2}$, where $\mathrm{Re}_x = U_0 x/\nu$. It will be noted that the relative error involved in this case is of order Re_x^{-1}, that is, $\nu/U_0 x$. Thus the solution is valid only for downstream flow, where the distance x is much larger than the viscous length $2\nu/U_0$.

6.8.2 The Two-Dimensional Jet

A symmetrical jet issues from a thin slit which lies along the z axis. The boundary conditions are, at $y = 0$, the axis of the jet, $\partial u/\partial y = 0$, and at the edges of the jet $u = 0$. The equations of motion will be assumed to be the boundary layer equations. This appears paradoxical in view of the fact that no solid boundary is present. However, the validity of the boundary layer equations depends basically on the fact that $\partial/\partial x \ll \partial/\partial y$. The validity of this assumption may be verified a posteriori. The equations of motion are (6.25)–(6.27). (Pressure again drops out because the pressure outside the jet is constant.) Again, by virtue of the geometrical simplicity of the flow, the argument given in the flat plate problem leads to the "similar" solution

$$\psi = x^p f\left(\frac{y}{x^q}\right). \tag{6.30}$$

This assumes that x is sufficiently large with respect to the finite width of the slit to be neglected. One condition on the values of p and q is obtained by substituting (6.30) in (6.27). A second condition follows from the fact that the momentum flux is constant along all sections of the jet. This condition follows from integration of (6.25); thus,

$$\frac{d}{dx}\int_0^\infty u^2\, dy + [uv]_0^\infty = \left[\nu \frac{\partial u}{\partial y}\right]_0^\infty .$$

Since the terms in square brackets vanish, the result

$$\int_0^\infty u^2\, dy = \text{constant} = \frac{1}{2}\,\frac{M}{\varrho} \tag{6.31}$$

follows, where M is the initial momentum flux. Taken together, the conditions fix p and q at $\frac{1}{3}$ and $\frac{2}{3}$, respectively. A simple form of the equation for f is obtained by putting

$$\psi = 6a\nu x^{1/3} F(\xi),$$

where $\xi = ay/x^{2/3}$ and a is a constant. Then F satisfies the equation

$$F''' + 2FF'' + 2F'^2 = 0. \tag{6.32}$$

The solution of this equation is

$$F = \tanh \xi. \tag{6.33}$$

The constant a is fixed by (6.31) and is equal to $(M/48\varrho\nu^2)^{1/3}$, and $u = 6a^2\nu x^{-1/3}\,\text{sech}^2\xi$. The Reynolds number in this case is given by $\text{Re}_x = u_{\text{axis}}x/\nu = 6a^2x^{2/3}$. Thus the error involved in the boundary layer equations is proportional to $x^{-2/3}$. The solution is therefore valid at large distances from the origin. The transverse component of velocity at the edge of the jet is $-2\nu a x^{-2/3}$. The jet acts therefore as a sink on the fluid which surrounds it.

6.8.3 Axially Symmetric Boundary Layers

In axial coordinates (x, r), the boundary layer equation is

$$u\frac{\partial u}{\partial x} + v\frac{\partial u}{\partial r} = -\frac{1}{\varrho}\frac{\partial p}{\partial r} + \nu\left(\frac{\partial^2 u}{\partial r^2} + \frac{1}{r}\frac{\partial u}{\partial r}\right), \tag{6.34}$$

and the equation of continuity is

$$\frac{\partial}{\partial x}(ur) + \frac{\partial}{\partial r}(vr) = 0. \tag{6.35}$$

The equation of continuity may be satisfied by introducing the stream function ψ defined by

$$ur = \frac{\partial \psi}{\partial r}, \qquad vr = -\frac{\partial \psi}{\partial x}.$$

An interesting problem in axially symmetric flow is that of a uniform flow of speed U_0 past a semi-infinite circular cylinder of radius a whose axis is parallel to the flow and which lies along the positive x axis. The boundary layer then grows in thickness with increasing x; at first its thickness is very much smaller than the radius of the cylinder, then it becomes comparable with the cylinder radius, and finally far down the cylinder it becomes very much larger than the radius.

In this problem there are therefore two limiting cases, namely, at small values of x and at large values of x.

Case (i) For small values of x, the boundary layer is small in thickness and is very much smaller than the radius of the cylinder. This means that the flow is approximately that over a flat plate. In this case the boundary layer thickness has an order of magnitude $(4\nu x/U_0)^{1/2}$. The flat plate profile will clearly be a good approximation when the boundary layer thickness is very much smaller than the radius of the cylinder; this may be expressed in terms of the dimensionless quantity

$$\frac{4\nu x}{U_0 a^2}, \qquad \text{since} \qquad \frac{4\nu x}{U_0 a^2} \ll 1.$$

This suggests the change of variables from (x, r) to (X, Y), where

$$X = 2\left(\frac{4\nu x}{U_0 a^2}\right)^{1/2}, \qquad Y = \left(\frac{U_0 a^2}{4\nu x}\right)^{1/2}\left[\frac{r^2 - a^2}{2a^2}\right]. \tag{6.36}$$

It will be noted that Y reduces approximately to

$$\left(\frac{U_0(r-a)^2}{4\nu x}\right)^{1/2}$$

when the boundary layer thickness is very small compared with a. This form clearly corresponds to the Blasius flat plate variable $(U_0 y^2/4\nu x)^{1/2}$ with $y = r - a$.

The stream function ψ is now written

$$\psi = 2\nu x\left(\frac{U_0 a^2}{4\nu x}\right)^{1/2} F(X, Y). \tag{6.37}$$

This implies that

$$u = \frac{1}{2} U_0 \frac{\partial F}{\partial Y}. \tag{6.38}$$

The boundary layer equation now becomes

$$\frac{\partial}{\partial y}\left[(1 + 2YX)\frac{\partial^2 F}{\partial Y^2}\right] + F\frac{\partial^2 F}{\partial Y^2} + X\left[\frac{\partial F}{\partial X}\frac{\partial^2 F}{\partial Y^2} - \frac{\partial F}{\partial Y}\frac{\partial^2 F}{\partial Y\,\partial X}\right] = 0 \tag{6.39}$$

with boundary conditions

$$\frac{\partial F}{\partial Y} = 0, \quad F + X\frac{\partial F}{\partial X} = 0, \quad Y = 0 \quad \text{and} \quad \frac{\partial F}{\partial Y} \to 2 \quad \text{as} \quad Y \to \infty. \tag{6.40}$$

A first approximation F_0 to F may be obtained by putting $X = 0$ in the above equation. Then F_0 satisfies the ordinary differential equation

$$\frac{d^3 F_0}{dY^3} + F_0\frac{d^2 F_0}{dY^2} = 0 \tag{6.41}$$

with boundary conditions

$$F_0 = 0, \quad \frac{\partial F_0}{\partial Y} = 0, \quad Y = 0 \quad \text{and} \quad \frac{\partial F_0}{\partial Y} \to 2 \quad \text{as} \quad Y \to \infty. \tag{6.42}$$

This equation and the boundary condition for F_0 are exactly the same as for the Blasius flat plate case. Now the error involved in the above approximation is of order X. This suggests that a better approximation to F is

$$F \doteq F_0 + XF_1(Y). \tag{6.43}$$

When this approximation is inserted in the boundary layer equation and the terms in X^2 are neglected, it is found that

$$F_1''' + F_0 F_1'' + F_0' F_1' + 2F_0'' F_1 + YF_0''' + F_0'' = 0. \qquad (6.44)$$

where primes denote differentiation with respect to Y. The boundary conditions satisfied by F_1 are

$$F_1 = 0, \quad F_1' = 0, \quad Y = 0 \quad \text{and} \quad F_1' \to 0 \quad \text{as} \quad Y \to \infty. \qquad (6.45)$$

The above equation may be solved numerically. This process can obviously be continued with

$$F = F_0 + XF_1 + X^2 F_2 + \cdots.$$

Case (*ii*) In this case the boundary layer is very much thicker than the radius of the cylinder. Since the cylinder is now a line singularity on the axis of the flow, the geometry of the problem is simple and a similarity-type solution,

$$\psi = x^p f(rx^q), \qquad (6.46)$$

is sought. The equation in this case reduces to an ordinary differential equation in rx^q only if $p = 1$, and the boundary condition $u \to U_0$ as $r \to \infty$ can be satisfied only if $q = -\frac{1}{2}$. This suggests the following dimensionless variables:

$$\xi = \frac{U_0 r^2}{4\nu x}, \qquad \beta = \log\left(\frac{4\nu x}{U_0 a^2}\right). \qquad (6.47)$$

Then

$$\psi = \nu x f(\xi), \qquad u = \tfrac{1}{2} U_0 f', \qquad v = \frac{\nu}{r}(\xi f' - f). \qquad (6.48)$$

The boundary conditions are

$$f' \to 2 \quad \text{as} \quad \xi \to \infty, \quad f = f' = 0, \quad \xi = e^{-\beta}. \qquad (6.49)$$

The equation for f is

$$\xi f''' + f'' + \tfrac{1}{2} f f'' = 0. \qquad (6.50)$$

Close to the wire, that is, for $\xi \sim e^{-\beta}$, f is small and this equation becomes approximately

$$\xi f''' + f'' = 0, \qquad (6.51)$$

and the approximate solution near the wire is

$$f' = A + B \log \xi, \qquad f = (A - B)\xi + B(\xi \log \xi) + C, \qquad (6.52)$$

and hence $B = A/\beta$, $C = Ae^{-\beta}/\beta$. Thus a solution of the above equation which is independent of β can only exist with an error of order $1/\beta$ in the velocity condition at the cylinder. This error may be corrected by expanding the stream function in the series

$$\psi = \nu x\left[f_0(\xi) + \frac{1}{\beta} f_1(\xi) + \cdots\right]. \qquad (6.53)$$

The equation for f_0 is

$$\xi f_0''' + f_0'' + \tfrac{1}{2} f_0 f_0'' = 0. \qquad (6.54)$$

This equation has the simple solution

$$f_0 = 2\xi. \qquad (6.55)$$

This satisfies all the conditions except for the velocity condition on the wire.

The equation for f_1 may be found by inserting the series expansion for ψ in the boundary layer equation and equating the coefficients of $1/\beta$. Thus

$$\xi f_1''' + f_1'' + \tfrac{1}{2} f_0 f_1'' + \tfrac{1}{2} f_1 f_0'' = 0. \qquad (6.56)$$

This simplifies to

$$\xi f_1''' + f_1''(1 + \xi) = 0 \qquad (6.57)$$

when $f_0 = 2\xi$ and has the solution

$$f_1' = A \int_\infty^\xi \frac{e^{-\xi}}{\xi}\, d\xi \equiv A\, \mathrm{Ei}(-\xi). \qquad (6.58)$$

Now $\mathrm{Ei}(-\xi)$ is a well-known function, which behaves like $\log \xi$ near $\xi = 0$; near $\xi = 0$ it has in fact the asymptotic expansion $\log \xi + \gamma$, where γ is Euler's constant 0.5772.

Now the velocity profile is given by

$$\frac{u}{U_0} = \frac{1}{2}\left[f_0' + \frac{1}{\beta} f_1' + \cdots\right]. \qquad (6.59)$$

Hence on the surface of the wire, $\xi = e^{-\beta}$,

$$\frac{u}{U_0} \doteq \frac{1}{2}\left[2 + \frac{1}{\beta} A(-\beta + \gamma)\right] \tag{6.60}$$

when β is large. It follows that if $A = 2$, the velocity condition on the surface of the cylinder is satisfied with an error of order $1/\beta$. This error may be reduced to order $1/\beta^2$ by evaluating the function f_2.

Thus, to within an error of order $1/\beta$, the velocity profile is

$$\frac{u}{U_0} \doteq 1 + \frac{1}{\beta}\int_{\infty}^{\xi} \frac{e^{-\xi}}{\xi}\, d\xi. \tag{6.61}$$

6.9 Approximate Methods of Solution of the Boundary Layer Equations

For most problems of practical interest, it is almost impossible to find exact solutions of the boundary layer equations. However, in these cases only a few bulk properties of the solution may be of interest. These are

skin friction coefficient: $\dfrac{(\mu\, \partial u/\partial y)_{y=0}}{\varrho U_0^2} = \dfrac{\tau}{\varrho U_0^2}$,

displacement thickness: $\delta_1 = \displaystyle\int_0^{\delta} \left(1 - \frac{u}{U_0}\right) dy$,

momentum thickness: $\vartheta = \displaystyle\int_0^{\delta} \left(1 - \frac{u}{U_0}\right)\left(\frac{u}{U_0}\right) dy$,

where δ is the thickness of the boundary layer and U_0 is the outer stream velocity.

The physical meaning of the displacement thickness is as follows: the volume carried in the boundary layer is equal to that of a uniform stream of speed U_0 and thickness $\delta - \delta_1$. A similar meaning may be attached to the momentum thickness ϑ.

Approximations to these coefficients may be obtained with fair accuracy by the approximate methods to be explained. Two main types of method will be dealt with; namely, the Rayleigh analogy and the Pohlhausen method.

6.9.1 The Rayleigh Analogy

The chief difficulty in solving the boundary layer equations exactly lies in the nonlinear differential operator

$$u \frac{\partial}{\partial x} + v \frac{\partial}{\partial y}.$$

This operator gives the transport rate through the fluid. One method of simplifying the boundary layer equations while at the same time taking some account of these terms lies in the approximation

$$u \frac{\partial}{\partial x} + v \frac{\partial}{\partial y} \doteq CU_0 \frac{\partial}{\partial x}, \tag{6.62}$$

where U_0 is the outer velocity and C is a constant lying between 0 and 1. This approximation is clearly useful in flows which are parallel to the x axis; in this case the operator becomes $u\, \partial/\partial x$ and the approximation means that the variable velocity u is replaced throughout a cross section of the boundary layer by a suitable value of u at some point in this cross section; obviously $C = 1$ gives an overestimate of this effect. Now boundary layer flows are sensibly unidirectional, thus the above approximation is clearly useful.

The boundary layer equation therefore becomes

$$CU_0 \frac{\partial u}{\partial x} = -\frac{1}{\varrho} \frac{dp_0}{dx} + \nu \frac{\partial^2 u}{\partial y^2}, \tag{6.63}$$

a linear heat-conduction-type equation for which standard methods of solution have been developed.

As a simple example, take the flow over a flat plate with constant outer velocity U_0. Then the Rayleigh equation is

$$\frac{\partial u}{\partial x} = \frac{\nu}{CU_0} \frac{\partial^2 u}{\partial y^2} \tag{6.64}$$

with boundary conditions $u = 0$ on $y = 0$, $u \to U_0$ as $y \to \infty$. This problem is well known in the theory of heat conduction; it has the solution

$$\frac{u}{U_0} = \operatorname{erf}\left[\frac{1}{2} y \left(\frac{CU_0}{x\nu}\right)^{1/2}\right]. \tag{6.65}$$

The skin friction coefficient on the plate is

$$\frac{(\mu\,\partial u/\partial y)_{y=0}}{\varrho U_0^2} = \left(\frac{\nu C}{\pi x U_0}\right)^{1/2} = 0.563\left(\frac{\nu C}{U_0 x}\right)^{1/2}. \tag{6.66}$$

This may be compared with the result of 6.8.1:

$$\frac{(\mu\,\partial u/\partial y)_{y=0}}{\varrho U_0^2} = 0.332\left(\frac{\nu}{U_0 x}\right)^{1/2}.$$

The Rayleigh analogy gives the correct x dependence, and the two methods agree if $C \sim 0.35$.

This suggests that fair results may be obtained by replacing the boundary layer equation by

$$0.35 U_0 \frac{\partial u}{\partial x} = -\frac{1}{\varrho}\frac{dp_0}{dx} + \nu \frac{\partial^2 u}{\partial y^2}. \tag{6.67}$$

6.9.2 The Pohlhausen Method

Consider the boundary layer on a flat plate; let the thickness of the boundary layer be $\delta(x)$. Then on integrating the continuity equation between $y = 0$ and $y = \delta$, the y component of velocity at the edge of the boundary layer, $v(\delta)$, is found to be

$$-\int_0^\delta \frac{\partial u}{\partial x}\,dy = \frac{d}{dx}(U_0\,\delta_1) - \frac{dU_0}{dx}\,\delta. \tag{6.68}$$

The boundary layer equation may be integrated in like manner to give

$$\int_0^\delta \frac{\partial}{\partial x}(u^2)\,dy + U_0 v(\delta) = -\frac{1}{\varrho}\frac{dp_0}{dx}\,\delta - \frac{\tau}{\varrho}, \tag{6.69}$$

where τ is the skin friction of the plate and is equal to $(u\,\partial u/\partial y)_{y=0}$. Now

$$\int_0^\delta \frac{\partial}{\partial x}(u^2)\,dy = \frac{d}{dx}\int_0^\delta u^2\,dy - U_0^2\,\frac{d\delta}{dx}$$

and

$$\int_0^\delta u^2\,dy = U_0^2(\delta - \delta_1 - \vartheta).$$

Further, just outside the boundary layer,

$$U_0 \frac{dU_0}{dx} = -\frac{1}{\varrho}\frac{dp_0}{dx}.$$

Thus the integrated momentum equation may be written

$$\frac{\tau}{\varrho_0 U_0^2} = \frac{d\vartheta}{dx} + \frac{2\vartheta + \delta_1}{U_0}\frac{dU_0}{dx}. \tag{6.70}$$

This equation is known as the Kármán–Pohlhausen equation.

As an introduction to the use of this equation, consider the case of uniform flow over a flat plate. In this case the equation becomes

$$\frac{\tau}{\varrho U_0^2} = \frac{d\vartheta}{dx}. \tag{6.71}$$

The basic idea behind this method is that the quantities δ_1 and ϑ are not critically dependent on the exact shape of the velocity profile, since they involve integrations of the velocity profile, while τ involves the properties of the profile just next to the wall. Thus, estimates of these quantities may be found from a suitably chosen velocity profile.

As an example, consider the profile

$$\frac{u}{U_0} = f(y, \delta) = \frac{y}{\delta}. \tag{6.72}$$

This satisfies the conditions $u = 0$ at $y = 0$ and $u = U_0$ at $y = \delta$, and is the simplest possible profile. In terms of this profile

$$\tau = \frac{\mu U_0}{\delta} \qquad \text{and} \qquad \vartheta = \int_0^\delta \frac{y}{\delta}\left(1 - \frac{y}{\delta}\right) dy = \tfrac{1}{6}\delta.$$

Thus, the Kármán–Pohlhausen equation becomes

$$\frac{\nu}{U_0 \delta} = \frac{1}{3}\frac{d\delta}{dx}. \tag{6.73}$$

This equation may easily be integrated to give

$$\delta^2 = \frac{6\nu x}{U_0}, \tag{6.74}$$

assuming that the boundary layer has zero thickness at the leading edge $x = 0$.

The skin friction coefficient may now be calculated: it is

$$\frac{\tau}{\varrho U_0^2} = \frac{\nu}{U_0 \delta} = \left(\frac{\nu}{6U_0 x}\right)^{1/2} = 0.41\left(\frac{\nu}{U_0 x}\right)^{1/2},$$

a value which is in remarkably good agreement with the exact value $0.332\ (\nu/U_0 x)^{1/2}$.

This result can be improved easily by choosing a more suitable profile which satisfies the conditions at the edge rather better than the straight line. Thus the profile

$$\frac{u}{U_0} = \sin\left(\frac{\pi}{2}\,\frac{y}{\delta}\right), \tag{6.75}$$

which satisfies the additional condition at $y = \delta$ of $\partial u/\partial y = 0$, gives improved results. In fact, this profile yields a skin friction coefficient of $0.325\ (\nu/U_0 x)^{1/2}$, which is in excellent agreement with the exact value $0.332\ (\nu/U_0 x)^{1/2}$ just given.

6.10 Steady Compressible Boundary Layers (in Two-Dimensional Flow)

When the density is variable, the continuity equation is

$$\operatorname{div}(\varrho \mathbf{u}) = 0. \tag{6.76}$$

The momentum equation is

$$u\frac{\partial u}{\partial x} + v\frac{\partial u}{\partial y} = -\frac{1}{\varrho}\frac{dp_0}{dx} + \frac{1}{\varrho}\frac{\partial}{\partial y}\left(\mu\frac{\partial u}{\partial y}\right), \tag{6.77}$$

where $p_0(x)$ is the pressure just outside the boundary layer. In this equation, account has to be taken of the variability of μ and ϱ with the flow. When the fluid is compressible, μ and ϱ are variables which depend on the pressure p and the temperature T. (p, ϱ, and T are related by the equation of state; in the case of a perfect gas, this is just $p = \varrho RT$ where R is the gas constant, while μ is sensibly a function of T only.) For a complete solution to a problem in compressible flow, these quantities have to be taken into account. Inside the boundary layer, the pressure p is equal to the outer pressure p_0 which is known.

Thus, to determine μ and ϱ it is necessary to find T, the absolute temperature. The dependent variables are u, v, and T. Since there are so far only two equations of motion, a third equation is necessary to determine the flow completely. This third equation is the equation of energy; this is, for steady flow in two dimensions,

$$\varrho \frac{D}{Dt}(C_{\mathrm{p}}T) - \frac{p}{\varrho}\frac{Dp}{Dt} = \tau_{11}\frac{\partial u}{\partial x} + \tau_{12}\frac{\partial u}{\partial y} + \tau_{21}\frac{\partial v}{\partial x} + \tau_{22}\frac{\partial v}{\partial y} + \frac{\partial}{\partial x}\left(K_H\frac{\partial T}{\partial x}\right) + \frac{\partial}{\partial y}\left(K_H\frac{\partial T}{\partial y}\right), \quad (6.78)$$

where $D/Dt = u\,\partial/\partial x + v\,\partial/\partial y$, and τ_{ij} are the stresses, K_H is the thermal conductivity, and C_{p} is the specific heat at constant pressure. When the equation is multiplied by an element of volume, it states the fact that the total energy contained in this element is constant. The four main factors affecting the energy are: the rate of convection of internal energy of the gas into the volume given by the first term on the left side of the equation; the rate at which the work is being done on the volume by the external pressure, represented by the second term on the left side of the equation; the rate of loss of mechanical energy through having to overcome the viscous stresses given by the terms τ_{ij}; and lastly the terms

$$\frac{\partial}{\partial x}\left(K_H\frac{\partial T}{\partial x}\right) + \frac{\partial}{\partial y}\left(K_H\frac{\partial T}{\partial y}\right),$$

which give the rate at which heat is being conducted out of the volume. This energy equation is too complicated to be useful in any problem of practical interest. For these regions in which the boundary layer assumptions hold it may be simplified by neglecting terms of order δ/L or higher (δ and L are defined earlier in this chapter). It is assumed that in the boundary layer T varies in much the same way as the main velocity of flow. Thus, if x is taken in the main direction of flow, then

$$\frac{\partial T/\partial x}{\partial T/\partial y} \sim \frac{\delta}{L}.$$

When the orders of magnitude of the terms whose coefficients are τ_{ij} are considered, it is seen that their dominant part is the element $\tau_{12}\,\partial u/\partial y$, namely $\mu(\partial u/\partial y)^2$—the other terms in τ_{ij} being of higher order. Next, the term $(\partial/\partial x)(K_H\,\partial T/\partial x)$ may be neglected in comparison with $(\partial/\partial y)(K_H\,\partial T/\partial y)$. Thus the right side of the energy equation may be approximated by

$$\mu\left(\frac{\partial u}{\partial y}\right)^2 + \frac{\partial}{\partial y}\left(K_H \frac{\partial T}{\partial y}\right).$$

The energy equation therefore simplifies to

$$\varrho \frac{D}{Dt}(C_p T) - \frac{p_0}{\varrho}\frac{D\varrho}{Dt} = \mu\left(\frac{\partial u}{\partial y}\right)^2 + \frac{\partial}{\partial y}\left(K_H \frac{\partial T}{\partial y}\right). \quad (6.79)$$

It is necessary, now, to explain the *relative* importance of the parts of the simplified equation. These parts are: the terms due to convection, namely $\varrho(D/Dt)(C_p T)$; the term due to viscous stress, $\mu(\partial u/\partial y)^2$; and the term due to the molecular conduction of heat, $(\partial/\partial y)(K_H\, \partial T/\partial y)$. The convection term is of order $\varrho u C_p\, \Delta T/L$, where ΔT is the order of magnitude of the variation in T in the flow; the viscous term $\mu(\partial u/\partial y)^2$ is of order $\mu u^2/\delta^2$, and the heat conduction term $(\partial/\partial y)(K_H\, \partial T/\partial y)$ is of order $K_H\, \Delta T/\delta^2$. Introduce now the dimensionless quantities: M, the local Mach number defined as the ratio of the velocity at a point in the flow to the local speed of sound, and Pr, the Prandtl number defined as $C_p\mu/K_H$. In the case of gases, Pr is of order unity; for air, Pr $= 0.733$. Then these three terms may be shown to have orders of magnitude in the ratios

$$\Delta T : M^2 T : \Delta T,$$

respectively, when terms of order unity such as Pr and the ratio of the specific heats are neglected. Thus, the term $\mu(\partial u/\partial y)^2$ is of importance only when M^2 is of the same order as $\Delta T/T$. This will be the case when temperature differences are brought about by changes in velocity. When flows at low Mach numbers are considered in which there are large sources of heat, then the viscous heating term $\mu(\partial u/\partial y)^2$ may be omitted and the energy equation reduces to one of diffusion type.

Before considering problems in compressible flow, it is necessary to consider the properties of the fluid, which in this case is a gas, in more detail.

The equation of state As mentioned above, only perfect gases are considered here. Perfect gases have the simple equation of state $p = \varrho RT$, where R is the gas constant. Most gases behave like perfect gases when conditions are far from the liquefication point of the gas concerned. The gases in ordinary use, such as air, require extremely low temperatures before entering the liquid state and may therefore be treated as though they were perfect.

Variation in density Since the pressure p_0 is known, the density p may be found from the equation of state when the temperature T is known.

Variation in viscosity It is well known that for perfect gases, viscosity is independent of pressure and density and therefore depends only on T. Von Kármán and Tsien have proposed an empirical law

$$\frac{\mu}{\mu_0} = \left(\frac{T}{T_0}\right)^n,$$

where μ_0 is the value of μ at temperature T_0 and n is approximately 1.24 in the case of air.

Variation in Prandtl number The dimensionless number Pr determines the ratio of heat generated by viscosity to heat conducted by molecular action in an element of volume through which fluid is in motion. Other things being equal, the larger Pr is, the smaller the amount of heat conducted and vice versa. Its importance as a parameter lies mainly in heat transfer theory in which temperature profiles are of interest. From the point of view of velocity profiles, if Pr is taken to be unity, instead of its actual value for air of about 0.733, then there is very little difference in the profiles. This was brought out in the case of boundary layer flow over a flat plate by Emmons and Brainserd. They solved the equations of motion using Pr $=$ 0.733 and Pr $=$ 1, and showed that the difference in the velocity profiles was at most 1 or 2%. It is therefore reasonable in the case of air to assume that Pr $=$ 1 when the velocity profiles are of primary interest. This approximation of taking Pr as unity is of some importance, because when the Prandtl number is unity it is possible to replace the boundary layer energy equation in certain cases by a simple algebraic relation.

To summarize, the boundary layer equations of steady two-dimensional compressible flow are

$$\frac{\partial}{\partial x}(\varrho u) + \frac{\partial}{\partial y}(\varrho v) = 0, \tag{6.80}$$

$$u\frac{\partial u}{\partial x} + v\frac{\partial u}{\partial y} = -\frac{1}{\varrho}\frac{dp_0}{dx} + \frac{1}{\varrho}\frac{\partial}{\partial y}\left(\mu\frac{\partial u}{\partial y}\right), \tag{6.81}$$

$$\varrho\frac{D}{Dt}(C_p T) - \frac{p_0}{\varrho}\frac{D\varrho}{Dt} = \mu\left(\frac{\partial u}{\partial y}\right)^2 + \frac{\partial}{\partial y}\left(K_H\frac{\partial T}{\partial y}\right), \tag{6.82}$$

and the equation of state $f(p, \varrho, T) = 0$, for perfect gases, is

$$p = \varrho RT.$$

6.11 Problem of Two-Dimensional Steady Compressible Flow

The equation of continuity is

$$\frac{\partial}{\partial x}(\varrho u) + \frac{\partial}{\partial y}(\varrho v) = 0.$$

From this equation it follows that a stream function exists such that

$$\varrho u = \frac{\partial \psi}{\partial y}, \qquad \varrho v = -\frac{\partial \psi}{\partial x}.$$

6.11.1 The Howarth Transformation

Let the independent variables be changed from (x, y) to (x, z), where $z = \int_0^y \varrho\, dy$. Then

$$\varrho u = \left(\frac{\partial \psi}{\partial y}\right)_x = \left(\frac{\partial \psi}{\partial z}\right)_x \left(\frac{\partial z}{\partial y}\right) = \varrho \frac{\partial \psi}{\partial z}, \tag{6.83}$$

so that $u = \partial\psi/\partial z$ and

$$-\varrho v = \left(\frac{\partial \psi}{\partial x}\right)_z + \left(\frac{\partial \psi}{\partial z}\right)_x \left(\frac{\partial z}{\partial x}\right)_y. \tag{6.84}$$

When these expressions are inserted into the momentum equation, the terms involving $\partial z/\partial x$ cancel and the momentum equation becomes

$$\frac{\partial \psi}{\partial z}\frac{\partial^2 \psi}{\partial x\, \partial z} - \frac{\partial \psi}{\partial x}\frac{\partial^2 \psi}{\partial z^2} = -\frac{1}{\varrho}\frac{dp_0}{dx} + \frac{\partial}{\partial z}\left[\mu\varrho \frac{\partial^2 \psi}{\partial z^2}\right]. \tag{6.85}$$

6.11.2 Simple Case of the Howarth Transformation

The Howarth transformation is most useful when the outer pressure gradient is zero; in this case the term $-\varrho^{-1}\, dp_0/dx$ cancels out. Now the coefficient of viscosity μ is in general a function of temperature T. Its dependence on T may be assumed to be of the form

$$\frac{\mu}{\mu_0} = \left(\frac{T}{T_0}\right)^n,$$

where μ_0 and T_0 refer to the gas at the edge of the boundary layer; in the case of air, $n = 0.76$. This law of variation is satisfactory so long as the variation in T is not too large. Next, ϱ depends in general on temperature and pressure, but since the latter is constant over the boundary layer, the dependence is on T alone. If the gas is perfect, it follows that

$$\frac{\varrho}{\varrho_0} = \left(\frac{T}{T_0}\right)^{-1}.$$

Thus, the Howarth equation becomes

$$\frac{\partial \psi}{\partial z} \frac{\partial^2 \psi}{\partial x \, \partial z} - \frac{\partial \psi}{\partial x} \frac{\partial^2 \psi}{\partial z^2} = \mu_0 \varrho_0 \frac{\partial}{\partial z}\left[\left(\frac{T}{T_0}\right)^{n-1} \frac{\partial^2 \psi}{\partial z^2}\right]. \tag{6.86}$$

An especially simple case which illustrates the general nature of the flow occurs when $n = 1$; in this case, T does not enter explicitly into the Howarth equation, which becomes

$$\frac{\partial \psi}{\partial z} \frac{\partial^2 \psi}{\partial x \, \partial z} - \frac{\partial \psi}{\partial x} \frac{\partial^2 \psi}{\partial z^2} = \mu_0 \varrho_0 \frac{\partial^3 \psi}{\partial z^3}. \tag{6.87}$$

Consider now the case of incompressible flow at constant temperature T_0. In this case, $z = \varrho_0 y$; if the solution in this case is $\psi = f(x, y)$, then the solution for the compressible case is

$$\psi = f\left(x, \frac{z}{\varrho_0}\right). \tag{6.88}$$

This means that the x dependence in the flow is unaltered. However, the velocity at a point (x, y) in the compressible case is the same as the velocity at a point (x, y_{inc}) in the incompressible case, where

$$y_{\text{inc}} = \int_0^y \varrho^* \, dy = \int_0^y \frac{1}{T^*} \, dy$$

and where

$$\varrho^* = \frac{\varrho}{\varrho_0}, \qquad T^* = \frac{T}{T_0}.$$

This means that the velocity profile is related to the incompressible velocity profile by an as yet undetermined change of scale in the y direction. To determine the flow completely, it is necessary to know the temperature distribution in the boundary layer. This is found from the energy equation which is, in the case of zero pressure gradient,

$$\varrho \frac{D}{Dt}(C_p T) = \mu \left(\frac{\partial u}{\partial y}\right)^2 + \frac{\partial}{\partial y}\left(K_H \frac{\partial T}{\partial y}\right). \tag{6.89}$$

When the Prandtl number $C_p\mu/K_H$ is unity, this equation may be reduced to an algebraic relation between T and u, known as the Crocco relation. Let

$$P = C_p T + \tfrac{1}{2}u^2;$$

P is known as the thermodynamic free energy. Then the energy equation may be written

$$\varrho \frac{DP}{Dt} = \mu \left(\frac{\partial u}{\partial y}\right)^2 + \frac{\partial}{\partial y}\left(K_H \frac{\partial T}{\partial y}\right) + \varrho u^2 \frac{\partial u}{\partial x} + \varrho uv \frac{\partial u}{\partial y}. \tag{6.90}$$

This is obtained by adding $\varrho u\, Du/Dt$ to both sides of the energy equation and using the continuity equation to modify the left side. The right side may be written

$$\frac{\partial}{\partial y}\left(\mu \frac{\partial P}{\partial y}\right) + \frac{\partial}{\partial y}\left[K_H \frac{\partial T}{\partial y} - \mu \frac{\partial}{\partial y}(C_p T)\right].$$

When C_p is constant and $\Pr = C_p\mu/K_H = 1$, the second bracket vanishes and the energy equation becomes

$$\varrho \frac{DP}{Dt} = \frac{\partial}{\partial y}\left(\mu \frac{\partial P}{\partial y}\right) \tag{6.91}$$

When P is replaced by u in this equation, the equation becomes identical with the momentum equation. This means that the general solution of the energy equation is, in this case,

$$P \equiv C_p T + \tfrac{1}{2}u^2 = A + Bu, \tag{6.92}$$

where A and B are constants related to the boundary conditions.

Now in these cases for which the vorticity is zero outside the boundary layer, P is constant† and therefore the relation becomes

$$C_p T + \tfrac{1}{2}u^2 = A = C_p T_0 + \tfrac{1}{2}U_0^2 = C_p T_{00}, \tag{6.93}$$

where U_0 is the speed just outside the boundary layer and where $C_p T_{00}$ is the stagnation enthalpy.

† See Chapter 9 on Compressible Fluid Flow.

Thus, in this simple case the thermodynamic free energy remains constant both inside and outside the boundary layer and its value is equal to the stagnation point enthalpy.

6.11.3 The Two-Dimensional Compressible Jet

A simple example of the above theory is the two-dimensional compressible jet. It has already been shown that the solution for the incompressible case is

$$\psi = 6a\mu_0 x^{1/3} \tanh\left(\frac{ay}{x^{2/3}}\right). \tag{6.94}$$

In this case the definition ψ has been modified so as to fit in with the definition given at the beginning of this section. The compressible jet then has the stream function

$$\psi = 6a\mu_0 x^{1/3} \tanh\left(\frac{az}{\varrho_0 x^{2/3}}\right). \tag{6.95}$$

In this case the Crocco relation becomes

$$C_p T + \tfrac{1}{2}u^2 = C_p T_0,$$

that is,

$$T^* = \frac{T}{T_0} = 1 - \frac{u^2}{2C_p T_0}. \tag{6.96}$$

Physically this means that T^* is less than unity by a quantity which is of the same order of magnitude as the square of the local Mach number. Furthermore, it follows from the definition of z that $z/\varrho_0 > y$. This means that the velocity profile in the compressible case is obtained from that in the incompressible case by shrinking in the y direction, that is to say, the effect of compressibility is to sharpen the jet.

The precise relation between y and z/ϱ_0 is obtained by inverting the definition of z; thus

$$\varrho_0 y = \int_0^z T^* \, dz. \tag{6.97}$$

Thus

$$\xi = \int_0^{\xi_1} \left[1 - \frac{18a^4 \nu_0^2}{x^{4/3} C_p T_0} \operatorname{sech}^4 \xi_1\right] d\xi_1,$$

where

$$\xi_1 = \frac{az}{\varrho_0 x^{2/3}}.$$

This integrates to

$$\xi = \xi_1 - \frac{18a^4 v_0^2}{x^{4/3} C_p T_0}\left(t_1 - \frac{t_1^3}{3}\right), \tag{6.98}$$

where $t_1 = \tanh \xi_1$. Near the axis of the jet where ξ_1 is small, this relation becomes approximately

$$\xi \doteq \xi_1\left(1 - \frac{18a^4 v_0^2}{x^{4/3} C_p T_0}\right) \doteq \xi_1(1 - \alpha M_{\text{axis}}^2), \tag{6.99}$$

where M_{axis} is the Mach number on the axis of the jet and $\alpha = 1/2(\gamma - 1)$, γ being the gas constant, using Eq. (6.96) and the relation

$$M^2 = \frac{u^2}{2C_p T_0(\gamma - 1)}.$$

6.12 The General Case

In the general case when $\Pr \neq 1$ and $\mu \propto T^n$, the full equations need to be solved. It has been shown, however, in a number of calculations that for air the difference between the profiles generated by the approximations $\Pr = 1$ and $\mu \propto T$ is at most a few percent. Of course, in calculations where the heat transfer between, say, a plate and the boundary layer is of importance, it is necessary to use the accurate value of Pr. There are, however, no simple analytical methods of solution in these cases, and methods of treatment are in the main of the Pohlhausen type.

The boundary layer then is recognized as being a thin layer in which the effect of viscosity is important however high the Reynolds number of the flow may be.

It has been indicated that a solid boundary over which a fluid is flowing acts as a vorticity source which is then diffused away by viscosity and convected downstream with the fluid (the vorticity changing, of course, due to rotation and stretching of the vortex lines). Prandtl's boundary layer hypothesis postulates that viscous effects are comparable with inertial effects only in layers adjoining solid boundaries and that the thickness of such a layer goes to zero as the Reynolds number of the flow approaches infinity, and are very small outside such layers. Jets and wakes also are regions of significant viscous effects.

In this chapter it was shown how this concept simplifies the Navier–Stokes equations into a form known as the boundary layer equation. Some methods of solution of these equations were described. The reader is referred to the literature for more detailed accounts of laminar boundary layers and for numerical methods of solution. The important topics of unsteady boundary layers and transition to turbulent boundary layer flow will be dealt with in subsequent chapters.

PROBLEMS

6.1 Calculate the drag on a length L (and unit width) of a flat plate immersed in a uniform stream, given that $F''(0) = 0.47$ [see Eq. (6.29)].

Answer: Drag $= 1.33(U_0{}^3 \varrho\mu L)^{1/2}$.

6.2 Write the boundary layer equations for the flow past a plane wall $y = 0$ with the prescribed external velocity distribution $U_0(x)$.

Show that when $U_0 = cx^m$, where c, m are constants, a similarity solution may be found whose stream function is given by

$$\psi = (U_0 \nu x)^{1/2} f(\eta), \qquad \eta = \left(\frac{U_0}{\nu x}\right)^{1/2} y,$$

and which satisfies the differential equation

$$m(f')^2 - \tfrac{1}{2}(m+1)ff'' = m + f'''.$$

Show that if the further change of variables

$$Y = [\tfrac{1}{2}(m+1)]^{1/2}\eta, \qquad F = [\tfrac{1}{2}(m+1)]^{1/2} f, \qquad \beta = \frac{2m}{m+1}$$

are made, then the above equation becomes

$$\frac{d^3F}{dY^3} + F\frac{d^2F}{dY^2} = \beta\left[\left(\frac{dF}{dY}\right)^2 - 1\right]$$

with boundary conditions

$$F = 0, \quad \frac{dF}{dY} = 0 \quad \text{on} \quad Y = 0, \qquad \frac{dF}{dY} \to 1 \quad \text{as} \quad Y \to \infty.$$

6.3 Evaluate the skin friction coefficient in Problem 6.2 in terms of $F''(0)$.

6.4 State the two-dimensional, incompressible laminar boundary layer equations. Show that a similarity solution is possible for a mainstream $U(x) = cx$, where c is a positive constant, and find the ordinary differential equation to be solved and the boundary conditions to be satisfied.

Show that this velocity field also leads to a solution of the two-dimensional Navier–Stokes equations in Cartesian coordinates, provided that the pressure is suitably chosen. Find this pressure distribution. (Manchester †)

6.5 A uniform stream U of incompressible fluid of kinematic viscosity ν flows in the x direction over the semi-infinite plate $y = 0$, $x \geq 0$. The plate is porous and suction is applied through the surface so that the boundary condition on the plate is $u = 0$, $v = -S(x)$, where $S \ll U$.

If $S(x) = ax^{-1/2}$, where a is a constant, verify that the flow in the boundary layer on the plate continues to be governed by Blasius' equation

$$f''' + \tfrac{1}{2}ff'' = 0$$

with boundary conditions $f(0) = 2a(U\nu)^{-1/2}$, $f'(0) = 0$, $f'(\infty) = 1$, where the stream function $\psi = (U\nu x)^{1/2}f(\eta)$, and $\eta = (U/\nu x)^{1/2}y$.

If S is constant, show that the boundary layer equation has a series solution for small x in the form

$$\psi = (U\nu x)^{1/2}f(\eta) + Sxg(\eta) + \cdots,$$

where $f(\eta)$ is the Blasius function. Find the differential equation and boundary conditions satisfied by $g(\eta)$. (E. Anglia ‡)

6.6 Discuss the case of a flat sheet which issues from a thin slit: $x = 0$, $y = 0$ and moves along the plane $y = 0$ with speed $u = U_0(x)$.

Show that an equation similar to that in Problem 6.2 may be derived.

Show that in the case $U_0 = cx$ the solution is of the form $u = cxe^{-\gamma y}$, where γ is to be determined.

Show that in this case the boundary layer has a constant thickness. This problem corresponds to that of a stretching plastic sheet.

† From Manchester University, Honours in Mathematics (1971); by permission of the University of Manchester, Manchester, England.

‡ From the University of East Anglia, Honours in Mathematics (1971); by permission of the University of East Anglia, Norwich, England.

6.7 Show that the v component of the velocity on the edges of the boundary layer flow in the flat plate problem (Fig. 6.1) are

$$v = \frac{d}{x^{1/2}} \qquad \text{on} \quad \vartheta = 0_{\pm},$$

where (r, ϑ) are polar coordinates and d is to be determined. Find the outer potential flow brought about by the velocity.

$$\textit{Answer} \quad u \doteq U_0 - \frac{d}{r^{1/2}} \sin \tfrac{1}{2}\vartheta,$$

$$v \doteq \frac{d}{r^{1/2}} \cos \tfrac{1}{2}\vartheta.$$

6.8 Assuming the axially symmetric boundary layer equations for a round jet in still air,

$$u \frac{\partial u}{\partial x} + v \frac{\partial u}{\partial r} = \frac{\nu}{r} \frac{\partial}{\partial r}\left(r \frac{\partial u}{\partial r}\right),$$

show that a similarity solution of the form

$$\psi = \nu x F(\eta),$$

where $\eta = r/x$, is possible where

$$\frac{FF'}{\eta^2} - \frac{(F')^2}{\eta} - \frac{FF''}{\eta} = \frac{d}{d\eta}\left(F'' - \frac{F'}{\eta}\right).$$

Show that this equation may be integrated to give

$$FF' = F' - \eta F''.$$

Show finally that

$$F = \frac{\xi^2}{1 + \frac{1}{4}\xi^2},$$

where $\xi = \gamma\eta$ and γ is a constant determined from the fact that the total momentum flux is constant and equal to $\frac{16}{3}\pi\varrho\gamma^2/\nu^2$.

6.9 A body is held fixed in a stream of viscous incompressible fluid of uniform speed U at infinity. The motion is everywhere steady and symmetric about the axis Ox. In the wake far downstream of the body, where x is large and positive, the velocity of the fluid is approximately unidirectional with magnitude $U - u$, where $u \ll U$ and u vanishes outside the wake, and the pressure is approximately uniform. Show that

$$u = \frac{AU}{\nu x} \exp\left(-\frac{Ur^2}{4\nu x}\right),$$

where r measures distance from Ox, A is a constant, and ν is the kinematic viscosity of the fluid.

Show that the rate at which fluid volume is drawn by the wake across a plane normal to Ox and fixed relative to the fluid at infinity is $4\pi A$. (Cambridge†)

6.10 Give a simple argument to obtain the boundary layer equations for the flow of an incompressible stream of velocity U past a semi-infinite flat plate parallel to the stream.

Obtain an approximate solution of these equations by assuming that in the nonlinear inertia terms the velocity can be approximated by cU, where c is an undetermined constant (modified Oseen technique). Show that this gives the boundary layer velocity profile in the form $u = U$ $\operatorname{erf}(y\gamma(x))$, where

$$\gamma(x) = \frac{1}{2}\left(\frac{cU}{\nu x}\right)^{1/2} \qquad \text{and} \qquad \operatorname{erf} z = \frac{2}{\pi^{1/2}} \int_0^z \exp(-t^2)\, dt.$$

Determine c by requiring the exact momentum equation to be satisfied on the average across the layer. Show that this gives a value for the skin friction on the plate of

$$\frac{2^{1/4}}{\pi^{1/2}} \varrho U^2 \left(\frac{\nu}{Ux}\right)^{1/2}. \qquad \text{(Oxford ‡)}$$

6.11 Fluid flows past a plane wall $Y = 0$ with prescribed external velocity $U(x)$. Assuming the velocity profile $u/U = \tanh(y/\delta(x))$, where $\delta(x)$ is a measure of boundary layer thickness, U is constant, and $v = 0$ on the boundary, show that the skin friction is given by

$$\frac{\tau_w}{\varrho U^2} = \left(\frac{1 - \ln 2}{2}\right)^{1/2} \left(\frac{Ux}{\nu}\right)^{-1/2}. \qquad \text{(Manchester §)}$$

(*Hint*: use the Pohlhausen equation.)

† From Cambridge University, Mathematics Tripos (1970); by permission of Cambridge University Press, London.

‡ From Oxford University, Second Public Examination (1970); by permission of the Clarendon Press, Oxford.

§ From Manchester University, Honours in Mathematics (1969); by permission of the University of Manchester, Manchester, England.

6.12 Show that the MacLaurin expansion of the boundary layer flow past a flat plate is

$$u \doteqdot -(\zeta_w)y, \qquad v \doteqdot \tfrac{1}{2}y^2\left(\frac{\partial \zeta_w}{\partial x}\right),$$

where $\zeta_w \doteqdot -(\partial u/\partial y)_0$ is the vorticity on the wall.

Hence show that the angle of inclination of the streamlines close to the wall will be large when ζ_w is small.

(*Note*: Separation of the boundary layer from the wall takes place when $\zeta_w = 0$.)

6.13 Show that separation occurs only in decelerating flows, that is, when $dU/dx < 0$.

6.14 Write down the Navier–Stokes equations for two-dimensional motion of a viscous incompressible fluid on a plane wall. Show by means of a careful argument that the boundary layer equations provide an asymptotic approximation to these equations, indicating carefully any assumptions and limitations. By differentiating the boundary layer equation with respect to y, or otherwise, show that *according to boundary layer theory*, the skin friction τ_w just upstream of a point of separation x_s is given by

$$\tau_w = \varrho U^2\left[\frac{2A}{\mathrm{Re}_L}\left(\frac{x_s - x}{l}\right)\right]^{1/2},$$

where ϱ, U, l are reference values of density, velocity, and length, Re_L is the Reynolds number based on these quantities and on the coefficient of viscosity μ, and

$$A = -\frac{\nu^2 l^2}{U^3}\left(\frac{\partial^4 u}{\partial y^4}\right)_{x=x_s,y=0},$$

which may be assumed to differ from zero. (Oxford†)

BIBLIOGRAPHY

L. Rosenhead, "Laminar Boundary Layers." Oxford Univ. Press, London and New York, 1963.

H. Schlichting, "Boundary Layer Theory," 4th ed. McGraw-Hill, NewYork, 1960.

S. I. Pai, "Fluid Dynamics of Jets." Van Nostrand-Reinhold, Princeton, New Jersey, 1954.

† From Oxford University, Second Public Examination (1967) ; by permission of the Clarendon Press, Oxford.

CHAPTER SEVEN *SLOW VISCOUS FLOW*

7.1 Introduction

This chapter deals with flows at low Reynolds numbers. In the first part, namely that dealing with Stokes flows, it follows that the inertia or transport terms in the Navier–Stokes equation can be neglected. In other words, vorticity is transferred essentially by diffusion. A linearization of the equations of motion is effected, which reduces them to the biharmonic equation. This equation has been studied extensively in classical elasticity. It is shown that, in problems involving infinite regions, the solutions of the Stokes equations give rise to nonnegligible inertia terms, which means that in these cases the Stokes equations are not consistent.

The difficulty is resolved in the case when there is a uniform flow at large distances from an obstacle (see Section 7.6 on Oseen flows). In this case it is shown that the inertia terms may be approximated by assuming that vorticity is transported with speed U_0 in the x direction only.

This approximation gives good results at large distances; at small distances from the obstacle the solutions of the Oseen flows and Stokes flows differ by a negligible amount, of the order of the Reynolds number.

7.2 Stokes Flows

At Reynolds numbers very much smaller than unity, the transport terms in the Navier–Stokes equations are very much smaller than the diffusion terms. This may be seen by considering the orders of magnitude of representative terms of these types. Thus, a typical transport term is $u\ \partial u/\partial x$ which has order u^2/L, where L is the length over which significant changes in the flow take place (it might for example be the size of an obstacle or the diameter of a pipe). In the same way a typical diffusion term is $\nu\ \partial^2 u/\partial x^2$, which is of order $\nu U/L^2$. The ratio

$$\frac{\text{transport terms}}{\text{diffusion terms}} \sim \frac{u^2/L}{\nu u/L^2} \sim \frac{UL}{\nu} \sim \mathrm{Re}_L, \tag{7.1}$$

where $\mathrm{Re}_L = UL/\nu$ is the Reynolds number of the flow and where U is a representative velocity in the flow. Thus when Re_L is small, the Navier–Stokes equations reduce to

$$\operatorname{grad} p = \mu\, \nabla^2 \mathbf{u} \tag{7.2}$$

in rectangular coordinates or, in a general coordinate system,

$$\operatorname{grad} p = -\mu \operatorname{curl} \boldsymbol{\zeta}, \tag{7.3}$$

where $\boldsymbol{\zeta}$ is the vorticity vector.

This equation is known as the Stokes equation. Mathematically it has the advantage over the Navier–Stokes equations of being linear, and thus the principle of superposition holds, a fact which greatly facilitates the solution of problems.

7.2.1 Properties of Stokes's Equation

Take the divergence of Stokes's equation (7.3). Then

$$\operatorname{div} \operatorname{grad} p = -\mu \operatorname{div} \operatorname{curl} \boldsymbol{\zeta} \equiv 0. \tag{7.4}$$

Thus the pressure p is a harmonic function satisfying the equation

$$\nabla^2 p = 0. \tag{7.5}$$

On the other hand, take the curl of Stokes's equation (7.3). Then

$$\text{curl grad } p \equiv \mathbf{0} = -\mu \text{ curl curl } \boldsymbol{\zeta}.$$

Thus

$$\text{curl curl } \boldsymbol{\zeta} = \mathbf{0} \qquad \text{or} \qquad \text{curl curl curl } \mathbf{u} = \mathbf{0}. \tag{7.6}$$

7.3 Two-Dimensional Flows

Before considering three-dimensional motions, it is of interest to consider problems in two dimensions, which have several simplifying factors. First, in a two-dimensional motion in the xy plane,

$$\boldsymbol{\zeta} = \zeta(x, y)\mathbf{k}.$$

Now, in rectangular coordinates, the equation

$$\text{curl curl } \boldsymbol{\zeta} = \mathbf{0}$$

implies that

$$\nabla^2 \boldsymbol{\zeta} = \mathbf{0} \qquad \text{or} \qquad \nabla^2 \zeta = 0. \tag{7.7}$$

Thus, in two-dimensional problems the vorticity vector ζ is also a harmonic function.

Consider now the equation $\text{grad } p = -\mu \text{ curl } \boldsymbol{\zeta}$, which in two dimensions becomes

$$\frac{\partial p}{\partial x} = -\mu \frac{\partial \zeta}{\partial y}, \qquad \frac{\partial p}{\partial y} = \mu \frac{\partial \zeta}{\partial x}. \tag{7.8}$$

These are just the Cauchy–Riemann equations for the function of a complex variable $f(z)$ $(z = x + iy)$, whose real and imaginary parts are ζ and p/μ, respectively, that is

$$\zeta + i \frac{p}{\mu} = f(z). \tag{7.9}$$

This result gives a powerful method of determining possible distributions of pressure and vorticity by considering suitable functions $f(z)$.

7.3.1 Drag on a Two-Dimensional Body

It can be shown that the force exerted on a surface element ds of a body at rest in a viscous fluid is

$$[-p\hat{\mathbf{n}} + \mu(\mathbf{n} \times \boldsymbol{\zeta})]\, ds, \tag{7.10}$$

where $\hat{\mathbf{n}}$ is the outward drawn normal. In two dimensions this reduces to

$$-p\, d\mathbf{s} - \mu\boldsymbol{\zeta}\mathbf{k} \times d\mathbf{s}, \tag{7.11}$$

where $d\mathbf{s} = \hat{\mathbf{n}}\, ds$; $d\mathbf{s} = \mathbf{k} \times d\mathbf{r}$ is the surface element bounded by $d\mathbf{r} = dx\, \mathbf{i} + dy\, \mathbf{j}$ and unit length in the z direction.

On simplification this reduces to

$$\mathbf{i}[-p\, dy + \mu\zeta\, dx] + \mathbf{j}[p\, dx + \mu\zeta\, dy] \equiv \mathbf{i}\, dR_x + \mathbf{j}\, dR_y,$$

where dR_x and dR_y are the elements of force on $d\mathbf{s}$ in the x and y directions, respectively. This result may be expressed in the neat form

$$dR_x + i\, dR_y = \mu f(z)\, dz, \tag{7.12}$$

where

$$f(z) = \zeta + i\frac{p}{\mu}.$$

Thus, the total force on the body is given by

$$R_x + iR_y = \int_C \mu f(z)\, dz, \tag{7.13}$$

where C is the contour of the body. This result is comparable in simplicity to that of Blasius for two-dimensional irrotational flows.

7.3.2 The Biharmonic Equation for Two-Dimensional Flows

In two-dimensional flows a stream function $\psi(x, y)$ may be introduced such that

$$u = \frac{\partial \psi}{\partial y}, \qquad v = -\frac{\partial \psi}{\partial x}.$$

Now it can be shown that

$$\nabla^2\psi = -\zeta. \tag{7.14}$$

Further, since $\nabla^2\zeta = 0$, it follows therefore that

$$\nabla^2(\nabla^2\psi) = 0 \tag{7.15}$$

or

$$\frac{\partial^4\psi}{\partial x^4} + 2\frac{\partial^4\psi}{\partial x^2\,\partial y^2} + \frac{\partial^4\psi}{\partial y^2} = 0. \tag{7.16}$$

This equation is known as the *biharmonic* equation. It is the equation of two-dimensional elasticity, and many of the results of that field may be translated into slow viscous flow.

It can be shown that any solution of the biharmonic equation may be expressed as

$$\psi = \mathcal{R}[\bar{z}F(z) + G(z)], \tag{7.17}$$

where $\bar{z} \equiv x - iy$ and $\mathcal{R}$ denotes the real part. It follows that

$$u - iv = \bar{z}F' + G' + \bar{F}, \tag{7.18}$$

where $\bar{F}$ is the complex conjugate of F, and that

$$\zeta = -\mathcal{R}[4F'(z)]. \tag{7.19}$$

Thus $f(z) = -4F'(z)$, and it follows that

$$\frac{p}{\mu} = -4\mathcal{I}[F'(z)], \tag{7.20}$$

where $\mathcal{I}$ denotes the imaginary part. It will be noted that the flow may be split into a rotational part and an irrotational part corresponding to the complex functions F and G, respectively.

To sum up, solutions of the Stokes equations in two dimensions may be generated by the pair of analytic functions $F(z)$ and $G(z)$, giving a stream function

$$\psi = \mathcal{R}[\bar{z}F + G].$$

7.3.3 Examples

A number of simple but important examples may be solved by considering certain combinations of the potentials F and G. Consider for example the set of rotational stream functions given by

$$F = (A + iB) \log z \qquad \text{and} \qquad G = (A + iB)z \log z. \tag{7.21}$$

Then

$$\psi = 2x(A \log r - B\vartheta). \tag{7.22}$$

Again

$$F = (A + iB) \log z \qquad \text{and} \qquad G = -(A + iB)z \log z \tag{7.23}$$

leads to

$$\psi = 2y(B \log r + A\vartheta). \tag{7.24}$$

Another simple case is given by

$$F = z \log z \qquad \text{and} \qquad G = 0.$$

This gives

$$\psi = r^2 \log r.$$

Thus, $x \log r$, $y \log r$, $x\vartheta$, $y\vartheta$, and $r^2 \log r$ are all possible stream functions, as is any linear combination of them.

To these may be added any irrotational flow which is simply derived from

$$F = 0, \qquad G = G(z).$$

Example (i) Consider the case of a sheet of material issuing from a slit in a wall with speed U_0 (Fig. 7.1). The positive x and y axes form

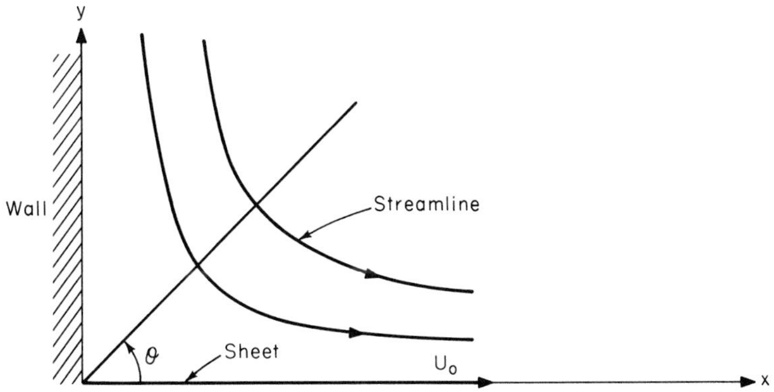

Fig. 7.1

the streamline $\psi = 0$. Now take

$$\psi = r\vartheta(B\cos\vartheta + C\sin\vartheta) + Ar\sin\vartheta \tag{7.25}$$

in polar coordinates. This automatically satisfies $\psi = 0$ on $\vartheta = 0$, and satisfies the same condition on $\vartheta = \pi/2$ if

$$A = -\frac{\pi}{2}C.$$

Next, the velocity components are given by

$$\begin{aligned} u_r &= \frac{1}{r}\frac{\partial\psi}{\partial\vartheta} \\ &= (B\cos\vartheta + C\sin\vartheta) + \vartheta(-B\sin\vartheta + C\cos\vartheta) + A\cos\vartheta, \end{aligned}$$

$$u_r = U_0 \quad \text{on} \quad \vartheta = 0 \quad \text{if} \quad U_0 = A + B, \tag{7.26}$$

$$u_r = 0 \quad \text{on} \quad \vartheta = \frac{\pi}{2} \quad \text{if} \quad C = \frac{\pi}{2}B.$$

Thus the stream function which satisfies the conditions of the problem is

$$\psi = \frac{rU_0}{\alpha}\left[\vartheta\left\{\frac{2}{\pi}\cos\vartheta + \sin\vartheta\right\} - \frac{\pi}{2}\sin\vartheta\right], \tag{7.27}$$

where

$$\alpha = \frac{2}{\pi} - \frac{\pi}{2}.$$

Example (ii) Flow in an annulus Consider the flow in the annulus bounded by the arcs $r = a$ and $r = b$ (Fig. 7.2). Suppose the flow is

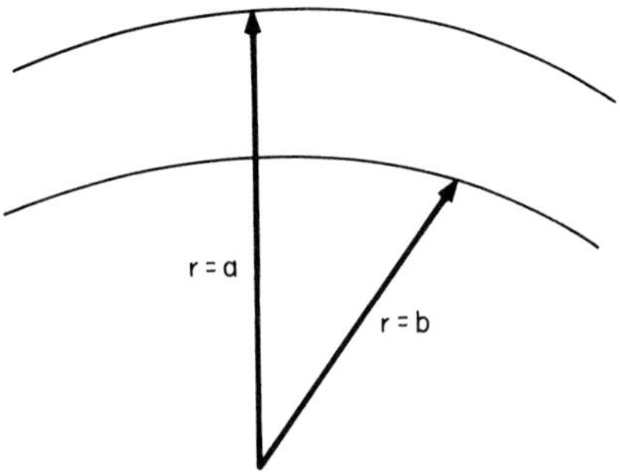

Fig. 7.2

independent of ϑ. Then the simplest rotational ψ is

$$\psi = Br^2 + Dr^2 \log r + C \log r, \tag{7.28}$$

the first two terms being rotational and the last term being irrotational.

This may be made to fit the boundary conditions in the arcs if

$$B = -\left[\frac{1}{2} + \frac{b^2 \log b - a^2 \log a}{b^2 - a^2}\right] D \tag{7.29}$$

and

$$C = \frac{2a^2b^2}{b^2 - a^2} \log\left(\frac{b}{a}\right) D. \tag{7.30}$$

The pressure may be derived by noting that

$$F = Bz + Dz \log z. \tag{7.31}$$

Thus,

$$f(z) = -4[B + D \log z + D], \tag{7.32}$$

and therefore since $p/\mu = \mathscr{I}(f(z))$,

$$p = -4\mu D\vartheta.$$

The total volume flux Q is given by

$$Q = \frac{2D}{b^2 - a^2}\left[\left(ab \log \frac{b}{a}\right)^2 - \left(\frac{b^2 - a^2}{2}\right)^2\right]. \tag{7.33}$$

Thus,

$$p = -\frac{12\mu Q}{(b - a)^3} \frac{1}{2} (a + b)\vartheta,$$

that is, the pressure gradient along the mean arc $r = \frac{1}{2}(a + b)$ is

$$\frac{12\mu Q}{(b - a)^3}.$$

This may be contrasted with the flow between two plane walls a distance h apart, for which the pressure gradient is

$$\frac{12\mu Q}{h^3}.$$

Example (iii) Consider

$$F = \frac{i}{a} \log\left(\frac{z}{a}\right), \qquad G = -\frac{i}{a} \log\left(\frac{z}{a}\right).$$

This gives

$$\psi = \frac{2r}{a} \sin \vartheta \log\left(\frac{r}{a}\right); \tag{7.34}$$

therefore, $\psi = 0$, on the cylinder $r = a$. Consider a possible flow past a stationary cylinder:

$$u_\vartheta = 0 \qquad \text{on} \quad r = a.$$

In this case

$$\begin{aligned} u &= -\frac{\partial \psi}{\partial r} = -\left[\frac{2}{a} \sin \vartheta \log\left(\frac{r}{a}\right) + \frac{2}{a} \sin \vartheta\right] \\ &= -\frac{2}{a} \sin \vartheta \qquad \text{on} \quad r = a. \end{aligned} \tag{7.35}$$

This may be modified to give $u_\vartheta = 0$ on $r = a$ by changing G to

$$G = -\frac{iz}{a} \log\left(\frac{z}{a}\right) + i\left(\frac{z}{a} + \frac{a}{z}\right). \tag{7.36}$$

This gives a stream function

$$\psi = A \sin \vartheta \left[2\frac{r}{a} \log \frac{r}{a} - \frac{r}{a} + \frac{a}{r}\right], \tag{7.37}$$

where A is an arbitrary constant. The drag on the cylinder may be evaluated to give

$$\text{drag} = \frac{8\pi\mu A}{a} \tag{7.38}$$

in the x direction. One disadvantage of this solution is that it yields a logarithmically infinite velocity at $r = \infty$. It is not, therefore, a valid solution at large radial distances. Later in this chapter it will be indicated how this solution may be related to a uniform flow at large radial distances.

7.4 Three-Dimensional Stokes Flows

In the case of three-dimensional flows, the starting point is the equation

$$\nabla^2 p = 0. \tag{7.39}$$

The general solution of this equation may be expressed as

$$p = \sum p_n, \tag{7.40}$$

where n takes integral values from $-\infty$ to $+\infty$, except for $n = -1$, and p_n is a solid spherical harmonic of degree n in x, y, z, and is a solution of Laplace's equation. Simple examples of the p_n's are

$$p_0 = \text{constant}, \quad p_1 = Ax + By + Cz, \quad p_2 = A(x^2 - y^2) + Bxy, \quad \text{etc.}$$

Note that the harmonics corresponding to negative values of n may be deduced from the p_n (n positive) by using the result that $(r^{2n+1})^{-1}p_n$ is a solid spherical harmonic. Now

$$\sum \frac{\partial p_n}{\partial x} = \mu \nabla^2 u. \tag{7.41}$$

It can be verified that this equation has the particular integral

$$u = \frac{1}{\mu} \sum \frac{1}{2(n+1)(2n+3)} \left[(n+3)r^2 \frac{\partial p_n}{\partial x} - 2np_n x \right], \tag{7.42}$$

with similar forms for the other components. This yields

$$\boldsymbol{\zeta} = \sum \frac{1}{\mu(n+1)} \mathbf{r} \times \operatorname{grad} p_n. \tag{7.43}$$

Now the above expressions are not complete integrals. Thus, to $\boldsymbol{\zeta}$ we can add

$$\sum \operatorname{grad} V_n,$$

where V_n is any solid spherical harmonic of degree n, without altering the value of p. The general form of $\boldsymbol{\zeta}$ is thus

$$\boldsymbol{\zeta} = \sum \left[\frac{1}{\mu(n+1)} (\mathbf{r} \times \operatorname{grad} p_n) + \operatorname{grad} V_n \right]. \tag{7.44}$$

Since $\boldsymbol{\zeta} = \operatorname{curl} \mathbf{u}$, this leads to

$$\mathbf{u} = \sum \frac{1}{2\mu(n+1)(2n+3)} \left[(n+3)r^2 \operatorname{grad} p_n - 2np_n \mathbf{r} \right] - \sum \frac{1}{(n+1)} \mathbf{r} \times \operatorname{grad} V_n + \sum \operatorname{grad} W_n, \tag{7.45}$$

where W_n is any solid spherical harmonic of degree n, and where the summation extends over all integral values of n in the case of W_n; but $n = -1$ is excluded in the first two summations.

Example (*i*) An interesting case arises where there is a pressure doublet at the origin. In this case

$$p = \frac{1}{r^2} P_1(\vartheta) \equiv \frac{1}{r^2} \cos \vartheta, \tag{7.46}$$

where P_1 is the *Legendre polynomial* of degree 1. This corresponds to a solid spherical harmonic with $n = -2$. Then the flow due to this doublet is given by

$$\mathbf{u} = \frac{1}{2\mu r} (2 \cos \vartheta \, \hat{\mathbf{r}} - \sin \vartheta \, \hat{\boldsymbol{\vartheta}}). \tag{7.47}$$

This flow has been called a *Stokeslet.*

Example (*ii*) Consider the case of slow flow in the z direction past a sphere of radius a with center at the origin. This flow is clearly axially symmetrical about the x axis. The boundary conditions are

$$\mathbf{u} = \mathbf{0} \quad \text{as} \quad r = a \qquad \text{and} \qquad \mathbf{u} \to U_0 \mathbf{i} \quad \text{as} \quad r \to \infty.$$

The last condition implies that

$$\mathbf{u} = U_0 r \sin \vartheta \, \hat{\mathbf{r}} - U_0 r \sin \vartheta \, \hat{\boldsymbol{\vartheta}}. \tag{7.48}$$

The above is clearly a solution of the Stokes equations. Express the complete solution to the problem in the term

$$\mathbf{u} = U_0 \mathbf{i} + \mathbf{u}. \tag{7.49}$$

Now clearly if $\mathbf{u}_1$ is a multiple of the Stokeslet (7.47), then either the radial or the transverse components of velocity can be eliminated on the sphere but not both. However, the irrotational motion due to a potential doublet gives

$$\mathbf{u} = \frac{2}{r^3} \cos \vartheta \, \hat{\mathbf{r}} + \frac{\sin \vartheta}{r^3} \, \hat{\boldsymbol{\vartheta}}. \tag{7.50}$$

Both the Stokeslet and the doublet give velocities which tend to zero as $r \to \infty$.

A suitable combination of these solutions with the uniform flow enables the condition $\mathbf{u} = \mathbf{0}$ on $r = a$ to be satisfied. Thus,

$$\mathbf{u} = U_0 \mathbf{i} - \frac{3}{4} \frac{U_0 a^2}{r} (2 \cos \vartheta \, \hat{\mathbf{r}} - \sin \vartheta \, \hat{\boldsymbol{\vartheta}})$$

$$+ \frac{1}{4} \frac{U_0 a^4}{r^3} (2 \cos \vartheta \, \hat{\mathbf{r}} + \sin \vartheta \, \hat{\boldsymbol{\vartheta}}). \tag{7.51}$$

The pressure is given by

$$p = p_\infty - \frac{3}{2}\mu U_0 a^2 \frac{\cos\vartheta}{r}, \tag{7.52}$$

where p_∞ is the uniform pressure at infinity, while

$$\boldsymbol{\zeta} = -\frac{3}{2}\frac{U_0 a^2}{r^4}\sin\vartheta\,\hat{\boldsymbol{\varphi}}, \tag{7.53}$$

where $\hat{\boldsymbol{\varphi}}$ is unit vector in the φ direction. The drag on the sphere may be obtained from the formula

$$[-p\hat{\mathbf{n}} + \mu(\hat{\mathbf{n}} \times \boldsymbol{\zeta})]\, ds. \tag{7.54}$$

The drag due to p on the sphere is zero; the total drag is therefore due to $\boldsymbol{\zeta}$ and therefore to the Stokeslet and is in fact equal to

$$6\pi\mu U_0 a. \tag{7.55}$$

This formula has been confirmed experimentally and has been shown to be correct for values of $U_0 a/\nu$, up to 0.5.

7.5 Criticism of Stokes's Solution

Even though the solution (7.51) satisfies the boundary conditions at infinity, nevertheless it is not consistent with the approximations which must be satisfied if the Stokes equations are to be validly derived from the Navier–Stokes equations. This may be seen by considering the ratio of the typical transport and viscous terms

$$u_r \frac{\partial u_r}{\partial r} \quad \text{and} \quad \nu \frac{\partial^2 u_r}{\partial r^2}.$$

At large radial distances these are of the order of

$$\frac{3}{2}\frac{U_0^2 a^2}{r^2} \quad \text{and} \quad \frac{3}{4}\frac{\nu U_0 a^2}{r^3},$$

respectively. The are therefore in the ratio $U_0 r/\nu$. Thus, at very large radial distances the transport terms dominate and the Stokes approximations break down. The Stokes solution is therefore not a valid approximate solution to the Navier–Stokes equations. An order of magnitude of the distance at which this solution is invalid is provided by observing that

$$\frac{U_0 r}{\nu} = \frac{r}{\nu/U_0}.$$

Thus, when r is comparable to the viscous length ν/U_0, the solution becomes invalid.

The question of the how this difficulty is to be resolved is deferred to the next section.

7.6 The Oseen Equations

Consider the slow flow past an obstacle; close to the obstacle, that is, at radial distances much smaller than the viscous length ν/U_0, the appropriate equations are those of Stokes. At large distances from the obstacle, or in other words distances comparable to ν/U_0, the Stokes equation no longer holds (as has been observed in the previous section in the particular case of the sphere). It is the purpose of this section to examine the region of flow.

To particularize the problem, at large distances from the obstacle the flow is essentially uniform in the x direction. Thus

$$\mathbf{u} = U_0\mathbf{i} + \mathbf{u}', \tag{7.56}$$

where $|\mathbf{u}'| \ll U_0$. Then the Navier–Stokes equations may be linearized by inserting (7.56) and neglecting quadratic terms in $\mathbf{u}'$.

Thus, in Cartesian coordinates,

$$U_0 \frac{\partial \mathbf{u}'}{\partial x} = -\frac{1}{\varrho}\operatorname{grad} p + \nu \nabla^2 \mathbf{u}', \tag{7.57}$$

$$\operatorname{div} \mathbf{u}' = 0. \tag{7.58}$$

These equations are called the *Oseen equations*. It will be noted that these equations are valid even in the region where the Stokes equations hold because in this region

$$U_0 \frac{\partial \mathbf{u}'}{\partial x}$$

is negligibly small, being of order Re_L compared with the other terms. Thus even though

$$U_0 \frac{\partial \mathbf{u}'}{\partial x}$$

is clearly not equal to the convection terms

$$u \frac{\partial \mathbf{u}}{\partial x} + v \frac{\partial \mathbf{u}}{\partial y},$$

its presence does not affect the solution when Re_L is small.

An immediate consequence of these equations, which follows by taking the divergence of the Oseen equation (7.57), is that p is a harmonic function, just as for Stokes flow.

The velocity $\mathbf{u}'$ is such that $\mathbf{u}' \to 0$ at large distances and $p \to$ constant there.

It is convenient to split $\mathbf{u}'$ into two components, which separately tend to zero at infinity; these are (a) a rotational component $\mathbf{u}_\mathrm{R}$ which contains all the vorticity in the flow and (b) a potential component $\mathbf{u}_\mathrm{P}$ which satisfies the condition curl $\mathbf{u}_\mathrm{P} = 0$.

Then $\mathbf{u}_\mathrm{P}$ satisfies the equation

$$U_0 \frac{\partial \mathbf{u}_\mathrm{P}}{\partial x} = -\frac{1}{\varrho} \operatorname{grad} p, \tag{7.59}$$

where p is the total pressure field. This may be integrated to give

$$\mathbf{u}_\mathrm{P} = -\operatorname{grad}\left(\int \frac{p}{\varrho U_0}\, dx\right) + f(y, z), \tag{7.60}$$

where f is an arbitrary function; the boundary condition $\mathbf{u}_\mathrm{P} \to \mathbf{0}$ at ∞ implies $f \equiv 0$.

Thus, $\mathbf{u}_\mathrm{P} = \operatorname{grad} \varphi$, where

$$\varphi = -\int \frac{p}{\varrho U_0}\, dx,$$

that is,

$$p - p_\infty = -\varrho U_0 \frac{\partial \varphi}{\partial x},$$

where p_∞ is the constant pressure at ∞. This equation is in fact the linearized form of Bernoulli's equation, when quadratic terms in $\mathbf{u}'$ are neglected.

Since the entire pressure field may be attributed to $\mathbf{u}_P$, it follows that the equation for $\mathbf{u}_R$ is

$$U_0 \frac{\partial \mathbf{u}_R}{\partial x} = \nu \nabla^2 \mathbf{u}_R \tag{7.61}$$

in rectangular coordinates.

It will be noted that (7.61) simplifies to

$$(\nabla^2 - k^2)\bar{\mathbf{u}}_R = \mathbf{0} \tag{7.62}$$

in rectangular coordinates, where $k = U_0/2\nu$ and where $\mathbf{u}_R = e^{kx}\bar{\mathbf{u}}_R$. This equation is known as the Helmholtz equation and its properties have been extensively studied.

7.6.1 Two-Dimensional Problems

Consider first the rotational part $\mathbf{u}_R$. Then

$$\mathbf{u}_R \equiv u_R \mathbf{i} + v_R \mathbf{j} \equiv e^{kx}[\bar{u}_R \mathbf{i} + \bar{v}_R \mathbf{j}]. \tag{7.63}$$

Then

$$(\nabla^2 - k^2)\bar{u}_R = 0, \qquad (\nabla^2 - k^2)\bar{v}_R = 0.$$

Let $\bar{v}_R = \partial G/\partial y$, where G satisfies

$$(\nabla^2 - k^2)G = 0. \tag{7.64}$$

Then from the continuity equation

$$\frac{\partial u_R}{\partial x} = -e^{kx} \frac{\partial^2 G}{\partial y^2} = -e^{kx}\left[k^2 G - \frac{\partial^2 G}{\partial x^2}\right]. \tag{7.65}$$

After a straightforward integration by parts,

$$u_R = e^{kx}\left[\frac{\partial G}{\partial x} - kG\right] + h(y); \tag{7.66}$$

$h(y)$ (which is an arbitrary function of y) is obviously zero by virtue of the fact that $u_R \to 0$ at infinity.

In summary, the solution of the two-dimensional Oseen equations consists of an irrotational part, which gives the pressure field, and a rotational part $\mathbf{u}_{\mathrm{R}}$, given by

$$\mathbf{u}_{\mathrm{R}} = e^{kx}[\operatorname{grad} G - kG\mathbf{i}], \tag{7.67}$$

where G is a solution of

$$(\nabla^2 - k^2)G = 0.$$

The simple result

$$\boldsymbol{\zeta} = 2ke^{kx}\frac{\partial G}{\partial y}\mathbf{k} \tag{7.68}$$

will be noted.

7.6.2 Simple Two-Dimensional Solutions

Since two-dimensional irrotational flows have been considered elsewhere in this book, attention in this section is focused on the rotational component $\mathbf{u}_{\mathrm{R}}$.

Consider first the set of possible solutions of

$$(\nabla^2 - k^2)G = 0,$$

which tend to zero at infinity. This equation has the set of solutions, in polar coordinates (r, ϑ),

$$G = \sum_{n=0}^{\infty} K_n(kr)[A_n \cos n\vartheta + B_n \sin n\vartheta], \tag{7.69}$$

where K_n is the modified Bessel function of the second kind; K_n is singular at $r = 0$, and has the asymptotic behavior at $kr = \infty$:

$$K_n(kr) \sim \left(\frac{\pi}{2kr}\right)^{1/2} e^{-kr}. \tag{7.70}$$

Case (i) $G = K_n(kr) \cos n\vartheta$ In this case

$$\mathbf{u}_{\mathrm{R}} = e^{kx}[kK_n' \cos n\vartheta\,\hat{\mathbf{r}} - nK_n \sin n\vartheta\,\hat{\boldsymbol{\vartheta}} - kK_n \cos n\vartheta\,\mathbf{i}], \tag{7.71}$$

where $\hat{\mathbf{r}}$ and $\hat{\boldsymbol{\vartheta}}$ are unit vectors in the radial and transverse directions, respectively. When kr is large,

$$\mathbf{u}_{\mathrm{R}} \sim e^{-k(r-x)}\left(\frac{\pi}{2kr}\right)^{1/2}[k \cos n\vartheta\,\hat{\mathbf{r}} + n \sin n\vartheta\,\hat{\boldsymbol{\vartheta}} + k \cos n\vartheta\,\mathbf{i}]; \tag{7.72}$$

when ϑ is not small, $\mathbf{u}_R$ tends exponentially to zero. When, however, ϑ is small,

$$\mathbf{u}_R \sim \exp\left(-\frac{kr\vartheta^2}{2}\right)\left(\frac{\pi}{2kr}\right)^{1/2}[k \cos n\vartheta\, \hat{\mathbf{r}} + n \sin n\vartheta\, \hat{\boldsymbol{\vartheta}} + k \cos n\vartheta\, \mathbf{i}]$$

$$\sim \exp(-\eta^2)\left(\frac{\pi}{2kx}\right)^{1/2}[2k\mathbf{i}], \tag{7.73}$$

where

$$\eta^2 = \frac{kr\vartheta^2}{2} \sim \frac{ky^2}{2x}.$$

This approximation holds provided ϑ^4 can be neglected. The region $ky^2/2x$ finite defines roughly a parabolic region in the neighborhood of the x axis. This region is known as the *wake*; in it the velocity decays like $x^{-1/2}$. It will be noted that there is a backward mass flux in the wake given by

$$\int_{-\infty}^{\infty} -2k \exp(-\eta^2)\left(\frac{\pi}{2kx}\right)^{1/2} d\eta \left(\frac{2x}{k}\right)^{1/2} = -2\pi,$$

that is, this flow behaves like a sink of unit strength. It can be shown that this solution may be interpreted as being caused by a point force $-\pi\varrho U_0\mathbf{i}$ at the origin.

Case (ii) $G = K_n(kr) \sin n\vartheta$ Again $\mathbf{u}_R$ is exponentially small except in the *wake*, where

$$\mathbf{u}_R \sim \exp(-\eta^2)\left(\frac{\pi}{2kx}\right)^{1/2} n\mathbf{j}.$$

The flow in this case is essentially transverse in the wake. There is a finite flux of transverse momentum down the wake, equal to

$$\int_{-\infty}^{\infty} \varrho U_0 n \exp(-\eta^2)\left(\frac{\pi}{2kx}\right)^{1/2} d\eta\left(\frac{2x}{k}\right)^{1/2} = \frac{\varrho U_0 n\pi}{k}.$$

This is in fact equal to $\varrho U_0 \Gamma$, where Γ is the circulation around a very large circle due to the rotational flow. This flow may be interpreted as due to a force at the origin on the fluid equal to

$$\varrho U_0 \frac{n\pi}{k}.$$

Case (iii) Consider the wake past a symmetric body. Here, by symmetry,

$$G = \sum_{n=0}^{\infty} A_n K_n(kr) \cos n\vartheta, \qquad \varphi = C_0 \log r + \sum_{n=1}^{\infty} \frac{C_n \cos n\vartheta}{r}.$$

Then, in the wake there is a net inflow, equal to

$$2\pi \sum_{n=0}^{\infty} A_n,$$

due to the G. This inflow must be balanced by the outflow due to the potential flow, which is equal to

$$2\pi C_0.$$

Thus

$$C_0 = 2\pi \sum_{n=0}^{\infty} A_n.$$

The drag on the symmetric body is equal to

$$\varrho U_0 C_0.$$

This drag is composed of two equal parts from the rotational and potential parts of the flow, respectively.

Case (iv) Consider the flow with speed U_0 past an infinite set of small similar symmetrical obstacles placed a distance $2d$ apart on the y axis (Fig. 7.3). The symmetry conditions imply that the lines $y = \pm d$ are streamlines and that the flow is periodic of period $2d$ in the y direction.

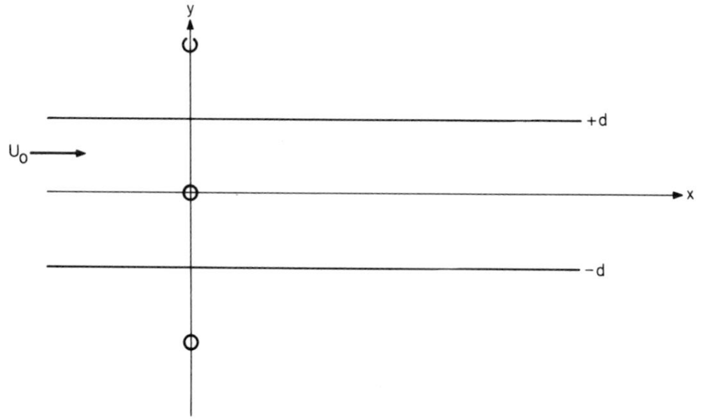

Fig. 7.3

Thus the rotational part of the flow satisfies the equation

$$\frac{\partial^2 G}{\partial x^2} + \frac{\partial^2 G}{\partial y^2} - k^2 G = 0. \tag{7.74}$$

There are two regions of flow, namely $x > 0$ and $x < 0$, for which the appropriate $G's$ are

$$G = B_0 e^{-kx} + \sum_{n=0}^{\infty} A_n \cos \frac{n\pi y}{d} e^{-\alpha_n x} \qquad \text{for} \quad x > 0, \tag{7.75}$$

$$G = B_1 e^{kx} + \sum_{n=0}^{\infty} C_n \cos \frac{n\pi y}{d} e^{\alpha_n x} \qquad \text{for} \quad x < 0, \tag{7.76}$$

where

$$\alpha_n = \left(\frac{n^2 \pi^2}{d^2} + k^2 \right)^{1/2}.$$

These equations lead to

$$u_{\rm R} = -e^{kx}\left[\sum_{n=1}^{\infty} (\alpha_n + k) A_n \cos \frac{n\pi y}{d} e^{-\alpha_n x} \right] - 2kB_0, \qquad x > 0, \tag{7.77}$$

$$u_{\rm R} = -e^{kx}\left[\sum_{n=1}^{\infty} (-\alpha_n + k) C_n \cos \frac{n\pi y}{d} e^{\alpha_n x} \right], \qquad x < 0. \tag{7.78}$$

The rotational flow thus gives rise to a finite velocity

$$u_{\rm R} = -2kB_0 \qquad \text{at} \quad x = +\infty;$$

at $x = -\infty$, the rotational part of the flow decays exponentially with x. The rotational flow thus gives rise to a sink at the origin of strength $-2kB_0$. This must be balanced by a potential source at the origin, also of strength $-2kB_0$. Thus, for $x > 0$,

$$\varphi = kB_0 x + \sum_{n=1}^{\infty} D_n \cos \frac{n\pi y}{d} e^{-n\pi x/d}, \tag{7.79}$$

and for $x < 0$,

$$\varphi = -kB_0 x + \sum_{n=1}^{\infty} E_n \cos \frac{n\pi y}{d} e^{+n\pi x/d}. \tag{7.80}$$

The pressure at $x = +\infty$ is $-\varrho U_0 k B_0$, and at $x = -\infty$ it is $\varrho U_0 k B_0$. Thus there is a pressure difference of about $2\varrho U_0 k B_0$. The drag exerted by each obstacle is just $4\varrho U_0\, dkB_0$. Upstream of the flow, the obstacles are manifested by means of the potential part of the flow, downstream by the rotational and potential parts of the flow. It is interesting to note that, in this case, the vorticity is uniformly distributed at large positive distances; this is due to the fact that the wakes due to each obstacle separately merge into each other.

There are two simple possibilities for the value of α_n. Now

$$\alpha_n = \left(\frac{n^2\pi^2}{d^2} + k^2\right)^{1/2} = k\left(1 + \frac{n^2\pi^2}{k^2d^2}\right)^{1/2}.$$

Now $kd = U_0 d/2\nu$ is the Reynolds number based on the distance separating the obstacles. When $kd \gg 1$, that is, when the separation distance $\gg$ the viscous length ν/U_0,

$$\alpha_n \doteqdot k$$

and

$$u_{\mathrm{R}} = -\sum_{n=1}^{\infty} 2kA_n \cos\frac{n\pi y}{d} - 2kB_0 \qquad \text{for} \quad x > 0; \qquad (7.81)$$

when $kd \ll 1$, that is, when the separation distance $\ll$ the viscous length ν/U_0, and

$$\alpha_n \doteqdot \frac{n\pi}{d}.$$

7.7 Three-Dimensional Oseen Flows

Just as in the two-dimensional case, the flow may be separated into two components—a rotational velocity $\mathbf{u}_{\mathrm{R}}$ and a potential velocity $\mathbf{u}_{\mathrm{P}}$. The rotational part has a similar form to the earlier case,

$$\mathbf{u}_{\mathrm{R}} = e^{kx}[\operatorname{grad} G - kG\mathbf{i}],$$

where G satisfies the equation

$$(\nabla^2 - k^2)G = 0;$$

the pressure field due to the rotational flow is again zero. In what follows attention will be confined to axially symmetric flows. Then it can be shown that

$$G = (kr)^{-1/2} \sum_{n=0}^{\infty} A_n K_{(n+\frac{1}{2})}(kr) P_n(\cos\vartheta), \qquad (7.82)$$

where K is the modified Bessel function of the second kind, and P_n is the *Legendre polynomial* of order n.

In this case the Bessel functions have a simple form. Thus,

$$K_{\frac{1}{2}}(t) = \left(\frac{\pi}{2t}\right)^{\frac{1}{2}} e^{-t}, \tag{7.83}$$

and in general

$$t^{-n-\frac{1}{2}} K_{(n+\frac{1}{2})}(t) = (-2)^n \left(\frac{\pi}{2}\right)^{\frac{1}{2}} \frac{d^n}{d(t^2)^n} \left(\frac{e^{-t}}{t}\right). \tag{7.84}$$

Example (*i*) Consider the solution corresponding to $n = 0$, that is,

$$G = \frac{1}{(kr)^{1/2}} K_{\frac{1}{2}}(kr) = \left(\frac{\pi}{2}\right)^{\frac{1}{2}} \frac{e^{-kr}}{kr}. \tag{7.85}$$

This gives, deleting the constant $(\pi/2)^{1/2}$, for simplicity,

$$\mathbf{u}_{\mathrm{R}} = e^{kx}\left[\frac{d}{dr}\left(\frac{e^{-kr}}{kr}\right)\hat{\mathbf{r}} - \frac{1}{r} e^{-kr}\,\mathbf{i}\right], \tag{7.86}$$

where $\hat{\mathbf{r}}$ is unit vector in the radial direction.

In this case, as in the two-dimensional cases already considered, the decay is exponential except within a thin wake region defined by $k(r - x)$ small, that is, $k\vartheta^2/2r$ small, where ϑ is the polar angle. This wake region is essentially a paraboloid of revolution about the x axis.

It can be shown that there is a net influx of fluid of an amount $4\pi/k$ along the wake.

If the flow is required to be such that there is no source of fluid at the origin, then a potential flow is required. This may be provided by a simple source, also of strength 4π, that is,

$$\varphi = -\frac{1}{r}.$$

Thus the net velocity is

$$\mathbf{u} = e^{kx}\left[\frac{d}{dr}\left(\frac{e^{-kr}}{kr}\right)\hat{\mathbf{r}} - \frac{1}{r} e^{-kr}\,\mathbf{i}\right] + \frac{\hat{\mathbf{r}}}{kr^2}. \tag{7.87}$$

The pressure associated with the flow is given by

$$p - p_\infty = -\varrho U_0 \frac{\partial \varphi}{\partial x} = -\varrho U_0 \frac{\cos\vartheta}{r^2}. \tag{7.88}$$

It can be shown that this flow may be generated by a point force

$$-\frac{4\pi\varrho U_0\mathbf{i}}{k}$$

applied at the origin.

It is instructive to consider the approximate value of $\mathbf{u}_R$ when kr is small, that is, close to the origin. Then

$$\begin{aligned}\mathbf{u}_R &\doteqdot k\left[-\left(\frac{1}{kr}+\frac{1}{k^2r^2}+\frac{\cos\vartheta}{kr}\right)\hat{\mathbf{r}}+\frac{1}{kr}\sin\vartheta\,\hat{\boldsymbol{\vartheta}}\right]\\ &\quad-\frac{k(x-r)}{k^2r^2}\,\hat{\mathbf{r}}+O(1) \qquad (7.89)\\ &= k\left[-\left(\frac{1}{k^2r^2}+\frac{2\cos\vartheta}{kr}\right)\hat{\mathbf{r}}+\frac{1}{kr}\sin\vartheta\,\hat{\boldsymbol{\vartheta}}\right]. \qquad (7.90)\end{aligned}$$

The sum of the potential and rotational flows then gives

$$\mathbf{u}=\frac{1}{r}\left(-2\cos\vartheta\,\hat{\mathbf{r}}+\sin\vartheta\,\hat{\boldsymbol{\vartheta}}\right) \qquad (7.91)$$

close to the origin.

Example (ii) Consider the total flow due to a point force

$$-\frac{4\pi\varrho U_0}{k}A\mathbf{i}$$

at the origin, that is,

$$\mathbf{u}=U_0\mathbf{i}+Ae^{kx}\left[\frac{d}{dr}\left(\frac{e^{kr}}{kr}\right)\hat{\mathbf{r}}-\frac{1}{r}e^{-kr}\,\mathbf{i}\right]+\frac{A}{kr^2}\,\hat{\mathbf{r}}. \qquad (7.92)$$

For small values of kr, this becomes

$$\mathbf{u}=\frac{A}{r}\left(-2\cos\vartheta\,\hat{\mathbf{r}}+\sin\vartheta\,\hat{\boldsymbol{\vartheta}}\right)+U_0\left(r\cos\vartheta\,\hat{\mathbf{r}}-r\sin\vartheta\,\hat{\boldsymbol{\vartheta}}\right). \qquad (7.93)$$

It will be noted that it is rather like the Stokes solution (7.51) for the sphere. Indeed, by adding a potential doublet

$$\mathbf{u}=\frac{B}{r^3}\left(2\cos\vartheta\,\hat{\mathbf{r}}+\sin\vartheta\,\hat{\boldsymbol{\vartheta}}\right)$$

to (7.93), and suitably choosing A and B, it is possible to satisfy the condition $\mathbf{u} = 0$ on $r = a$ approximately, with an error of order ka, $\sim \mathrm{Re}_L$. Thus this solution approximates to the Stokes solution (7.93) for the sphere given earlier in the chapter, with

$$A = \tfrac{5}{4} U_0 a^2, \qquad B = \tfrac{1}{4} U_0 a^4.$$

To sum up, the Oseen flow for the sphere consists of (in units of velocity)

a uniform flow: $U_0 \mathbf{i}$,

a potential source: $\dfrac{3}{4} \dfrac{U_0 a^2}{k} \hat{\mathbf{r}}$,

a potential dipole: $\dfrac{1}{4} \dfrac{U_0 a^4}{r^3} [2 \cos \vartheta \, \hat{\mathbf{r}} + \sin \vartheta \, \hat{\boldsymbol{\vartheta}}]$,

and

a rotational flow (Oseenlet):

$$\frac{3}{4} U_0 a^2 e^{kx} \left[\frac{d}{dr} \left(\frac{e^{-kr}}{kr} \right) - \frac{1}{r} e^{-kr} \, \mathbf{i} \right]. \tag{7.94}$$

This flow satisfies the boundary conditions on the sphere approximately and at infinity exactly. As is to be expected, it gives the same value as the Stokes equation for the drag. However, it strongly differentiates between the upstream and downstream regions of flow and predicts a wake—features which are absent from the Stokes solution.

PROBLEMS

7.1 A two-dimensional flat plate is defined by $y = 0$, $-a < x < a$, and is immersed in viscous fluid which is rotating far from the plate, like a solid body, with angular velocity ω. Show that a solution of the slow-motion equations of the above form can be constructed in which

$$f(z) = \tfrac{1}{2} i\omega (a^2 - z^2)^{1/2},$$

where $(a^2 - z^2)^{1/2}$ is made single-valued by a branch cut from $z = -a$ to $+a$ and takes the value $+iz$ as $|z| \to \infty$. Hence show that a couple $2\pi\mu\omega a^2$ acts on the plate. (Oxford†)

† From Oxford University, Second Public Examination (1969) ; by permission of the Clarendon Press, Oxford.

7.2 The stream function ψ for the slow two-dimensional flow of a viscous incompressible fluid satisfies $\nabla^4\psi = 0$. Show that ψ can be represented as the real part of $\bar{z}f(z) + g(z)$, where $z = x + iy$ and f and g are analytic functions. Hence show that the pressure and vorticity can be represented by

$$p = 4i\mu\{f'(z) - \overline{f'(z)}\}, \qquad \zeta = -4\{f'(z) + \overline{f'(z)}\}.$$

A curve is defined by $z = z(s)$. Prove that on it

$$\tau_{nn} + p + i\tau_{ns} = -4\mu i\left(\frac{dz}{ds}\right)^2 (\bar{z}f''(z) + g''(z)),$$

where τ_{ij} is the viscous stress tensor and (s, n) are coordinates parallel and perpendicular to the curve. Hence show that on a free boundary to a slow viscous flow

$$\bar{z}f(z) + g(z) = 0 \quad \text{and} \quad \overline{f(z)}\frac{dz}{ds} + f(z)\frac{d\bar{z}}{ds} = 0. \qquad \text{(Oxford}^{\dagger}\text{)}$$

7.3 Given the Navier–Stokes and continuity equations for plane incompressible fluid flow, show that slow steady motions are governed by

$$\nabla^4\psi = \left(\frac{\partial^2}{\partial r^2} + \frac{1}{r}\frac{\partial}{\partial r} + \frac{1}{r^2}\frac{\partial^2}{\partial \vartheta^2}\right)^2 \psi(r, \vartheta) = 0,$$

where r, ϑ are polar coordinates and ψ is the stream function.

Derive solutions in the form $\psi = r^{n+1}f(\vartheta)$ and show that, for flow in a corner between two plane rigid boundaries at an angle β, n must satisfy

$$n \sin\beta = \pm\sin n\beta. \qquad \text{(London}^{\ddagger}\text{)}$$

7.4 A viscous fluid, coefficient of viscosity μ, flows steadily down a long cylindrical pipe, whose cross section is defined by $f(x, y) = 0$. If there is a pressure gradient p' down the pipe and no other forces act, show that the velocity at any point is

$$\frac{1}{4}\frac{p'}{\mu}(x^2 + y^2) + \operatorname{Re} F(z),$$

where F is an analytic function of $x + iy$ which is nonsingular inside $f(x, y) = 0$ and whose real part equals $-\frac{1}{4}(p'/\mu)z\bar{z}$ on $f(x, y) = 0$.

† From Oxford University, Second Public Examination (1967); by permission of the Clarendon Press, Oxford.

‡ From University of London (Imperial College), B.Sc. (1969); by permission of the Senate of the University.

Show that the transformation $\zeta = z/(1 + az)$, where a is a real constant, maps the region between a pair of eccentric circles in the z plane onto the region between $|\zeta| = \varrho_1$ and $|\zeta| = \varrho_2$ in the ζ plane ($\varrho_2 < \varrho_1 < 1/a$). Hence show that the solution for $F(z)$, when $f(x, y) = 0$ defines a pair of eccentric circles, is given by $F(z) = G(\zeta)$, where $G(\zeta)$ is non-singular in $\varrho_1 > |\zeta| > \varrho_2$ and

$$\text{Re}\, G(\zeta) = \frac{1}{4} \frac{p'\zeta\zeta}{\mu(1 - a^2\zeta\zeta)} \left\{1 + \frac{a\zeta}{1 - a\zeta} + \frac{a\zeta}{1 - a\zeta}\right\}$$

on $|\zeta| = \varrho_1$ and $|\zeta| = \varrho_2$.

By writing $\zeta = \varrho e^{i0}$, or otherwise, find an expression for $G(\zeta)$ in the form of an infinite series of powers of ζ and $1/\zeta$ with real coefficients defined in terms of ϱ_1, ϱ_2 and a. (Oxford †)

7.5 Define Stokes's stream function ψ for an axisymmetric flow in spherical polar coordinates (r, ϑ, φ). Show that the "slow motion" equations, in which inertia terms are neglected, can be reduced to $(\text{curl})^4 (0, 0, \psi/r \sin\vartheta) = 0$. Obtain a solution for ψ in the form $\mathrm{f}(r) \sin^2 \vartheta$.

Find the "slow motion" flow between two concentric spheres which are rotating with different angular velocities about a common diameter.

Show that the couple on a sphere of radius a which is rotating slowly about a diameter with angular velocity ω in an unbounded viscous fluid is $8\pi\mu\omega a^3$.

[*Hint*: Assume that the component of stress $p_{r\varphi}$ is

$$\mu\left(\frac{\partial V_\varphi}{\partial r} - \frac{V_\varphi}{r} + \frac{1}{r \sin\vartheta} \frac{\partial V_r}{\partial \varphi}\right).\Big] \qquad \text{(Oxford †)}$$

7.6 Taking as a starting point the equation

$$\text{curl curl curl}\, \mathbf{u} = 0,$$

satisfied by the velocity field $\mathbf{u}$ in a steady flow of a viscous incompressible fluid at very small Reynolds number, show that in the case of axisymmetric motion the stream function ψ must satisfy the equation

$$D^4\psi = 0,$$

where

$$D^2 = \frac{\partial^2}{\partial r^2} + \frac{\sin\vartheta}{r^2} \frac{\partial}{\partial\vartheta} \frac{1}{\sin\vartheta} \frac{\partial}{\partial\vartheta}.$$

† From Oxford University, Second Public Examination (1965); by permission of the Clarendon Press, Oxford.

Here r is the length of the vector from a fixed origin to a typical point and ϑ is the angle made by this vector with the axis of symmetry, so that r, ϑ, and an azimuthal angle φ are spherical polar coordinates.

A rigid sphere of radius a is placed in a uniform stream of speed U. Show that if the axis of the spherical polar coordinates is parallel to the stream and their origin is at the center of the sphere, then

$$\psi = \frac{1}{2} U \sin^2 \vartheta \left(r^2 - \frac{3}{2} ar + \frac{1}{2} \frac{a^3}{r} \right).$$

[In spherical polar coordinates (r, ϑ, φ), for any vector with components $(u_r, u_\vartheta, u_\varphi)$,

$$\operatorname{div} \mathbf{u} = \frac{1}{r^2} \frac{\partial}{\partial r} (r^2 u_r) + \frac{1}{r \sin \vartheta} \frac{\partial}{\partial \vartheta} (\sin \vartheta u_\vartheta) + \frac{1}{r \sin \vartheta} \frac{\partial u \varphi}{\partial \varphi},$$

$$\operatorname{curl} \mathbf{u} = \left(\frac{1}{r \sin \vartheta} \left\{ \frac{\partial}{\partial \vartheta} (\sin \vartheta u_\varphi) - \frac{\partial u_\vartheta}{\partial \varphi} \right\}, \right.$$

$$\left. \frac{1}{r \sin \vartheta} \left\{ \frac{\partial u_r}{\partial \varphi} - \sin \vartheta \frac{\partial}{\partial r} (r u_\varphi) \right\}, \frac{1}{r} \frac{\partial}{\partial r} (r u_\vartheta) - \frac{1}{r} \frac{\partial u_r}{\partial \vartheta} \right).]$$

(London †)

7.7 Consider the two-dimensional motion generated in a viscous fluid by the wavy motion of an inextensible sheet (whose equilibrium position is in the plane $y = 0$). The displacement of this sheet is

$$y = h \sin k(x - Ut).$$

Assuming that the motion which is generated satisfies Stokes's equations, show that the sheet moves with velocity $\frac{1}{2}(kh)^2 U$ when quantities of order $(kh)^4$ are neglected.

This problem was first considered by G. I. Taylor [*Proc. Roy. Soc.* **A209**, 447–461 (1951)] and has applications to the propulsion of certain microorganisms.

7.8 Discuss the following‡:

"Let me list just what we ought to take pains to forget in these circumstances when pressure gradients are purely balanced by viscous forces.

† From University of London (Imperial College), B.Sc. (1971) ; by permission of the Senate of the University.

‡ From M. J. Lighthill, *J. Fluid Mech.* **52**, 486 (1972) ; by permission of Cambridge University Press, London.

We must forget about Bernoulli's equation: there is no measurable difference between static and dynamic pressures. We must forget about centrifugal forces: fluid can now negotiate sharp bends without any difficulty at all and without setting up any kind of secondary flow. Generally speaking, in fact, motions are much less sensitive to vessel geometry: there is practically no tendency for flow separation. Indeed, in a certain subset of cases (fluid satisfying Newton's viscosity law flowing in rigid vessels) the flow is completely 'reversible'."

7.9 A viscous incompressible fluid of kinematic viscosity ν is flowing steadily in the xy plane in such a manner that the velocity $(U + \partial\psi/\partial y, -\partial\psi/\partial x)$ at any point differs only slightly from the uniform stream $(U, 0)$. Here $U\,(>0)$ is the constant speed of the stream and ψ is a disturbance stream function.

Show that ψ satisfies the equation

$$U \frac{\partial}{\partial x} (\nabla^2\psi) = \nu\, \nabla^4\psi,$$

provided that terms involving products of derivatives of ψ can be neglected.

The y axis is occupied by a gauze which causes the velocity at the point $(0, y)$ of the gauze to have the value $(U + cU \cos(y/L), 0)$, where ε and L are constants and $\varepsilon \ll 1$. By seeking a solution of the form

$$\psi = f(x) \sin(y/L),$$

find ψ both upstream and downstream of the gauze.

When the Reynolds number $\mathrm{Re}_L = UL/\nu$ is large, show that the vorticity is confined to a region of streamwise extent $O(L/R)$ upstream of the gauze, but extends downstream a distance $O(LR)$. (London†)

BIBLIOGRAPHY

W. E. LANGLOIS, "Slow Viscous Flow." Macmillan, New York, 1964.

H. LAMB, "Hydrodynamics," 6th ed. Cambridge Univ. Press, London and New York, 1932.

C. TRUESDELL, *in* "Handbuch der Physik" (S. Flügge, ed.), Vol. VIII/2. Springer-Verlag, Berlin, 1959.

R. BERKER, *in* "Handbuch der Physik" (S. Flügge, ed.). Springer-Verlag, Berlin, 1959.

† From University of London (Imperial College), B.Sc. (1971); by permission of the Senate of the University.

CHAPTER EIGHT *UNSTEADY FLOWS, STABILITY, AND TURBULENCE*

8.1 Introduction

This chapter is divided into four parts, namely: Part I on unsteady boundary layer flows; Part II on the instability of laminar boundary layers; Part III on fully developed turbulent boundary layers; Part IV on the structure of turbulence.

Part I considers the effect on a laminar boundary layer of an outer flow which varies sinusoidally with time. An important case is when the frequency is large; here an inner boundary layer develops, close to the walls in which the bulk of the transition of the unsteady component occurs. It is shown that the unsteady terms have a second-order effect on the mean flow, by means of the so-called Reynolds stresses.

Part II considers the effect of introducing a small disturbance on a steady laminar flow. It is shown that at sufficiently small Reynolds numbers all laminar flows are stable, that is the disturbance dies out. Two basic types of instability are treated, namely: Tollmien–Schlichting and Taylor–Görtler. The former is in the plane of the flow, the latter is a three-dimensional disturbance.

Part III deals with certain aspects of fully developed

turbulent flows. A feature of this type of flow is that the viscous stresses are very much smaller than the Reynolds stresses, except in the neighborhood of solid boundaries. The equations of mean flow are derived and some theories of Reynolds stress considered. Some problems on jet type flows, which have no solid boundaries, are also treated.

Part IV deals largely with the structure of locally isotropic and homogeneous turbulence.

PART I UNSTEADY FLOWS

8.2 Simple Examples

Some simple examples of unsteady flow have already been considered. One simple case is the flow generated by a flat plate (immersed in a viscous fluid which is at rest) oscillating in its own plane. Then the solution of the problem is

$$u = U_0 \exp(-\alpha y) \cos(\alpha y - nt),$$

where the velocity of the plate, which coincides with the x, z plane, is $U_0 \cos nt$ and $\alpha = (n/2\nu)^{1/2}$. One interesting feature of this flow is that the unsteady motion is appreciable only within a distance $\delta_0 = (2\nu/n)^{1/2}$ of the wall. This means that at high frequencies the transition region is very thin. The shear stress at the wall,

$$\tau_{xy} = -\varrho(\nu n)^{1/2} U_0 \cos\left(nt + \frac{\pi}{4}\right),$$

is out of phase with the velocity. The rate at which work is done by the wall on the fluid is

$$-\tau_{xy} U_0 = \frac{\varrho(\nu n)^{1/2} U_0^2 \cos nt \cos[nt + (\pi/4)]}{\text{unit area of wall.}}$$

This has a mean value

$$\frac{1}{2}\varrho U_0^2\left(\frac{\nu n}{2}\right)^{1/2},$$

a result which indicates that at high frequencies considerable energy is lost by viscous heating.

As a second example, consider the two-dimensional flow between two parallel planes a distance h apart, in which a vibrating piston causes a fluctuating pressure gradient. Let the planes bounding the flow be $y = \pm h/2$ and let the pressure gradient term be

$$-\frac{1}{\varrho}\frac{\partial p}{\partial x} = \Re\{Ge^{int}\}.$$

Then the equation of momentum is

$$\frac{\partial u}{\partial t} = Ge^{int} + \nu\frac{\partial^2 u}{\partial y^2}, \tag{8.1}$$

assuming that motion is parallel to the x axis. It is clear that u is of the form

$$u = \Re\{f(y)e^{int}\}.$$

Thus

$$inf = G + \nu f'',$$

from which

$$f = \frac{G}{in} + A\cosh\alpha(1+i)y,$$

where $\alpha = (n/2\nu)^{1/2}$ and A is a constant of integration.

Finally, imposing the no-slip condition on $y = \pm h/2$,

$$f = \frac{G}{in}\left[1 - \frac{\cosh\alpha(1+i)y}{\cosh\alpha(1+i)(h/2)}\right]. \tag{8.2}$$

The required velocity profile is then

$$u = \Re\left\{\frac{G}{in}e^{int}\left[1 - \frac{\cosh\alpha(1+i)y}{\cosh\alpha(1+i)(h/2)}\right]\right\}. \tag{8.3}$$

There are two interesting limiting cases, namely when $\alpha h/2 \ll 1$ and when $\alpha h/2 \gg 1$. In the former case, which corresponds to low frequencies, cosh may be replaced by its quadratic approximation. This gives

$$u \doteqdot \frac{G}{n}\alpha^2(h^2 - y^2)e^{int}.$$

Thus, at low frequencies, the usual quadratic profile is maintained; furthermore, the motion is everywhere in phase with the pressure.

Consider next the high frequency case $\alpha h/2 \gg 1$. Then, when $y > 0$,

$$f \doteq \frac{G}{in}\left\{1 - \exp\left[\alpha(1+i)\left(y - \frac{h}{2}\right)\right]\right\}. \tag{8.4}$$

Thus, $f \doteq G/in$ when $\alpha(y - (h/2))$ is large, and in the central core of the flow,

$$u \doteq \frac{G}{n} \sin nt.$$

Near the edge of the flow, $y = h/2$, where $y - h/2 \sim \delta_0 = 1/\alpha$,

$$u = \frac{G}{n}\left\{\sin nt - \exp\left[\alpha\left(y - \frac{h}{2}\right)\right] \sin\left[\alpha\left(y - \frac{h}{2}\right) + nt\right]\right\}. \tag{8.5}$$

Thus, in the high frequency case, transition again takes place in a region of thickness δ_0; this region contains the fluctuating vorticity of the flow, while the central core is irrotational. Note that in this case there are considerable phase changes in the flow.

8.3 Boundary Layer Flows with Outer Fluctuating Velocity (High Frequency)

Consider now the general case of a two-dimensional boundary layer flow along a flat plate (which coincides with the positive x axis). The outer flow is

$$u = U_0(x) + U_1'(x, t), \qquad p = P_0(x) + P_1'(x, t).$$

In what follows the mean motion is signified by the suffix 0 and the fluctuating part by a prime. The flow inside the boundary layer is

$$u = u_0(x, y) + u_1'(x, y, t),$$
$$v = v_0(x, y) + v_1'(x, y, t),$$
$$p = p_0(x) + p_1'(x, t).$$

Assume now that the fluctuating part is sinusoidal with the factor e^{int}. Then

$$\frac{\partial}{\partial t} = in,$$

and the operator

$$u \frac{\partial}{\partial x} + v \frac{\partial}{\partial y} \sim \frac{U_0}{L},$$

where L is the characteristic length of the mean boundary layer flow in the x direction. The L satisfies the relation

$$\frac{\delta}{L} = (\mathrm{Re}_L)^{-1/2} = \left(\frac{\nu}{U_0 L}\right)^{1/2}$$

(see Chapter 6 on Vorticity and the Laminar Boundary Layer); thus $L = U_0\, \delta^2/\nu$, where δ is the thickness of the mean boundary layer flow U_0.

Now the outer flow equation is

$$\frac{\partial U_1'}{\partial t} + (U_0 + U_1')\frac{\partial}{\partial x}(U_0 + U_1') = -\frac{1}{\varrho}\frac{\partial P_0}{\partial x} - \frac{1}{\varrho}\frac{\partial P_1}{\partial x}. \tag{8.6}$$

Take the mean of Eq. (8.6) over one period. Then†

$$U_0 \frac{\partial U_0}{\partial x} + \overline{U_1' \frac{\partial U_1'}{\partial x}} = -\frac{1}{\varrho}\frac{\partial P_0}{\partial x}. \tag{8.7}$$

If this equation is substracted from the original equation, then

$$\frac{\partial U_1'}{\partial t} + U_1' \frac{\partial U_0}{\partial x} + U_0 \frac{\partial U_1'}{\partial x} - \overline{U_1' \frac{\partial U_1'}{\partial x}} = -\frac{1}{\varrho}\frac{\partial P_1}{\partial x}. \tag{8.8}$$

Now at high frequencies this equation reduces to

$$\frac{\partial U_1'}{\partial t} = -\frac{1}{\varrho}\frac{\partial P_1}{\partial x}, \tag{8.9}$$

since the ratio of the neglected terms to $\partial U_1'/\partial t$ is U_0/nL, which is small if n is large.

Consider now the boundary layer equation:

$$\frac{\partial u}{\partial t} + u \frac{\partial u}{\partial x} + v \frac{\partial u}{\partial y} = -\frac{1}{\varrho}\frac{\partial p}{\partial x} + \nu \frac{\partial^2 u}{\partial y^2}. \tag{8.10}$$

Taking the mean value of this equation:

$$u_0 \frac{\partial u_0}{\partial x} + v_0 \frac{\partial u_0}{\partial y} + \overline{u_1' \frac{\partial u_1'}{\partial x}} + \overline{v_1' \frac{\partial u_1'}{\partial y}} = -\frac{1}{\varrho}\frac{\partial p_0}{\partial x} + \nu \frac{\partial^2 u_0}{\partial y^2}. \tag{8.11}$$

† Overbars denote mean quantities.

This may be arranged to give

$$u_0 \frac{\partial u_0}{\partial x} + v_0 \frac{\partial u_0}{\partial y} = U_0 \frac{dU_0}{dx} + \nu \frac{\partial^2 u_0}{\partial y^2} + F, \tag{8.12}$$

where

$$\begin{aligned} F &= \overline{U_1' \frac{\partial U_1'}{\partial x}} - \overline{u_1' \frac{\partial u_1'}{\partial x}} - \overline{v_1' \frac{\partial u_1'}{\partial y}} \\ &= \overline{U_1' \frac{\partial U_1'}{\partial x}} - \frac{\partial}{\partial x} \overline{(u_1')^2} - \frac{\partial}{\partial y} \overline{(u_1' v_1')}. \end{aligned} \tag{8.13}$$

The mean flow equation (8.12) is the same as that of a steady boundary layer flow apart from the body force ϱF. The ϱF consists of two parts—one is due to the acceleration of the flow, the other to the fluctuating part of the flow, namely

$$-\varrho\left[\frac{\partial}{\partial x} \overline{(u_1')^2} + \frac{\partial}{\partial y} \overline{(u_1' v_1')}\right].$$

The quantities $-\varrho\overline{(u_1')^2}$ and $-\varrho\overline{u_1' v_1'}$ are examples of Reynolds stresses which occur in all fluctuating flows. For example, the quantity $-\varrho\overline{u_1' v_1'}$ may be interpreted as a shear stress and is due to the transfer in the y direction of the momentum $\varrho u_1' \mathbf{i}$ by the velocity $v_1' \mathbf{j}$.

Consider now the full equation for flow inside the boundary layer, If the equation for mean flow is substracted, the remaining terms can be simplified at high frequencies to

$$\frac{\partial u_1'}{\partial t} = \frac{\partial U_1'}{\partial t} + \nu \frac{\partial^2 u_1'}{\partial y^2}.$$

The error in this equation is of order U_0/nL. Now if

$$U_1' = U_1(x)\mathcal{I}(e^{int}),$$

then it follows that [see Example (ii) of Section 8.2, p. 331]

$$u_1' = U_1(x)[\sin nt - e^{-\alpha y} \sin(\alpha y + nt)], \tag{8.14}$$

where $\alpha = (n/2\nu)^{1/2}$. The fluctuating part of the boundary layer flow thus attains its full value within a distance of order δ_0 of the wall. Now the basic approximation used in this theory is

$$\frac{nU_0}{L} \gg 1;$$

this may be rearranged to give

$$\frac{\delta}{\delta_0} \gg 1,$$

that is, the boundary layer thickness $\gg \delta_0$. Thus the Reynolds stresses are of importance only within a thickness δ_0 of the wall. The effect of $U_1\, \partial U_1/\partial x$ is felt throughout the boundary layer. One interesting deduction that may be made is that $F = 0$ when U_1' is independent of x; that is to say, no matter how large U_1' may be, if it is independent of x, it will have no effect on the *mean* flow or on the shear stress at the wall.

PART II INSTABILITY OF BOUNDARY LAYER FLOWS

It is well known that laminar flows tend in certain circumstances to develop into an irregular flow. Thus when smoke from a stationary cigarette rises in still air, there is at first a thin pencil-like column to about a heighth of one foot; this column then tends to become wavy and finally becomes completely irregular.

At very low Reynolds numbers when viscous forces far outweigh pressure forces, viscosity is sufficient to damp out any disturbance in the flow. It will be shown that at a sufficiently low Reynolds number, all two-dimensional parallel flows are stable, but that as the Reynolds number is increased, energy is transferred from the basic flow to the disturbance and the flow therefore is unstable.

In Part II, we treat the instability of two-dimensional flows. The destabilizing effects of heat, gravity, etc. are not considered. Two main types of instability are treated—Tollmien–Schlichting and Taylor–Görtler. The former consists of a wavy disturbance propagated in the plane of the flow; the latter is a vortex motion with an axis parallel to the basic flow and is therefore three-dimensional.

8.4 The Energy Equation for Disturbances in Parallel Flows

Consider a parallel flow $u = \bar{u}(y)$ between two parallel planes $y = y_1$ and $y = y_2$. Let primes denote a wavy disturbance which is periodic in the x direction so that

$$u = \bar{u}(y) + u_1'(x, y, t),$$
$$v = \qquad v_1'(x, y, t),$$
$$p = p_0(x) + p_1'(x, y, t).$$

Then u_1', v_1' are zero and $y = y_1$, $y = y_2$. The kinetic energy of the disturbance between y_1 and y_2 over one wavelength is

$$\tfrac{1}{2} \iint (u_1'^2 + v_1'^2)\, dx\, dy = E.$$

Now

$$\frac{\partial E}{\partial t} = \iint \left(u_1' \frac{\partial u_1'}{\partial t} + v_1' \frac{\partial v_1'}{\partial t} \right) dx\, dy, \tag{8.15}$$

and

$$\frac{\partial u_1'}{\partial t} + \bar{u} \frac{\partial u_1'}{\partial x} + \begin{matrix}\text{quadratic}\\ \text{terms in}\end{matrix}\ u_1',\ v_1' = -\frac{1}{\varrho} \frac{\partial p_1'}{\partial x} + \nu \nabla^2 u_1', \tag{8.16}$$

$$\frac{\partial v_1'}{\partial t} + v_1' \frac{\partial \bar{u}}{\partial y} + \begin{matrix}\text{quadratic}\\ \text{terms in}\end{matrix}\ u_1',\ v_1' = -\frac{1}{\varrho} \frac{\partial p_1'}{\partial y} + \nu \nabla^2 v_1'. \tag{8.17}$$

Equations (8.16) and (8.17) are derived from the unsteady flow equations by canceling out the basic flow equations:

$$0 = -\frac{1}{\varrho} \frac{\partial p_0}{\partial x} + \nu \frac{\partial^2 \bar{u}}{\partial y^2}. \tag{8.18}$$

A simple calculation then yields, using the fact that the flow is periodic, and neglecting quadratic terms,

$$\frac{\partial E}{\partial t} = -\varrho \iint u_1' v_1' \frac{d\bar{u}}{dy}\, dx\, dy - \mu \iint \zeta_1'^2\, dx\, dy; \tag{8.19}$$

ζ_1' is the vorticity of the disturbance and is equal to

$$\frac{\partial v_1'}{\partial x} - \frac{\partial u_1'}{\partial y}.$$

Now the term

$$-\mu \iint \zeta_1'^2 \, dx \, dy$$

is clearly negative and is in fact the loss in energy of the disturbance in viscous heating; it has, therefore, a stabilizing effect on the flow. The term

$$-\varrho \iint u_1'v_1' \frac{d\bar{u}}{dy} \, dx \, dy$$

is positive only when $d\bar{u}/dy$ and $-\varrho u_1'v_1'$ have the same sign over a dominant part of the period. This term gives the transfer of the kinetic energy from the basic flow to the disturbance by means of the Reynolds stress $-\varrho u_1'v_1'$.

Thus the condition for the flow to be unstable is

$$-\varrho \iint u_1'v_1' \frac{d\bar{u}}{dy} \, dx \, dy > \mu \iint \zeta_1'^2 \, dx \, dy. \tag{8.20}$$

It is of interest to introduce dimensionless variables so that lengths are made nondimensional by dividing by the width of the channel h, and velocities by dividing by the maximum velocity in the channel U_0. The above condition then becomes

$$\frac{U_0 h}{\nu} > \text{dimensionless constant} = R_{\text{crit}}. \tag{8.21}$$

The dimensionless constant R_{crit} depends on the velocity profile. The exact value of R_{crit} depends on the solution of the disturbance equations; however, useful upper bounds for R_{crit} may be obtained by inserting suitable disturbance u_1' and v_1' in the above inequality. These disturbances must satisfy the kinematic conditions of the flow but not necessarily the momentum equations.

8.5 Stability of Parallel Flows

In this section attention is focused on obtaining the basic equation for wavy disturbances in parallel flows.

It is convenient to nondimensionalize the variables; thus, lengths and velocities are nondimensionalized as above, and pressure is nondimensionalized by dividing by ϱU_0^2. The variables as now stated are thus in nondimensional form.

The basic flow is $u = \bar{u}(y)$ and primes denote the disturbance. Thus the equation for the nondimensional vorticity of small disturbances is

$$\frac{\partial \zeta'}{\partial t} + \bar{u}\frac{\partial \zeta'}{\partial x} = -\frac{\partial p'}{\partial x} + \frac{1}{\mathrm{Re}_L}\nabla^2\zeta', \tag{8.22}$$

where

$$\zeta' = \frac{\partial v'}{\partial x} - \frac{\partial u'}{\partial y}, \qquad \frac{\partial u'}{\partial x} + \frac{\partial v'}{\partial y} = 0,$$

and Re_L is the Reynolds number $U_0 h/\nu$. It is convenient to introduce a disturbance stream function ψ' such that

$$u' = \frac{\partial \psi'}{\partial y}, \qquad v' = -\frac{\partial \psi'}{\partial x}.$$

Then $\zeta' = -\nabla^2\psi'$. Consider now a wavy disturbance which travels downstream such that

$$\psi' = \varphi(y)e^{i\alpha(x-ct)}, \tag{8.23}$$

where α is the wave number of the disturbance and is real, c may be real or complex; if real, the disturbance is neutral; if complex, the disturbance grows or decays according as

$$\mathscr{I}(c) > 0 \qquad \text{or} \qquad \mathscr{I}(c) < 0.$$

Then

$$\zeta' = -[\varphi'' - \alpha^2\varphi]e^{i\alpha(x-ct)},$$

and the equation for φ is

$$(\bar{u} - c)(\varphi'' - \alpha^2\varphi) - \bar{u}''\varphi = \frac{i}{\alpha\,\mathrm{Re}_L}(\varphi^{\mathrm{iv}} - 2\alpha^2\varphi'' + \alpha^4\varphi). \tag{8.24}$$

Equation (8.24) is known as the Orr–Sommerfeld equation (O–S equation). It will be noted that the O–S equation depends not only on a knowledge of the velocity profile but also on its second derivative $\bar{u}''$. Thus an accurate knowledge of the profile is important.

8.5.1 Solutions of the O–S Equations When $\alpha\,\mathrm{Re}_L$ is Large

When $\alpha\,\mathrm{Re}_L$ is large, the O–S equation becomes approximately

$$\varphi'' - \alpha^2\varphi - \left(\frac{\bar{u}''}{\bar{u} - c}\right)\varphi = 0. \tag{8.25}$$

This has the effect of reducing the order of the O–S equation from 4 to 2. Thus the no-slip condition at the walls has to be relaxed in this case. Thus $\varphi = 0$ at the walls $y = y_1$ and $y = y_2$.

It will be shown now that a necessary condition for instability (at large values of $\alpha\,\mathrm{Re}_L$) is that $\bar{u}(y)$ should have a point of inflection. The proof is as follows.

Let

$$L \equiv \frac{d^2}{dy^2} - \alpha^2 - \frac{\bar{u}''}{\bar{u} - c},$$

and let $\tilde{L}$ denote the complex conjugate of L, that is,

$$\tilde{L} \equiv \frac{d^2}{dy^2} - \alpha^2 - \frac{\bar{u}''}{\bar{u} - \tilde{c}}.$$

Also, let $\tilde{\varphi}$ be the complex conjugate of φ. Then consider

$$\tilde{\varphi}L(\varphi) - \varphi\tilde{L}(\tilde{\varphi}) = \tilde{\varphi}\varphi'' - \varphi\tilde{\varphi}'' - \frac{2ic_i\bar{u}''\varphi\tilde{\varphi}}{|\bar{u} - c|^2},$$

where $c_i = \mathcal{I}(c)$. Then

$$\int_{y_1}^{y_2} [\tilde{\varphi}L(\varphi) - \varphi\tilde{L}(\tilde{\varphi})]\,dy = \int_{y_1}^{y_2} (\tilde{\varphi}\varphi'' - \varphi\tilde{\varphi}'')\,dy - 2ic_i \int_{y_1}^{y_2} \frac{\bar{u}''\varphi\tilde{\varphi}\,dy}{|\bar{u} - c|^2}. \tag{8.26}$$

Since $L(\varphi) = 0$ and $\tilde{L}(\tilde{\varphi}) = 0$, the integral on the left side is zero. Next, the first integral on the right side of (8.26) is zero. This may be proved by integrating by parts and imposing the boundary conditions. This leaves

$$2ic_i \int_{y_1}^{y_2} \bar{u}'' \frac{\varphi\tilde{\varphi}}{|\bar{u} - c|^2}\,dy = 0. \tag{8.27}$$

Now in the case of an increasing disturbance,

$$c_i > 0 \qquad \text{and} \qquad \frac{\varphi\tilde{\varphi}}{|\bar{u} - c|^2} > 0.$$

Thus $\bar{u}''$ must change sign in the interval. Since $\bar{u}''$ is continuous, it follows that

$$\bar{u}'' = 0$$

at a point $y = y_c$ in the interval. The profile has therefore at least one point of inflection, at $y = y_c$, for amplification of the disturbance. Further, in the case of a neutral disturbance, that is, in the case $c_i \to 0$, it is necessary that

$$\bar{u}(y_c) = c,$$

otherwise the integral does not exist at $y = y_c$; that is to say, at large Reynolds numbers, amplified disturbances are possible only if a point of inflection exists in the profile; further, the neutral disturbance is propagated with the speed of the basic velocity profile at the point of inflection $y = y_c$.

8.6 Stability of the Schlichting Jet

A simple example of the above analysis is that of the two-dimensional jet. In this case the spreading of the jet is neglected. This assumption is correct provided the wavelength of the disturbance is very much smaller than the characteristic boundary layer length in the x direction. Since the wavelength of interest is usually of the same order of magnitude as the boundary layer thickness, this assumption is clearly valid.

Thus the nondimensional velocity profile is

$$\bar{u} = \operatorname{sech}^2 y, \tag{8.28}$$

where the velocity on the axis of the jet is taken as the reference velocity and where the reference length has the same order of magnitude as the boundary layer thickness.

This profile has two points of inflection at $y = \pm\frac{2}{3}$. The limiting case of the O–S equation becomes, for $\alpha \operatorname{Re}_L$ large,

$$\varphi'' - \alpha^2\varphi + 6\varphi \operatorname{sech}^2 y = 0. \tag{8.29}$$

It is of interest to separate two types of solution, namely symmetrical oscillation with $\varphi = 0$ at $y = 0$, $y = \infty$ and antisymmetrical oscillations with $\varphi' = 0$ at $y = 0$, $\varphi = 0$ at $y = \infty$. It may be verified that the symmetrical solution is

$$\varphi = A \operatorname{sech} y \tanh y \qquad \text{with} \quad \alpha = 1,$$

and that the antisymmetrical oscillations is

$$\varphi = A \operatorname{sech}^2 y \qquad \text{with} \quad \alpha = 2.$$

In both cases the critical wavelength $2\pi/\alpha$ is in fact of the same order of magnitude as the boundary layer thickness.

These solutions represent a set of eddies which travel downstream with a speed $\frac{2}{3}$ that of the main speed of the flow.

8.7 General Method of Solution in Inviscid Case

In this section velocity profiles with a point of inflection at $y = y_c$ are considered. In the limit, as $\alpha \operatorname{Re}_L \to \infty$, the O–S equation is

$$(\bar{u} - c)(\varphi'' - \alpha^2\varphi) - \bar{u}''\varphi = 0. \tag{8.30}$$

Consider first the special case when $\alpha^2 = 0$. It can be verified that a solution in this case is

$$\varphi_{10} = \bar{u} - c. \tag{8.31}$$

A second solution can be obtained for $\alpha^2 = 0$ by writing $\varphi_{20} = (\bar{u} - c)\Phi$. It can be verified that

$$\varphi_{20} = (\bar{u} - c) \int_{y_c}^{y} \frac{dy}{(\bar{u} - c)^2}, \tag{8.32}$$

the lower limit being taken at $y = y_c$ to avoid convergence difficulty.

These results may be used as a starting point in the calculation of general solutions in ascending powers of α^2; thus,

$$\varphi_1 = \varphi_{10} + \alpha^2\varphi_{11} + \alpha^4\varphi_{12} + \cdots, \qquad \varphi_2 = \varphi_{20} + \alpha^2\varphi_{21} + \alpha^4\varphi_{22} + \cdots. \tag{8.33}$$

These series converge provided α^2 is small. The equation for φ_{11} is obtained by substituting φ_1 in the inviscid O–S equation and equating coefficients of α^2. Thus

$$(\bar{u} - c)\varphi_{11} - \bar{u}''\varphi_{11} = (\bar{u} - c)\varphi_{10} = (\bar{u} - c)^2.$$

The solution of this equation is

$$\varphi_{11} = (\bar{u} - c) \int_{y_c}^{y} \frac{dy}{(\bar{u} - c)^2} \int_{y_c}^{y} (\bar{u} - c)\varphi_{10}\, dy.$$

In general,

$$\varphi_{1,n+1} = (\bar{u} - c) \int_{y_c}^{y} \frac{dy}{(\bar{u} - c)^2} \int_{y_c}^{y} (\bar{u} - c)\varphi_{1,n}\, dy, \tag{8.34}$$

$$\varphi_{2,n+1} = (\bar{u} - c) \int_{y_c}^{y} \frac{dy}{(\bar{u} - c)^3} \int_{y_c}^{y} (\bar{u} - c)^2\varphi_{2,n}\, dy. \tag{8.35}$$

It will be noted that both φ_1 and φ_2 are regular at $y = y_c$. The required solution is a linear combination of φ_1 and φ_2, which satisfies $\varphi = 0$ at $y = y_1$ and $y = y_2$, that is,

$$a\varphi_1(y_1) + b\varphi_2(y_1) = 0, \qquad a\varphi_1(y_2) + b\varphi_2(y_2) = 0.$$

For a nontrivial solution,

$$\begin{vmatrix} \varphi_1(y_1) & \varphi_2(y_1) \\ \varphi_1(y_2) & \varphi_2(y_2) \end{vmatrix} = 0. \tag{8.36}$$

This determinantal condition is satisfied only when α has a certain value, the value of c being fixed by the point of inflection.

8.8 Stability at Finite Reynolds Numbers

Thus far the effect of a finite Reynolds number has been neglected in the stability problem. When the Reynolds number is finite, it is necessary to consider all four solutions of the O–S equation. It can be shown that two of these solutions tend, as $\alpha \, \mathrm{Re}_L \to \infty$, to the inviscid solutions φ_1 and φ_2 obtained in the previous section. The second pair of solutions, φ_3 and φ_4, belong to the complete O–S equation and are called the viscous solutions. A complete solution which satisfies the no-slip conditions at y_1 and y_2 then requires consideration of all four solutions. The mathematics of the stability problem is, in this case, a matter of some difficulty and will not be attempted here. Instead, the results will be briefly stated.

There are two main types of parallel flow differentiated by whether or not there is a point of inflection.

Flows with a point of inflection are unstable at infinite Reynolds number. In this type of flow, viscosity can be shown to have a purely stabilizing effect. The neutral stability cure is shown in Fig. 8.1.

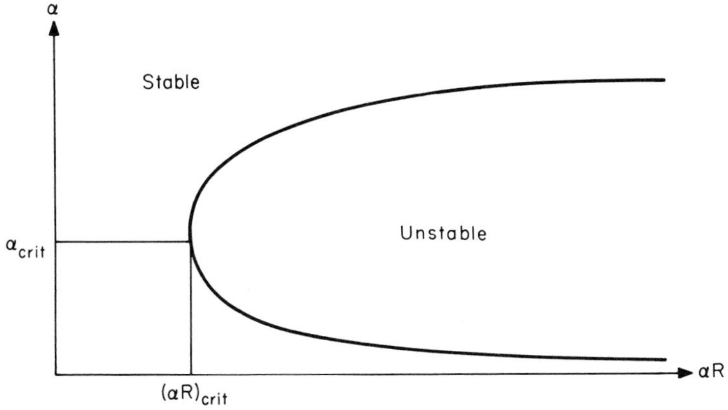

Fig. 8.1 Neutral stability curve for flows with a point of inflection.

This type of flow occurs in practice in boundary layer flows over a flat plate with adverse pressure gradient and in free flows such as jets and wakes. The latter flows are very unstable, being unstable at values of R_{crit} as low as 10.

Flows without a point of inflection have been shown to be stable at infinite Reynolds number. A typical neutral stability curve in this case is shown in Fig. 8.2.

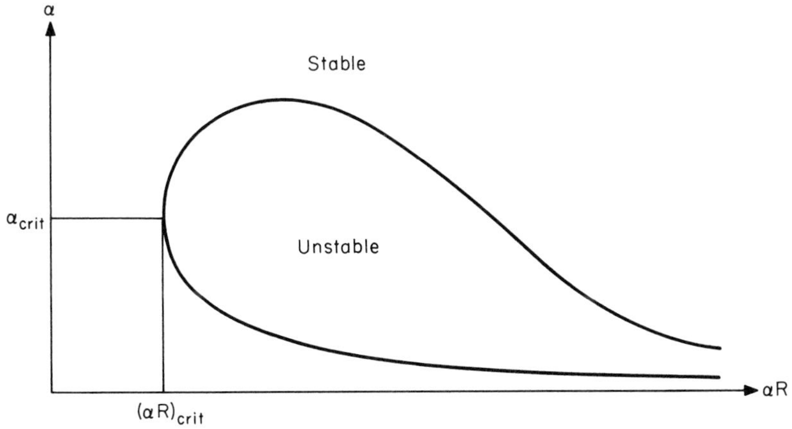

Fig. 8.2 Neutral stability curve for flows without a point of inflection.

In this case viscosity plays a dual role. It brings the disturbance into being and damps it. This type of flow is very much more stable than flows with a point of inflection. Instability first occurs in this type of flow at a Reynolds number of order 10^4. A typical example of flow without a point of inflection is the Blasius flow over a flat plate.

8.8.1 Taylor–Görtler Instability

Consider the boundary layer flow over a curved wall whose radius of curvature R is very much larger than the boundary layer thickness δ (Fig. 8.3). Let curvilinear coordinates (x, y) be taken: x is the arc length along the wall, y is in the radial direction. Then the curvature of the flow manifests itself essentially in a body force

$$-\frac{u^2}{(R-y)}\,\mathbf{j} \doteq -\frac{u^2}{R}\,\mathbf{j}$$

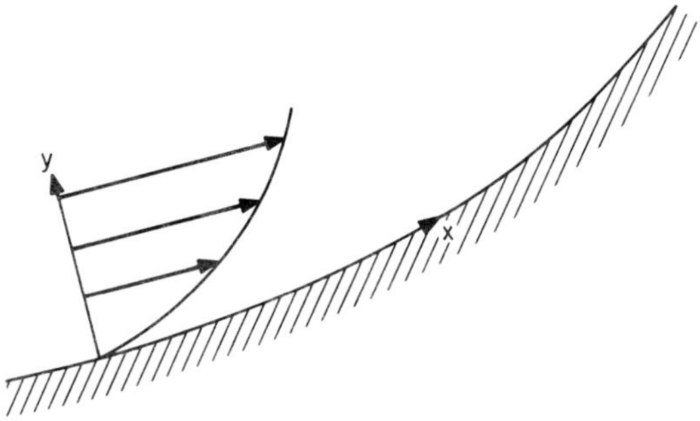

Fig. 8.3

per unit mass of the fluid, where u, v are velocity components in the x, y directions. Then it is readily seen that there is a tendency for the outside particles, with relatively large kinetic energy, to migrate under this centrifugal force toward the wall, where the kinetic energy is small.

In this motion the flow displaces particles of low kinetic energy toward the edge of the boundary layer. Thus the possibility of secondary motion exists in which the fluid particles migrate in spirals whose axes are in the x direction. This secondary flow is produced basically by the centrifugal forces u^2/R; it is opposed by viscous forces of order $\nu u/\delta^2$ per unit mass. The ratio of these forces gives

$$\frac{U_0{}^2\delta^2}{\nu^2}\,\frac{\delta}{R} = \frac{\bar{\mu}}{2}.$$

Thus it is to be expected that instability will occur for $\bar{\mu}$ sufficiently large. The exact value of $\bar{\mu}_{\text{crit}}$ at which the secondary flow first occurs depends on the velocity profile.

The equations of momentum are

$$\frac{Du}{Dt} = -\frac{1}{\varrho}\frac{\partial p}{\partial x} + \nu\,\nabla^2 u, \tag{8.37}$$

$$\frac{Dv}{Dt} = -\frac{1}{\varrho}\frac{\partial p}{\partial y} + \nu\,\nabla^2 v - \frac{u^2}{R}, \tag{8.38}$$

$$\frac{Dw}{Dt} = -\frac{1}{\varrho}\frac{\partial p}{\partial z} + \nu\,\nabla^2 w, \tag{8.39}$$

and

$$\frac{\partial u}{\partial x} + \frac{\partial v}{\partial y} + \frac{\partial w}{\partial z} = 0. \tag{8.40}$$

These equations are approximate and hold only if $\delta/R \ll 1$, where δ is the boundary layer thickness.

Let the basic flow be parallel to the x direction and be denoted by the subscript 0. Then

$$\begin{aligned} u_0 &= u_0(y), & w_0 &= 0, \\ v_0 &= 0, & p_0 &= p_0(y). \end{aligned} \tag{8.41}$$

Let U_0 be the constant velocity on the outside of the boundary layer. Let primes denote a small perturbation to this flow of the form

$$\begin{aligned} u' &= u_1(y)\cos\alpha z\, e^{\beta t}, & w' &= w_1(y)\sin\alpha z\, e^{\beta t}, \\ v' &= v_1(y)\cos\alpha z\, e^{\beta t}, & p' &= p_1(y)\cos\alpha z\, e^{\beta t}. \end{aligned} \tag{8.42}$$

In these expressions $2\pi/\alpha$ is the spacing between successive vortices in the z direction and β is the amplification factor of the disturbances. Then u_1, v_1, and w_1 satisfy the equations

$$\begin{aligned} \beta u_1 + v_1 u_0' &= \nu(u_1'' - \alpha^2 u_1), \\ \beta v_1 + \frac{2u_0}{R} u_1 &= -\frac{1}{\varrho} p_1' + \nu(v_1'' - \alpha^2 v_1), \\ \beta w_1 &= -\frac{1}{\varrho}\alpha p_1 + \nu(w_1'' - \alpha^2 w_1), \\ v_1' + \alpha w_1 &= 0, \end{aligned} \tag{8.43}$$

when quadratic terms are neglected.

It is convenient to eliminate w_1 and p_1; thus

$$\beta u_1 + v_1 u_0' = \nu(u_1'' - \alpha^2 u_1), \tag{8.44}$$

$$\beta v_1 + \frac{2u_0}{R} u_1 = \frac{\beta}{\alpha^2} v_1'' - \frac{\nu}{\alpha^2}(v_1^{\text{iv}} - 2\alpha^2 v_1'' + \alpha^4 v_1). \tag{8.45}$$

Introduce now the dimensionless variables

$$\eta = \frac{y}{\delta}, \quad U = \frac{u_0}{U_0}, \quad \sigma = \alpha\delta, \quad \bar{u} = \frac{u_1}{U_0\delta/\nu}.$$

Then the disturbance equations become, in the case of neutral stability $\beta = 0$,

$$L(\bar{u}) = v_1 \frac{dU}{d\eta}, \qquad L^2(v_1) = -\sigma^2 \bar{\mu} U \bar{u},$$

where

$$L = \frac{d^2}{d\eta^2} - \sigma^2, \qquad \bar{\mu} = 2\left(\frac{U_0\delta}{\nu}\right)^2 \frac{\delta}{R}.$$

The boundary conditions are

$$\bar{u} = 0, \quad v_1 = 0, \quad v_1' = 0 \qquad \text{on} \quad \eta = 0,$$
$$\bar{u} = 0, \quad v_1 = 0, \quad v_1' = 0 \qquad \text{at} \quad \eta = \infty.$$

The method of solution is to convert the above equations into a pair of integral equations:

$$\bar{u} = -\int_0^\infty G(\eta, \eta_0) \frac{dU_0(\eta_0)}{d\eta_0} v(\eta_0)\, d\eta_0, \tag{8.46}$$

$$v_1 = \sigma^2 \mu \int_0^\infty H(\eta, \eta_0) U(\eta_0) \bar{u}(\eta_0)\, d\eta_0, \tag{8.47}$$

where G and H are the Green's functions of the operators L and L^2, respectively, which satisfy the appropriate boundary conditions. This pair of integral equations may be solved by converting the integrals into finite sums. This results in a set of linear algebraic equations for the unknown values of $\bar{u}$ and v_1. The condition that this set should have a nontrivial solution then yields a determinantal equation which gives a relation between $\bar{u}$ and σ. It has been shown that this relation is remarkably independent of the particular basic velocity profile. In fact, the lowest value of $\bar{\mu}$ for which this secondary flow occurs is about 0.1. This type of flow is then remarkably sensitive to quite a small curvature in the wall.

8.9 Transition to Turbulent Flows

Consider a laminar flow at sufficiently low Reynolds number when this flow is stable. At R just greater than R_{crit}, one and only one mode of disturbance is amplified. As the disturbance increases (exponentially at first), it affects the basic mean flow pattern. In fact, the mean flow pattern can develop regions in which the streamlines are curved. Where the curvature is of the right sign, Taylor–Görtler vortices develop. This then converts the two-dimensional disturbances into three-dimensional flow. On this new secondary flow, further Tollmien–Schlichting eddies form until the flow becomes completely random. The subject of turbulence is treated in Part III.

PART III TURBULENT FLOWS

Fully developed turbulent flows will be considered here. In Part II the basic mechanism for transition from a laminar flow to a random-type turbulent flows has been explained. Now it will be assumed that the turbulent eddies no longer have any discernible periodic structure.

Section 8.10 deals with the equations of steady mean flow. Then the boundary layer equations of turbulent flow are developed. There follows a discussion on the Reynolds shear stress in free flows such as jets and wakes. The turbulent jet is then treated as an example.

Section 8.15 deals with flows involving solid boundaries, for example, flows in channels and along a flat plate.

8.10 The Equations of Mean Flow

In this section the equations of steady mean flow are derived. The mean flow will be denoted by a bar, and the fluctuations from the mean value by primes. Means will be taken over a sufficiently long period of time. Thus

$$\begin{aligned} \mathbf{u}(x, y, z, t) &= \bar{\mathbf{u}}(x, y, z) + \mathbf{u}'(x, y, z, t), \\ p(x, y, z, t) &= \bar{p}(x, y, z) + p'(x, y, z, t). \end{aligned} \tag{8.48}$$

It follows from the definition [Eq. (8.48)] that

$$\bar{\mathbf{u}}' = 0, \qquad \overline{\frac{\partial \mathbf{u}}{\partial t}} = 0, \qquad \overline{\frac{\partial \mathbf{u}'}{\partial x}} = 0, \qquad \text{etc.,}$$

with similar results for p'.

The continuity equation is now

$$\operatorname{div} \bar{\mathbf{u}} + \operatorname{div} \mathbf{u}' = 0.$$

When means are taken, this equation becomes

$$\operatorname{div} \bar{\mathbf{u}} = 0,$$

and it follows that

$$\operatorname{div} \mathbf{u}' = 0.$$

The momentum equation is

$$\frac{\partial \mathbf{u}}{\partial t} + u \frac{\partial \mathbf{v}}{\partial x} + v \frac{\partial \mathbf{u}}{\partial y} + w \frac{\partial \mathbf{u}}{\partial z} = -\frac{1}{\varrho} \operatorname{grad} p + \nu \nabla^2 \mathbf{u} \tag{8.49}$$

in rectangular coordinates, with

$$\mathbf{u} = u\mathbf{i} + v\mathbf{j} + w\mathbf{k}.$$

When means are taken and the above mentioned properties of the means are used, we have

$$\begin{aligned} \bar{u} \frac{\partial \bar{\mathbf{u}}}{\partial x} + \bar{v} \frac{\partial \bar{\mathbf{u}}}{\partial y} + \bar{w} \frac{\partial \bar{\mathbf{u}}}{\partial z} + \overline{u' \frac{\partial \mathbf{u}'}{\partial x} + v' \frac{\partial \mathbf{u}'}{\partial y} + w' \frac{\partial \mathbf{v}'}{\partial z}} \\ = -\frac{1}{\varrho} \operatorname{grad} \bar{p} + \nu \nabla^2 \bar{\mathbf{u}}. \end{aligned} \tag{8.50}$$

The mean flow equation is identical to that of steady laminar flow, except that a term

$$\begin{aligned} &-\overline{\left[u' \frac{\partial \mathbf{u}'}{\partial x} + v' \frac{\partial \mathbf{u}'}{\partial y} + w' \frac{\partial \mathbf{u}'}{\partial z}\right]} \\ &= -\overline{\left[\frac{\partial}{\partial x} (u'\mathbf{u}') + \frac{\partial}{\partial y} (v'\mathbf{u}') + \frac{\partial}{\partial z} (w'\mathbf{u}')\right]} \end{aligned}$$

must be added to the right side of the momentum equation. This term multiplied by ϱ may be considered as a body force term due to the Reynolds stresses.

It follows that a stress tensor may be defined for the mean turbulent flow. This is equal to

$$\begin{bmatrix} -\bar{p} + \frac{2}{3}\mu\frac{\partial \bar{u}}{\partial x}, & \mu\left(\frac{\partial \bar{u}}{\partial y} + \frac{\partial \bar{v}}{\partial x}\right), & \mu\left(\frac{\partial \bar{w}}{\partial x} + \frac{\partial \bar{u}}{\partial z}\right) \\ & \text{etc.} & \end{bmatrix}$$

$$+\varrho\begin{bmatrix} \frac{\partial}{\partial x}(-\overline{(u')^2}), & \frac{\partial}{\partial y}\overline{(u'v')}, & \frac{\partial}{\partial z}\overline{(u'w')} \\ & \text{etc.} & \end{bmatrix};$$

that is to say, the stress tensor is made up of the usual viscous stress tensor and a Reynolds stress tensor.

In general the Reynolds stresses are very much larger than the viscous stresses except close to a wall where the Reynolds stresses vanish by virtue of the no-slip condition,

$$\mathbf{u}' = \mathbf{0}.$$

Apart however from this region (known as the sublaminar region), the Reynolds stresses outweigh the viscous stresses by a factor of order 100.

Thus the momentum equations of steady mean turbulent flow are, in rectangular coordinates,

$$\begin{aligned} \frac{D\bar{u}}{Dt} &= -\frac{1}{\varrho}\frac{\partial \bar{p}}{\partial x} - \left[\frac{\partial}{\partial x}\overline{(u')^2} + \frac{\partial}{\partial y}\overline{(u'v')} + \frac{\partial}{\partial z}\overline{(u'w')}\right], \\ \frac{D\bar{v}}{Dt} &= -\frac{1}{\varrho}\frac{\partial \bar{p}}{\partial y} - \left[\frac{\partial}{\partial x}\overline{(u'v')} + \frac{\partial}{\partial y}\overline{(v')^2} + \frac{\partial}{\partial z}\overline{(v'w')}\right], \\ \frac{D\bar{w}}{Dt} &= -\frac{1}{\varrho}\frac{\partial \bar{p}}{\partial z} - \left[\frac{\partial}{\partial x}\overline{(u'w')} + \frac{\partial}{\partial y}\overline{(v'w')} + \frac{\partial}{\partial z}\overline{(w')^2}\right], \end{aligned} \tag{8.51}$$

where

$$\frac{D}{Dt} = \bar{u}\frac{\partial}{\partial x} + \bar{v}\frac{\partial}{\partial y} + \bar{w}\frac{\partial}{\partial z}.$$

8.11 Turbulent Boundary Layers

In this section the simple case of a two-dimensional boundary layer flow is treated. The basic assumption is the same as that for laminar flow, namely

$$\frac{\partial}{\partial x} \ll \frac{\partial}{\partial y}.$$

Experimental results suggest that the turbulent fluctuations u', v', w' are all of the same order of magnitude. Thus the most important Reynolds stress term in the x-momentum equation is $(\partial/\partial y)\overline{(u', v')}$ and the x-momentum boundary layer equation is

$$\bar{u}\frac{\partial \bar{u}}{\partial x} + \bar{v}\frac{\partial \bar{u}}{\partial y} = -\frac{1}{\varrho}\frac{\partial \bar{p}}{\partial x} - \frac{\partial}{\partial y}\overline{(u'v')}. \tag{8.52}$$

Next, the y-momentum equation is approximately

$$\bar{u}\frac{\partial \bar{v}}{\partial x} + \bar{v}\frac{\partial \bar{v}}{\partial y} = -\frac{1}{\varrho}\frac{\partial \bar{p}}{\partial y} - \frac{\partial}{\partial y}\overline{(v')^2}. \tag{8.53}$$

Suppose the x axis is a flat plate on which $\bar{u} = 0$, $\bar{v} = 0$, $v' = 0$. Let the boundary layer have thickness δ; then

$$\int_0^\delta \bar{u}\frac{\partial \bar{v}}{\partial x}\,dy + \frac{1}{2}[\bar{v}^2]_0^\delta = -\left[\frac{p}{\varrho}\right]_0^\delta - [\overline{(v')^2}]_0^\delta. \tag{8.54}$$

Now $v' = 0$ at the edge of the boundary layer, where the flow is no longer turbulent; thus

$$\frac{p}{\varrho\bar{u}^2} \sim \left(\frac{\bar{v}}{\bar{u}}\right)^2 \sim \left(\frac{\delta}{L}\right)^2,$$

where L is the characteristic boundary layer length in the x direction. It follows that the pressure is sensibly constant over the boundary layer just as in the laminar case.

Thus the equations of steady turbulent boundary layer flow are

$$\bar{u}\frac{\partial \bar{u}}{\partial x} + \bar{v}\frac{\partial \bar{v}}{\partial y} = -\frac{1}{\varrho}\frac{\partial \bar{p}}{\partial x} - \frac{\partial}{\partial y}\overline{(u'v')}, \tag{8.55}$$

with $\bar{p} = \bar{p}(x)$.

The above equation is valid except in the sublaminar region close to the wall where an additional term

$$\nu \frac{\partial^2 \bar{u}}{\partial y^2}$$

has to be added to the right side. The thickness of this region may be estimated by using the results of Part I. Thus, if n is the order of magnitude of the frequency of the dominant eddies ($1/n$ is known as the time scale of the turbulence), then the thickness of the sublaminar region is of order $(2\nu/n)^{1/2}$.

Outside the boundary layer the Reynolds stresses vanish and the equation of mean flow is

$$\bar{u}\frac{\partial \bar{u}}{\partial x} + \bar{v}\frac{\partial \bar{u}}{\partial y} = -\frac{1}{\varrho}\frac{\partial \bar{p}}{\partial x}, \qquad \bar{u}\frac{\partial \bar{v}}{\partial x} + \bar{v}\frac{\partial \bar{v}}{\partial y} = -\frac{1}{\varrho}\frac{\partial \bar{p}}{\partial y}. \tag{8.56}$$

Just outside the boundary layer, flow is sensibly in the x direction, that is $\mathbf{u} = U_0(x)\mathbf{i}$, so that

$$-\frac{1}{\varrho}\frac{\partial \bar{p}}{\partial x} = U_0\frac{dU_0}{dx}.$$

The mean flow equation inside the boundary layer therefore becomes

$$\bar{u}\frac{\partial \bar{u}}{\partial x} + \bar{v}\frac{\partial \bar{u}}{\partial y} = U_0\frac{dU_0}{dx} - \frac{\partial}{\partial y}\overline{(u'v')}. \tag{8.57}$$

8.12 The Momentum Balance Equation

Consider the full turbulent boundary layer equation:

$$\bar{u}\frac{\partial \bar{u}}{\partial x} + \bar{v}\frac{\partial \bar{u}}{\partial y} = U_0\frac{dU_0}{dx} - \frac{\partial}{\partial y}\overline{(u'v')} + \nu\frac{\partial^2 \bar{u}}{\partial y^2}. \tag{8.58}$$

Then integration over the boundary layer thickness δ gives

$$\int_0^\delta \frac{\partial}{\partial x}(\bar{u}^2)\,dy + [\bar{u}\bar{v}]_0^\delta = U_0\frac{dU_0}{dx}\delta - [\overline{u'v'}]_0^\delta + \left[\nu\frac{\partial \bar{u}}{\partial y}\right]_0^\delta.$$

Since the turbulent fluctuations vanish at $y = 0$, $y = \delta$, the term

$$[\overline{u'v'}] = 0.$$

Also,

$$[\bar{u}\bar{v}]_0 = 0, \qquad \left[\frac{\partial \bar{u}}{\partial y}\right]^\delta = 0.$$

It follows therefore that the Reynolds stresses do not enter explicitly into the mean momentum equation, which is formally the same as in laminar flow (see Chapter 6 on Vorticity and the Laminar Boundary Layer). This apparent paradox is resolved by remembering that it is the turbulent stresses which determine the mean velocity profile.

8.13 The Turbulent Jet

As an example which is readily amenable to analysis, consider a two-dimensional jet which emerges from a slit. Let the axis of the jet coincide with the x axis and let the motion take place in the xy plane. The slit is then at the origin. Then the pressure outside the jet is constant, and the mean equation of motion is

$$\bar{u}\frac{\partial \bar{u}}{\partial x} + \bar{v}\frac{\partial \bar{u}}{\partial y} = -\frac{\partial}{\partial y}\overline{(u'v')}\,. \tag{8.59}$$

The momentum balance equation then gives

$$\int_0^\delta \frac{\partial}{\partial x}(\bar{u})^2\,dy = 0, \tag{8.60}$$

since $\partial\bar{u}/\partial y = 0$ on the axis of the jet. This result may be written

$$\frac{d}{dx}\int_0^\delta \bar{u}^2\,dy = 0$$

since $\bar{u} = 0$ on the edge of the jet. This implies

$$\int_0^\delta \bar{u}^2\,dy = \frac{1}{2}\,\frac{M_0}{\varrho} = \text{constant},$$

where M_0 is the (constant) momentum flux.

Now it is known from experiment that in the case of free flows, that is, flows in the absence of solid boundaries such as jets or wakes, the exchange coefficient ε defined by

$$\varepsilon = \frac{-\overline{u'v'}}{\partial\bar{u}/\partial y}$$

is almost independent of y, except near the edge of the jet where the flow is intermittently laminar and turbulent. If the intermittency is taken into account this result holds at the edges too. This result is known as the *constant exchange coefficient hypothesis.* It implies that

$$-\overline{u'v'} = \varepsilon(x)\frac{\partial\bar{u}}{\partial y}. \tag{8.61}$$

Thus the equation of mean motion becomes

$$\bar{u}\frac{\partial\bar{u}}{\partial x} + \bar{v}\frac{\partial\bar{u}}{\partial y} = \varepsilon(x)\frac{\partial^2\bar{u}}{\partial y^2}. \tag{8.62}$$

This equation may be written in terms of a mean stream function ψ, defined by

$$\bar{u} = \frac{\partial\psi}{\partial y}, \qquad \bar{v} = -\frac{\partial\psi}{\partial x},$$

as

$$\frac{\partial\psi}{\partial y}\frac{\partial^2\psi}{\partial x\,\partial y} - \frac{\partial\psi}{\partial x}\frac{\partial^2\psi}{\partial y^2} = \varepsilon(x)\frac{\partial^3\psi}{\partial y^3}.$$

Now change the independent variable x to X, where

$$X = \int_0^x \varepsilon(x)\,dx.$$

Then the equation for the stream function becomes

$$\frac{\partial\psi}{\partial y}\frac{\partial^2\psi}{\partial X\,\partial y} - \frac{\partial\psi}{\partial X}\frac{\partial^2\psi}{\partial y^2} = \frac{\partial^3\psi}{\partial y^3}. \tag{8.63}$$

This is exactly the same equation for the laminar jet with $\nu = 1$ and (x, y) replaced by (X, y). The solution is therefore

$$\psi = 2cX^{1/3}\tanh\left(\frac{cy}{3X^{2/3}}\right) \tag{8.64}$$

and

$$\bar{u} = \frac{2}{3}\frac{c^2}{X^{1/3}}\operatorname{sech}^2\xi, \qquad \text{where} \quad \xi = \frac{cy}{3X^{1/3}}.$$

It now remains to determine the quantity $\varepsilon(x)$. This may be obtained by using the following dimensional argument. The mean velocity component $\bar{u}$ may be written as the product of the axial velocity $\bar{u}_{\text{axis}}$ times a factor which depends on the spatial variables x, y and the physical variables ν, ϱ, M_0. Thus

$$\bar{u} = \bar{u}_{\text{axis}} f(x, y, \nu, \varrho, M_0).$$

If the velocity profiles are similar, then

$$f = f(\eta),$$

where $\eta = x^L y^M \nu^N \varrho^P M_0$. Now the motion is sensibly independent of viscosity, so $N = 0$. The only possible combination of the remaining variables is $\eta = y/x$.

It follows that

$$\bar{u} = \bar{u}_{\text{axis}}\, f\left(\frac{y}{x}\right).$$

Thus $X^{2/3} \propto x$, so that $\varepsilon(x) = \varepsilon_0 x^{1/2}$, where ε_0 is a constant.

Finally, the velocity profile may be written

$$\bar{u} = c^{3/2}\left(\frac{4\sigma}{3x}\right)^{1/2} \operatorname{sech}^2\left(\frac{\sigma y}{x}\right),$$

where

$$\sigma = \frac{c}{3(\frac{2}{3}\varepsilon_0)^{2/3}}$$

is a dimensionless constant, which is determined experimentally to be about 10. The constant c depends on the momentum flux M_0.

It is important to summarize the basic assumptions involved. These are

(i) that $\overline{u'v'}/(\partial\bar{u}/\partial y)$ is constant over a section of the jet,
(ii) that the flow is similar at large distances from the slit, and
(iii) that viscosity may be neglected as a factor in determining the mean flow pattern.

It may be noted that the same conclusions about $\varepsilon(x)$ are reached when Prandtl's hypothesis is used. Prandtl's hypothesis is

$$\varepsilon(x) \propto (U_{\max} - U_{\min})\, \delta,$$

where U_{max} and U_{min} are the maximum and minimum velocities in the section and δ is the boundary layer thickness.

In the case of the jet,

$$U_{min} = 0, \qquad U_{max} \propto \frac{1}{X^{1/3}}, \qquad \delta \propto X^{2/3},$$

so that $\varepsilon(x) \propto X^{1/3}$ and hence $X \propto x^{3/2}$ as before.

Prandtl's hypothesis is useful in determining $\varepsilon(x)$ in the case of free flows when dimensional arguments cannot be used.

8.14 Flows Involving Solid Boundaries

Problems involving solid boundaries are much more difficult than those in which boundaries are absent. The reason for this is that the presence of a solid boundary entails a region in which the viscous stresses are comparable with the turbulent stresses. Thus, in general, the velocity profiles will not be similar. Further, the exchange coefficient

$$\frac{-\overline{u'v'}}{\partial\bar{u}/\partial y}$$

is clearly not constant over a section of the flow, since at a wall $\overline{u'v'} = 0$, whereas $\partial\bar{u}/\partial y$ is not zero there. The quantity $\overline{u'v'}$ is thus in general a function of both x and y.

However, in simple cases some progress can be made in solving the exact equations of motion. Two such problems will be considered here. Both involve the flow in the x direction in the channel. The mean flow in both cases is assumed to be parallel to the x axis; this means that the mean transport operator

$$\bar{u}\frac{\partial}{\partial x} + \bar{v}\frac{\partial}{\partial y}$$

is zero.

Example (*i*) In this case the equations of mean flow reduce to

$$0 = \nu\frac{\partial^2\bar{u}}{\partial y^2} - \frac{\partial}{\partial y}(\overline{v'v'}), \qquad 0 = -\frac{1}{\varrho}\frac{\partial\bar{p}}{\partial y} - \frac{\partial}{\partial y}\overline{(v')^2}. \quad (8.65)$$

These equations may be integrated to give

$$\mu \frac{\partial \bar{u}}{\partial y} - \varrho\overline{u'v'} = \text{constant} = \left(\mu \frac{\partial \bar{u}}{\partial y}\right)_{\text{wall}} = \tau_{\text{w}}, \tag{8.66}$$

$$\bar{p} + \varrho\overline{(v')^2} = \text{constant} = \bar{p}_{\text{wall}}. \tag{8.67}$$

Equation (8.66) then implies that the *net* shear stress, that is, the sum of viscous and turbulent shear stresses, is constant over the channel and is equal to the drag on the wall. The pressure has its greatest value at the wall.

This problem has been studied experimentally by H. Reichardt.† A sketch of his results is shown in Fig. 8.4. At low Reynolds numbers,

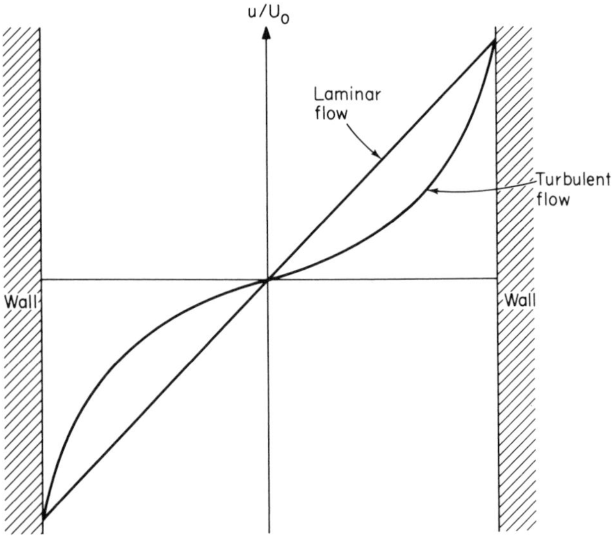

Fig. 8.4 From "Boundary Layer Theory" by H. Schlichting. Copyright 1966, McGraw-Hill, New York. Used with permission of McGraw-Hill Book Company.

say below 1500, the flow is completely laminar. It will be observed that as the Reynolds number increases, the velocity profile is flat in the central region and is very steep at the walls. This means that the flow region consists of two parts—a central region where turbulent stresses are dominant and a thin region close to the wall where the viscous stresses are dominant; this region is known as the viscous sublayer. An order of mag-

† See H. Schlichting, "Boundary Layer Theory," p. 492. McGraw-Hill, New York, 1966.

nitude of the thickness of this sublayer may be obtained from the first problem of Part I. In this case a single component of frequency was present giving a transition region of thickness $(\nu/n)^{1/2}$. In the turbulent case a spectrum of frequencies is present; nevertheless, one can estimate a thickness based on the frequency of the dominant eddies. Thus, if T is the time scale of the turbulence, that is, $1/T$ is the frequency of the dominant eddies, then the viscous sublayer has thickness $\sim(\nu T)^{1/2}$.

In the present case this means that over the central region of the flow $-\varrho u'v'$ is equal to the shear stress at the wall.

Example (ii) In this case both walls of the channel are fixed; a uniform mean flow is maintained by a pressure gradient in the x direction whose mean is constant, that is,

$$-\frac{1}{\varrho}\frac{\partial \bar{p}}{\partial x} = \text{constant} \equiv \frac{P}{\varrho}. \tag{8.68}$$

Thus the equations of the mean motion are

$$0 = \frac{P}{\varrho} + \nu\frac{\partial^2 \bar{u}}{\partial y^2} - \frac{\partial}{\partial y}\overline{(u'v')}, \tag{8.69}$$

$$0 = -\frac{1}{\varrho}\frac{\partial \bar{p}}{\partial y} - \frac{\partial}{\partial y}\overline{(v')^2}. \tag{8.70}$$

As in Example (i),

$$p + \varrho\overline{(v')^2} = p_{\text{wall}}.$$

However,

$$Py + \mu\frac{\partial \bar{u}}{\partial y} - \varrho\overline{u'v'} = \text{constant}. \tag{8.71}$$

Now the terms on the left side of this equation are clearly antisymmetric, so that the constant must be zero. Thus

$$\mu\frac{\partial \bar{u}}{\partial y} - \varrho\overline{u'v'} = -Py. \tag{8.72}$$

Thus the net shear stress is a linear function of y.

This problem has also been studied experimentally by Reichardt† (Fig. 8.5). His results give excellent agreement with his theory. They

† H. Schlichting, "Boundary Layer Theory," p. 466. McGraw-Hill, New York, 1966.

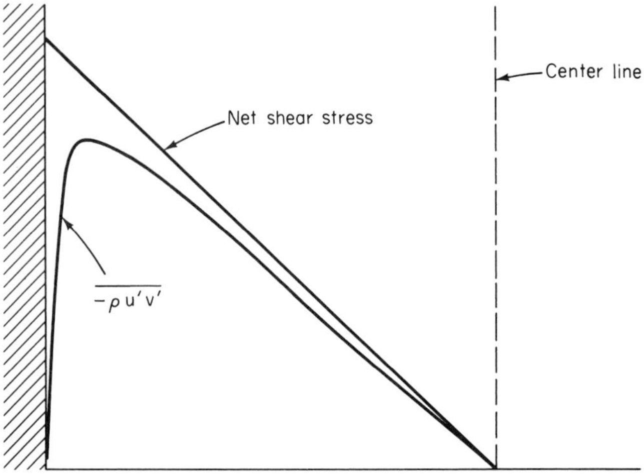

Fig. 8.5 From "Boundary Layer Theory" by H. Schlichting. Copyright 1966, McGraw-Hill, New York. Used with permission of McGraw-Hill Book Company.

also show that viscosity is important only in the viscous sublayer close to the wall.

To summarize, the last two examples show that in flows in a channel the turbulent stresses dominate away from the wall. Further, this central region is virtually frictionless—energy dissipation takes place in the viscous sublayer.

8.15 Boundary Layer Flow along a Flat Plate

In this section the boundary layer flow along a flat plate will be considered. (The outer mean flow is assumed to be constant.) Let the flat plate lie along the positive x axis and let the outer mean flow be in the x direction.

Consider now the flow close to the wall, that is, when y is small. It is clear from the previous examples that in the viscous sublayer the net shear stress

$$\mu \frac{\partial \bar{u}}{\partial y} - \varrho \overline{u'v'}$$

is approximately constant and equal to the wall stress τ_w. Now $\overline{u'v'}$ depends on the local properties of the fluid close to the wall; it must increase with increasing $\partial \bar{u}/\partial y$ since the larger $\partial \bar{u}/\partial y$ the greater the turbulent fluctuations u', v'; and it must vanish at the wall ($y = 0$). The simplest hypothesis is that $-\overline{u'v'} \propto y\, \partial \bar{u}/\partial x$. Now the proportionality factor in this hypothesis has the dimensions of a velocity. Close to the wall the only local quantity having the dimensions of a constant velocity is u^*, where

$$u^* = \left(\frac{\tau_{\mathrm{W}}}{\varrho}\right)^{1/2};$$

u^* is called the wall friction velocity. Thus

$$-\overline{u'v'} = Ku^*y \frac{\partial \bar{u}}{\partial y},$$

where K is a dimensionless constant. Thus, in the viscous sublayer

$$\mu \frac{\partial \bar{u}}{\partial y} + \varrho K u^* y \frac{\partial \bar{u}}{\partial y} = \tau_{\mathrm{W}}. \tag{8.73}$$

This equation may be integrated to give

$$\bar{u} = \frac{\tau_{\mathrm{W}}}{\varrho K u^*} \log\left(1 + \frac{Ku^*y}{\nu}\right).$$

When u^*y/ν is small,

$$\bar{u} \doteqdot \frac{\tau_{\mathrm{W}} y}{\mu}.$$

This is to be expected since $\overline{u'v'}$ is very small in this region. On the other hand, when $u^*y/\nu \gg 1$,

$$\bar{u} \doteqdot \frac{\tau_{\mathrm{W}}}{\varrho K u^*} \log y + \text{constant}. \tag{8.74}$$

Obviously the result given above breaks down at very large values of y because it predicts an infinite value of $\bar{u}$. However, the basic assumptions made above about $\overline{u'v'}$, particularly that

$$\overline{u'v'} \propto y,$$

clearly do not apply at large distances.

8.16 Summary

Very close to the wall, when $u^*y/\nu \ll 1$, a linear velocity profile is predicted. When $u^*y/\nu \gg 1$ but is still not too large, the mean velocity profile is logarithmic. This profile has been experimentally confirmed.†

Outside the laminar sublayer, that is for $u^*y/\nu \gg 1$, not much is known. The indications are that

$$\overline{u'v'} \bigg/ \frac{\partial \bar{u}}{\partial y}$$

is constant as in free flows.

The logarithmic profile has also been observed in turbulent flows in pipes.

A factor that has been omitted so far is the degree of roughness permissible. A criterion for a smooth wall is that the size of the protuberances be very much less than the thickness of the viscous sublayer. The case in which the roughness is comparable with the viscous sublayer is complicated and will not be treated here.

PART IV STRUCTURE OF TURBULENCE

8.17 The Microstructure of Turbulent Flow

The *dynamics* of turbulence, which has been dealt with in Parts I–III, is based on the hypothesis of the validity of the Navier–Stokes equations for turbulent flow.

But the mechanics, or structure, of turbulence is a statistical mechanics. Turbulent boundary layers are complicated by the fact that their structure is neither homogeneous nor isotropic. Little progress has been made in the theory of nonisotropic turbulence.

† See for example H. Schlichting, "Boundary Layer Theory," p. 538. McGraw-Hill, New York, 1966.

Considerable progress has been made, however, in the study of, at least locally, isotropic turbulence. The isotropy of turbulent flow downstream from a grid seems well established experimentally.

In 1941 a general theory of locally isotropic turbulence was proposed by Kolmogorov, which permitted the prediction of a number of laws of turbulent flow for large Reynolds numbers.

The most important of these laws are:

(1) the dependence of the mean square of the difference in velocities at two points on their distance apart;

(2) the dependence of the coefficient of turbulence diffusion on the scale of the phenomenon.

The fundamental physical concepts which are the basis of Kolmogorov's theory can be summarized as follows.

A turbulent flow at large Reynolds numbers is considered to be the result of the superposing of disturbances (vortices or eddies) of all possible sizes. Only the very largest of these disturbances, or vortices, are due directly to the instability of the mean flow. The scale L of these large vortices is of the same order as the distance over which the velocity of the mean flow changes. For example, in a turbulent boundary layer this is the distance from the wall. The length L corresponds to the length of the mixing path introduced in Prandtl's semiempirical theory of turbulence.

The motion of the large vortices is unstable and this produces smaller (second-order) vortices; these produce even smaller (third-order) vortices and so on. Since for all vortices, except the very smallest, the characteristic Reynolds number is large, the viscosity has no appreciable effect on their motion.

Note $\mathrm{Re}_L = uL/\nu$.

So the motion of all but the very small vortices is not associated with any marked dissipation of energy; the vortices of the nth order use nearly all the energy received from those of the $(n - 1)$th order to form vortices of the $(n + 1)$th order.

But the motion of the smallest vortices is "laminar" and depends basically on the molecular viscosity. The entire energy transferred to them along vortex cascade is transformed eventually into heat energy.

The motion of all the vortices except the largest can be assumed to be homogeneous and isotropic. Any directional effect of the mean flow is ineffective for low-order vortices. Moreover, the motion can be assumed to be quasistationary; that is, a change in the statistical characteristics of the vortices occurs on a time scale much larger than the periods of the vortices. Hence the motion of all vortices whose scales are much smaller than L (the microstructure or *local* structure of the flow) must be subject to general statistical laws which do not depend on the geometry of the flow or on the properties of the mean flow. The establishment of these general statistical laws constitutes the theory of local isotropic turbulence.

Considerations of similitude and dimensions are again of great value in establishing results. One must first identify those fundamental magnitudes on which the local structure of the flow may depend. Due to the homogeneous and isotropic character of the motion of the vortex system considered, the characteristics of the mean motion (length and velocity characteristics, etc.) do not enter into these fundamentals. So, only two magnitudes remain—the mean dissipation of energy per unit time per unit mass of the fluid, ε, and the kinematic viscosity ν. The former determines the intensity of the energy flow transferred along a cascade of vortices of different scales, and the latter plays a basic role in the process of energy dissipation.

In a compressible fluid, or where the structure of the pressure field is of interest, the density of the fluid ϱ must be added as a fundamental property.

The dimensions of ε and ν are

$$[\varepsilon] = L^2T^{-3}, \qquad [\nu] = L^2T^{-1}.$$

From these two quantities, one can form a single quantity with the dimensions of length:

$$\eta = \left(\frac{\nu^3}{\varepsilon}\right)^{1/4}. \tag{8.75}$$

Isotropic turbulence also implies that the mean velocity fluctuations in the three coordinate directions at any point in the domain of isotropy are equal:

$$\overline{u'^2} = \overline{v'^2} = \overline{w'^2}.$$

In such cases the intensity of turbulence can be described by the quantity

$$\frac{[\frac{1}{3}(\overline{u'^2} + \overline{v'^2} + \overline{w'^2}]^{1/2}}{U_\infty} = \frac{(\overline{u'^2})^{1/2}}{U_\infty}.$$

The length η determines an internal scale characteristic of the local structure (in contrast to the external scale L).

Thus, Kolmogorov's postulate of similarity is: in a domain where the turbulence is locally homogeneous and isotropic, the laws of probability depend only on ε and ν. It is possible to identify η with the scale of the smallest vortices in which a dissipation of energy occurs.

8.18 The Velocity Correlation

Consider the fluid velocities $\mathbf{v}_2$ and $\mathbf{v}_1$ at two nearby points and let the radius vector between these two points be $\mathbf{r}$. It is supposed that $|\mathbf{r}| \ll L$ but not necessarily large with respect to η. Consider the tensor with components

$$B_{ik} = \overline{(v_{2i} - v_{1i})(v_{2k} - v_{1k})}, \tag{8.76}$$

where the bar implies an average with respect to time.

For isotropic turbulence the tensor B_{ik} cannot depend on any direction in space. The only vector that can appear in the expression for B_{ik} is the radius vector $\mathbf{r}$. Thus B_{ik} can contain only $|\mathbf{r}|$, the unit tensor δ_{ik}, and the unit vector $\mathbf{n}$ in the direction of $\mathbf{r}$. The most general form of such a tensor of rank 2 is

$$B_{ik} = A(r)\delta_{ik} + B(r)n_i n_k. \tag{8.77}$$

One takes the coordinate axes so that one of them is in the direction of $\mathbf{n}$, denoting the velocity component along this axis by v_r and the component perpendicular to $\mathbf{n}$ by v_t.

Then B_{rr} is the mean square relative velocity of two neighboring particles along the line joining them. Similarly, B_{tt} is the mean square transverse velocity of one particle relative to the other.

Now $n_r = 1$, $n_t = 0$, and so

$$B_{rr} = A + B, \qquad B_{tt} = A. \tag{8.78}$$

Expanding Eq. (8.76), one obtains

$$B_{ik} = \overline{v_{1i}v_{1k}} - \overline{v_{2i}v_{2k}} - \overline{v_{1i}v_{2k}} - \overline{v_{1k}v_{2i}} .$$

Since the flow is homogeneous and isotropic,

$$\overline{v_{1i}v_{1k}} = \overline{v_{2i}v_{2k}}\,, \qquad \overline{v_{2i}v_{2k}} = \overline{v_{1k}v_{2i}}\,.$$

Therefore

$$B_{ik} = 2\,\overline{v_{1i}v_{1k}} - 2\overline{v_{1k}v_{2k}}. \tag{8.79}$$

It can now easily† be shown that

(i) for distances $r \gg \eta$,

$$B_{tt} = \tfrac{4}{3}B_{rr}; \tag{8.80}$$

(ii) for distances $r \ll \eta$,

$$B_{tt} = 2B_{rr}. \tag{8.81}$$

For distances $r \ll \eta$, velocity differences are proportional to r and hence B_{rr} and B_{tt} are proportional to r^2. At these distances they can also be expressed in terms of the mean energy dissipation ε. Writing $B_{rr}=ar^2$ (a is a constant) and combining Eqs. (8.77)–(8.79), one gets

$$\overline{v_{1i}v_{2k}} = \overline{v_{1i}v_{1k}} - ar^2\delta_{ik} + \tfrac{1}{2}ar^2 n_i n_k.$$

Differentiating this relation,

$$\overline{\frac{\partial v_{1i}}{\partial x_{1l}}\,\frac{\partial v_{2i}}{\partial x_{2l}}} = 15a, \qquad \overline{\frac{\partial v_{1i}}{\partial x_{1l}}\,\frac{\partial v_{2l}}{\partial x_{2i}}} = 0. \tag{8.82}$$

Since this holds for arbitrarily small r, one can put $x_{1i} = x_{2i}$. Therefore

$$\overline{\left(\frac{\partial v_i}{\partial x_l}\right)^2} = 15a, \qquad \overline{\frac{\partial v_i}{\partial x_l}\,\frac{\partial v_l}{\partial x_i}} = 0. \tag{8.83}$$

Now the general formula for the mean energy dissipation is

$$\varepsilon = \tfrac{1}{2}\nu\overline{\left(\frac{\partial v_i}{\partial x_l} + \frac{\partial v_l}{\partial x_i}\right)^2} = \nu\left[\overline{\left(\frac{\partial v_i}{\partial x_l}\right)^2} + \overline{\frac{\partial v_i}{\partial x_l}\,\frac{\partial v_l}{\partial x_i}}\right],$$

and using Eq. (8.83), one obtains

$$\varepsilon = 15a\nu, \qquad a = \frac{\varepsilon}{15\nu}, \tag{8.84}$$

† L. D. Landau and E. M. Lifschitz, "Fluid Mechanics," p. 125. Pergamon, Oxford, 1959.

whence

$$B_{tt} = \frac{2}{15} \frac{\varepsilon r^2}{\nu}, \qquad B_{rr} = \frac{1}{15} \frac{\varepsilon r^2}{\nu}. \tag{8.85}$$

One can discuss *triple* correlation (B_{ikl}) similarly.

8.19 Spectrum of Turbulence

An alternative description of the structure of turbulence is obtained when a frequency analysis of the motion is provided instead of the correlation function. This leads one to the concept of the spectrum of a turbulent stream.

Let n denote the frequency and $F(n)\,dn$ the fractional content of the root-mean-square value, $\overline{u'^2}$, of the longitudinal fluctuation (or any one of the three components in isotropic turbulence) which belongs to the frequency interval from n to $n + dn$.

The function $F(n)$, which represents the density of the distribution of $\overline{u'^2}$ in n, is known as the spectral distribution of $\overline{u'^2}$. Obviously

$$\int_0^\infty F(n)\,dn = 1; \tag{8.86}$$

$F(n)$, in fact, is the Fourier transform of the autocorrelation function.

Figure 8.6 displays experimentally determined spectra for turbulent flow on a flat plate.† Except for the measurements at the outer edge of the boundary layer ($y/\delta = 1$), the maximum value of $F(n)$ always occurs at the lowest measured frequency.

As the frequency n is increased,

$$F(n) \sim n^{-5/3},$$

agreeing with the theory developed by Kolmogorov and Heisenberg. As the frequency becomes larger, $F(n)$ decreases through viscous action at an even faster rate. The relation here is of the form

$$F(n) \sim n^{-7}.$$

† P. S. Klebenoff, NACA 1247 (1955).

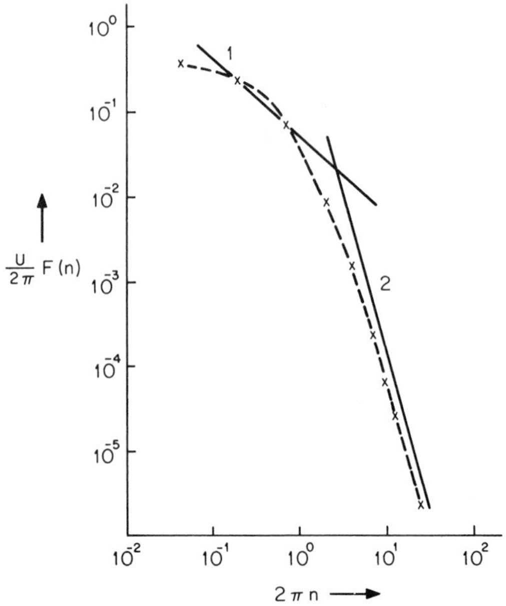

Fig. 8.6 Curve 1 → $n^{-5/3}$, curve 2 → n^{-7}.

8.20 Nonisotropic Turbulence

Whereas local isotropy is a reasonable assumption for fully developed turbulence downstream of grids, or in the far-field region of wakes and jets, this cannot be valid for the transition region of jets, wakes, and boundary layers in general.

Lumley, in 1966, formulated a theory in which he decomposes the correlation tensor into a sum of its eigenmodes and interprets the dominant mode as the large eddies. He therefore provides us with an objective definition of the large eddy, which can be used in the analysis of experimental data. However, his approach is still based on the Reynolds (1895) decomposition of the flow field into "mean" and "fluctuating" components. This involves the closure problem inherent in all treatments of turbulence to date, as such a decomposition can only be a crude approximation to the physics of the problem.

PROBLEMS

8.1 A reciprocating piston acts on fluid in a pipe of radius α whose axis coincides with the x axis. This causes a fluctuating pressure gradient

$$-\frac{1}{p}\frac{\delta p}{\delta x} = \Re(Ge^{int}).$$

Assuming that

$$u = \Re[f(r)e^{int}],$$

show that

$$f''(r) + \frac{1}{r}f'(r) - \frac{in}{\nu}f(r) = \frac{-G}{\nu}.$$

Show that the solution is

$$u = \Re\left[\frac{-iG}{n}e^{int}\left\{1 - \frac{J_0(\alpha r)}{J_0(\alpha a)}\right\}\right],$$

where $\alpha = (-in/\nu)^{1/2}$. Discuss the flow at high frequencies.

8.2 Fluid flows between parallel walls $y = y_1$ and $y = y_2$. An "inviscid" disturbance $u = \bar{u}(y)$ is superposed on the basic flow. The stream function of this disturbance is

$$\psi = \varphi(y)e^{|\alpha|x-ct}.$$

Assuming that φ satisfies (8.25), prove that

$$\int_{y_1}^{y_2} [\tilde{\varphi}L(\varphi) + \varphi\tilde{L}(\tilde{\varphi})]\, dy$$
$$= -2\int_{y_1}^{y_2}\left[\alpha^2 |\varphi|^2 + |\varphi'|^2 + \bar{u}''(\bar{u} - c_r)\left|\frac{\varphi}{\bar{u} - c}\right|^2\right] dy.$$

Deduce that

$$\int_{y_1}^{y_2} \bar{u}''(\bar{u} - \bar{u}_I)\left|\frac{\varphi}{\bar{u} - c}\right|^2 dy < 0,$$

where $\bar{u}_I = \bar{u}(y_c)$.

Deduce that for amplified oscillations in monotonically varying velocity profiles, the vorticity of the disturbance has a numerical maximum at $y = y_c$. [*Hint*: Add to the above result a multiple of (8.27).]

8.3 Determine the Reynolds stresses for the disturbances which satisfy Eq. (8.29) of the Schlichting jet.

8.4 Discuss the turbulent round jet by an argument similar to that involved for the plane jet. Given that the three basic assumptions as stated on p. 328 hold.

Answer: In this case the coefficient ε_0 is found to be constant and the solution is in fact the same as for the laminar round jet with ν replaced by ε_0. See problems in Chapter 6.

Note that the turbulent boundary layer equation is in cylindrical coordinates:

$$\bar{u}\frac{\partial \bar{u}}{\partial r} + \bar{v}\frac{\partial \bar{u}}{\partial r} = \frac{\varepsilon_0}{r}\frac{\partial}{\partial r}\left(r\frac{\partial \bar{u}}{\partial r}\right).$$

8.5 In free turbulent flow it is found that at the edge of the jet the flow is intermittently laminar and turbulent.

Show that small scale eddies at the edge of the boundary layer have a depth of penetration into the potential region whose order of magnitude is that of the scale of the eddies. Deduce that the eddies responsible for intermittency are large in scale.

8.6 Show that the Prandtl hypothesis and the assumption of similarity at large distances imply that the width of the two-dimensional wake increases as $x^{1/2}$ while the velocity deficit on the axis decays as $x^{-1/2}$.

Determine similar results for the circular turbulent wake.

Answer: Breadth $\propto x^{1/3}$; velocity defect $\propto x^{-2/3}$.

8.7 Consider the turbulent flow at large distances from a row of cylinders as in Chapter 7 on Slow Viscous Flows. Use the Prandtl hypothesis to show that the velocity defect at large distances is proportional to

$$\frac{1}{x}\cos\left(\frac{2\pi y}{n}\right).$$

8.8 Starting from the equation

$$U_1\frac{\partial U}{\partial x} + \frac{\partial \overline{u'v'}}{\partial y} = 0,$$

show that at large Reynolds numbers the width of a two-dimensional wake increases as $x^{1/2}$ and the velocity defect decreases as $x^{-1/2}$, where x is the distance downstream from a virtual origin, and U_1 is the steady velocity of the main stream. State clearly any assumptions you use.

If the Reynolds stress and the transverse gradient of mean velocity are related by means of an eddy viscosity which is constant across sections of the wake, show that the profile of the velocity defect is Gaussian.

8.9 The amplitude $\varphi(y)$ of a wavy disturbance in a parallel flow of an inviscid fluid satisfies the inviscid O–S equation

$$(U - c)(\varphi'' - \alpha^2\varphi) - U''\varphi = 0.$$

Derive an expression for the nondimensional Reynolds stress $\tau = -\overline{u'v'}$ in terms of $\varphi(y)$, and prove that in an amplified disturbance

$$\frac{d\tau}{dy} = \frac{c_i \bar{v}'^2 U''}{\alpha \mid U - c \mid^2}.$$

Deduce that the velocity profile must have an inflection.

Comment on the consequences of making $c_i \to 0$. (Manchester†)

BIBLIOGRAPHY

L. Rosenhead, "Laminar Boundary Layers." Oxford Univ. Press, London and New York, 1963.

H. Schlichting, "Boundary Layer Theory," McGraw-Hill, New York, 1968.

C. C. Lin, "Hydrodynamic Stability." Cambridge Univ. Press, London and New York, 1955.

S. I. Pai, "Fluid Dynamics of Jets." Van Nostrand-Reinhold, Princeton, New Jersey, 1954.

J. O. Hinze, "Turbulence." McGraw-Hill, New York, 1959.

R. Betchov and W. Criminale, "Stability of Parallel Flows." Academic Press, New York, 1967.

† From Manchester University, Honours in Mathematics (1969); by permission of the University of Manchester, Manchester, England.

CHAPTER NINE *COMPRESSIBLE FLUID FLOW*

Thus far, fluids have been treated largely as incompressible. In fact, no fluid is incompressible. But any fluid, in some range of flow conditions, is effectively incompressible.

At sufficiently low flow velocities, any gas simulates incompressible flow and, on the contrary, any liquid at high enough velocities exhibits compressibility effects. The Mach number of the flow of a gas (ratio of the flow velocity to the local speed of sound) provides a rough guide. It appears that at a Mach number of 0.2 the error involved in evaluating the stagnation pressure developed at the nose of a body in a fluid stream is 1%. This rises to 2.25% at $M = 0.3$, and 28% at $M = 1$.

Moreover, in compressible fluids, the speed of propagation of pressure (or density) waves is finite. For small pressure fluctuations there is the phenomenon of the sound wave, and for large pressure fluctuations one gets the phenomenon of shock, or finite amplitude, waves.

Thus, the study of compressible flow (often called gas dynamics) is an important part of fluid dynamics. In the study of compressible flow one must consider the variation of fluid density due to variations in pressure and temperature.

9.1 Thermodynamics and Fluid Flow

For any specific fluid, its density is related to pressure and temperature by the *equation of state*. Pressure, temperature, and density are called the *state variables*. An equation of state is formed by expressing any one of these variables in terms of the other two. For example,

$$p = p(\varrho, T). \tag{9.1}$$

Under ordinary conditions, the equation of state of most gases is of the form

$$p = \varrho RT, \tag{9.2}$$

where T is the absolute temperature and R is a constant.

By considering a definite mass M of gas, and since $\varrho = M/V$, Eq. (9.2) can be put in the form

$$pV = MRT. \tag{9.3}$$

If we define a dimensionless gas ratio, $\mu = M/m$, where m is the mass of one molecule, Eq. (9.3) assumes the form

$$\frac{pV}{\mu} = \mathscr{R}T, \tag{9.4}$$

where $\mathscr{R}$ is called the *universal gas constant*. If p is measured in dynes per square centimeter (dyn/cm^2) and ϱ in grams per cubic centimeter (gm/cm^3), then $\mathscr{R}$ has the value

$$\mathscr{R} = 8.314 \quad \text{ergs/mole}\ {}^0\text{K}.$$

At the molecular level a perfect gas is characterized by the absence of intermolecular attractions and so no real gas obeys the perfect gas law exactly. Nevertheless, the concept of a perfect gas is widely used in the dynamics of compressible gas flow.

9.1.1 The First Law of Thermodynamics

As seen in Chapter 4, the concept of energy must be introduced in the study of compressible flow.

Consider a small element of fluid in thermodynamic equilibrium in a state 1. If the surroundings add heat energy ΔQ to the element and the element does work ΔW on the surroundings, then the element changes to a state 2. The first law states that an entity ΔE exists such that

$$\Delta E_{12} = \Delta Q_{12} - \Delta W_{12}, \tag{9.5}$$

and ΔE_{12} depends only on states 1 and 2 (initial and final).

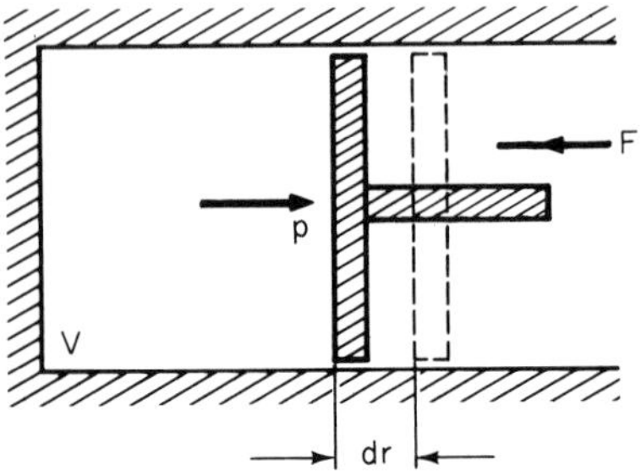

Fig. 9.1

Here E is a state variable and is called the *internal energy* of the fluid, and ΔE_{12} is then the change in internal energy. Consider the cylinder–piston arrangement shown in Fig. 9.1. The work done by the surroundings is given by

$$W = + \int pA\, dr = + \int p\, dV. \tag{9.6}$$

Therefore

$$dW = p\, dV$$

and

$$\Delta E_{12} = \Delta Q_{12} - p\, \Delta V_{12}.$$

In differential form, then,

$$dE = dQ - p\, dV$$

or, for unit mass,

$$de = dq - p\, dv. \tag{9.7}$$

If the process is reversible, then

$$\Delta E_{12} = -\Delta E_{21}.$$

The change in the above cylinder–piston system must be carried out in extremely small steps so that no fluxes or currents of heat, mass, or momentum are set up if the change is to be reversible. Irreversibility results from deviation from equilibrium during the change in state. A flux of heat flows if a finite temperature difference exists; a flux of mass flows if a finite concentration difference in a component exists; a flux of momentum flows if differences in velocity exist.

In fact, a system is in a state of equilibrium if no fluxes are present in it. So the work W must be done on and the heat Q added to the system in such a way that no fluxes are created, if the change is to be a reversible one. Again, the ideal of reversibility or equilibrium can only be approximated in practice. Reversibility and small departures from equilibrium are compatible for stable systems.

9.1.2 Specific Heats and Reversible Processes

The internal energy per unit mass, e, is a function of density and temperature in general:

$$e = e(\varrho, T). \tag{9.8}$$

Experimentally it is observed that under ordinary conditions e is quite insensitive to the density and so one can reasonably assume that $e = e(T)$. This relation can be written in the form

$$e = e_0 + \int_{T_0}^{T} C_v(T)\, dT,$$

where

$$C_v = \left(\frac{\partial e}{\partial T}\right)_v; \tag{9.9}$$

e has a constant value e_0 at some reference temperature T_0; C_v is called the *specific heat at constant volume* of the gas. From the First Law of Thermodynamics one sees that C_v can also be written

$$C_v = \left(\frac{dq}{dT}\right)_v. \tag{9.10}$$

A specific heat at constant pressure can also be defined:

$$C_p = \left(\frac{dq}{dT}\right)_p. \tag{9.11}$$

The value of the specific heat of a gas depends on the process by which the heat q is added.

Since $de = dq - p\,dv$,

$$C_p = \left(\frac{\partial e}{\partial T}\right)_p + \left[\left(\frac{\partial e}{\partial v}\right)_p + p\right]\left(\frac{\partial v}{\partial T}\right)_p. \tag{9.12}$$

Experimentally it is found that within reasonable ranges of temperature, C_v can be regarded as constant. Therefore,

$$e = (e_0 - C_v T_0) + C_v T = \text{constant} + C_v T. \tag{9.13}$$

It turns out that for monatomic gases, such as helium, C_v is very nearly $3R/2$, where $R = \mathscr{R}/m$. For diatomic gases, such as oxygen and nitrogen, it is nearly $5R/2$.

9.1.3 Specific Enthalpy

There is another very important state variable which is related to e. It is the specific enthalpy or heat function,

$$h = e + pv \tag{9.14}$$

or, for an arbitrary mass of fluid,

$$H = E + pV. \tag{9.15}$$

Thus

$$dh = de + p\,dv + v\,dp,$$

and the First Law of Thermodynamics can be written as

$$dh = dq + v\,dp. \tag{9.16}$$

Forming C_v and C_p as before, one has

$$C_v = \frac{\partial h}{\partial T} + \left(\frac{\partial h}{\partial p} - v\right)\left(\frac{\partial p}{\partial T}\right)_v, \tag{9.17}$$

$$C_p = \frac{\partial h}{\partial T}. \tag{9.18}$$

For a perfect gas, $e = e(T)$; therefore,

$$h = e(T) + pv = e(T) + RT = h(T).$$

From Eqs. (9.9) and (9.12),

$$C_v = \frac{de}{dT}, \qquad C_p = C_v + p\left(\frac{\partial v}{\partial T}\right)_p = C_v + R,$$

and from Eqs. (9.17) and (9.18),

$$C_p = \frac{dh}{dT}, \qquad C_v = C_p - v\left(\frac{\partial p}{\partial T}\right)_v = C_p - R.$$

So, for a perfect gas there are the important thermodynamic relations

$$C_p - C_v = R, \tag{9.19}$$

$$e(T) = \text{constant} + \int C_v\, dT, \tag{9.20}$$

$$h(T) = \text{constant} + \int C_p\, dT. \tag{9.21}$$

If C_p and C_v can be treated as constant (the gas is calorically perfect), then

$$e = C_v T + \text{constant}, \tag{9.22}$$

$$h = C_p T + \text{constant}. \tag{9.23}$$

9.1.4 Isentropic Relations

Processes in which no heat is transferred to, or removed from, the system and in which the work is done reversibly are known as isentropic, or adiabatic and reversible, processes. Under these conditions,

$$de = -p\, dv \tag{9.24}$$

or

$$dh = v\, dp. \tag{9.25}$$

Since $e = e(v, T)$, Eq. (9.24) can be expanded in partial differentials:

$$\frac{\partial e}{\partial v}\, dv + \frac{\partial e}{\partial T}\, dT = -p\, dv, \tag{9.26}$$

and from Eq. (9.25),

$$\frac{\partial h}{\partial p}\, dp + \frac{\partial h}{\partial T}\, dT = v\, dp, \tag{9.27}$$

since

$$h = h(p, T).$$

So, for an isentropic, or adiabatic and reversible, process, the following relations hold:

$$\frac{dT}{dv} = -\frac{1}{C_v}\left(\frac{\partial e}{\partial v} + p\right), \tag{9.28}$$

$$\frac{dT}{dp} = -\frac{1}{C_p}\left(\frac{\partial h}{\partial p} - v\right), \tag{9.29}$$

$$\frac{dp}{dv} = -\frac{p}{v}\frac{dh}{de}. \tag{9.30}$$

For a perfect gas ($pv = RT$), these relations yield

$$\frac{v}{T}\frac{dT}{dv} = -\frac{R}{C_v}, \tag{9.31}$$

$$\frac{p}{T}\frac{dT}{dp} = \frac{R}{C_p}, \tag{9.32}$$

$$\frac{v}{p}\frac{dp}{dv} = -\frac{C_p}{C_v}. \tag{9.33}$$

Now

$$R = C_p - C_v,$$

and

$$\gamma = \frac{C_p}{C_v}$$

is the specific heat ratio. Using this in Eq. (9.31) and integrating, one obtains

$$\ln v = -\int \frac{dT}{T(\gamma - 1)}.$$

If the gas is calorically perfect, then γ is not a function of T and thus

$$v = \text{constant} \cdot (T^{-1/(\gamma-1)}). \tag{9.34}$$

Similarly, one can derive the following:

$$p = \text{constant} \cdot (T^{\gamma/(\gamma-1)}) \tag{9.35}$$

or

$$p = \text{constant} \cdot v^{-\gamma}. \tag{9.36}$$

Equation (9.36) is called the isentropic relation for a perfect gas.

The conditions for isentropicity demand that the fluid element be involved in frictionless and non-heat-conducting flow. Thus P/ϱ^{γ} must be a state variable for such a flow.

9.1.5 The Second Law of Thermodynamics and Entropy

In compressible flows there are two additional unknowns—density and temperature. The continuity and momentum equations must now be augmented by the equation of state and the energy equation to render the system determinable. The Second Law of Thermodynamics asserts that all the thermodynamic systems possess a state variable called *entropy*, S, defined by

$$S_B - S_A = \Delta S = \int_{\text{rev}} \frac{dQ}{T}. \tag{9.37}$$

This implies that the change of entropy between two states A and B is the sum of all $\Delta Q/T$ produced by any reversible change which joins the two states. For a closed system—that is, one that exchanges neither heat nor work with the surroundings—S increases in any spontaneous process. The system has reached equilibrium if S has reached a maximum.

It can be shown that for any natural, that is, irreversible process, $\Delta S > \Delta Q/T$. So, for arbitrary processes, Eq. (9.37) can be written

$$S_B - S_A \geq \int_A^B \frac{dQ}{T}. \tag{9.38}$$

Now the First Law of Thermodynamics gave the relation

$$dE = -p\, dV + dQ, \qquad dQ = T\, dS.$$

Therefore,

$$S = \int \frac{dE + p\, dV}{T} = m \int \frac{de + p\, dv}{T} + \text{constant}. \tag{9.39}$$

For a perfect gas,

$$\frac{S}{m} = s = \int C_{\text{v}} \frac{dT}{T} + R \ln v + \text{constant}. \tag{9.40}$$

For a calorically perfect gas, $C_v = C_p =$ constant. Therefore,

$$s - s_1 = C_p \ln \frac{T}{T_0} - R \ln \frac{p}{p_0} \tag{9.41}$$

$$= C_v \ln \frac{T}{T_0} + R \ln \frac{v}{v_0}, \tag{9.42}$$

where p_0, v_0, and T_0 are reference conditions.

9.1.6 Canonical Equation of State

Now, for reversible changes of state,

$$dE = T\, dS - p\, dV, \tag{9.43}$$

$$dH = T\, dS + V\, dp. \tag{9.44}$$

The natural variables for E are obviously S and V, and for H, they are S and p, and so

$$E = E(S, V) \qquad \text{and} \qquad H = H(S, p) \tag{9.45}$$

are called *canonical equations of state*. Each describes a system completely. Useful relations are

$$\left(\frac{\partial E}{\partial S}\right)_V = T, \qquad \left(\frac{\partial E}{\partial V}\right)_S = -p, \tag{9.46}$$

$$\left(\frac{\partial H}{\partial S}\right)_p = T, \qquad \left(\frac{\partial H}{\partial p}\right)_S = V. \tag{9.47}$$

9.1.7 Free Energy and Free Enthalpy

Free energy and free enthalpy are two useful functions, related to E, S and H, which have V, T and p, T as natural variables.

The so-called *free energy* (available energy) is defined as

$$F = E - TS, \tag{9.48}$$

and the *free enthalpy* (or Gibbs free energy) as

$$G = H - TS. \tag{9.49}$$

Now

$$dF = dE - S\,dT - T\,dS = -S\,dT - p\,dV, \tag{9.50}$$

$$dG = dH - S\,dT - T\,dS = -S\,dT + V\,dp. \tag{9.51}$$

So, the natural variables for F are T, V and for G they are T, p.

Useful differential relations are

$$\left(\frac{\partial F}{\partial T}\right)_V = -S, \qquad \left(\frac{\partial F}{\partial V}\right)_T = -p, \tag{9.52}$$

$$\left(\frac{\partial G}{\partial T}\right)_p = -S, \qquad \left(\frac{\partial G}{\partial p}\right)_T = V. \tag{9.53}$$

9.1.8 Real Gas Effects

The behavior of real gases is bound to deviate from that of the perfect gas since there are significant intermolecular attractive forces.

The equation of state of a real gas may be written in terms of a "compressibility factor" Z:

$$\frac{pv}{RT} = Z(p, T). \tag{9.54}$$

If $Z = 1$, then this becomes the perfect gas law.

At low temperatures, and at high pressures, the intermolecular interactions become important. They are called van der Waals forces. At high temperatures dissociation and ionization of molecules occur and Z differs from unity because the number of particles in the system changes.

The van der Waals force effect can be expressed to a first approximation by expanding Z to the "second virial coefficient" $b(T)$:

$$\frac{pv}{RT} = Z = 1 + b(T)\frac{p}{RT}; \tag{9.55}$$

b/R is a characteristic function for a specific gas and varies from gas to gas.

The equation of state of a dissociating diatomic gas is

$$\frac{pv}{RT} = 1 + \alpha, \tag{9.56}$$

where α is the degree of dissociation and $\alpha = \alpha(p, T)$. Consider the dissociation of a diatomic gas such as oxygen. At high enough temperatures ($\approx 3000\,^{\circ}$K) the gas consists of both O_2 and O and is a reacting gas mixture. Suppose c_O and c_{O_2} are the masses of O and O_2 in the mixture and that a unit mass of mixture is involved, so that

$$c_O + c_{O_2} = 1.$$

Then the degree of dissociation α is defined by

$$\alpha = \frac{c_O}{c_O + c_{O_2}},$$
$$c_O = \alpha, \qquad c_{O_2} = 1 - \alpha. \tag{9.57}$$

If h_1 and h_2 are the specific enthalpies of O and O_2, respectively, then the heat of this reaction,

$$2O \rightleftharpoons O_2,$$

is given by

$$Q_R = h_1 - h_2. \tag{9.58}$$

A characteristic temperature for the reaction is given by

$$\vartheta_D = \frac{Q_R}{R}; \tag{9.59}$$

Q_R is related to the force with which the atoms are held within the molecule. Now, the heat of vaporization Q_V is related to the force between molecules. Thus, $Q_R \gg Q_V$, and the temperature range over which dissociation effects occur is usually much higher than that over which the van der Waals forces are important. It is rarely, then, in a gas dynamics problem that both these effects need be considered simultaneously.

In the case of ionization, the ionization energy Q_i is involved and this concerns the forces that bind the electrons in an atom. Even higher energies ($\gtrsim$10,000°K) are needed to produce appreciable degrees of ionization in gases.

9.2 Speed of Propagation of Sound Waves

The speed of sound a is the speed at which pressure disturbances are propagated in the form of small pressure waves through a compressible fluid. Compressible effects depend fundamentally on the role played by the speed of sound and so it is appropriate to study the latter before proceeding with a study of one-dimensional compressible flow.

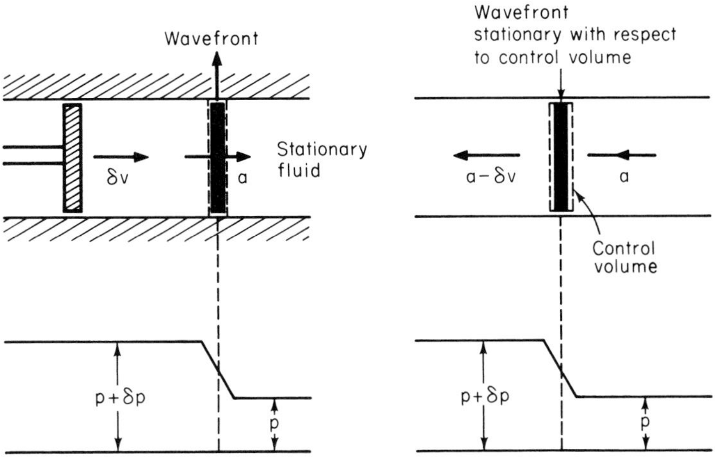

Fig. 9.2

An expression for the speed of propagation of small pressure waves will be derived by using the idealized model shown in Fig. 9.2.

The tube fitted with a piston is rigid and has a uniform cross-sectional area A. It is filled initially with fluid at rest. The piston gives an impulsive motion and produces a small pressure wave moving to the right at the speed a. The fluid conditions on either side of the wavefront are different. On the left side, through which the front has passed, the fluid moves to the right with velocity δv, and its pressure and density are $p + \delta p$ and $\varrho + \delta\varrho$, respectively. On the right side, the fluid is still undisturbed and the pressure and density are p and ϱ, respectively.

Now suppose a coordinate (reference) system is attached to the moving front, and consider a small (control) volume surrounding the front (Fig. 9.3). With respect to this coordinate system, the wavefront appears to be stationary. The fluid on the right appears to be moving to the left with velocity a; that on the left appears to be moving with velocity $a - \delta v$.

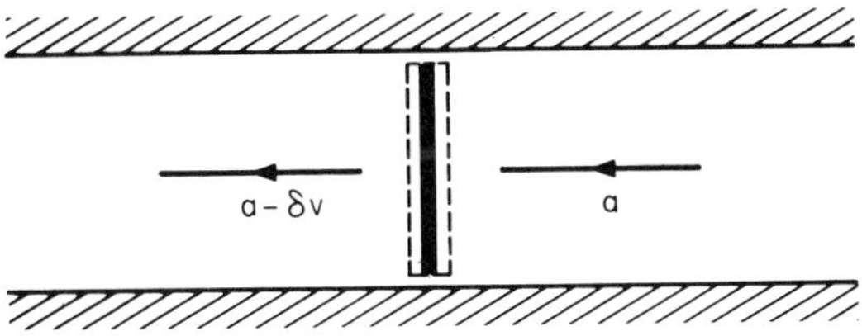

Fig. 9.3

The continuity equation for the control volume is

$$a\varrho A = (a - \delta v)(\varrho + \delta\varrho)A = (a\varrho + a\,\delta\varrho - \varrho\,\delta v - \delta v\,\delta\varrho)A.$$

Dropping the second-order term, one therefore has

$$a\,\delta\varrho = \varrho\,\delta v. \tag{9.60}$$

The shear force on the control volume is negligibly small, and the momentum equation can be written

$$(p + \delta p)A - pA = \varrho aA[-(a - \delta v)] - \varrho aA(-a).$$

Therefore

$$\delta p = \varrho a\,\delta v. \tag{9.61}$$

Eliminating δv from Eqs. (9.60) and (9.61), and going to the limit, one obtains

$$a = \left(\frac{dp}{d\varrho}\right)^{1/2} = \left(\frac{\partial p}{\partial\varrho}\right)_S^{1/2}. \tag{9.62}$$

Since the changes in density, pressure, and temperature in the pressure wave (sound) propagation are all infinitesimally small, the process is nearly reversible. Moreover, the pressure fluctuations are so rapid that the accompanying infinitesimal temperature changes make the process virtually adiabatic. So the process involved in a sound wave is very closely isentropic.

9.2.1 General Characteristics of Sound

Fluids generally have mass density and volume elasticity (macroscopic properties) and thus have many of the characteristics of linear arrays of masses and springs (Fig. 9.4). The masses are connected by linear springs.

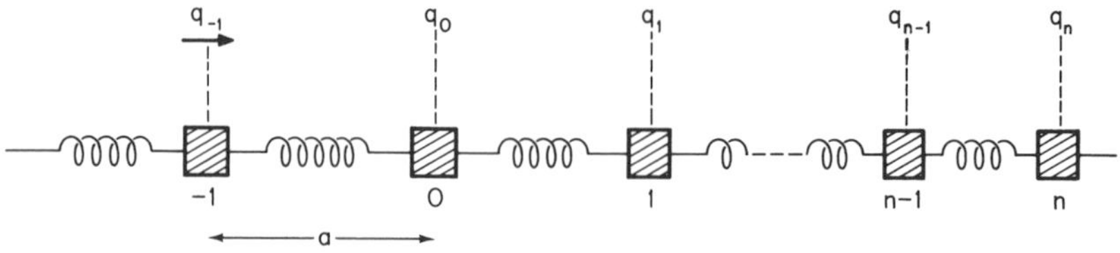

Fig. 9.4

The elasticity of the fluid enables it to resist compression and to tend to return to the equilibrium configuration. The inertia of the mass density causes "overshoot" in the movement and so wave motion can result.

At equilibrium, the fluid has density ϱ, in units of kilograms per cubic meter (mks system), is at a uniform pressure p in units of newtons per square meter, and is at a uniform temperature T°K. These three quantities are connected by the equation of state, either stated explicitly as in Eq. (9.2) or in terms of the partial derivatives:

$$K_T = -\frac{1}{V}\left(\frac{\partial V}{\partial p}\right)_T = \frac{1}{\varrho}\left(\frac{\partial \varrho}{\partial p}\right)_T, \tag{9.63}$$

$$\beta = \frac{1}{V}\left(\frac{\partial V}{\partial T}\right)_p = -\frac{1}{\varrho}\left(\frac{\partial \varrho}{\partial T}\right)_p; \tag{9.64}$$

K_T is the fractional rate of change of volume (or density) with pressure at constant temperature—called the *isothermal compressibility* of the fluid. It is interesting to note that the compressibility of a spring is

$$K = \frac{1}{l}\,\frac{\Delta x}{F},$$

where $F = K\,\Delta x$ is the force required to produce an extension Δx and K is the stiffness constant.

The fractional change in volume (or density) with temperature at constant pressure, β, is called the *coefficient of thermal expansion* of the fluid.

For a perfect gas, $pV = NRT$, and K_T and β take on very simple values,

$$K_T = \frac{1}{p}, \qquad \beta = \frac{1}{T}. \tag{9.65}$$

Now by kinetic theory,

$$p = \tfrac{1}{3}\varrho\langle v^2\rangle,$$

where $\langle v^2\rangle$ is the mean square molecular velocity. Hence, the compressibility of a perfect gas is inversely proportional to the mean kinetic energy of the gas molecules.

The velocity of propagation of waves along a mass–spring system is $(K\varepsilon)^{-1/2}$, where K is the spring compressibility and ε the mass per unit length of the chain. Analogously, the velocity of sound waves is $(K\varepsilon)^{-1/2}$. Since

$$K_T = \frac{1}{p} = \frac{3}{\varrho\langle v^2\rangle},$$

the speed of sound is

$$c = \left(\frac{1}{K\varrho}\right)^{1/2} \simeq \left(\frac{\langle v^2\rangle}{3}\right)^{1/2} \tag{9.66}$$

and is approximately equal to the root mean square (rms) of the molecular speed. In liquids and solids, where intermolecular forces are large, the speed of sound is considerably larger than the rms speed of vibration of an atom about its equilibrium position.

The amplitude of the pressure change in a barely audible sound wave of 1000 Hz is 0.00022 dyn/cm². For simple harmonic waves the pressure amplitude can be given in terms of the pressure level in decibels. This is 20 times the logarithm to the base 10 of the pressure amplitude in dynes per square centimeter above or below the level of 1 dyn/cm². Or it can be given as pressure level in decibels (dB) above or below 0.00022 dyn/cm². On this scale, for example, a pressure amplitude of 1 dyn/cm² rms amplitude is 74 dB.

In fluids with very large thermal conductivity, the fluid temperature is hardly changed by the passage of a sound wave. The isothermal compressibility K_T is directly applicable in this situation. If the pressure change caused by the sound is p_1 and the equilibrium pressure is p_0, then the density of the fluid is

$$\varrho_0 + \Delta = \varrho_0 + \left(\frac{\partial \varrho}{\partial p}\right)_T p = \varrho_0 + \varrho_0 K_T p.$$

Therefore,

$$\Delta = (K_T \varrho_0) p, \tag{9.67}$$

where ϱ_0 is the equilibrium density, and Δ the small increase in density caused by the sound. For sound at frequencies below $\sim 10^9$ Hz in gases, it is a better approximation to assume that the compression, as mentioned previously, is adiabatic or isentropic. The differential relations between density and pressure and temperature and pressure are

$$\left(\frac{\partial \varrho}{\partial p}\right)_S = \varrho K_S = \frac{1}{\gamma}\left(\frac{\partial \varrho}{\partial p}\right)_T,$$

therefore,

$$K_S = \frac{K_T}{\gamma}; \tag{9.68}$$

$$\left(\frac{\partial T}{\partial p}\right)_S = \frac{\gamma - 1}{\gamma}\left(\frac{\partial T}{\partial p}\right)_\varrho = (\gamma - 1)\frac{K_S}{\beta}, \tag{9.69}$$

therefore

$$\varrho_0 + \Delta = \varrho_0 + \varrho_0 K_S p, \tag{9.70}$$

$$T_0 + T_1 = T_0 + (\gamma - 1)\frac{K_S}{\beta} p \tag{9.71}$$

for adiabatic compression, where $\gamma = C_p/C_v$ and T_1 is the small change in temperature caused by the sound wave. For perfect gases, $mp = RT\varrho$ and so

$$K_S = \frac{1}{\gamma p}, \tag{9.72}$$

$$\left(\frac{\partial \varrho}{\partial p}\right)_S = \frac{\varrho}{\gamma p} = \frac{m}{\gamma RT}, \tag{9.73}$$

$$\left(\frac{\partial T}{\partial p}\right)_S = \frac{\gamma - 1}{\gamma}\frac{T}{p} = (\gamma - 1)\frac{m}{\gamma R \varrho}. \tag{9.74}$$

If the pressure in the presence of the sound is $p_0 + p_1$, then the density and temperature are

$$\varrho_0 + \varrho_1 = \varrho_0 + \left(\frac{\varrho_0}{\gamma p}\right) p_1,$$

therefore,

$$\varrho_1 = (\varrho_0 K_S) p_1 = \left(\frac{m}{\gamma RT}\right) p_1; \tag{9.75}$$

and

$$T_0 + T_1 = T_0 + \frac{(\gamma - 1)T_0}{\gamma p_0} p_1,$$

therefore,

$$T_1 = \frac{(\gamma - 1)m}{\gamma R \varrho_0} p_1 \tag{9.76}$$

for adiabatic compression of a perfect gas.

9.2.2 Mach Number and Mach Cone

The Mach number M at any point in a compressible flow field is given by the ratio of the local stream velocity U and the speed of sound a:

$$M = \frac{U}{a}. \tag{9.77}$$

The speed of sound is computed from the local thermodynamic conditions of the stream. For a perfect gas,

$$a = (\gamma RT)^{1/2}, \tag{9.78}$$

therefore

$$M = \frac{U}{(\gamma RT)^{1/2}}. \tag{9.79}$$

Since γ and R are constant for a perfect gas,

$$\frac{dM}{M} = \frac{dU}{U} - \frac{1}{2}\frac{dT}{T}. \tag{9.80}$$

There are four regimes of compressible flow, associated with various ranges of magnitude of M:

(i)	$M < 1$,	subsonic flow,
(ii)	$M \approx 1$,	transonic flow,
(iii)	$1 < M < 3$,	supersonic flow,
(iv)	$M > 3$,	hypersonic flow.

Consider a body moving from right to left in a stationary fluid, at a velocity smaller than the velocity of sound. As the body moves, pressure disturbances are generated, which propagate in all directions with the velocity of sound. If the Mach number is nearly zero, the spherical wave surfaces spreading out from the body are symmetrical about it, as indicated in Fig. 9.5a.

Three time intervals are shown, Δt, $2\,\Delta t$, and $3\,\Delta t$. The radius of the spherical wavefront at $3\,\Delta t$ is $3a\,\Delta t$. If the velocity of the body is now V (a finite Mach number), the resulting propagation of the disturbance is no longer symmetrical. The body moves a distance $3V\,\Delta t$, while the spherical wavefront travels $3a\,\Delta t$. If $V < a$, obviously the pressure disturbance moves ahead of the projectile and will eventually reach the whole fluid space. This is the case which defines subsonic flow (Fig. 9.5b).

When $V = a$, the disturbance cannot propagate ahead of the object and so the propagation of the pressure waves is restricted to the half space bounded by the plane perpendicular to the direction of the moving object. This is illustrated in Fig. 9.5c and characterizes the regime known as transonic flow.

When $V > a$, $3V\,\Delta t$ is larger than $3a\,\Delta t$. As shown in Fig. 9.5d,

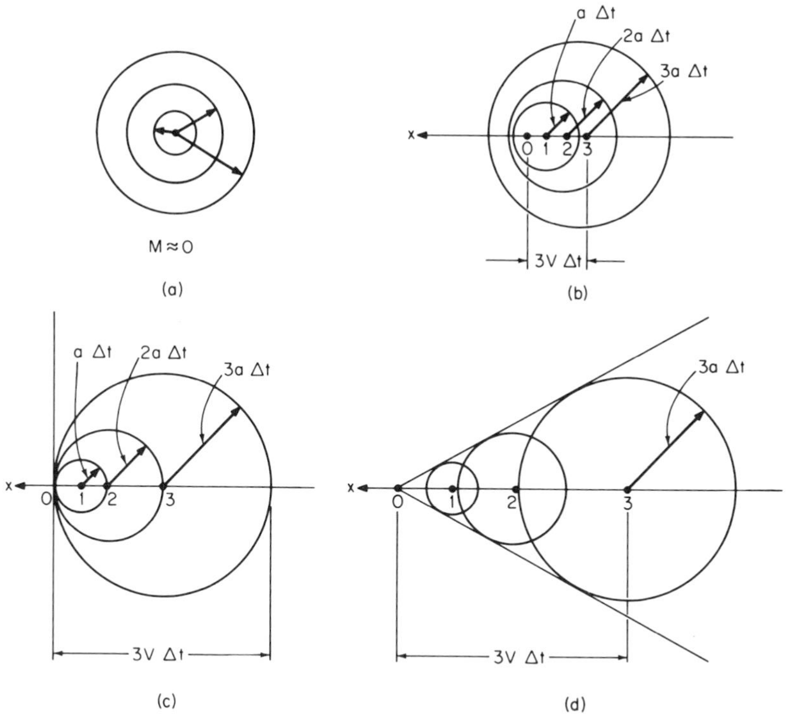

Fig. 9.5

the object at 0 is outside the spherical waves originating at positions 2 and 3. So the effect of the pressure disturbance is restricted to the inside of the envelope of these spherical surfaces. This is a cone whose vertex is at the position of the object, and whose semivertex angle is given by

$$\sin \vartheta = \frac{a}{V}. \tag{9.81}$$

This is called the Mach angle and the cone is called the Mach cone.

In plane (two-dimensional) flow, this cone is formed by a pair of intersecting lines, each of which is called a Mach line or Mach wave.

The Mach cone divides the fluid space into that region disturbed by the object motion and that which is undisturbed. Von Kármán called the former region the region of action, and the latter the region of silence. This phenomenon, regions of action and silence, characterizes the regime of supersonic flow.

9.3 One-Dimensional Compressible Flow

As a first step in developing the dynamics of compressible fluids, it is important to establish the controlling parameters in the problem. The Reynolds similarity principle was obtained in Eq. (4.26) from the Navier–Stokes equation and is the controlling parameter in viscous incompressible flow.

In viscous compressible flow, controlling parameters can be deduced from the energy equation. The relevant energy equation [Eq. (4.26)] is

$$\varrho \frac{D(C_p T)}{Dt} = \frac{Dp}{Dt} + \nabla \cdot (k \nabla T) + \mu\left[-\frac{2}{3}(\nabla \cdot \mathbf{v})^2 + \cdots\right]. \tag{9.82}$$

The last term on the right-hand side of Eq. (9.82) is the dissipation function.

The subscript ∞ will denote the value of a parameter at infinity. If L is a characteristic length in the flow system, then a characteristic time is given by

$$t_0 = \frac{L}{v_\infty}.$$

The following nondimensional quantities can now be introduced:

$$t^* = \frac{t_0 v_\infty}{L}, \quad \nabla^* = L\nabla, \quad \mathbf{v}^* = \frac{\mathbf{v}}{v_\infty}, \quad \varrho^* = \frac{\varrho}{\varrho_\infty}, \quad p^* = \frac{p}{\varrho_\infty {v_\infty}^2}, \tag{9.83}$$

$$T^* = \frac{T}{T_\infty}, \quad \mu^* = \frac{\mu}{\mu_\infty}, \quad k^* = \frac{k}{k_\infty}, \quad {C_p}^* = \frac{C_p}{C_{p\infty}}.$$

Substituting these nondimensional quantities into Eq. (9.82), one gets the nondimensional form of the energy equation:

$$\varrho^* \frac{D({C_p}^* T^*)}{Dt^*} = \left(\frac{{v_\infty}^2}{C_{p\infty} T_\infty}\right)\frac{Dp^*}{Dt^*} + \left(\frac{\mu_\infty}{\varrho_\infty v_\infty L}\right)\left(\frac{k_\infty}{C_{p\infty}\mu_\infty}\right)\nabla^* \cdot (k^* \nabla^* T^*)$$
$$+ \left(\frac{\mu_\infty}{\varrho_\infty v_\infty L}\right)\left(\frac{{v_\infty}^2}{C_{p\infty} T_\infty}\right)\mu^*\left[-\frac{2}{3}(\nabla^* \cdot \mathbf{v}^*)^2 + \cdots\right]. \tag{9.84}$$

Consider the bracketted coefficient terms in turn. In the first term,

$$\frac{{v_\infty}^2}{C_{p\infty} T_\infty} = \frac{\gamma R {v_\infty}^2}{C_{p\infty}\gamma R T_\infty} = (\gamma - 1)\frac{{v_\infty}^2}{{a_\infty}^2} = (\gamma - 1){M_\infty}^2. \tag{9.85}$$

This gives the Mach number and the ratio of the specific heats. The Mach number is a measure of the compressibility of the fluid, and γ is a measure of the molecular structure of the fluid.

The coefficients in the second term on the right-hand side of Eq. (9.84) can be written

$$\left(\frac{\mu_\infty}{\varrho_\infty v_\infty L}\right)\left(\frac{k_\infty}{C_{p\infty}\mu_\infty}\right) = \frac{1}{\text{Re Pr}}, \tag{9.86}$$

where Re is the Reynolds number and Pr is the Prandtl number, with

$$\text{Pr} = \frac{\mu_\infty C_{p\infty}}{k_\infty} = \frac{\mu_\infty/\varrho_\infty}{k_\infty/\varrho_\infty C_{p\infty}} = \frac{\text{kinematic viscosity}}{\text{thermal diffusivity}}; \tag{9.87}$$

Pr is a measure of the relative importance of heat conduction and viscosity of a fluid. For inviscid compressible flow, of course, the Reynolds number and Prandtl number are not relevant, while the Mach number and specific heat ratio are. For *viscous* compressible flows, all four parameters are relevant.

9.4 Basic Equations of Flow

9.4.1 The Continuity Equation

The continuity equation [Eq. (3.9)] is

$$\frac{\partial \varrho}{\partial t} + \boldsymbol{\nabla} \cdot (\varrho \mathbf{v}) = 0. \tag{9.88}$$

In terms of the material derivative,

$$\frac{D\varrho}{Dt} + \varrho(\boldsymbol{\nabla} \cdot \mathbf{v}) = 0. \tag{9.89}$$

In Cartesian coordinates

$$\frac{\partial \varrho}{\partial t} + \frac{\partial(\varrho u)}{\partial x} + \frac{\partial(\varrho y)}{\partial y} + \frac{\partial(\varrho z)}{\partial z} = 0. \tag{9.90}$$

For steady one-dimensional flow this becomes

$$\frac{d(\varrho u)}{dx} = 0. \tag{9.91}$$

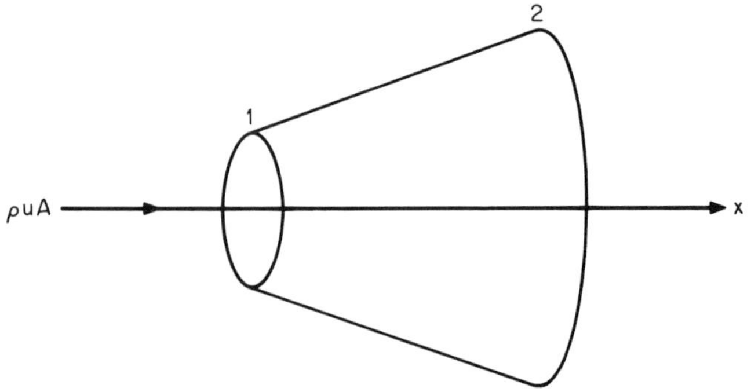

Fig. 9.6

If one considers fluid flowing along a rigid tube, as shown in Fig. 9.6, the mass flow is constant all along the tube, and so

$$\varrho_1 u_1 A_1 = \varrho_2 u_2 A_2$$

or

$$\varrho u A = \text{constant};$$

therefore

$$\frac{d}{dx}(\varrho u A) = 0. \tag{9.92}$$

The corresponding unsteady equation is

$$\frac{\partial}{\partial t}(\varrho u A) + \frac{\partial}{\partial x}(\varrho u A) = 0. \tag{9.93}$$

If the steady mass flow rate through the tube is $\dot{m}$, then

$$\dot{m} = \varrho u A = \text{constant}.$$

The differential equation of continuity for one-dimensional compressible flow is therefore

$$\frac{d\varrho}{\varrho} + \frac{du}{u} + \frac{dA}{A} = 0. \tag{9.94}$$

For a perfect gas,

$$\dot{m} = \varrho u A \frac{(\gamma RT)^{1/2}}{(\gamma RT)^{1/2}} = \varrho RT \frac{u}{(\gamma RT)^{1/2}} \frac{\gamma^{1/2}}{(RT)^{1/2}} A = pM \frac{\gamma^{1/2}}{(RT)^{1/2}} A. \tag{9.95}$$

So the differential form for a perfect gas in terms of the Mach number is

$$\frac{dp}{p} + \frac{dM}{M} - \frac{1}{2}\,\frac{dT}{T} + \frac{dA}{A} = 0. \tag{9.96}$$

9.4.2 The Energy Equation

For a flowing fluid, the basic thermodynamic quantity is the enthalpy rather than the internal energy. The difference is that the former contains the "flow work."

For adiabatic flow of an ideal fluid, the thermal energy equation (based on the law of conservation of energy) was given by Eq. (4.36) as

$$\frac{Dh}{Dt} = \frac{1}{\varrho}\,\frac{Dp}{Dt} = \frac{1}{\varrho}\,\frac{\partial p}{\partial t} + \mathbf{v}\cdot\frac{\nabla p}{\varrho}. \tag{9.97}$$

The Navier–Stokes equation (4.15), under the same conditions and for zero body force, gives, after scalar multiplication by $\mathbf{v}$.

$$\frac{1}{2}\,\frac{Dv^2}{Dt} = -\mathbf{v}\cdot\frac{\nabla p}{\varrho}. \tag{9.98}$$

Substituting Eq. (9.98) into Eq. (9.97), one obtains the energy equation in the form

$$\frac{D}{Dt}\left(h + \frac{v^2}{2}\right) = \frac{1}{\varrho}\,\frac{\partial p}{\partial t}. \tag{9.99}$$

This shows that a change in the quantity

$$h + \frac{v^2}{2} = h_0$$

is due to nonsteady pressure.

For a steady flow, then,

$$d\left(h + \frac{v^2}{2}\right) = 0 \tag{9.100}$$

or

$$h + \frac{v^2}{2} = C_pT + \frac{v^2}{2} = \text{constant} = h_0 = C_pT_0 \tag{9.101}$$

for a perfect gas, and T_0 is the absolute temperature at points where the fluid is at rest.

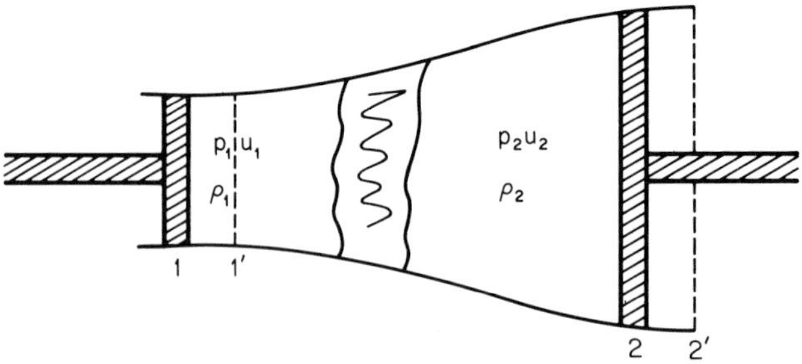

Fig. 9.7

Consider the "system" formed by a definite volume of fluid between sections 1 and 2 (Fig. 9.7). The pistons move from positions 1 and 2, to 1′ and 2′. The initial and final states are in equilibrium. Then the adiabatic energy equation given in Eq. (9.101) implies

$$h_2 + \tfrac{1}{2}u_2{}^2 = h_1 + \tfrac{1}{2}u_1{}^2. \tag{9.102}$$

This is true even if there are viscous stresses, heat transfer, or other nonequilibrium (transport) conditions between sections 1 and 2.

If equilibrium exists all along a flow, then

$$h + \frac{v^2}{2} = \text{constant},$$

continuously.

9.4.3 Reservoir Conditions

In Eq. (9.101) it was written that

$$h + \tfrac{1}{2}v^2 = h_0; \tag{9.103}$$

h_0 is called the *stagnation* or *reservoir enthalpy*. It will be the enthalpy value in a large reservoir where v is practically zero (Fig. 9.8). With

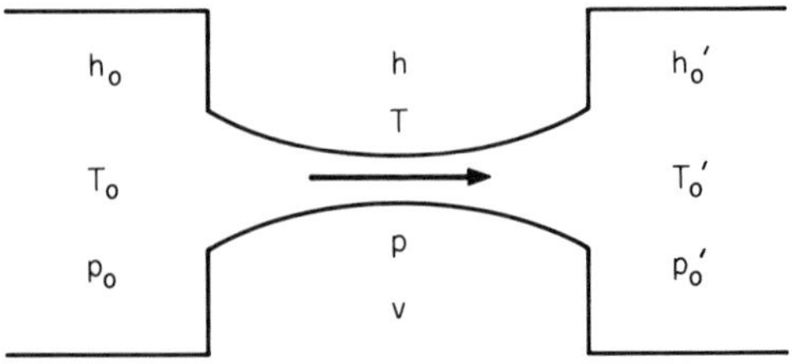

Fig. 9.8

no heat addition to the flow between the two reservoirs,

$$h_0' = h_0. \tag{i}$$

Also, for a perfect gas,

$$C_p T + \tfrac{1}{2}v^2 = C_p T_0,$$

and so

$$T_0' = T_0, \tag{ii}$$

independent of the two pressures.

According to the Second Law of Thermodynamics,

$$S_0' - S_0 \geq 0, \tag{iii}$$

and $S_0' = S_0$ only if the flow is isentropic (that is, it must be free from the energy dissipation which results from friction along the walls, or from the formation of turbulent eddies).

From thermodynamics, $S_0' - S_0$ for a perfect gas is

$$S_0' - S_0 = R \ln \frac{p_0}{p_0'} + C_p \ln \frac{T_0'}{T_0}, \tag{9.104}$$

and if $T_0' = T_0$, then

$$S_0' = S_0 - R \ln \frac{p_0'}{p_0}. \tag{9.105}$$

Now, h_0, T_0, etc. are the local total conditions in a fluid flow—that is, the maximum values obtained by bringing the flow to rest isentropically. At such a stagnation point in a flow (for example, where the flow impinges on a blunt body), the stagnation entropy $S_0' = S'$ is related to the local static entropy $S_0 = S$ by Eq. (9.105)—that is,

$$S' = S - R \ln \frac{p_0'}{p_0}, \tag{9.106}$$

since $T_0' = T_0$ again.

Thus, a measurement of the local total pressure provides a measure of the entropy of the flow (for example, using a pitot tube).

Note A measurement of stagnation temperature in this way is not possible, because thermal equilibrium is not established at the stagnation point of a thermometer.

9.4.4 Euler's Equation of Motion

The one-dimensional form of Euler's equation given in Eq. (4.16), with no body force, is

$$\frac{\partial u}{\partial t} + u\frac{\partial u}{\partial x} = -\frac{1}{u}\frac{\partial p}{\partial x}, \tag{9.107}$$

where u is the x component of velocity.

For steady flow,

$$u\,du + \frac{dp}{\varrho} = 0$$

or, in integral form,

$$\frac{u^2}{2} + \int \frac{dp}{\varrho} = \text{constant.} \tag{9.108}$$

The constant remains unchanged along a given streamline, but it may vary from streamline to streamline. This is Bernoulli's equation for compressible flow.

9.4.5 Entropy Variation along a Streamline

Returning to the general form of Euler's equation given in (4.16) and again omitting the body force, one obtains

$$\frac{\partial \mathbf{v}}{\partial t} + (\mathbf{v}\cdot\nabla)\mathbf{v} = -\frac{\nabla p}{\varrho}. \tag{9.109}$$

Now,

$$(\mathbf{v}\cdot\nabla)\mathbf{v} = \tfrac{1}{2}\nabla v^2 - \mathbf{v}\times\boldsymbol{\zeta},$$

where $\boldsymbol{\zeta}$ is the vorticity in the flow. So Eq. (9.109) can be written as

$$\frac{\partial \mathbf{v}}{\partial t} + \nabla\left(\frac{v^2}{2}\right) + \mathbf{v}\times\boldsymbol{\zeta} = -\frac{\nabla p}{\varrho}. \tag{9.110}$$

Now

$$d\mathbf{v}\cdot\frac{\nabla p}{\varrho} = \frac{dp}{\varrho} = d\int\frac{dp}{\varrho} = d\mathbf{r}\cdot\nabla\int\frac{dp}{\varrho}.$$

Therefore,

$$\frac{\nabla \mathbf{p}}{\varrho} = \nabla\int\frac{dp}{\varrho} \tag{9.111}$$

and

$$\frac{\partial \mathbf{v}}{\partial t} + \nabla\left(\int \frac{dp}{\varrho} + \frac{v^2}{2}\right) = \mathbf{v} \times \boldsymbol{\zeta}. \tag{9.112}$$

For steady flow, and after scalar multiplication by $d\mathbf{r}$,

$$d\left(\int \frac{dp}{\varrho} + \frac{v^2}{2}\right) = (\mathbf{v} \times \boldsymbol{\zeta}) \cdot d\mathbf{r}. \tag{9.113}$$

Now

$$dQ = T\,dS = dh - \frac{dp}{\varrho} = d\left(h - \int \frac{dp}{\varrho}\right),$$

and combining this with Eq. (9.113), one obtains

$$d\left(h + \frac{v^2}{2}\right) - T\,dS = (\mathbf{v} \times \boldsymbol{\zeta}) \cdot d\mathbf{r}. \tag{9.114}$$

But

$$d\left(h + \frac{v^2}{2}\right) = 0,$$

and so

$$T\,dS = (\mathbf{v} \times \boldsymbol{\zeta}) \cdot d\mathbf{r}. \tag{9.115}$$

Since $(\mathbf{v} \times \boldsymbol{\zeta})$ is perpendicular to $d\mathbf{r}$ and along a streamline, entropy is conserved along a streamline. Again, entropy can vary from streamline to streamline.

If the flow is irrotational everywhere, then from Eq. (9.115) the entropy must be constant and the flow is isentropic, and so

$$p = \text{constant} \cdot \varrho^\gamma \tag{9.116}$$

or

$$\frac{p}{p_0} = \left(\frac{\varrho}{\varrho_0}\right)^\gamma = \left(\frac{T}{T_0}\right)^{\gamma/(\gamma-1)}. \tag{9.117}$$

9.4.6 Pressure and Density Ratios

For a perfect gas, from Eq. (9.103),

$$\frac{v^2}{2} + C_p T = C_p T_0.$$

Using the thermodynamic expression for the speed of sound, one gets

$$a^2 = \gamma RT,$$

so then

$$\frac{v^2}{2} + \frac{a^2}{\gamma - 1} = \frac{a_0^2}{\gamma - 1}. \tag{9.118}$$

Multiplying across by $((\gamma - 1)/a^2)$, one has

$$\frac{a_0^2}{a^2} = \frac{T_0}{T} = 1 + \frac{\gamma - 1}{2} M^2, \tag{9.119}$$

and using the isentropic relations (9.117), one gets

$$\frac{p_0}{p} = \left(1 + \frac{\gamma - 1}{2} M^2\right)^{\gamma/(\gamma-1)}, \tag{9.120}$$

$$\frac{\varrho_0}{\varrho} = \left(1 + \frac{\gamma - 1}{2} M^2\right)^{1/(\gamma-1)}. \tag{9.121}$$

In Eqs. (9.118) and (9.119), T_0 and a_0 are constant through the flow and may be taken as actual reservoir values. In Eqs. (9.120) and (9.121), the values of p_0 and ϱ_0 are local reservoir values and are constant throughout if the flow is isentropic.

Instead of evaluating these constants in a reservoir, any other point in the flow could be used. For example, in a convergent/divergent duct, the throat is the point at which $M = 1$. The flow variables are called "sonic" and are denoted by an asterisk; for example, the flow speed v^*. Since $M = 1$, thus $v^* = a^*$. From Eq. (9.118),

$$\frac{v^2}{2} + \frac{a^2}{\gamma - 1} = \frac{v^{*2}}{2} + \frac{a^{*2}}{\gamma - 1} = \frac{1}{2}\,\frac{\gamma + 1}{\gamma - 1} a^{*2}. \tag{9.122}$$

Comparing this with Eq. (9.118), one obtains

$$\frac{a^{*2}}{a_0^2} = \frac{2}{\gamma + 1} = \frac{T^*}{T_0}. \tag{9.123}$$

This gives the relation between the speed of the fluid at the throat and in the reservoir.

For air,

$$\frac{T^*}{T_0} = 0.533, \qquad \frac{a^*}{a_0} = 0.913.$$

Setting $M = 1$ in Eqs. (9.120) and (9.121), one obtains

$$\frac{p^*}{p_0} = \left(\frac{2}{\gamma + 1}\right)^{\gamma/(\gamma-1)}, \tag{9.124}$$

$$\frac{\varrho^*}{\varrho_0} = \left(\frac{2}{\gamma + 1}\right)^{1/(\gamma-1)}. \tag{9.125}$$

Note It is not necessary that a throat exist in the flow to use the sonic values as a reference.

9.4.7 Dynamic Pressure

The energy equation for a perfect gas,

$$\frac{v^2}{2} + C_p T = C_p T_0,$$

can be written, using the gas law to eliminate T, as

$$\frac{v^2}{2} + \frac{\gamma}{\gamma - 1}\frac{p}{\varrho} = \frac{\gamma}{\gamma - 1}\frac{p_0}{\varrho_0}. \tag{9.126}$$

For isentropic flow, the isentropic relation $p/p_0 = (\varrho/\varrho_0)^\gamma$ can be used to eliminate ϱ:

$$\frac{v^2}{2} + \frac{\gamma}{\gamma - 1}\frac{p_0}{\varrho_0}\left(\frac{p}{p_0}\right)^{(\gamma-1)/\gamma} = \frac{\gamma}{\gamma - 1}\frac{p_0}{\varrho_0}. \tag{9.127}$$

The right-hand side is a constant and so this equation is another form of the integral form of the Bernoulli equation given in Eq. (9.108).

In compressible flow, the dynamic pressure $\frac{1}{2}\varrho U^2$ depends on both the Mach number and the static pressure p. For a perfect gas,

$$\frac{1}{2}\varrho v^2 = \frac{1}{2}\varrho a^2 M^2 = \frac{1}{2}\varrho\left(\frac{\gamma p}{\varrho}\right)M^2 = \frac{1}{2}\gamma p M^2. \tag{9.128}$$

9.5 Normal Shock Relations for a Perfect Gas

For constant area adiabatic flow through a nonequilibrium region, the equations of momentum [Eq. (9.107)], energy [Eq. (9.102)], and continuity [Eq. (9.93)] are

$$p_1 + \varrho_1 v_1^2 = p_2 + \varrho_2 v_2^2, \tag{9.129}$$

$$h_1 + \tfrac{1}{2} v_1^2 = h_2 + \tfrac{1}{2} v_2^2, \tag{9.130}$$

$$\varrho_1 v_1 = \varrho_2 v_2. \tag{9.131}$$

The index 1 indicates the region upstream and 2 the region downstream.

The distribution region can be of any shape, but the reference sections must lie outside it (Fig. 9.9). If the distribution region is regarded as a

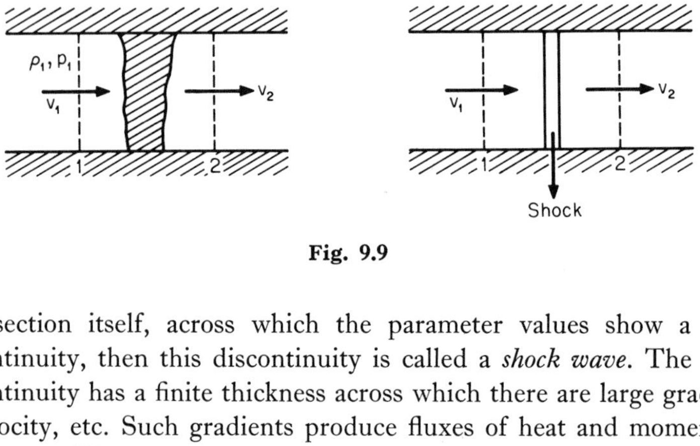

Fig. 9.9

thin section itself, across which the parameter values show a jump discontinuity, then this discontinuity is called a *shock wave*. The shock discontinuity has a finite thickness across which there are large gradients in velocity, etc. Such gradients produce fluxes of heat and momentum, and nonequilibrium conditions prevail inside a shock. For most aerodynamic purposes the shock can be regarded as a discontinuity.

Equations (9.129)–(9.131) are general equations for a normal shock wave, and usually can be solved only numerically. But for a thermally and calorically perfect gas one can derive explicit solutions in terms of the Mach number upstream of the shock.

Dividing the left-hand side of Eq. (9.129) by $\varrho_1 v_1$ and the right-hand side by $\varrho_2 v_2$ $(= \varrho_1 v_1)$, one obtains

$$v_1 - v_2 = \frac{p_2}{\varrho_2 v_2} - \frac{p_1}{\varrho_1 v_1} = \frac{a_2^2}{\gamma v_2} - \frac{a_1^2}{\gamma v_1}, \tag{9.132}$$

where $a^2 = \gamma p/\varrho$ for a perfect gas.

Using the form of the energy equation given in Eq. (9.122), one obtains

$$\frac{v_1^2}{2} + \frac{a_1^2}{\gamma - 1} = \frac{v_2^2}{2} + \frac{a_2^2}{\gamma - 1} = \frac{1}{2}\left(\frac{\gamma + 1}{\gamma - 1}\right)a^{*2}. \quad (9.133)$$

From these two equations it can be deduced that

$$v_1 v_2 = a^{*2}. \quad (9.134)$$

This is known as the Prandtl relation.

Let

$$M^* = \frac{v^*}{a}.$$

Therefore

$$M_2^* = \frac{1}{M_1^*}, \quad (9.135)$$

and so the Prandtl relation shows that the velocity variation across a normal shock wave can only be from supersonic to subsonic or vice versa (not from subsonic to subsonic, for example). In view of the energy dissipation inside the shock, obviously the first alternative is the only possible one. From Eq. (9.122) it is easily shown that

$$M^{*2} = \frac{\dfrac{\gamma + 1}{2} M^2}{1 + \dfrac{\gamma - 1}{2} M^2} = \frac{(\gamma + 1)M^2}{(\gamma - 1)M^2 + 2}, \quad (9.136)$$

where $M^* = v/a^*$.

Using Eq. (9.136) and the Prandtl relation, one obtains the relation between M_1 and M_2:

$$M_2^2 = \frac{1 + \dfrac{\gamma - 1}{2} M_1^2}{\gamma M_1^2 - \dfrac{\gamma - 1}{2}}. \quad (9.137)$$

Now,

$$\frac{v_1}{v_2} = \frac{v_1^2}{v_1 v_2} = \frac{v_1^2}{a^{*2}} = M_1^{*2}; \quad (9.138)$$

therefore,

$$\frac{\varrho_2}{\varrho_1} = \frac{v_1}{v_2} = \frac{(\gamma + 1)M_1^2}{(\gamma - 1)M_1^2 + 2}. \quad (9.139)$$

Then, from the momentum equation,

$$p_2 - p_1 = \varrho_1 v_1^2 - \varrho_2 v_2^2 = \varrho_1 v_1 (v_1 - v_2)$$

or

$$\frac{p_2 - p_1}{p_1} = \frac{\varrho_1 v_1^2}{p_1}\left(1 - \frac{v_2}{v_1}\right).$$

Now, $a_1^2 = \gamma p_1/\varrho_1$ and using (9.139) for v_1/v_2, one obtains the pressure jump:

$$\frac{p_2 - p_1}{p_1} = \frac{\Delta p_1}{p_1} = \frac{2\gamma}{\gamma + 1}(M_1^2 - 1); \tag{9.140}$$

$\Delta p/p_1$ is used to define the *shock strength.* The change in entropy is given by

$$\frac{S_2 - S_1}{R} = \ln\left[\left(\frac{p_2}{p_1}\right)^{1/(\gamma-1)}\left(\frac{\varrho_2}{\varrho_1}\right)^{-\gamma/(\gamma-1)}\right], \tag{9.141}$$

and using the above relations for ϱ_2/ϱ_1 and p_2/p_1, one obtains

$$\frac{S_2 - S_1}{R} = \ln\left[1 + \frac{2\gamma}{\gamma + 1}(M_1^2 - 1)\right]^{1/(\gamma-1)}\left[\frac{(\gamma + 1)M_1^2}{(\gamma - 1)M_1^2 + 2}\right]^{-\gamma/(\gamma-1)}. \tag{9.142}$$

Other relations, such as temperature and total pressure ratios, are obtained by similar manipulations. It can be seen that the jumps in ϱ, p, and T are from lower to higher values—the shock *compresses* the flow.

9.6 The Propagating Wave

In the preceding sections, the pressure, or shock, wave was assumed to be stationary. Actually a shock, or pressure wave, propagates through the fluid. Let us consider the propagating plane sinusoidal wave first.

The equation of continuity is

$$\frac{\partial \varrho}{\partial t} + \nabla \cdot (\varrho \mathbf{v}) = 0, \tag{9.143}$$

and the equation of motion for inviscid fluids with negligible body forces is

$$\frac{\partial \mathbf{v}}{\partial t} + (\mathbf{v} \cdot \boldsymbol{\nabla})\mathbf{v} = -\frac{\boldsymbol{\nabla} p}{\varrho}, \tag{9.144}$$

where p and ϱ are connected by the isentropic relation.

If p_∞ and ϱ_∞ are the pressure and density of the gas at infinity (where the gas is undisturbed), then the total flow parameter values are

$$p = p_\infty + p_1, \qquad \varrho = \varrho_\infty + \varrho_1, \qquad \mathbf{v} = \mathbf{v}_1. \tag{9.145}$$

For small disturbances, $p_1 \ll p_\infty$, $\varrho_1 \ll \varrho_\infty$, and $|\mathbf{v}_1| \ll a$. Moreover, Eq. (9.144) is valid only if the velocity gradients are small enough to ensure that friction is negligible. Thus the term $(\mathbf{v} \cdot \boldsymbol{\nabla})\mathbf{q}$ in Eq. (9.144) can be neglected, and

$$\frac{\boldsymbol{\nabla} p}{\varrho} = \frac{\boldsymbol{\nabla} p_1}{\varrho_\infty + \varrho_1} = \frac{1}{\varrho_\infty}\left(1 - \frac{\varrho_1}{\varrho_\infty} + \cdots\right)\boldsymbol{\nabla} p_1 \doteq \frac{\boldsymbol{\nabla} p_1}{\varrho_\infty}. \tag{9.146}$$

The equation of motion may thus be written

$$\frac{\partial \mathbf{v}_1}{\partial t} + \frac{\boldsymbol{\nabla} p_1}{\varrho_\infty} = 0, \tag{9.147a}$$

and the continuity equation is

$$\frac{\partial \varrho_1}{\partial t} + (\mathbf{v}_1 \cdot \boldsymbol{\nabla})\varrho_1 + (\varrho_\infty + \varrho_1)(\boldsymbol{\nabla} \cdot \mathbf{v}_1) = 0.$$

Neglecting second-order terms, one gets

$$\frac{\partial \varrho_1}{\partial t} + \varrho_\infty(\boldsymbol{\nabla} \cdot \mathbf{v}_1) = 0, \tag{9.147b}$$

and in terms of pressure,

$$\frac{\partial p_1}{\partial t} + a_\infty^2 \varrho_\infty(\boldsymbol{\nabla} \cdot \mathbf{v}_1) = 0, \tag{9.148}$$

where

$$\left(\frac{\partial p}{\partial \varrho}\right)_S = \left(\frac{\partial p_1}{\partial \varrho_1}\right)_S = a^2 = \left(a_\infty - \frac{\gamma - 1}{2} v_1\right)^2 \approx a_\infty^2.$$

Taking the divergence of Eq. (9.147a) and the derivative of Eq. (9.148) with respect to time, one obtains

$$\frac{\partial(\nabla \cdot \mathbf{v}_1)}{\partial t} + \frac{\nabla^2 p_1}{\varrho_\infty} = 0$$

and

$$\frac{\partial^2 p_1}{\partial t^2} + a_\infty{}^2 \varrho_\infty \frac{\partial}{\partial t} (\nabla \cdot \mathbf{v}_1) = 0.$$

Combining these two equations, one has

$$\frac{\partial^2 p_1}{\partial t^2} - a_\infty{}^2 \nabla^2 p_1 = 0. \tag{9.149}$$

This is a second-order linear partial differential equation for p_1—it is the *wave equation*. It is the same as the electromagnetic wave equation and shows that the pressure wave propagates with the speed of sound.

Similarly, $\mathbf{v}_1$ and $\mathbf{p}_1$ can be eliminated from Eqs. (9.147a), (9.147b), and (9.148) to give

$$\frac{\partial^2 \varrho_1}{\partial t^2} - a_\infty{}^2 \nabla^2 \varrho_1 = 0 \tag{9.150}$$

and

$$\frac{\partial^2 v_1}{\partial t^2} - a_\infty{}^2 \nabla^2 \mathbf{v}_1 = 0. \tag{9.151}$$

These are also wave equations, and they show that both the density and velocity variations follow the same wave pattern as does the pressure change, with a propagation speed equal to the same sound speed.

Because solutions of the three-dimensional wave equation are difficult, only one-dimensional propagation of plane waves will be considered here, the pressure equation being

$$\frac{\partial^2 p_1}{\partial t^2} - a_\infty{}^2 \frac{\partial^2 p}{\partial x^2} = 0. \tag{9.152}$$

The general form of the solution of this equation is

$$\frac{p_1}{p_\infty} = f_1(x - a_\infty t) + g_1(x + a_\infty t), \tag{9.153}$$

where f_1 and g_1 are arbitrary functions of their arguments.

Consider first the solution for $g_1 = 0$:

$$\frac{p_1}{p_\infty} = f_1(x - a_\infty t).$$

This represents a disturbance, or wave, which at time $t = 0$ had the arbitrary shape

$$\frac{p_1}{p_\infty} = f_1(x),$$

and which at time t has exactly the same shape but with corresponding points displaced a distance $a_\infty t$ to the right. Thus, the velocity of each point of the wave is a_∞, as is the wave velocity itself. If one takes $f_1 = 0$, then

$$\frac{p_1}{p_\infty} = g_1(x + a_\infty t),$$

and this represents a wave traveling to the left with velocity a_∞. In fact, any acoustic wave can be resolved into two simple waves—one going to the right and one to the left.

If the lines of constant p_1/p_∞ are plotted on an xt plane, as in Fig. 9.10, a family of straight lines with constant slopes, either $+a_\infty$ or $-a_\infty$, is obtained. The lines of constant $f_1(x - a_\infty t)$ and $g_1(x + a_\infty t)$ are called the characteristics of the wave equation (9.152).

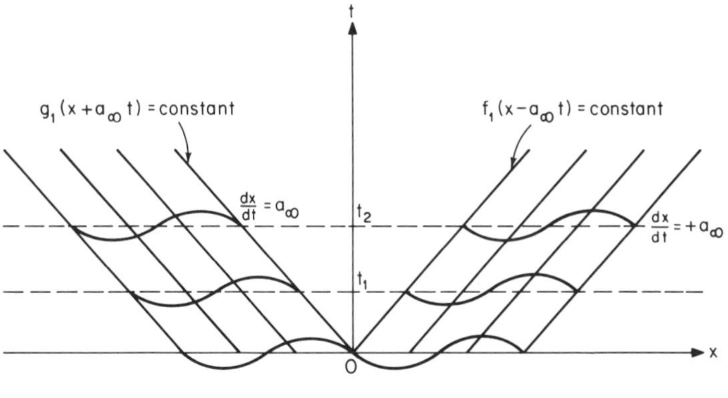

Fig. 9.10

In like manner, the wave solutions for the density ratio ϱ_1/ϱ_∞ and the velocity disturbance v_1 are

$$\frac{\varrho_1}{\varrho_\infty} = f_2(x - a_\infty t) + g_2(x + a_\infty t) \tag{9.154}$$

and

$$v_1 = f_3(x - a_\infty t) + g_3(x + a_\infty t). \tag{9.155}$$

9.6.1 The "Linearized" Shock Tube

The wave equations derived previously resulted from the linearizing of the equations of fluid motion, which restricts us to the propagation of waves of low intensity. Thus, the pressure difference across the diaphragm in the tube shown in Fig. 9.11 can only be small, so that when it is broken a weak wave motion results. In this sense it is an "acoustic" or "linearized" shock tube.

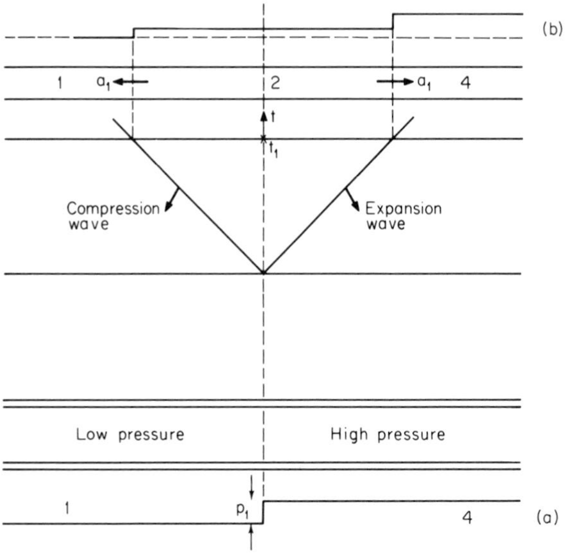

Fig. 9.11

For convenience, the pressure ratio (p_1/p_∞) will be designated as $s = s(x, t)$. At the instant of breaking the diaphragm, that is, $t = 0$, the wave shape is as shown in Fg. 9.11a. It is a step distribution in pressure. The equations then are

$$s(x, 0) = s_0 = f_1(x) + g_1(x) = \begin{cases} s_4, & x > 0, \\ 0, & x < 0, \end{cases} \tag{9.156}$$

$$v(x, 0) = 0.$$

Now

$$f_1 = g_1 = \tfrac{1}{2}s_0 = \begin{cases} \tfrac{1}{2}s_4, & x > 0, \\ 0, & x < 0. \end{cases} \tag{9.157}$$

At any subsequent time, the motion is given by

$$s(x, t) = \tfrac{1}{2}s_0(x - a_\infty t) + \tfrac{1}{2}s_0(x + a_\infty t)$$
$$= \begin{cases} s_4, & x > a_\infty t, \\ \tfrac{1}{2}s_4, & -a_\infty t < x < +a_\infty t, \\ 0, & x < -a_\infty t. \end{cases} \tag{9.158}$$

The pressure distribution is shown in Fig. 9.11b. A compression wave propagates into the low pressure side (1), and an expansion wave, of equal strength, into the high pressure side (2)—both with velocity a_∞.

9.6.2 Propagation of Plane Waves of Finite Amplitude—Formation of Shock Waves

If the assumption of infinitesimal amplitudes and gradients is removed, then the wave velocity will not be a constant and a simple wave will distort as it propagates. Special methods are required to solve the complete nonlinear equation of fluid motion. The reader may refer to more detailed texts on compressible flow for treatment of the method of characteristics which can be applied to solve such hyperbolic-type equations. The one-dimensional continuity and momentum equations are

$$\frac{\partial \varrho}{\partial t} + v\frac{\partial \varrho}{\partial x} + \varrho\frac{\partial v}{\partial x} = 0, \tag{9.159}$$

$$\frac{\partial v}{\partial t} + v\frac{\partial v}{\partial x} + \frac{1}{\varrho}\frac{\partial p}{\partial x} = 0. \tag{9.160}$$

To find a simple analytic solution of these equations, the following approach is adopted.

Now $\varrho = \varrho(v)$, and so these equations can be written

$$\frac{d\varrho}{dv}\frac{\partial v}{\partial t} + v\frac{d\varrho}{dv}\frac{\partial v}{\partial x} + \varrho\frac{\partial v}{\partial x} = 0, \tag{9.159a}$$

$$\frac{\partial v}{\partial t} + v\frac{\partial v}{\partial x} + \frac{1}{\varrho}\frac{dp}{d\varrho}\frac{d\varrho}{dv}\frac{\partial v}{\partial x} = 0. \tag{9.160a}$$

Dividing (9.159a) by $d\varrho/dv$ and combining the equations, one obtains

$$\frac{dv}{d\varrho} = \pm\frac{1}{\varrho}\left(\frac{dp}{d\varrho}\right)^{1/2} = \pm\frac{a}{\varrho}. \tag{9.161}$$

Using the isentropic relation

$$a^2 = \frac{dp}{d\varrho} = \gamma\left(\frac{p}{\varrho}\right)$$

and integrating Eq. (9.161), one obtains

$$v = \pm \int_{\varrho_\infty}^{\varrho} \left(\frac{dp}{d\varrho}\right)^{1/2} \frac{d\varrho}{\varrho} = \pm \frac{2}{\gamma - 1}(a - a_\infty). \tag{9.162}$$

This gives the relation between the particle velocity and the local speed of sound as

$$a = a_\infty \pm \frac{\gamma - 1}{2} v, \tag{9.163}$$

which was used in the derivation of Eq. (9.148).

Substituting Eq. (9.161) into Eqs. (9.159a) and (9.160a), one obtains, respectively,

$$\frac{\partial \bar{\varrho}}{\partial t} + (v \pm a)\frac{\partial \bar{\varrho}}{\partial x} = 0, \tag{9.164}$$

$$\frac{\partial v}{\partial t} + (v \pm a)\frac{\partial v}{\partial x} = 0, \tag{9.165}$$

where $\bar{\varrho} = (\varrho/\varrho_\infty) - 1$ and is a dimensionless quantity sometimes called the *condensation*. Now, along the characteristics,

$$v \pm a = \text{constant}, \tag{9.166}$$

and for characteristics, therefore, the solutions of (9.164) and (9.165), respectively, are

$$\bar{\varrho} = f_1(x - (v \pm a)t), \tag{9.167}$$

$$v = f_2(x - (v \pm a)t), \tag{9.168}$$

where f_1 and f_2 are arbitrary functions. These equations show that the disturbance is propagated at the *instantaneous* velocity $v + a$ or $v - a$, instantaneous because this velocity is now a function of time, in general. Substituting for a in Eq. (9.167) from Eq. (9.163), one therefore gets

$$\bar{\varrho} = f_1\left[x - \left(a_\infty + \frac{\gamma + 1}{2} v\right)t\right] = f_1(\xi). \tag{9.169}$$

The instantaneous velocity of propagation is thus

$$a_\infty + \left(\frac{\gamma - 1}{2}\right) v.$$

Suppose the initial condensation distribution is a sine function and that $x = 0$ at $t = 0$. Then

$$\bar{\varrho} = f_1(0) = 0, \tag{9.170a}$$

$$\bar{\varrho}_1 = f_1(x_1), \tag{9.170b}$$

$$\bar{\varrho}_2 = f_1(x_2) = 0. \tag{9.170c}$$

The distribution is shown in Fig. 9.12a.

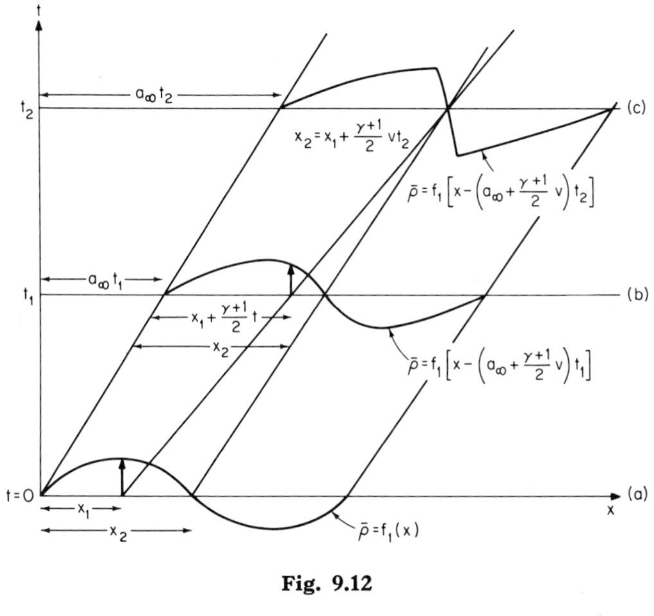

Fig. 9.12

These values of the condensation distribution can be maintained if the corresponding locations on the x axis now become

$$x = a_\infty t_1, \tag{9.171a}$$

$$x = x_1 + a_\infty t_1 + \frac{\gamma + 1}{2} v t_1, \tag{9.171b}$$

$$x = x_2 + a_\infty t_1. \tag{9.171c}$$

If the increment in velocity v is $\ll a_\infty$, then the situation is the case of the sound wave and the curve $\bar{\varrho} = f_1(\xi)$ does not change its shape as the disturbance propagates. But, if v is not small, then the value of $\bar{\varrho} = \varrho_1$ will no longer occur at $x_1 + a_\infty t_1$, but will appear further on at a distance $((\gamma + 1)/2)vt_1$. The points of zero $\bar{\varrho}$ remain at the same positions [see Eqs. (9.171a) and (9.171c)]. So the curve $\bar{\varrho} = f_1(\xi)$ changes its shape as it propagates. The resulting distortions in the condensation distribution are shown in Figs. 9.12b and 9.12c. The physical limit of the change in shape occurs when $d\bar{\varrho}/dx$ becomes infinite (otherwise the velocity and density would become multivalued functions of position).

The regions of higher condensation tend to overtake those of lower condensation. Before $d\bar{\varrho}/dx$ becomes infinite, the velocity and temperature gradients in the compression regions become so large that friction and heat transfer effects become significant. These have a diffusive action which opposes the steepening tendency and a balance is achieved—the compression part of the wave becomes "stationary" in so much as it propagates without further distortion. It is then known as a *compression* or *shock wave*.

Across this surface of discontinuity the velocity and all the thermodynamic properties change suddenly and the gas no longer undergoes isentropic changes.

Before discussing nonisentropic flow through a shock front, the Fanno and Rayleigh lines for steady compressible flow through a duct of constant cross section will be considered. These lines are extremely useful in gaining a physical understanding of nonisentropic flow.

9.7 Flow through a Constant-Area Tube with Heat Transfer

The case of flow through variable cross-sectional area tubes, with no frictional heat losses but with exchange of heat between the gas and the surroundings, has a basic relevance to the study of shock wave formation.

The flow situation involved here is reversible but nonadiabatic and thus nonisentropic. Heat exchangers and combustion chambers are examples of flow through tubes with heat transfer.

To simplify analysis it will be assumed that there is a uniform cross section, that the gas is perfect, and that there is no change in gas composition. The continuity and momentum equations for this situation are, respectively,

$$\varrho v = \text{constant}, \tag{9.172}$$

$$\varrho v\, dv = -dp. \tag{9.173}$$

The energy equation and the equation of state (for a perfect gas) may be written, respectively, as

$$dQ = C_p\, dT + p\, d\left(\frac{1}{\varrho}\right), \tag{9.174}$$

$$p = \varrho RT, \tag{9.175}$$

where dQ is the heat addition to the system.

Differentiating Eqs. (9.172) and (9.175), one obtains the relations

$$-\frac{d\varrho}{\varrho} = \frac{dv}{v}, \tag{9.176}$$

$$\frac{dp}{p} = \frac{R}{p}(\varrho\, dT + T\, d\varrho) = \frac{d\varrho}{\varrho} + \frac{dT}{T}. \tag{9.177}$$

From Eqs. (9.173) and (9.174),

$$-\frac{dp}{p} = \frac{\varrho}{p} v\, dv = \frac{v^2}{RT}\frac{dv}{v} = \gamma M^2 \frac{dv}{v}, \tag{9.178}$$

and

$$dQ = C_p\, dT - \frac{dp}{\varrho} = C_p\, dT + v\, dv. \tag{9.179}$$

Equations (9.176) and (9.179) are the basic equations for derivation of the differential equations of the velocity, density, pressure, temperature, and Mach number as a function of the heat addition.

Now, from Eqs. (9.176)–(9.178),

$$\frac{dT}{T} = \frac{dp}{p} - \frac{d\varrho}{\varrho} = (1 - \gamma M^2)\frac{dv}{v}, \tag{9.180}$$

and from (9.179),

$$\frac{dv}{v} = \frac{dQ}{v^2} - \frac{C_p\, dT}{v^2} = \frac{1}{M^2}\left(\frac{C_p}{\gamma RT}\right)\frac{dQ}{C_p} - \frac{1}{M^2}\frac{C_p}{\gamma RT}\, dT$$

$$= \frac{1}{(\gamma - 1)M^2}\left(\frac{dQ}{h} - \frac{dT}{T}\right) \tag{9.181}$$

or, using Eq. (9.180),

$$\frac{dv}{v} = \frac{1}{1 - M^2} \frac{dQ}{h}, \tag{9.182}$$

where h is the enthalpy of the fluid per unit mass.

Using Eq. (9.182) in Eqs. (9.176), (9.178), and (9.180), one gets the following differential relations for the density, pressure, and temperature:

$$\frac{d\varrho}{\varrho} = -\frac{1}{1 - M^2} \frac{dQ}{h}, \tag{9.183}$$

$$\frac{dp}{p} = -\frac{\gamma M^2}{1 - M^2} \frac{dQ}{h}, \tag{9.184}$$

$$\frac{dT}{T} = \frac{1 - \gamma M^2}{1 - M^2} \frac{dQ}{h}. \tag{9.185}$$

Finally, since

$$\frac{dM^2}{M^2} = \frac{d(v^2/a^2)}{v^2/a^2} = \frac{T}{v^2} d\left(\frac{v^2}{T}\right) = 2\frac{dv}{v} - \frac{dT}{T},$$

using Eqs. (9.182) and (9.180), one therefore obtains

$$\frac{dM^2}{M^2} = \frac{1 + \gamma M^2}{1 - M^2} \frac{dQ}{h} = \frac{1 + \gamma M^2}{1 - M^2} \frac{dS}{C_p}, \tag{9.186}$$

where dS is the increase in entropy.

Equations (9.182)–(9.186) can be used to explain the effect of heat addition on the thermodynamic and flow properties of flows in the subsonic and supersonic regions.

9.7.1 Heat Added to Subsonic Flow

$\frac{dv}{v}$ is positive, or v increases,

$\frac{d\varrho}{\varrho}$ is negative, or ϱ decreases,

$\frac{dp}{p}$ is negative, or p decreases,

$$\frac{dM^2}{M^2} \text{ is positive, or } M \text{ increases,}$$
$$\frac{dT}{T} \text{ is positive, or } T \text{ increases if } M^2 < \frac{1}{\gamma},$$
$$\frac{dT}{T} \text{ is negative, or } T \text{ decreases if } M^2 > \frac{1}{\gamma}.$$

In the Mach range, $1/\gamma < M^2 < 1$; therefore the heat added to the flow will result in an increase in the kinetic energy and the internal energy of the mean flow. Thus, the temperature will actually decrease when heat is added, provided M lies in the subsonic range.

As the internal energy increases, so will the entropy, and this is shown in the T–S diagram in Fig. 9.13.

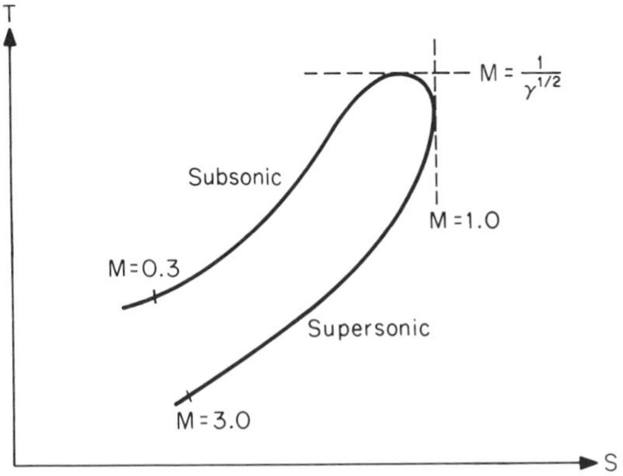

Fig. 9.13

9.7.2 Heat Added to Supersonic Flow

In this case,

$$\frac{dv}{v} \text{ is negative, or } v \text{ decreases,}$$
$$\frac{d\varrho}{\varrho} \text{ is positive, or } \varrho \text{ increases,}$$
$$\frac{dT}{T} \text{ is positive, or } T \text{ increases,}$$
$$\frac{dM^2}{M^2} \text{ is negative, or } M \text{ decreases.}$$

Thus, the supersonic flow is decelerated when heat is added.

It must be noted that Eqs. (9.182)–(9.186) break down when $M = 1$ for $dQ \neq 0$. All the variables, v, p, etc., have an infinite gradient at this point. This is indicated in the T–S diagram (Fig. 9.13). It implies that a subsonic flow in a constant cross-section tube cannot become supersonic, nor can a supersonic flow become subsonic (without some external influence). They both approach the sonic state in the limit.

9.8 The Equation of the Rayleigh Line

The curve shown in the T–S diagram is called the Rayleigh line. Consider the conditions in the duct at positions 1 and 2 between which an amount of heat Q has been added (Fig. 9.14). Then, using the energy equation for a perfect gas, one gets

$$Q = C_p(T_2 - T_1) + \frac{v_2^2 - v_1^2}{2} = C_p(T_{02} - T_{01}), \qquad (9.187)$$

where T_0 is a stagnation temperature.

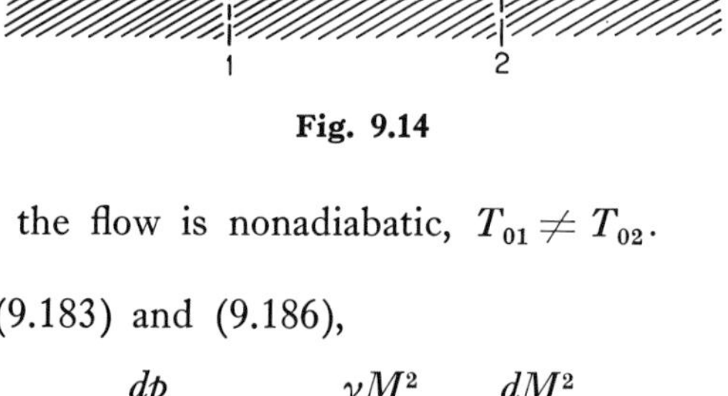

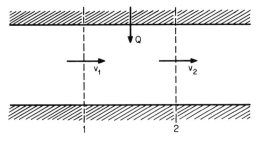

Fig. 9.14

Note Since the flow is nonadiabatic, $T_{01} \neq T_{02}$.

From Eqs. (9.183) and (9.186),

$$\frac{dp}{p} = -\frac{\gamma M^2}{1 + \gamma M^2} \frac{dM^2}{M^2}, \qquad (9.188)$$

and integrating between states 1 and 2, one obtains

$$\ln \frac{p_2}{p_1} = \ln \frac{1 + \gamma M_1^2}{1 + \gamma M_2^2}$$

or

$$\frac{p_2}{p_1} = \frac{1 + \gamma M_1^2}{1 + \gamma M_2^2}. \tag{9.189}$$

Now, from Eq. (9.120),

$$p_0 = p\left(1 + \frac{\gamma - 1}{2} M^2\right)^{\gamma/(\gamma-1)};$$

therefore

$$\frac{p_{02}}{p_{01}} = \frac{1 + \gamma M_1^2}{1 + \gamma M_2^2}\left[\frac{1 + \frac{\gamma - 1}{2} M_2^2}{1 + \frac{\gamma - 1}{2} M_1^2}\right]^{\gamma/(\gamma-1)}. \tag{9.190}$$

Now

$$\frac{T_2}{T_1} = \frac{M_2^2}{M_1^2}\left[\frac{1 + \gamma M_1^2}{1 + \gamma M_2^2}\right], \tag{9.191}$$

and it follows that

$$\frac{T_{02}}{T_{01}} = \frac{M_2^2}{M_1^2}\left(\frac{1 + \gamma M_1^2}{1 + \gamma M_2^2}\right)^2\left[\frac{1 + \frac{\gamma - 1}{2} M_2^2}{1 + \frac{\gamma - 1}{2} M_1^2}\right]. \tag{9.192}$$

Also, from Eqs. (9.189) and the ratio T_2/T_1,

$$\frac{\varrho_2}{\varrho_1} = \frac{v_1}{v_2} = \frac{p_2 T_1}{p_1 T_2} = \frac{M_1^2}{M_2^2}\left(\frac{1 + \gamma M_2^2}{1 + \gamma M_1^2}\right). \tag{9.193}$$

From Eq. (9.186),

$$dS = C_p\left(\frac{1 - M^2}{1 + \gamma M^2}\right)\frac{dM^2}{M^2},$$

and integrating between states 1 and 2, one obtains

$$S_2 - S_1 = \frac{\gamma R}{\gamma - 1}\ln\left[\frac{M_2^2}{M_1^2}\left(\frac{1 + \gamma M_2^2}{1 + \gamma M_1^2}\right)^{(\gamma+1)/\gamma}\right], \tag{9.194}$$

where

$$C_p = \frac{\gamma R}{\gamma - 1}.$$

Equations (9.191) and (9.194) can be used to construct the loci of the Mach number for Rayleigh flow in the T–S diagram. It is seen from Eq.

(9.194) that the entropy is maximized when $M = 1$. Because entropy increases with heat addition, it is not possible to pass from subsonic to supersonic flow or vice versa. The Second Law of Thermodynamics is rigorously obeyed.

9.9 The Fanno Line—Flow with Friction

In this case the flow of a fluid through a constant area tube with no heat addition (or work done) is considered. But there is friction with the wall of the tube. If the axis of the tube is the x axis, then the continuity equation is

$$\varrho v = \dot{m} = \text{constant}$$

or

$$\frac{d(\varrho v)}{dx} = 0. \tag{9.195}$$

The energy equation (no work done) is

$$\frac{d}{dx}\left(C_{\mathrm{p}}T + \frac{v^2}{2}\right) = 0. \tag{9.196}$$

From Eq. (9.195),

$$\frac{dv}{dx} = -\frac{\dot{m}}{\varrho^2}\frac{d\varrho}{dx}. \tag{9.197}$$

Substituting this in Eq. (9.196), one obtains

$$C_{\mathrm{p}}\frac{dT}{dx} - \frac{\dot{m}^2}{\varrho^3}\frac{d\varrho}{dx} = 0. \tag{9.198}$$

Integrating (9.198), with the index 1 indicating some reference state, and converting to enthalpy, one gets

$$h = h_1 - \frac{\dot{m}^2}{2}\left(\frac{1}{\varrho^2} - \frac{1}{\varrho_1{}^2}\right). \tag{9.199}$$

This is the equation of the Fanno line and relates enthalpy and density. A family of lines, for various flow rates ($\dot{m}$), can be drawn, as shown in Fig. 9.15. This is an adiabatic (no work done or heat transfer) but

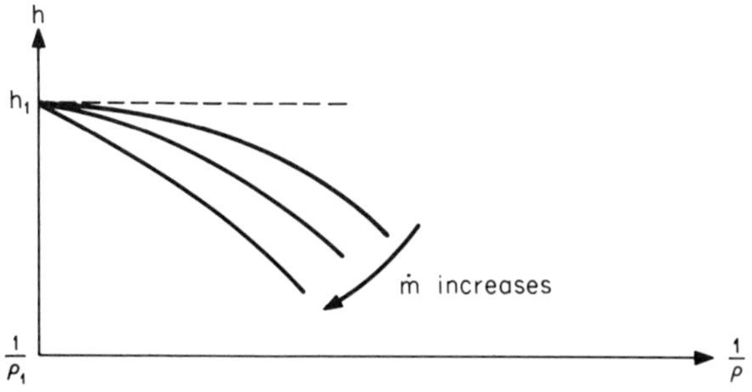

Fig. 9.15

irreversible process—irreversible because there is an increase in entropy caused by the friction with the walls. In such a process,

$$T\,dS = dh - \frac{dp}{\varrho} \tag{9.200}$$

or

$$dS = R\left[\frac{\gamma}{\gamma - 1}\,\frac{dT}{T} - \frac{dp}{\varrho}\right]. \tag{9.201}$$

Now

$$\frac{dT}{T} = (1 - \gamma M^2)\,\frac{dv}{v} \qquad \text{[see Eq. (9.180)].} \tag{9.202}$$

Also,

$$M^2 = \frac{v^2}{\gamma RT};$$

therefore

$$\frac{dM}{M} = \frac{dv}{v} - \frac{1}{2}\,\frac{dT}{T}.$$

Using Eq. (9.202) for dT/T, one obtains

$$\frac{dv}{v} = \frac{dM/M}{\dfrac{\gamma - 1}{2}M^2 + 1} = -\frac{d\varrho}{\varrho}; \tag{9.203}$$

therefore,

$$\frac{dT}{T} = \frac{(1 - \gamma)M\,dM}{\dfrac{\gamma - 1}{2}M^2 + 1} = \frac{-dM^2}{M^2 + \dfrac{2}{\gamma - 1}}. \tag{9.204}$$

Using these expressions in the differential form of the perfect gas law, one gets the differential pressure relation

$$\frac{dp}{p} = \frac{\left[-\dfrac{1}{M} + (1-\gamma)M\right] dM}{\left(\dfrac{\gamma-1}{2}\right)M^2 + 1} = -\left[\frac{1}{M} + \frac{M}{M^2 + \dfrac{2}{\gamma-1}}\right] dM. \tag{9.205}$$

Equations (9.204) and (9.205) can be used in Eq. (9.201) for dS:

$$dS = \frac{R}{\dfrac{\gamma-1}{2}M^2 + 1}\left[-\gamma M + \frac{1}{M} - (1-\gamma)M\right] dM \tag{9.206}$$

or

$$\frac{dS}{dM} = \frac{R(1-M^2)}{M\left(\dfrac{\gamma-1}{2}M^2 + 1\right)}. \tag{9.207}$$

Thus, S will be a maximum when $M = 1$:

$$\frac{dS}{R} = \left[\frac{1}{M} + \frac{\dfrac{\gamma+1}{2}M}{\dfrac{\gamma-1}{2}M^2 + 1}\right] dM,$$

and on integrating, one obtains

$$\frac{S}{R} = \ln M - \frac{1}{2}\left(\frac{\gamma+1}{\gamma-1}\right)\ln\left(M^2 + \frac{2}{\gamma-1}\right) + C \tag{9.208}$$

or

$$\frac{S-S_1}{R} = \ln \frac{M}{M_1}\left[\frac{1 + \dfrac{\gamma-1}{2}M_1^2}{1 + \dfrac{\gamma-1}{2}M^2}\right]^{(\gamma+1)/(\gamma-1)}, \tag{9.209}$$

the change of entropy from some reference state S_1.

Equations (9.209) and (9.191) can be used to construct a T–S (or h–S) diagram for various Mach numbers. The general shape of the Fanno line is shown in Fig. 9.16.

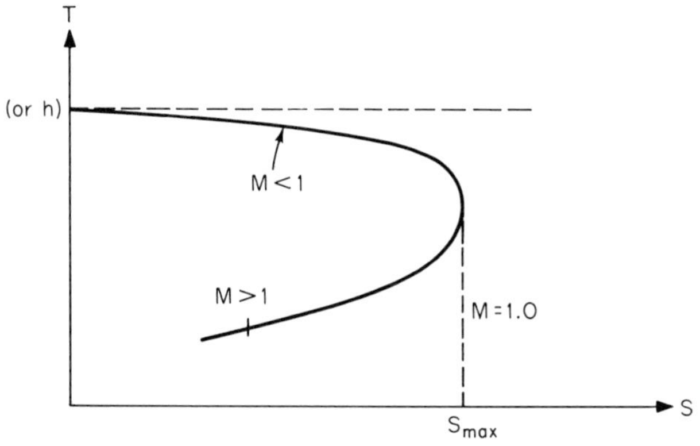

Fig. 9.16

When the entropy S becomes a maximum, $M = 1$. It is seen that the effect of friction is to accelerate a flow when it is subsonic and to decelerate it when it is supersonic. If there is enough friction in the tube, the flow Mach number in both cases will approach the limit value of 1. At this point the flow is said to be choked. In the supersonic case a shock wave would be formed before choking occurs.

Again, as in the Rayleigh case, for adiabatic constant-area conditions, a subsonic flow can never become supersonic and, in the absence of a discontinuity, a supersonic flow cannot become subsonic.

9.10 Shock Waves

A one-dimensional normal shock is essentially a plane surface of discontinuity normal to the direction of flow.

The thickness of the normal shock is usually very small—for monatomic gases it is of the order of a few mean free paths of a molecule. Actually, complex irreversible dissipative phenomena occur in the interior of the shock. But for practical purposes one is interested only in the net changes in fluid properties across the shock.

Across the shock wave, the pressure, density, and temperature rise, while the flow speed relative to the shock wave decreases.

In the case of a supersonic flow past a wedge (or a cone), the shock waves which form can either be attached or detached from the wedge, as shown in Fig. 9.17. They are oblique shock waves.

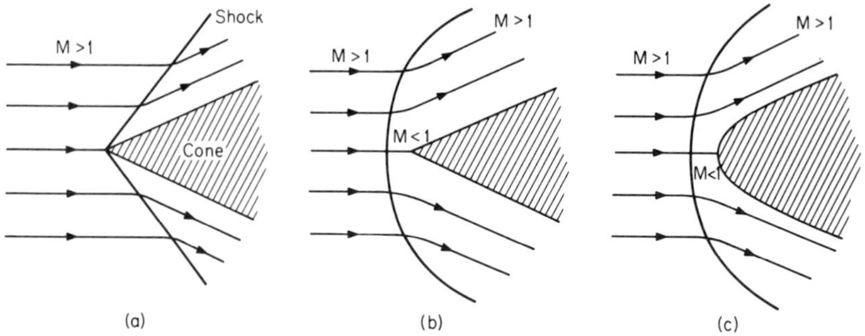

Fig. 9.17 (a) Attached shock. (b) Detached (bow) shock. (c) Detached shock, blunt body.

The Mach number of the flow ahead of a stationary shock is *always* larger than unity (that is, the flow is supersonic relative to the shock). Downstream of the shock the flow can be either supersonic or subsonic, depending on the angle between the upstream flow velocity and the line normal to the shock wave. Downstream of a normal shock the flow is always subsonic relative to the shock.

Whether the shock wave is attached or detached from a wedge (or cone) depends on the wedge (or cone) angle and the Mach number of the undisturbed flow ahead. When the solid body is blunt-nosed, the shock wave is always detached. Detached shocks are usually called *bow shock waves*.

9.10.1 Normal Shock Wave in a Duct

Consider a shock wave moving through a stationary gas with a velocity v_1 in a direction normal to the shock front (Fig. 9.18).

Conditions in front of the shock will be denoted by the index 1 and

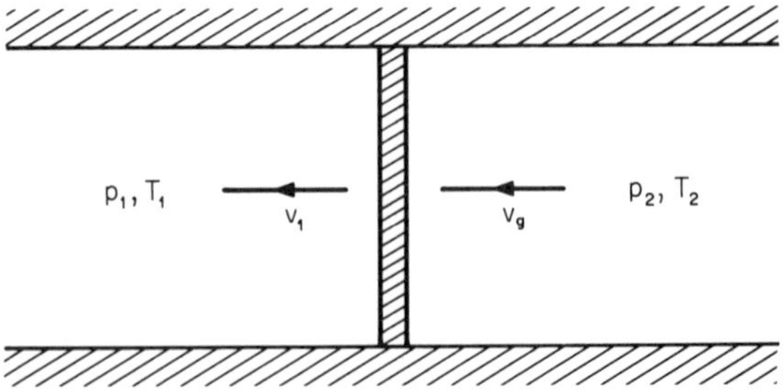

Fig. 9.18

those behind by the index 2. The gas velocity behind the front will be denoted by v_g.

A coordinate system is chosen which moves with velocity v_1 to the right (Fig. 9.19). In this frame the shock front is at rest and the gas behind

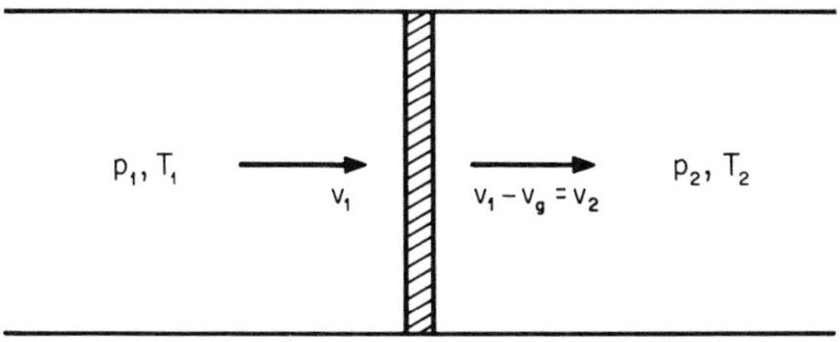

Fig. 9.19

the front has a velocity $v_2 = v_1 - v_g$. The conditions behind the shock can be determined from the various conservation laws. But first it is useful to note that points representing states at sections 1 and 2 must lie at the intersections of the Fanno and Rayleigh lines drawn on an h–S diagram (Fig. 9.20). Both lines are constructed to pass through point 1,

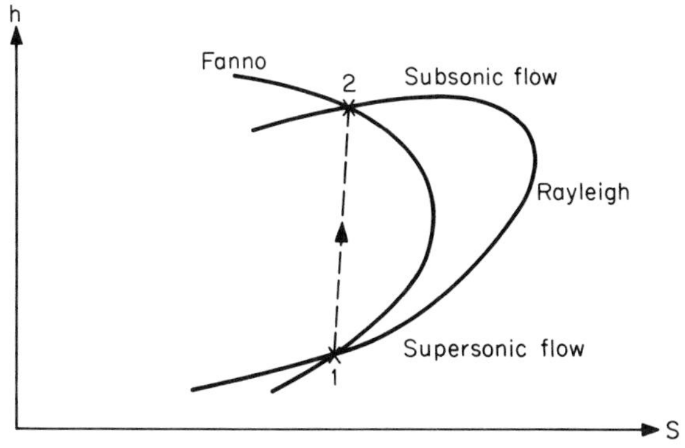

Fig. 9.20

which represents the known fluid state at section 1. Points on the Fanno line represent various possible fluid states in an adiabatic flow which starts with the fluid state corresponding to section 1, whereas points on the Rayleigh line represent various fluid states in a flow with no wall friction which starts with the same fluid state. As the flow through a shock is both adiabatic and with no friction, the fluid state at section 2 must be at the second intersection of the Fanno and Rayleigh lines.

Thus, in all compressible flows, if 1 lies on the supersonic branch, 2 appears on the subsonic one. Moreover, point 2 is always to the right of point 1. This leads to the following conclusions:

(a) from the second law of thermodynamics, the flow passing a normal shock must always proceed from point 1 to point 2, since the entropy of an irreversible process must increase;
(b) a normal shock can occur only in supersonic flow.

9.10.2 Basic equations

These have been written down previously and will be requoted here. The pressure ratio across the shock [Eq. (9.140)] is

$$\frac{p_2}{p_1} = 1 + \frac{2\gamma}{\gamma+1}(M_1^2 - 1), \tag{9.210}$$

and the density ratio [Eq. (9.139)] is

$$\frac{\varrho_2}{\varrho_1} = \frac{(\gamma+1)M_1^2}{(\gamma-1)M_1^2+2}. \tag{9.211}$$

Equations (9.210) and (9.211) are one form of the Rankine–Hugoniot relations governing the change of state across a shock wave. The velocity, or Prandtl, relation [Eq. (9.134)] was

$$v_1 v_2 = a^{*2}, \tag{9.212}$$

where a^* is the critical sonic velocity. It follows from this equation that if the flow is supersonic in front of a normal shock wave ($M_1 > 1$), then the flow behind must be subsonic ($M_2 < 1$).

Finally, the entropy change across the shock is [Eq. (9.144)]

$$\frac{S_2 - S_1}{R} = \ln\left[1 + \frac{2\gamma}{\gamma+1}(M_1^2 - 1)\right]^{1/(\gamma-1)}\left[\frac{(\gamma+1)M_1^2}{(\gamma-1)M_1^2+2}\right]^{-\gamma/(\gamma-1)}. \tag{9.213}$$

The entropy change is positive only if $M_1 > 1$, again indicating that a normal shock can occur only in supersonic flow.

9.10.3 Oblique Shock Waves

An oblique shock, as described previously, is formed when supersonic flow approaches a shock at an angle β, $\beta \neq \pi/2$ (Fig. 9.21). The coordinate system is attached to the body and so the shock appears stationary in the moving fluid. The angle δ is called the turning, or wedge, angle.

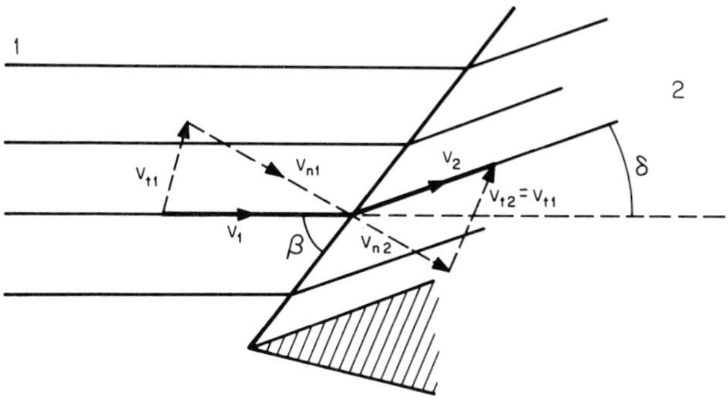

Fig. 9.21

Again, let the parameters upstream of the shock be designated by the index 1 and those downstream by the index 2.

The velocity v is resolved in both cases into tangential and normal components, v_t and v_n, respectively. The relations previously established for the normal shocks will, of course, hold for the normal component v_n. Moreover, the tangential component v_{t1} does not change; $v_{t1} = v_{t2}$. Since $v_{n2} < v_{n1}$, the final velocity v_2 must turn toward the shock (through the angle δ).

The Rankine–Hugoniot relations are

$$M_{1n} = \frac{v_{1n}}{a_1} = \frac{v_1}{a_1} \sin \beta = M_1 \sin \beta, \tag{9.214}$$

and therefore

$$\frac{p_2}{p_1} = \frac{2\gamma}{\gamma + 1} M_1^2 \sin^2 \beta - \frac{\gamma - 1}{\gamma + 1}, \tag{9.215}$$

$$\frac{\varrho_2}{\varrho_1} = \frac{(\gamma + 1)M_1^2 \sin^2 \beta}{(\gamma - 1)M_1^2 \sin^2 \beta + 2}. \tag{9.216}$$

The strength of an oblique shock is given by

$$\frac{\Delta p}{p_1} = \frac{p_2 - p_1}{p_1} = \frac{2\gamma}{\gamma + 1} (M_1^2 \sin^2 \beta - 1). \tag{9.217}$$

The entropy change is given by

$$\frac{S_2 - S_1}{R} = \frac{1}{\gamma - 1} \ln\left[\frac{2\gamma}{\gamma + 1} p_1{}^2 \sin^2 \beta - \frac{\gamma - 1}{\gamma + 1}\right] \times \left[\frac{(\gamma - 1)M_1{}^2 \sin^2 \beta + 2}{(\gamma + 1)M_1{}^2 \sin^2 \beta}\right]^{\gamma}, \tag{9.218}$$

where some slight rearrangement has been made from Eq. (9.144). The existence of an oblique shock, then, requires that

$$M_1 \sin \beta > 1.$$

An oblique shock becomes a normal one when $\beta = \pi/2$, and so the range of possible wave angles is

$$\sin^{-1} \frac{1}{M_1} < \beta < \frac{\pi}{2}. \tag{9.219}$$

The phenomena of oblique shocks are useful in studying supersonic flows over corners. The oblique shock formed in a concave corner of angle δ is shown in Fig. 9.22.

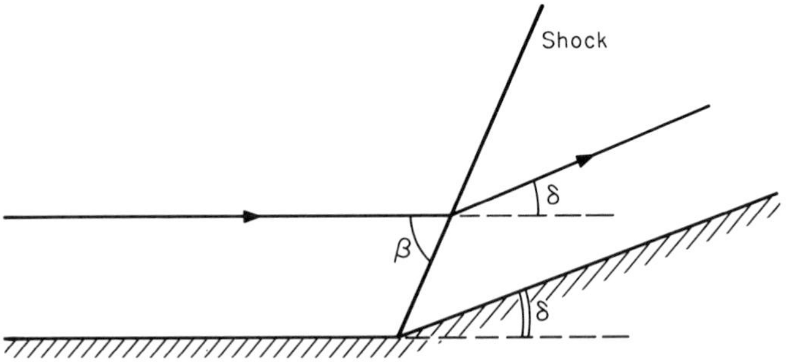

Fig. 9.22

The fluid density, pressure, and temperature will be increased as the fluid flows through the shock—it is compressed nonisentropically.

When the concave corner is a smooth and continuous curve, the individual Mach lines converge to form a shock some distance off the surface. In this case, the fluid is compressed isentropically (Fig. 9.23). At a convex corner, the Mach lines diverge and the fluid expands. Thus,

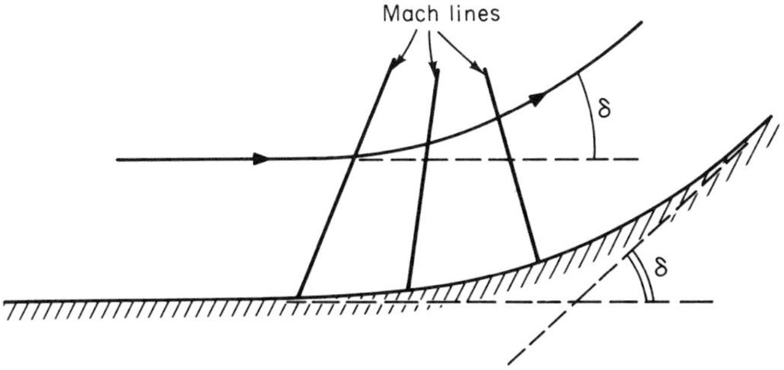

Fig. 9.23

no oblique shock can be formed in an expanding supersonic flow (Fig. 9.24). Such oblique expansion waves are called *Prandtl–Meyer expansions*.

The reader is referred to other textbooks for more detailed treatments of shock wave phenomena, such as the formation of detached shocks, shock polars, compression, and expansion waves.

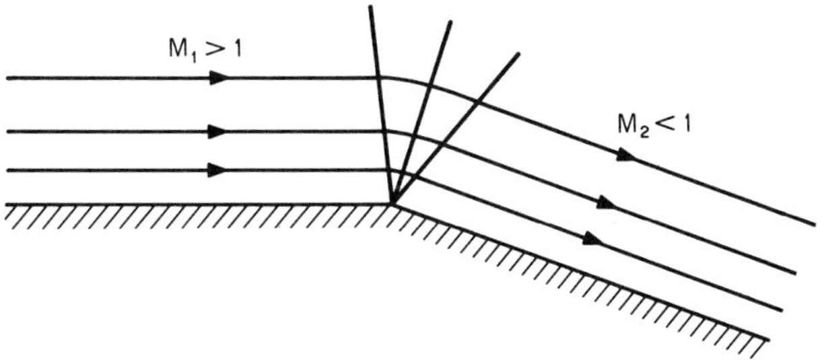

Fig. 9.24

9.11 The Use of Shock Tubes for the Study of Kinetics in Chemical Physics

A strong shock wave can raise the temperature of a gas to several thousand degrees Kelvin. At such temperatures, molecular vibrations are excited in the gas, molecular dissociation may be enhanced, chemical reactions induced, and even ionization produced.

When the rate of the gas dynamic process is very high (as in shock compression), insufficient time is available for the establishment of thermodynamic equilibrium and the gas particles remain essentially in a state of nonequilibrium. Gas kinetics then becomes a predominant factor.

Such a situation arises in practice in the entry of space vehicles into the earth's atmosphere, in supersonic flows in high power jet engine combustion chambers, and in strong explosions.

In contrast to thermodynamic parameters (which can be calculated theoretically), most gas kinetic information (such as cross sections for various types of molecular collisions or rates of various chemical reactions) is obtained by experiment. The shock tube is a widely used device for obtaining high temperatures in the laboratory and for studying the chemical physics of gases.

The use of controlled shocks can heat the gas (or mixture) to the required temperature. Following the rapid heating, various processes take place in the hot gas, such as molecular vibrations and the others already mentioned. Their relative importance and rates depend on the temperature and density of the gas. The nonequilibrium layer behind the compression shock is where the relaxation processes occur, and this is the layer studied experimentally. The density and temperature distributions in the relaxation layer can be theoretically related to the reaction rates. In some cases it is possible to record the reaction kinetics directly.

The reader is referred to the literature for details on the design and operation of shock tubes and the associated methods for the measurement of the various quantities.

PROBLEMS

9.1 Prove that for a perfect gas the isentropic equation is

$$pv^{\gamma} = C,$$

where γ is the specific heat ratio and C is a constant.

9.2 (a) For a perfect gas undergoing isothermal changes, show that the speed of sound is

$$a = (RT)^{1/2}.$$

(b) For a liquid with bulk modulus φ and density ϱ, show that the speed of sound is

$$a = \left(\frac{\varphi}{\varrho}\right)^{1/2}.$$

9.3 Illustrate the propagation of a sound wave in terms of the relative pressure change.

9.4 Derive the energy equation for a fluid in which the energy variations are due only to the work done by pressure forces (no body forces) in the form

$$\frac{\partial}{\partial t}\left[\varrho(\tfrac{1}{2}v^2 + E)\right] + \nabla \cdot \left[p\mathbf{v}\left(\tfrac{1}{2}v^2 + E + \frac{p}{\varrho}\right)\right] = 0,$$

where $\mathbf{v}$ is the fluid velocity, p the fluid pressure, and E the external energy per unit mass.

Show that for a polytropic fluid and for a steady flow

$$\tfrac{1}{2}v^2 + C_p T = \text{constant}$$

along a streamline where T is the fluid temperature and C_p the specific heat at constant pressure.

Also show that if heat conduction is important, a term $-\nabla \cdot (-k\,\nabla T)$, where k is the coefficient of thermal conductivity, must be added to the energy equation. If inertial terms are negligible with respect to heat conduction terms, then show that this equation becomes

$$\varrho C_p \frac{\partial T}{\partial t} + \varrho C_p \mathbf{v} \cdot \nabla T - \nabla \cdot (k\,\nabla T) = 0.$$

9.5 Prove that for an isentropic flow

(a) $$a^{*2} = \left(\frac{2}{\gamma + 1}\right)^{1/2} a^2 + \left(\frac{\gamma - 1}{\gamma + 1}\right)v^2,$$

where a^* is the critical velocity, and

(b) $$M^* = \left(\frac{(\gamma + 1)M^2}{(\gamma - 1)M^2 + 2}\right)^{1/2},$$

where $M^* = v/a^*$.

9.6 The Rankine–Hugoniot relation for normal shock waves is

$$\frac{p_2 - p_1}{\varrho_2 - \varrho_1} = \gamma\left(\frac{p_1 + p_2}{\varrho_1 + \varrho_2}\right).$$

Show that this can be written in the form

$$\tfrac{1}{2}(p_1 + p_2)\left(\frac{1}{\varrho_1} - \frac{1}{\varrho_2}\right) = -\frac{1}{\gamma - 1}\left(\frac{p_1}{\varrho_1} - \frac{p_2}{\varrho_2}\right).$$

9.7 Describe three quantities that are continuous through a steady normal plane shock, and hence deduce the Prandtl relation

$$v_1 v_2 = a^{*2},$$

where v_1 and v_2 are the velocities with which the gas enters and leaves the shock front and a^* is the critical (or sonic) velocity.

9.8 Starting from the one-dimensional continuity and momentum equations, derive the relation between the particle velocity and the local speed of sound.

Show that the solutions of these equations can be written as

$$\bar{\varrho} = f_1(x - (v \pm a)t), \qquad v = f_2(x - (v \pm a)t),$$

where $\bar{\varrho}$ is the condensation.

Explain the significance of the term $(v \pm a)$ in relation to the original hyperbolic equations.

9.9 Air enters a constant-area pipe with a velocity of 200 m/sec, at a temperature of 80°C, and a pressure of 1.0 MN/m². If 200 kJ/kg of heat is added to the pipe, determine (i) the final stagnation temperature, (ii) the final temperature, (iii) the final pressure, (iv) the final Mach number, and (v) the final velocity.

9.10 Air at a Mach number of 4.0 is decelerated in an insulated pipe (of inside diameter 100 mm) to $M = 2.0$. If the friction constant is 0.003, find the length of the pipe over which the deceleration occurs.

9.11 Construct the temperature–entropy diagram

(a) for air undergoing Fanno flow at a mass flow of 40 kg/sec, which is discharged from a reservoir at standard conditions;

(b) for air undergoing Rayleigh flow when the mass flow is 6 kg/sec and the reservoir is at standard conditions.

9.12 The Rankine–Hugoniot relations for normal shocks are given by

$$\frac{p_2}{p_1} = \frac{\dfrac{\gamma+1}{\gamma-1}\dfrac{\varrho_2}{\varrho_1} - 1}{\dfrac{\gamma+1}{\gamma-1} - \dfrac{\varrho_2}{\varrho_1}} = \frac{1 + \dfrac{\gamma+1}{2}\,\dfrac{\varrho_2 - \varrho_1}{\varrho_1}}{1 - \dfrac{\gamma-1}{2}\,\dfrac{\varrho_2 - \varrho_1}{\varrho_1}}.$$

Show that

(a) for weak shocks

$$\frac{\Delta \varrho}{\varrho_1} \doteqdot \frac{1}{\gamma} \frac{\Delta p}{p_1}, \qquad \frac{\Delta T}{T_1} \doteqdot \left(\frac{\gamma - 1}{\gamma}\right) \frac{\Delta p}{p_1};$$

(b) for strong shocks

$$\frac{\varrho_2}{\varrho_1} \to \left(\frac{\gamma + 1}{\gamma - 1}\right), \qquad \frac{T_2}{T_1} \to \left(\frac{\gamma - 1}{\gamma + 1}\right) \frac{p_2}{p_1}.$$

9.13 In an oblique shock wave the incident velocity is U along Ox, the angle the incident stream makes with the shock front is β, and u, v are the velocity components behind the front. Show that

$$Uu = a^{*2} + \frac{2}{\gamma + 1} U^2 \cos^2 \beta, \qquad Uv = U(U - u) \cot \beta,$$

where a^* is the sonic velocity.

9.14 Starting from the equation of motion, derive the Bernoulli equation in the form

$$1 + \tfrac{1}{2}(\gamma - 1)M^2 = \frac{a_0^2}{a^2}$$

for the irrotational steady adiabatic flow of a gas, with no body forces and when pressure and density are related by $p = C\varrho^\gamma$; $a = (\gamma p/\varrho)^{1/2}$ is the local speed of sound, a_0 the value of a when the fluid speed is zero, and $M = v/c$ the Mach number.

Such a gas is flowing down a convergent tube. Show that the fluid speed increases or decreases according as $M <$ or $>$ unity.

9.15 A wedge with a total angle of 30° is introduced into a flow with Mach number 3.0. What wave angles can the shock have if the incident Mach number is given by

$$\frac{1}{M_1^2} = \sin^2\beta - \frac{\gamma + 1}{2} \left(\frac{1}{1 + \cot \beta \cot \vartheta}\right),$$

where ϑ is the wedge angle, β is the shock angle. [Use the inequality in Eq. (9.217).]

BIBLIOGRAPHY

R. Pao, "Fluid Dynamics," Chapter 9. Merrill, Columbus, Ohio, 1967.

K. Oswatitisch, "Gas Dynamics." Academic Press, New York, 1956.

H. Liepman and A. Roshko, "Elements of Gas Dynamics." Wiley, New York, 1957.

S. W. Yuan, "Foundations of Fluid Mechanics." Chapters 11 and 12. Prentice-Hall, Englewood Cliffs, New Jersey, 1970.

H. Schlichting, "Boundary Layer Theory," Chapter 13. McGraw-Hill, New York, 1960.

W. Elmore and M. Heald, "The Physics of Waves," Chapter 5. McGraw-Hill, New York, 1969.

Y. Zel'dovich and Y. Raizer, "Physics of Shock Waves and High Temperature Hydrodynamic Phenomena," Vol. 1. Academic Press, New York, 1966.

CHAPTER TEN *PARTICLE FLUID DYNAMICS*

Only single-phase fluid flow has been dealt with so far. When a particulate phase is also present, then one has a multiphase system. If the fluid is a gas, the particulate phase may consist of solid particles, liquid droplets, or both. If it is a liquid, the particulate phase may consist of solid particles, gas bubbles, or liquid droplets which do not mix with the fluid phase.

Examples of practical multiphase systems are as follows:

(i) *Gas–solid particle* Dust collectors, fluidized beds, metallized propellant rockets, cosmic dusts, and nuclear fallout problems.

(ii) *Gas–liquid droplets* Atomizers, rocket engine injectors, clouds, factory stack effluents and evaporators.

(iii) *Liquid–solid particles* Fluidized beds, sedimentation, and factory waste products.

Studies of the dynamics of multiphase systems have fallen into two main categories:

(a) the dynamics of single particles is established and extended to a multiple-particle system in an analogous way to that used in molecular kinetic theory based on the Boltzmann equation;

(b) modification of the continuum mechanics of single-phase fluids so as to account for the presence of the particle "cloud."

The problem is immensely difficult and it is only in the last few years that any sort of acceptable general treatment has emerged—and this only by considerable idealization of the situation. For example, only particles of uniform size and perfectly rigid spherical shape are considered, dispersed in a viscous incompressible fluid. The particle "cloud" is regarded as a continuum and its interaction with the fluid phase is analyzed. Thus, approach (b), described above, is followed. It is in this context that a discipline known as *particle fluid dynamics* is at present in the process of formulation.

10.1 The Particle Continuum

Consider solid particles in a gas. When they are metallic, the ratio of the mass density of the solids to the mass density of the gas at standard conditions, ϱ_s/ϱ, is of the order of 10^3. Interest generally lies in the range in which the total mass content of the particles is of the same order as the total mass content of the gas in a unit volume of the mixture—that is, $\varrho_p/\varrho = O(1)$. Then the ratio of the total volume occupied by the particles to that of the gas in the unit volume of mixture is of the order of 10^{-3}. So one can speak of a quantity in terms of per unit volume of gas, as the solid volume is negligible. When $\varrho_p/\varrho = O(1)$, if all the particles are of radius 0.1 micron, then the number of particles in a cubic millimeter is about 10^5, and the interparticle distance is about 10^{-2} mm. When $r_p = 10$ microns, these values are about 10^3 and 10^{-1}, respectively. So in the range $0.1 \lesssim r_p \lesssim 10$, one has a macroscopic "point" of the order of a fraction of a millimeter over which an average quantity of the particle cloud can be suitably defined. At the same time, the interparticle distance is large enough compared to the particle size so that the particle–particle interaction (collisions) is small compared to the particle–fluid interaction. The latter interaction is a continuous one since the mean free path of a gas at standard conditions is about 5×10^{-5} mm. So the particle cloud can be regarded as a continuum.

To understand the complicated phenomenon of a cloud of solid particles in a viscous fluid, it is essential that the phenomenon involving a single particle first be fairly well understood, and this must now be examined.

10.2 Forces on a Rigid Spherical Particle in a Fluid

10.2.1 Drag Coefficient

From Newton's experiments in 1710, and later observations, the magnitude of the drag force on spheres in steady motion of a viscous fluid was given as

$$\mathscr{D} = 0.22\pi a^2 \varrho v^2, \tag{10.1}$$

where v is the relative velocity between particle and fluid, a is the particle radius, and ϱ is the fluid density. This relation is for large values of v, for which inertial effects are important.

Stokes [1], in 1850, suggested that at very low relative velocities the inertial effects are so small that they can be omitted from the Navier–Stokes equations. Under these conditions, the total drag on a sphere has been derived in Chapter 7 to be

$$\mathscr{D} = 6\pi a \mu v, \tag{10.2}$$

where μ is the viscosity of the fluid. This is known as the Stokes formula for the drag of a sphere.

The drag coefficient of a sphere is given by

$$C_{\mathrm{D}} = \frac{\mathscr{D}}{(\pi a^2)(\frac{1}{2}\varrho v^2)}, \tag{10.3}$$

and so for the Stokes law regime (small Reynolds numbers)

$$C_{\mathrm{D}} = \frac{24}{(2av\varrho/\mu)} = \frac{24}{N_{\mathrm{Re}}}, \tag{10.4}$$

where

$$N_{\mathrm{Re}} = (2a)v\frac{\varrho}{\mu} \tag{10.5}$$

is the Reynolds number for a sphere, based on its diameter (N_{Re} must be <1 for the Stokes law regime).

For the Newton regime,

$$C_{\mathrm{D}} = 0.44. \tag{10.6}$$

Oseen improved the Stokes relation by taking inertial-force terms partly into consideration. The improved equation for the drag coefficient was

$$C_D = \frac{24}{\text{Re}}\left(1 + \frac{3}{16}\,\text{Re}\right); \tag{10.7}$$

C_D versus Re is plotted in Fig. 10.1 for the Stokes and Oseen relations and for the experimentally observed results. It is seen that Oseen's formula is good up to a Reynolds number of about 5.

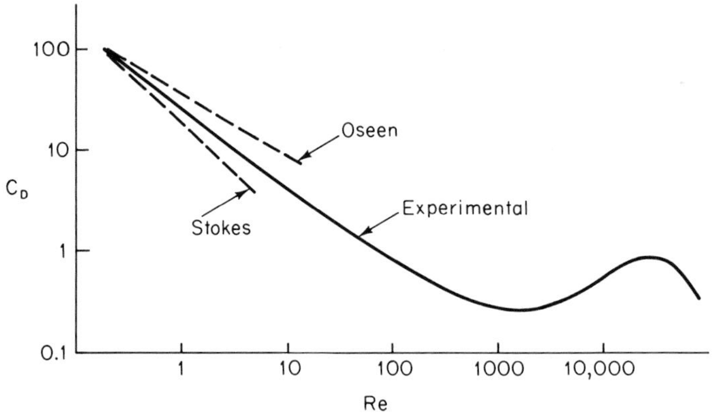

Fig. 10.1

At N_{Re} of about 10, separation of the boundary layer on the sphere occurs. The streamlines in the wake curl up to form a stationary vortex ring at the rear of the sphere. Further increase of N_{Re} leads to growth of the vortex in size and strength. In a laminar stream at $N_{Re} \sim 500$, the vortex system separates from the body and a wake is formed. The vortex rings continue to form and shed from the sphere, producing an instantaneous wake drag force. A very complex situation now arises, especially when dealing with a cloud of particles. So only Re < 5 will be considered here.

10.2.2 Transverse Forces on a Particle

A particle can experience two kinds of transverse force due to the presence of a velocity gradient in the fluid, such as the shear layer near a wall.

(i) *Magnus Force* The velocity gradient can cause a solid particle to rotate. At low Reynolds numbers rotation causes fluid entrainment, increasing the velocity on one side of the body and lowering the velocity on the other side. This is known as the Magnus effect and tends to move the particle toward the region of higher velocity [2]. This would suggest that particles tend to migrate to the center of the pipe in Poiseuille flow. However, Segre and Silberberg [3, 4] showed by experiment that particles tend to concentrate in an annular region at about 0.6 times the pipe radius from the axis. Their experiments were carried out in a glass tube of 11.2 $\pm$ 0.2 mm internal diameter, using polymethylmethacrylate spheres of diameters close to 0.32, 0.8, 1.21, and 1.71 mm in a media of equal density. The particle concentration ranged from 0.33 to 4 particles per cubic centimeter (this hardly satisfies the continuum requirement, however). The concentration distributions were determined by optical scanning.

Rubinow and Keller [5] computed the Magnus lift force on a rotating sphere as

$$\mathfrak{F}_{\mathrm{L}} = \pi a^3 \varrho \boldsymbol{\omega} \times \mathbf{v}[1 + O(N_{\mathrm{Re}})], \tag{10.8}$$

where $\mathbf{v}$ is the fluid velocity and $\boldsymbol{\omega}$ is the angular velocity of the sphere. The torque on the sphere was determined to be

$$\mathfrak{T} = -8\pi\mu a^3 \boldsymbol{\omega}[1 + O(N_{\mathrm{Re}})]. \tag{10.9}$$

The equations of motion were determined also and are

$$\begin{aligned} m\frac{d\mathbf{v}}{dt} &= -6\pi\mu a[1 + \tfrac{3}{8} N_{\mathrm{Re}}]\mathbf{v} + \pi a^3 \varrho \boldsymbol{\omega} \times \mathbf{v}, \\ I\frac{d\boldsymbol{\omega}}{dt} &= -8\pi\mu a^3 \boldsymbol{\omega}, \end{aligned} \tag{10.10}$$

where m is the mass of the sphere and I is the moment of inertia of the sphere.

(ii) *Slip-Shear Force* Saffman [6] showed that when a particle is in a fluid velocity gradient it experiences a transverse lift force even when it is not rotating. Saffman derived the net force on a small translating sphere which is simultaneously rotating in an unbounded, uniform simple shear flow field. The translation is along the streamlines.

Three independent particle Reynolds numbers arise in the analysis;

$$\text{shear:} \qquad (\mathrm{Re})_k = \frac{4a^2k}{\nu}, \tag{10.11}$$

$$\text{rotation:} \quad (\mathrm{Re})_\Omega = \frac{4a^2\Omega}{\nu}, \tag{10.12}$$

$$\text{slip:} \qquad (\mathrm{Re})_\mathrm{p} = \frac{2a(u_\mathrm{p} - u)}{\nu}, \tag{10.13}$$

where the particle relative (slip) velocity $(u_\mathrm{p} - u)$ is measured at its center, k is the magnitude of the velocity gradient ($|\,\partial u/\partial y\,|$, say), and Ω is the magnitude of the angular velocity of the particle.

The analysis showed that *in addition* to the Stokes drag force

$$\mathscr{D} = 6\pi\mu a(u_\mathrm{p} - u), \tag{10.14}$$

the particle experiences a transverse force given by

$$\mathfrak{F}_\mathrm{T} = 81.2\, a^2(\varrho\mu)^{1/2}k^{1/2}(u_\mathrm{p} - u), \tag{10.15}$$

which is due to the combination of slip and shear, and also a lift force due to the rotation given by

$$\mathfrak{F}_\mathrm{L} = \pi\nu a^3\Omega(u_\mathrm{p} - u). \tag{10.16}$$

The latter is the same result as obtained by Rubinow and Keller for a rotating sphere. The analysis is valid when

$$(\mathrm{Re})_\mathrm{p},\ (\mathrm{Re})_k,\ (\mathrm{Re})_\Omega \ll 1 \qquad \text{and} \qquad (\mathrm{Re})_k,\ (\mathrm{Re})_\Omega \gg (\mathrm{Re})_\mathrm{p}^{\,2}.$$

Saffman's analysis showed, however, that unless the angular velocity of the particle is much greater than the rate of shear, and for a freely rotating particle $\omega = \frac{1}{2}(\partial u/\partial y)$, the lift force due to particle rotation is less by an order of magnitude than that due to slip-shear when the Reynolds number is small.

Thus, under normal circumstances the Magnus effect can be ignored. This is not true in particles in a jet emerging from an orifice, where due to friction at the walls the particles emerge with initially high angular velocities and this can change their subsequent trajectories significantly [7]. One notes also that the slip-shear force tends to move the particle to *lower* velocity regions (to reduce the relative velocity between the particle and the fluid).

Eichorn and Small [8] experimentally measured the transverse force by the lift, drag, and rotating speed of small spheres (0.061 to 0.126 in. diameter) in the shear region of Poiseuille flow. The relation they derived for the lift coefficient on the sphere was

$$C_{\mathrm{L}} \sim 7 \times 10^4 \left[\frac{2a}{u} \left(\frac{du}{dy} \right) \frac{a}{R} \frac{1}{(N_{\mathrm{Re}})_{\mathrm{p}}} \right], \tag{10.17}$$

where u is the velocity and R the tube radius. In the experiments, $(N_{\mathrm{Re}})_{\mathrm{p}}$ ranged from 75 to 230 and $(2a/R)(\partial u/\partial y)$ ranged from 0.4 to 1. From Saffman's results, one should get

$$C_{\mathrm{L}} = \frac{2K}{\pi} \left[\frac{(2a/u)(\partial u/\partial y)}{(N_{\mathrm{Re}})_{\mathrm{p}}} \right]. \tag{10.18}$$

This is believed to be more accurate at low Reynolds numbers ($K = 81.2$).

The above relations show that the ratio of lift force to drag force is

$$\frac{|\mathcal{F}_{\mathrm{L}}|}{|\mathcal{D}|} = \frac{1}{6} \frac{a^2\omega}{\nu} \tag{10.19}$$

for the rotational effect, and

$$\frac{|\mathcal{F}_{\mathrm{T}}|}{|\mathcal{D}|} = \frac{K}{6\pi} \left[\frac{a^2(\partial u/\partial y)}{\nu} \right]^{1/2} \tag{10.20}$$

for the slip-shear effect.

Hence, for small particles, the transverse force due to its passage through a shear layer can be neglected when

(a) $a^2\omega/\nu$ is small or
(b) $a^2(\partial u/\partial y)/\nu$ is small.

The time required for the particle to reduce its slip velocity by e^{-1} is called the *slip relaxation time*. If $\bar{u}_{\mathrm{p}}$ is the mean particle velocity, then a slip relaxation length is defined by

$$\lambda = \tau_{\mathrm{p}} \bar{u}_{\mathrm{p}}. \tag{10.21}$$

10.2.3 Pressure Gradient Force

The pressure gradient in the fluid will also produce a force on a particle. For a particle of radius a in a pressure gradient $\partial P/\partial x$ as shown in Fig. 10.2,

$$d\mathcal{F}_{\mathrm{p}} = -a \cos \vartheta \left(\frac{\partial P}{\partial x} \right) 2\pi a \sin \vartheta \; a \, d\vartheta \cos \vartheta.$$

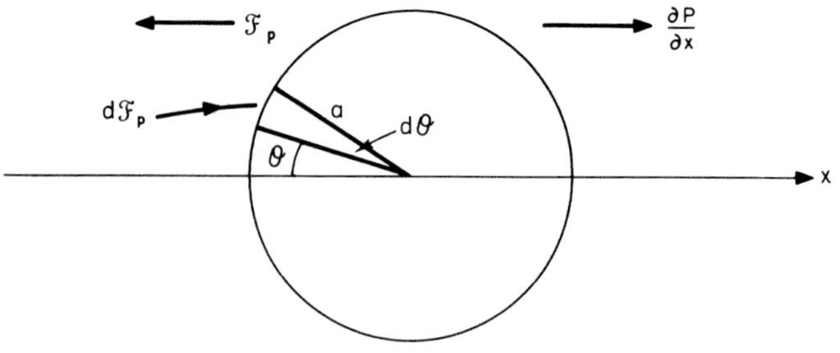

Fig. 10.2

Integration gives

$$\mathfrak{F}_{\mathrm{p}} = -\frac{\partial P}{\partial x} 2\pi a^3 \int_0^{\pi} \cos^2\vartheta \sin\vartheta \, d\vartheta = -\frac{\partial P}{\partial x} \frac{4\pi a^3}{3}. \quad (10.22)$$

The force is in the opposite direction to the pressure gradient. For small particles, unless there is a very large pressure gradient, this force will be very small.

Forces can also arise on a particle due to the temperature gradient in the gas (thermophoresis) and to nonuniform radiation (photophoresis). However, it is only in the submicron particle size range that the effect of temperature gradients on the motion of a particle starts to become significant.

10.2.4 Basset Acceleration Force

Basset [9] estimated the effect on the particle due to the deviation of the flow pattern from the steady state. It is an instantaneous flow resistance and accounts for the acceleration history of the particle. The expression he derived was

$$\mathfrak{F}_{\mathrm{a}} = 6a^2(\pi\varrho\mu)^{1/2} \int_{t_{\mathrm{p}0}}^{t_{\mathrm{p}}} \frac{(d/d\tau)(\mathbf{u} - \mathbf{u}_{\mathrm{p}})}{(t_{\mathrm{p}} - \tau)^{1/2}}, \quad (10.23)$$

where τ is the time constant for the effect. The $|\mathfrak{F}_{\mathrm{a}}|$ becomes significant only when the solid particle is accelerated at high rate, when the observed drag force becomes many times the steady state drag coefficient [10]. So only in cases of extreme particle accelerations need this effect be considered. In the case of passage of a particle through a shock front, both the pressure gradient and Basset effects would be significant.

In the absence of extreme conditions, then, it would appear reasonable to consider only the Stokesian drag force in general, and also the slip-stream force when a substantial shear flow field is present.

10.3 Viscosity of a Dilute Suspension of Small Particles

Again, the fluid will be considered as incompressible, and the particles to be exactly spherical and their number density relatively small. Under such conditions the motion of the fluid on which the disturbance flow due to the presence of a specific particle is superposed consists approximately of a uniform translation, uniform rotation, and a uniform pure straining motion.

The particle translates and rotates with the surrounding fluid and so only the fluid straining motion gives rise to a disturbance flow. As the straining motion is changed by the presence of the particle, one would expect an accompanying increase in the rate of energy dissipation and that the effective viscosity (shear) of the suspension would be greater than the viscosity of the pure fluid.

Since the particle will also be assumed to be incompressible, only the shear viscosity is relevant.

The disturbance flow due to a single incompressible particle must first be established.

10.3.1 Disturbance Flow Due to a Sphere Present in a Pure Shear Flow (or Straining Motion)

Consider a fluid of viscosity μ and density ϱ surrounding a sphere of radius a. Far from the sphere the motion is a pure shear flow specified by the rate-of-strain tensor e_{ij} with $e_{ii} = 0$.

The velocity and pressure in the fluid may be written as

$$v_i = v_i' + e_{ij}x_j \qquad \text{and} \qquad p = p' + P, \tag{10.24}$$

where P is the pressure in the pure shear flow represented by e_{ij} in the absence of the particle. The v_i' and p_i' are the perturbations due to the presence of the sphere and

$$v_i' \to 0 \qquad \text{and} \qquad p' \to 0 \qquad \text{as} \quad r \to \infty. \tag{10.25}$$

The center of the sphere is chosen as the origin and there is no tendency for the sphere to translate. The surface of the sphere, then, is always given by $r = a$, and so

$$\mathbf{n} \cdot \mathbf{v} = 0 \qquad \text{at} \quad r = a. \tag{10.26}$$

If the material in the sphere can flow, then there are other conditions to be satisfied at its surface. The velocity must be continuous across the interface between the sphere and the surrounding fluid, and the tangential component of stress must be also (assuming that the interface has no mechanical properties other than a uniform surface tension). Using the superscript s for the fluid *in* the sphere, one gets, at $r = a$,

$$\begin{aligned} v_i &= v_i' + e_{ij}x_j = v_i^{\mathrm{s}}, \\ \varepsilon_{kli}n_l n_j(\sigma_{ij} &- \sigma_{ij}^{\mathrm{s}}) = 0, \end{aligned} \tag{10.27}$$

where x_j denotes the position vector and n_j the normal to the interface. The viscosity of the fluid in the sphere will be denoted by μ^{s}, and if the sphere is rigid, then $\mu^{\mathrm{s}}/\mu \to \infty$. The velocities $\mathbf{v}$ and $\mathbf{v}^{\mathrm{s}}$ satisfy the Navier–Stokes equation.

Substituting from Eq. (10.24) into this equation, one gets

$$\varrho\left\{\frac{\partial v_i'}{\partial t} + (v_j' + e_{jk}x_k)\frac{\partial v_i'}{\partial x_j} + e_{ij}v_j'\right\} = -\frac{\partial p'}{\partial x_i} + \mu \nabla^2 v_i'. \tag{10.28}$$

For a small disturbance velocity v_i' (and thus negligible inertial effects), the left-hand side of Eq. (10.28) can be ignored; thus the flow near the particle is governed by the equation

$$\nabla p' = \mu \nabla^2 \mathbf{v}'. \tag{10.29}$$

Under the same conditions, the velocity $\mathbf{v}^{\mathrm{s}}$ and pressure p^{s} inside the sphere also satisfy such an equation,

$$\nabla p^{\mathrm{s}} = \mu^{\mathrm{s}} \nabla^2 \mathbf{v}^{\mathrm{s}}. \tag{10.30}$$

From mass conservation, finally,

$$\nabla \cdot \mathbf{v}' = \nabla \cdot \mathbf{v} = 0, \qquad \nabla \cdot \mathbf{v}^{\mathrm{s}} = 0, \tag{10.31}$$

Equations (10.29)–(10.31), with boundary conditions (10.25)–(10.27), govern the disturbance motion. They are homogeneous and linear in $\mathbf{v}', p', \mathbf{v}^{\mathrm{s}}, p^{\mathrm{s}}$, and e_{ij}. The details will not be given here, but it can be shown [11] that the pressures and velocities have the form

$$\begin{aligned} p' &= \frac{C\mu e_{ij}x_i x_j}{r^5}, \\ p^s - p_0^s &= C^s\mu^s e_{ij}x_i x_j, \\ v_i' &= e_{ij}x_j M + e_{jk}x_i x_j x_k Q, \\ v_i^s &= e_{ij}x_j M^s + e_{jk}x_i x_k x_j Q^s, \end{aligned} \tag{10.32}$$

where M, Q, M^s, and Q^s are functions of r alone, and C, C^s, and p_0^s are constants. The form of the functions that satisfies the governing equations and the conditions far from the particle and at $r = 0$ are found to be

$$\begin{aligned} M &= \frac{D}{r^5}, & Q &= \frac{C}{2r^5} - \frac{5D}{2r^7}, \\ M^s &= D^s + \frac{5}{21}C^s r^2, & Q^s &= -\frac{2}{21}C^s, \end{aligned} \tag{10.33}$$

and the conditions at the interface, $r = a$, are then satisfied if

$$\frac{C}{(2\mu + 5\mu^s)} = \frac{D}{\mu^2 a^5} = -\frac{2C^s a^2}{21\mu} = \frac{2D^s}{3\mu} = -\frac{1}{\mu + \mu^s}. \tag{10.34}$$

Note At large distances from the particle,

$$v_i' = \tfrac{1}{2}Ce_{jk}\frac{x_i x_j x_k}{r^5} + O(r^{-4}). \tag{10.35}$$

Thus, the disturbance velocity is one order smaller than that due to a sphere translational motion (see Batchelor [11]).

The above results can now be used to calculate the effective shear viscosity of a suspension of small incompressible spherical particles executing a prescribed bulk motion.

Consider a volume $\mathscr{V}$ of the suspension bounded by a flexible surface $\mathscr{A}$. The velocity *on* this surface is taken to be an exact linear function of position $(\mathbf{x})$.

As the rotational part of the motion is irrelevant to the analysis, the velocity at the boundary is chosen to be $e_{ij}x_j$, where e_{ij} is the symmetric tensor with $e_{ii} = 0$. The suspension moves compatibly with the boundary, and the fluid velocity is

$$e_{ij}x_j + v_i'.$$

Similarly, the pressure in the fluid is changed by the presence of the particles and is $P + p'$, where P would be the pressure in the absence of the particle.

The particles are far apart and so each is embedded in a pure straining motion characterized by the rate-of-strain tensor e_{ij}.

The stress at any point in the fluid of viscosity μ is then

$$\sigma_{ij} = -P\,\delta_{ij} + 2\mu e_{ij} + \sigma'_{ij},$$

where

$$\sigma'_{ij} = -p'\,\delta_{ij} + \mu\left(\frac{\partial u_j'}{\partial x_i} + \frac{\partial u_i'}{\partial x_j}\right). \qquad (10.36)$$

If the fluid had no particles present but had the same density and viscosity μ^*, then the stress tensor would be

$$-P\,\delta_{ij} + 2\mu^* e_{ij}.$$

The problem is to choose μ^* so that it represents the effect of the disturbance flows due to the particles in the suspension. The rate of dissipation of mechanical energy in the volume $\mathscr{V}$ is sensitive to the presence of the particles, and by determining the dissipation a value for μ^* can be obtained.

The rate at which work is being done by forces at the boundary $\mathscr{A}$ is

$$\int_{\mathscr{A}} e_{ij} x_k \sigma_{ij} n_j \, d\mathscr{A} = e_{ik} \int_{\mathscr{A}} (-P\,\delta_{ij} + 2\mu e_{ij} + \sigma'_{ij}) x_k n_j \, d\mathscr{A}.$$

If the suspension were a uniform fluid of the same density and viscosity μ^*, then the rate of doing work at the boundary would be

$$e_{ik} \int_{\mathscr{A}} (-P\,\delta_{ij} + 2\mu^* e_{ij}) x_k n_j \, d\mathscr{A}.$$

The P term in both expressions accounts for any increase in kinetic energy associated with the linear velocity field. The remaining parts of the two expressions represent the rate of dissipation of energy in $\mathscr{V}$. So the effective viscosity μ^* will be defined such that both expressions are equal. After using the divergence theorem on the e_{ij} terms,

$$2\mu^* e_{ij} e_{ij} \mathscr{V} = 2\mu e_{ij} e_{ij} \mathscr{V} + e_{ik} \int_{\mathscr{A}} \sigma'_{ij} x_k n_j \, d\mathscr{A}. \qquad (10.37)$$

The last term on the right-hand side represents the additional rate of dissipation in $\mathscr{V}$ due to the particles. It can be transformed to an integral over the surfaces of the particles as follows:

$$e_{ik} \int_{\mathscr{A}} \sigma'_{ij} x_k n_j \, d\mathscr{A} = e_{ik} \int_{\mathscr{V} - \Sigma \mathscr{V}_{\mathrm{p}}} \left(\frac{\partial \sigma'_{ij}}{\partial x_j} x_k + \sigma'_{ik} \right) d\mathscr{V}$$
$$+ e_{ik} \sum \int_{\mathscr{A}_{\mathrm{p}}} \sigma'_{ij} x_k n_j \, d\mathscr{A},$$

where $\mathscr{A}_{\mathrm{p}}$ and $\mathscr{V}_{\mathrm{p}}$ are the surface and volume of one particle, respectively, and the summation is over all particles in $\mathscr{V}$.

Now the disturbance flow is governed by Eq. (10.29), and so

$$\frac{\partial \sigma'_{ik}}{\partial x_k} = -\nabla p' + \mu \nabla^3 \mathbf{v}' = 0.$$

Also,

$$e_{ik} \int_{\mathscr{V} - \Sigma \mathscr{V}_{\mathrm{p}}} \sigma'_{ik} \, d\mathscr{V} = e_{ik} \int_{\mathscr{V} - \Sigma \mathscr{V}_{\mathrm{p}}} 2\mu \frac{\partial v_i'}{\partial x_k} \, d\mathscr{V}$$
$$= -e_{ik} \sum \int_{\mathscr{A}_{\mathrm{p}}} 2\mu v_i' n_k \, d\mathscr{A},$$

since $\mathbf{v}' = 0$ on the boundary $\mathscr{A}$. Thus, Eq. (10.37) becomes

$$2(\mu^* - \mu) e_{ij} e_{ij} = \frac{e_{ik}}{\mathscr{V}} \sum \int_{\mathscr{A}_{\mathrm{p}}} (\sigma'_{ij} x_k n_j - 2\mu v_i' n_k) \, d\mathscr{A}. \quad (10.38)$$

The right-hand side represents the (average) additional rate of energy dissipation per unit volume due to the particles. This is valid for any spacing of particles in $\mathscr{V}$. Now suppose the particles are spherical and spaced at distances much greater than their diameters. The disturbance flows around each particle do not interact with one another, and the results developed earlier can be used to evaluate the integral in Eq. (10.38). From Eqs. (10.32), (10.33), and (10.36), it follows that

$$\sigma'_{ij} x_j x_k - 2\mu v_i' x_k = \mu e_{ij} x_j x_k \left(\frac{C}{r^3} - \frac{10D}{r^5} \right)$$
$$+ \mu e_{jl} x_i x_j x_k x_l \left(-\frac{5C}{r^5} + \frac{25D}{r^7} \right). \quad (10.39)$$

The surface $\mathscr{A}$ is of a sphere of radius a, and using the well-known relations

$$\int n_j n_k \, d\Omega = \tfrac{4}{3} \pi \delta_{jk},$$
$$\int n_i n_j n_k n_l \, d\Omega = \tfrac{4}{15} \pi (\delta_{ij} \delta_{kl} + \delta_{ik} \delta_{jl} + \delta_{il} \delta_{jk}), \quad (10.40)$$

where the integration is over the complete solid angle subtended at the sphere center, one gets

$$\int_{\mathscr{A}_{\mathrm{p}}} (\sigma'_{ij}x_k n_j - 2\mu v_i' n_k)\, d\mathscr{A} = -\tfrac{4}{3}\pi\mu C e_{ik}.$$

Hence, from Eq. (10.34),

$$\frac{\mu^*}{\mu} = 1 - \frac{2\pi}{3\mathscr{V}} \sum C = 1 + \frac{1}{\mathscr{V}} \sum \left(\frac{\mu + \frac{5}{2}\mu^{\mathrm{s}}}{\mu + \mu^{\mathrm{s}}}\right) \mathscr{V}_{\mathrm{p}}. \tag{10.41}$$

If all the particles have the same internal viscosity (μ^{s}), then

$$\frac{\mu^*}{\mu} = 1 + \alpha\left(\frac{\mu + \frac{5}{2}\mu^{\mathrm{s}}}{\mu + \mu^{\mathrm{s}}}\right), \tag{10.42}$$

where $\alpha = \sum \mathscr{V}_{\mathrm{p}}/\mathscr{V}$ is the particle concentration by volume.

For a suspension of rigid particles, the effective viscosity is greater than the fluid viscosity by a fraction $\frac{5}{2}\alpha$. This result was first obtained by Einstein in 1906. For a suspension of gas bubbles ($\mu^{\mathrm{s}} = 0$), it is interesting to note that the fraction is just α.

Equations (10.41) and (10.42) are subject to the restriction $\alpha \ll 1$. Experiment indicates that the Einstein formula $\mu(1 + \frac{5}{2}\alpha)$ represents the viscosity of a suspension of small rigid spheres for values of $\alpha \lesssim 0.02$ [12]. Jeffrey (in 1922) determined the effective viscosity for rigid particles of ellipsoidal shape (assuming sufficiently strong Brownian motion to make all orientations of the ellipsoids equally likely). He showed that μ^*/μ increases with the departure from sphericity, because velocity gradients in the fluid are larger in magnitude for "sharper" particles. But the variation of the coefficient of viscosity is found to be small until the ratio of the semimajor to the semiminor axes is about 3.

In 1932, Taylor dealt with the problem of the viscosity of a fluid containing small drops of another liquid. Again the method used was essentially that of Einstein. The so-called "method of Stokes–Einstein" has been outlined in great detail by Sadron [13] and Frisch and Simha [14]. In 1938 Burgers adopted a substantially different approach—it is also described in references [13] and [14].

It must be mentioned that the very basic physical mechanism of energy dissipation is still a matter lacking general agreement, and so the "viscosity of a suspension" is very much a subject of present day research.

10.4 Macroscopic Continuum Description of Particle–Fluid Flow

Again, solid spheres in a viscous incompressible fluid will be considered—spheres small enough and numerous enough so that a particle continuum can be envisaged but without significant particle–particle interaction. Finally, the particles will be of uniform size unless otherwise specified.

Viscous interaction between neighboring spheres (due to the intervening fluid) tends to smooth out possible velocity dispersion among adjacent particles. So, to a first approximation, the velocity difference can be ignored. This implies that particle–particle collisions are not frequent. For slow viscous motion, direct contact collision between particles in which significant momentum exchange takes place is very rare. In this respect, then, the particle cloud can be regarded as collisionless.

The motion of the particle cloud can be described by defining a distribution function $f(\mathbf{x}, \mathbf{v}, t)$ for the particles and then deriving the conservation equations by taking moments of the collisionless Boltzmann equation [15]. The nature of the force acting on each particle is still under investigation and so such a procedure will not be dealt with here.

A simpler procedure is adopted. Under the conditions of large particle number density, small particle size, and negligible velocity dispersion, the particle cloud can be considered as a continuum, with a particle continuum velocity $\mathbf{v}_p$. The field equations for the particle continuum can then be written down as follows.

$$\text{continuity:} \qquad \frac{\partial n}{\partial t} + \nabla \cdot (n\mathbf{v}_p) = 0, \tag{10.43}$$

$$\text{linear momentum:} \qquad m_p n\left(\frac{\partial \mathbf{v}_p}{\partial t} + \mathbf{v}_p \cdot \nabla \mathbf{v}_p\right) = \mathfrak{F}_p, \tag{10.44}$$

$$\text{angular momentum:} \qquad I_p\left(\frac{\partial \boldsymbol{\omega}_p}{\partial t} + (\mathbf{v}_p \cdot \nabla)\boldsymbol{\omega}_p\right) = \mathfrak{T}, \tag{10.45}$$

where n is the particle number density, $\mathbf{v}_p$ the particle continuum velocity, m_p the single particle mass, I_p the moment of inertia of a single particle, $\boldsymbol{\omega}_p$ the particle angular velocity, $\mathfrak{F}_p$ the force exerted on the particle continuum by the surrounding fluid, per unit volume of space, and $\mathfrak{T}$ the torque exerted on a particle by the surrounding field. In a nonisothermal system there would also be a conservation of energy equation.

10.4.1 Bulk Fluid Equations

The fluid phase, as mentioned before, is assumed to be a viscous, incompressible fluid, the motion of which is described by the Navier–Stokes equations. For regions exterior to the solid spheres, the following equations apply:

$$\text{continuity:} \quad \frac{\partial \varrho}{\partial t} + \nabla \cdot (\varrho \mathbf{v}) = 0, \tag{10.46}$$

$$\text{momentum:} \quad \frac{\partial (\varrho \mathbf{v})}{\partial t} + \nabla \cdot (\varrho \mathbf{v}\mathbf{v}) = \nabla \cdot \boldsymbol{\sigma}, \tag{10.47}$$

$$\sigma_{ij} = -p\,\delta_{ij} + \mu\left(\frac{\partial v_i}{\partial x_j} + \frac{\partial v_j}{\partial x_i}\right),$$

where ϱ is the fluid density, μ the fluid viscosity, $\mathbf{v}$ the fluid velocity, $\boldsymbol{\sigma}$ the stress tensor, and p the fluid pressure. In principle, Eqs. (10.43)–(10.47) give a complete description of the two-phase system. A solution of these sets of equations with appropriate boundary conditions would give all available information. This is a virtually impossible task. Consider 10-micron particles spaced 100 microns apart. Then in 1 cubic millimeter there are about 1000 particles and 1000 isolated surfaces on which the solution has to satisfy the no-slip boundary condition.

For many practical purposes the detailed behavior of the fluid particle system is not required. A gross averaged description is enough to meet requirements.

Of course the "microscopic approach" involving a detailed account of the behavior of a typical particle and its surrounding fluid is indispensable in determining the transport properties of the two-phase system.

To describe the system macroscopically, one must first define bulk, or averaged, quantities of the fluid. Let the symbol $\langle\ \rangle$ denote an averaging operation which involves the following steps.

(a) the space is divided into small cubic volumes v, large with respect to the size of a particle but small with respect to macroscopic scales;

(b) the physical quantities of the fluid are averaged over the volume occupied by the fluid, v_f, in v. Thus, for any quantity Q,

$$\langle Q \rangle \equiv \frac{1}{v} \int_{v_f} Q \, dv. \tag{10.48}$$

Using this averaging operator, one can define the following bulk fluid quantities:

density: $\varrho_f = \langle \varrho \rangle$,

velocity: $\mathbf{v}_f = \dfrac{1}{\varrho_f} \langle \varrho \mathbf{v} \rangle$,

pressure: $p_f = \dfrac{1}{1-c} \langle p \rangle$,

where $c = \frac{4}{3}\pi a^3 n$ is the volume concentration of solid particles.

Obviously, for systems in which n is small, the presence of the particles causes only small changes in the fluid field. On the macroscale these local variations can be ignored, so that variations on the macroscale can be represented well enough by variations of the averages.

Another assumption that can be made is that on the macroscale the two operators ∇ and $\langle \ \rangle$ commute, or

$$\langle \nabla \ \rangle = \nabla \langle \ \rangle;$$

that is, the average of the derivative equals the derivative of the average (true in the limit $v \to 0$). Using this assumption and the average operator, one can write the bulk fluid equations (10.46) and (10.47) as

$$\frac{\partial \varrho_f}{\partial t} + \nabla \cdot (\varrho_f \mathbf{v}_f) = 0 \tag{10.49}$$

and

$$\frac{\partial}{\partial t} (\varrho_f \mathbf{v}_f) + \nabla \cdot (\varrho_f \mathbf{v}_f \mathbf{v}_f) = \nabla \cdot \boldsymbol{\sigma}_f + \mathfrak{F}_f, \tag{10.50}$$

where

$$\nabla \cdot \boldsymbol{\sigma}_f + \mathfrak{F}_f = \nabla \cdot (\varrho_f \mathbf{v}_f \mathbf{v}_f) - \langle \nabla \cdot (\varrho \mathbf{v}\mathbf{v}) \rangle + \langle \nabla \cdot \boldsymbol{\sigma} \rangle.$$

If the particles have negligible inertia, then the particles and fluid will move together as a homogeneous fluid, to which one can assign a stress tensor $\boldsymbol{\sigma}_f$ to describe its mechanical behavior. When the particles possess inertia and slip can occur, one can still retain the tensor $\boldsymbol{\sigma}_f$ but account for the difference by the term $\mathfrak{F}_f$. This signifies a body force due to particle–fluid interaction:

$$\langle \nabla \cdot \boldsymbol{\sigma} \rangle = \frac{1}{v} \int_{v_f} \nabla \cdot \boldsymbol{\sigma} \, dv = \frac{1}{v} \int_{S_f} \mathbf{n} \cdot \boldsymbol{\sigma} \, dS$$

and represents the total force exerted on the fluid by the particles in the volume v.

The difference term

$$\nabla \cdot (\varrho_f \mathbf{v}_f \mathbf{v}_f) - \langle \nabla \cdot (\varrho \mathbf{v}\mathbf{v}) \rangle$$

represents the momentum associated with the disturbances produced in the fluid due to the presence of the particles.

Equations (10.49) and (10.50) are the field equations for the description of an isothermal bulk fluid. They do not form a closed set, however. It is necessary to supplement them with phenomenological formulas relating $\boldsymbol{\sigma}_f$ and $\mathfrak{F}_f$ to the field variables.

The stress tensor components can be written as

$$\sigma_{fij} = -p_f \delta_{ij} + \mu_f \left(\frac{\partial v_{fi}}{\partial x_j} + \frac{\partial v_{fj}}{\partial x_i} \right), \tag{10.51}$$

where μ_f in the coefficient of viscosity of the bulk fluid (discussed previously—it depends on the homogeneous fluid viscosity μ and the particle concentration n).

As already mentioned in evaluating $\mathfrak{F}_f$, the interaction force between the phases, only the Stokesian drag force [Eq. (10.14)] and the transverse slip-shear force [Eq. (10.15)] need be considered under normal circumstances. In a linear shear field, then, and in terms of a force per unit volume of suspension, the interaction between the particulate and fluid phases becomes

$$\mathfrak{F}_f = n \left[\frac{\mathbf{v} - \mathbf{v}_p}{\tau_p} + 4.5 \left(\frac{\varrho/\varrho_s}{\tau_p (\text{curl } \mathbf{v})} \right)^{1/2} \text{curl } \mathbf{v} \times (\mathbf{v} - \mathbf{v}_p) \right], \tag{10.52}$$

where

$$\tau_{\mathrm{p}} = \frac{2}{9}\left(\frac{\varrho_{\mathrm{s}} a^2}{\mu}\right) = \text{slip relaxation time for the particulate phase}$$

and ϱ_s is the particle material density.

10.5 Two-Dimensional Suspension Boundary Layer Equations

The only specific particle–fluid system which will be treated in detail here is the two-dimensional boundary layer flow.

The general conservation equations, in indicial form, are as follows:

fluid phase:

$$\frac{\partial v_i}{\partial x_i} = 0, \tag{10.53}$$

$$\varrho\left(\frac{\partial v_i}{\partial t} + v_j \frac{\partial v_i}{\partial x_i}\right) = -\frac{\partial p}{\partial x_i} + \mu \nabla^2 v_i + \varrho f_i - \mathfrak{F}_{\mathrm{f}i}; \tag{10.54}$$

solid phase:

$$\frac{\partial n}{\partial t} + \frac{\partial}{\partial x_i}(n v_{\mathrm{p}})_i = 0, \tag{10.55}$$

$$n\left(\frac{\partial v_{\mathrm{p}i}}{\partial t} + v_{\mathrm{p}j} \frac{\partial v_{\mathrm{p}i}}{\partial x_j}\right) = n f_{\mathrm{p}i} + \mathfrak{F}_{\mathrm{f}i}; \tag{10.56}$$

where n is the particle number density, f the body force per unit mass, and $\mathfrak{F}_{\mathrm{f}i}$ is the interaction force per unit volume between the phases, and is given by Eq. (10.52).

Note The particle phase is regarded as compressible.

The above equations are in the forms originally derived by Soo [16], Hinze [17], and Marble [18].

It has been shown [19] that the standard boundary layer approximations are valid for the fluid phase; that is,

$$\frac{\partial p}{\partial y} = 0, \qquad p = p(x, t),$$

provided n/ϱ is of the order of 1. But simplification of the particulate phase momentum equations does not occur. Within the restrictions listed previously, the continuity and boundary layer momentum equations for the two phases become

$$\frac{\partial u}{\partial x} + \frac{\partial v}{\partial y} = 0, \tag{10.57}$$

$$\frac{\partial u}{\partial t} + u\frac{\partial u}{\partial x} + v\frac{\partial u}{\partial y} = -\frac{1}{\varrho}\frac{\partial p}{\partial x} + \nu\frac{\partial^2 u}{\partial y^2} + \frac{\varrho(u_p - u)}{\varrho_p \tau_p}, \tag{10.58}$$

$$\frac{\partial n}{\partial t} + \frac{\partial}{\partial x}(nu_p) + \frac{\partial}{\partial y}(nv_p) = 0, \tag{10.59}$$

$$\frac{\partial u_p}{\partial t} + u_p\frac{\partial u_p}{\partial x} + v_p\frac{\partial u_p}{\partial y} = \frac{(u - u_p)}{\tau_p}, \tag{10.60}$$

$$\frac{\partial v_p}{\partial t} + u_p\frac{\partial v_p}{\partial x} + v_p\frac{\partial v_p}{\partial y} = \frac{v - v_p}{\tau_p} + \frac{4.5(\varrho/\varrho_s)^{1/2}(\partial u/\partial y)^{1/2}(u - u_p)}{\tau_p^{1/2}}. \tag{10.61}$$

It is appropriate next to introduce the following dimensionless parameters:

$$\varphi = \frac{t}{\tau_p}, \quad x^* = \frac{x}{\lambda}, \quad y^* = \frac{y}{(\tau_p \nu)^{1/2}}, \quad u^* = \frac{u}{\bar{u}}, \quad v^* = v\left(\frac{\tau_p}{\nu}\right)^{1/2}, \tag{10.62}$$

$$p^* = \frac{p}{\varrho \bar{u}^2}, \qquad u_p{}^* = \frac{u_p}{\bar{u}}, \qquad v_p{}^* = v_p\left(\frac{\tau_p}{\nu}\right)^{1/2}, \qquad n^* = \frac{n}{n_\infty},$$

$$W = \left(\frac{\varrho}{n}\right)_\infty = \text{constant}, \qquad Z = 4.5\left(\frac{\varrho}{\varrho_s}\right)^{1/2}\left(\frac{\bar{u}\lambda}{\nu}\right)^{3/4} = \text{constant},$$

where

$$\lambda = \tau_p \bar{u} \tag{10.63}$$

is the slip relaxation length and $\bar{u}$ a characteristic velocity.

If $x^* \ll 1$, the particles have not had time to adjust to the gas flow and take on large velocity slips. But if $x^* \gg 1$, the particles have moved a large distance and show little velocity slip. In nonisothermal situations, analogous thermal relaxation lengths can be developed.

Substitution of these parameters into Eqs. (10.57)–(10.61) produces the following dimensionless equations:

$$\frac{\partial u^*}{\partial x^*} + \frac{\partial v^*}{\partial y^*} = 0, \tag{10.64}$$

$$\frac{\partial u^*}{\partial \varphi} + u^* \frac{\partial u^*}{\partial x^*} + v^* \frac{\partial u^*}{\partial y^*} = -\frac{dp^*}{dx^*} + \frac{\partial^2 u^*}{\partial y^{*2}} + W\varrho_{\mathrm{p}}^*(u_{\mathrm{p}}^* - u), \tag{10.65}$$

$$\frac{\partial n^*}{\partial \varphi} + \frac{\partial}{\partial x^*}(n^* u_{\mathrm{p}}^*) + \frac{\partial}{\partial y^*}(n^* v_{\mathrm{p}}^*) = 0, \tag{10.66}$$

$$\frac{\partial u_{\mathrm{p}}^*}{\partial \varphi} + u_{\mathrm{p}}^* \frac{\partial u_{\mathrm{p}}^*}{\partial x^*} + v_{\mathrm{p}}^* \frac{\partial u_{\mathrm{p}}^*}{\partial y^*} = (u^* - u_{\mathrm{p}}^*), \tag{10.67}$$

$$\frac{\partial v_{\mathrm{p}}^*}{\partial \varphi} + u_{\mathrm{p}}^* \frac{\partial v_{\mathrm{p}}^*}{\partial x^*} + v_{\mathrm{p}}^* \frac{\partial v_{\mathrm{p}}^*}{\partial y^*} = (v^* - v_{\mathrm{p}}^*) + Z(u^* - u_{\mathrm{p}}^*) \times \left| \frac{\partial u^*}{\partial y^*} \right|^{1/2}. \tag{10.68}$$

In these equations all the physical parameters governing the motion of a viscous suspension have been combined into two dimensionless constants W and Z. Equations (10.64)–(10.68) are a "universal" set of equations which describe the basic fluid mechanics of suspension boundary layer flows. They have been successfully applied to the case of laminar mixing of a suspension with a pure fluid.

10.6 Laminar Mixing [20]

Consider two streams moving parallel to each other with velocities u_1 and u_2, respectively, in the positive x direction (Fig. 10.3). The stream with velocity u_2 has particles (uniform size, etc.) suspended in it. At the position $x = 0$ the streams begin to mix. The object is to compute the steady state growth of the mixing region, the slip between

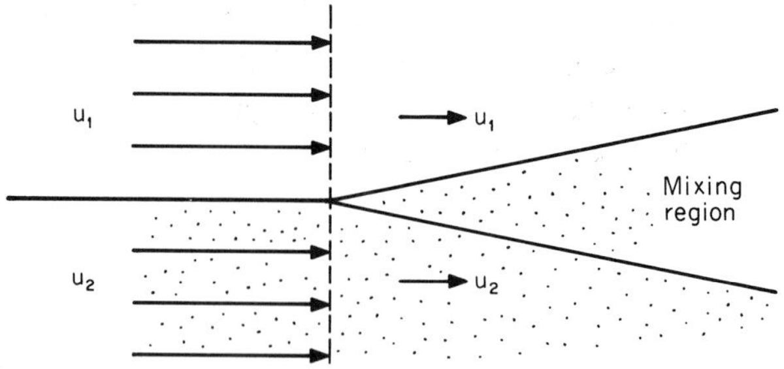

Fig. 10.3

the phases, and the particle distribution in the mixing (or boundary layer) region.

The governing equations are (10.64)–(10.68) with the time-dependent terms removed. It is also assumed that mixing occurs at constant pressure, that is,

$$\frac{dp}{dx} = 0.$$

The characteristic velocity $\bar{u}$ is defined as

$$\bar{u} = \frac{u_1 + u_2}{2}, \tag{10.69}$$

and the average velocity difference of the two streams is

$$\Lambda = \frac{u_1 - u_2}{u_1 + u_2}. \tag{10.70}$$

The boundary conditions are as follows:

$$\begin{aligned} u^*(0, y^*) &= u_p^*(0, y^*) = 1 + \Lambda, \\ u^*(x^*, \infty) &= u_p^*(x^*, \infty) = 1 + \Lambda, \\ u^*(x^*, -\infty) &= 1 - \Lambda, \\ \varrho_p^*(x^*, \infty) &= 1. \end{aligned} \tag{10.71}$$

Equations (10.64)–(10.68) and these boundary conditions form a boundary value problem for which a closed-form solution for the entire range of x^* cannot be found. Solutions applicable in the initial, or near-field, mixing region, and the far-field region can be found separately. Only the solution for the near-field region will be described here.

The solution is obtained in terms of the transformed variables

$$\xi = x^*, \qquad \eta = \frac{y^*}{(x^*)^{1/2}}.$$

10.6.1 The Near-Field Region

At $x \sim 0$, to zeroth order, the two phases flow independently of each other. This suggests that a separate stream function be defined for each phase ψ^* and $\psi_p{}^*$, where

$$u^* = \frac{\partial \psi^*}{\partial y^*}, \qquad v^* = -\frac{\partial \psi^*}{\partial x^*}, \tag{10.72}$$

$$n^* u_p{}^* = \frac{\partial \psi_p{}^*}{\partial y^*}, \qquad n^* v_p{}^* = -\frac{\partial \psi_p{}^*}{\partial x^*}. \tag{10.73}$$

These stream functions satisfy respectively the fluid and particulate phase continuity equations. Substituting (10.72) and (10.73) into Eqs. (10.65), (10.67), and (10.68), one gets

$$\frac{\partial \psi^*}{\partial y^*} \frac{\partial^2 \psi^*}{\partial x^* \partial y^*} - \frac{\partial \psi^*}{\partial x^*} \frac{\partial^2 \psi^*}{\partial y^{*2}} = \frac{\partial^3 \psi^*}{\partial y^{*3}} + W\left(\frac{\partial \psi_p{}^*}{\partial y^*} - n^* \frac{\partial \psi^*}{\partial y^*}\right), \tag{10.74}$$

$$\frac{\partial \psi_p{}^*}{\partial y^*}\left(n^* \frac{\partial^2 \psi_p{}^*}{\partial x^* \partial y^*} - \frac{\partial n^*}{\partial x^*} \frac{\partial \psi_p{}^*}{\partial y^*}\right) - \frac{\partial \psi_p{}^*}{\partial x^*}\left(n^* \frac{\partial^2 \psi_p{}^*}{\partial y^{*2}} - \frac{\partial n^*}{\partial y^*} \frac{\partial \psi_p{}^*}{\partial y^*}\right)$$
$$= \left(n^{*3} \frac{\partial \psi^*}{\partial y^*} - n^{*2} \frac{\partial \psi_p{}^*}{\partial y^*}\right), \tag{10.75}$$

$$\frac{\partial \psi_p{}^*}{\partial y^*}\left(n^* \frac{\partial^2 \psi_p{}^*}{\partial x^{*2}} - \frac{\partial n^*}{\partial x^*} \frac{\partial \psi_p{}^*}{\partial x^*}\right) - \frac{\partial \psi_p{}^*}{\partial x^*}\left(n^* \frac{\partial^2 \psi_p{}^*}{\partial y^* \partial x^*} - \frac{\partial n^*}{\partial y^*} \frac{\partial \psi_p{}^*}{\partial x^*}\right)$$
$$= \left(n^{*3} \frac{\partial \psi^*}{\partial x^*} - n^{*2} \frac{\partial \psi_p{}^*}{\partial x^*}\right) + Z\left(n^{*2} \frac{\partial \psi_p{}^*}{\partial y^*} - n^{*3} \frac{\partial \psi^*}{\partial y^*}\right) \left| \frac{\partial^2 \psi^*}{\partial y^{*2}} \right|^{1/2}. \tag{10.76}$$

Introducing the variables ξ and η, one has

$$\frac{\partial\psi^*}{\partial\eta}\left(\frac{\partial^2\psi^*}{\partial\xi\,\partial\eta}-\frac{1}{2\xi}\frac{\partial\psi^*}{\partial\eta}\right)-\frac{\partial\psi^*}{\partial\xi}\frac{\partial^2\psi^*}{\partial\eta^2}$$
$$=\frac{1}{\xi^{1/2}}\frac{\partial^3\psi^*}{\partial\eta^3}+\xi^{1/2}W\left(\frac{\partial\psi_p^*}{\partial\eta}-n^*\frac{\partial\psi^*}{\partial\eta}\right), \tag{10.77}$$

$$n^*\frac{\partial\psi_p^*}{\partial\eta}\left(\frac{\partial^2\psi_p^*}{\partial\xi\,\partial\eta}-\frac{1}{2\xi}\frac{\partial\psi_p^*}{\partial\eta}\right)$$
$$-\left(\frac{\partial\psi_p^*}{\partial\eta}\right)^2\frac{\partial n^*}{\partial\xi}+\frac{\partial\psi_p^*}{\partial\xi}\left(\frac{\partial n^*}{\partial\eta}\frac{\partial\psi_p^*}{\partial\eta}-n^*\frac{\partial^2\psi_p^*}{\partial\eta^2}\right)$$
$$=\xi\left(n^{*3}\frac{\partial\psi^*}{\partial\eta}-n^{*2}\frac{\partial\psi_p^*}{\partial\eta}\right), \tag{10.78}$$

$$\xi^{1/2}\left[\frac{\partial\psi_p^*}{\partial\eta}\frac{\partial n^*}{\partial\xi}\left(\frac{\partial\psi_p^*}{\partial\xi}-\frac{\eta}{2\xi}\frac{\partial\psi_p^*}{\partial\eta}\right)\right]$$
$$-\xi^{1/2}\left[\frac{\partial n^*}{\partial\eta}\frac{\partial\psi_p^*}{\partial\xi}-\left(\frac{\partial\psi_p^*}{\partial\xi}-\frac{\eta}{2\xi}\frac{\partial\psi_p^*}{\partial\eta}\right)\right]$$
$$-\xi^{1/2}\left[n^*\frac{\partial\psi_p^*}{\partial\eta}\left(\frac{\partial^2\psi_p^*}{\partial\xi^2}-\frac{\eta}{2\xi}\frac{\partial^2\psi_p^*}{\partial\eta\,\partial\xi}+\frac{\eta}{2\xi^2}\frac{\partial\psi_p^*}{\partial\eta}\right)\right]$$
$$+\xi^{1/2}\left[n^*\frac{\partial\psi_p^*}{\partial\xi}\left(\frac{\partial^2\psi_p^*}{\partial\eta\,\partial\xi}-\frac{\eta}{2\xi}\frac{\partial^2\psi_p^*}{\partial\eta^2}-\frac{1}{2\xi}\frac{\partial\psi_p^*}{\partial\eta}\right)\right]$$
$$=\xi\left[n^{*2}\left(\frac{\partial\psi_p^*}{\partial\xi}-\frac{\eta}{2\xi}\frac{\partial\psi_p}{\partial\eta}-n^*\frac{\partial\psi^*}{\partial\xi}-n^*\frac{\eta}{2\xi}\frac{\partial\psi^*}{\partial\eta}\right)\right]$$
$$+Zn^{*2}\left(n^*\frac{\partial\psi^*}{\partial\eta}-\frac{\partial\psi_p^*}{\partial\eta}\right)\left(\frac{\partial^2\psi^*}{\partial\eta^2}\right)^{1/2}. \tag{10.79}$$

The near-field region is characterized by $\xi \ll 1$. Thus the following expansions are assumed for ψ^*, ψ_p^*, and n^*:

$$\begin{aligned}\psi^* &= \xi^{1/2}[f_0(\eta)+\xi f_1(\eta)+\xi^{5/4}f_2(\eta)+\cdots],\\ \psi_p^* &= \xi^{1/2}[g_0(\eta)+\xi g_1(\eta)+\xi^{5/4}g_2(\eta)+\cdots],\\ n^* &= 1+\xi h_1(\eta)+\xi^{5/4}h_2(\eta)+\cdots.\end{aligned} \tag{10.80}$$

These expansions are now substituted into Eqs. (10.77) and (10.79), and equating coefficients of ξ^n to zero, one obtains the following.

Equation (10.77) gives to the zeroth order

$$f_0''' + \tfrac{1}{2} f_0 f_0'' = 0, \tag{10.81}$$

where the prime implies differentiation with respect to η.

Equations (10.78) and (10.79) give, respectively,

$$\tfrac{1}{2} g_0 g_0'' = 0,$$

$$\eta g_0'' g_0 - \eta (g_0')^2 + g_0' g_0 = 0,$$

and so

$$g_0'(g_0 - \eta g_0') = 0. \tag{10.82}$$

Now $g_0' \neq 0$, and so $g_0 = A\eta$, where A is a proportionality constant. Since $g_0'(\infty) = 1 + \Lambda$, therefore,

$$g_0 = (1 + \Lambda)\eta. \tag{10.83}$$

So, to the zeroth order the two phases flow independently of each other. Equation (10.81) is the Blasius equation and governs the flow of the fluid phase, subject to the boundary conditions

$$f_0'(+\infty) = 1 + \Lambda, \qquad f_0'(-\infty) = 1 - \Lambda. \tag{10.84}$$

The equation is third order, however, and to obtain an exact solution one can assume that $\eta = 0$ is a streamline and so

$$f_0(0) = 0. \tag{10.85}$$

This assumption implies that the interface between the two streams remains located along this line (the boundary layer develops symmetrically about $y = 0$). It probably will deviate from $y = 0$, but to first order, and in the absence of any significant disturbance, this assumption should be a reasonable one.

Using the Görtler series expansion method, let

$$f_0 = 2 \sum_{n=0}^{\infty} \Lambda^n V_n(\gamma), \tag{10.86}$$

where $\gamma = \eta/2 = V_0$.

Substituting this in Eq. (10.81), one has

$$\begin{aligned} V_1''' + 2\gamma V_1'' &= 0, \\ V_2''' + 2\gamma V_2'' &= -2V_1 V_1'', \\ V_3''' + 2\gamma V_3'' &= -2(V_1 V_2'' + V_2 V_1''), \end{aligned} \tag{10.87}$$

The boundary conditions on $V(\gamma)$ are

$$V_1(0) = 0, \tag{10.88}$$

$$\begin{aligned} V_1(\gamma) &= 1, \qquad \gamma \to \infty, \\ &= -1, \qquad \gamma \to -\infty, \end{aligned} \tag{10.89}$$

$$V_n(\gamma) = 0, \qquad \gamma = \pm\infty, \quad n \geq 2, \tag{10.90}$$

$$V_n(0) = 0, \qquad n \geq 2. \tag{10.91}$$

The solution of the first of Eqs. (10.87), with boundary conditions (10.88) and (10.89), is

$$V_1(\gamma) = \int_0^\gamma X(x)\, dx, \tag{10.92}$$

where

$$X(x) = \frac{2}{\pi^{1/2}} \int_0^x \exp(-z^2)\, dz = \text{erf}(x).$$

The fluid phase velocity can be approximated by the first two terms of the series expansion, that is,

$$f_0' = 1 + \Lambda\, \text{erf}\left(\frac{\eta}{2}\right). \tag{10.93}$$

Figure 10.4 shows the dimensionless fluid phase velocity profile for various values of Λ.

The first-order modification to the fluid phase stream function $f_1(\eta)$ is obtained from Eq. (10.77), which yields

$$f_1''' + \tfrac{1}{2} f_0 f_1'' - f_0' f_1' + \tfrac{3}{2} f_0'' f_1 = W(f_0' - (1 + \Lambda)). \tag{10.94}$$

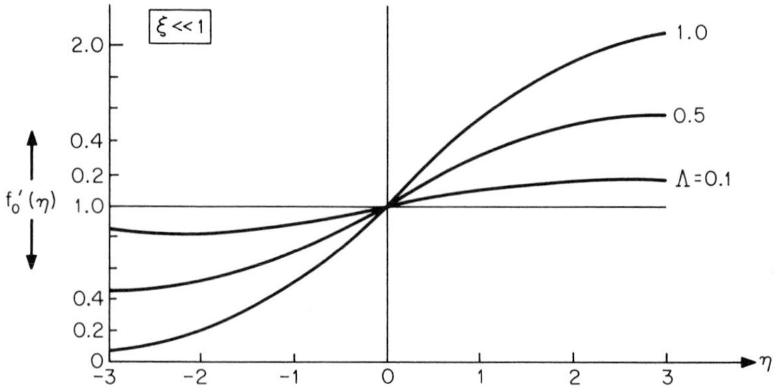

Fig. 10.4

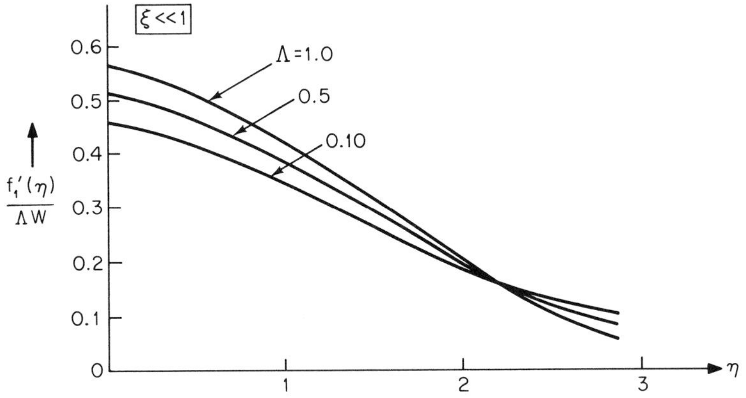

Fig. 10.5

This equation cannot be solved in closed form, however, and one must resort to numerical methods. The solution is plotted in Fig. 10.5.

The first-order functions $g_1(\eta)$ and $h_1(\eta)$ which relate to the particulate phase velocity are obtained from Eqs. (10.78) and (10.79):

$$\eta^2 G_1''(\eta) + 3\eta G_1'(\eta) + 3G_1(\eta) = 2(f_0 - \eta f_0'), \tag{10.95}$$

$$(1 + \Lambda)^2(\eta h_1'(\eta) - 2h_1(\eta)) = \eta G_1''(\eta) - 2G_1(\eta) + 2(f_0' - (1+\Lambda)), \tag{10.96}$$

where

$$G_1(\eta) = (1 + \Lambda)g_1(\eta). \tag{10.97}$$

The complementary solution of Eq. (10.95) is

$$C_1\eta + C_2\eta^3.$$

By the method of variation of parameters, that is,

$$G_1(\eta) = \eta^3 A(\eta) + \eta B(\eta),$$

one obtains for the particular solution

$$A(\eta) = -\frac{f_0(\eta)}{\eta^3} + 2\int_\eta^\infty \frac{f_0(x)}{x^4}\,dx, \tag{10.98}$$

$$B(\eta) = \frac{f_0(\eta)}{\eta}. \tag{10.99}$$

The complete solution is therefore

$$G_1(\eta) = 2\eta^3 \int_\eta^\infty \frac{f_0(x)}{x^4}\, dx + C_1\eta + C_3\eta^3, \qquad (10.100)$$

$$G_1'(\eta) = 6\eta^2 \int_\eta^\infty \frac{f_0(x)}{x^4}\, dx - \frac{2f_0(\eta)}{\eta} + C_1 + 3C_2\eta^2. \qquad (10.101)$$

Note

$$\lim_{\eta\to\infty} \frac{f_0(\eta)}{\eta} = 1 + \Lambda, \qquad (10.102)$$

$$\lim_{\eta\to-\infty} \frac{f_0(\eta)}{\eta} = 1 - \Lambda. \qquad (10.103)$$

Since $G_1'(\infty) = 0$, therefore $C_2 = 0$ and $C_1 = -(1 + \Lambda)$, and so finally

$$G_1'(\eta) = 6\eta^2 \int_\eta^\infty \frac{f_0(x)}{x^4}\, dx - \frac{2f_0(\eta)}{\eta} - (1 + \Lambda). \qquad (10.104)$$

Then

$$\psi_p^{*\prime} = \xi^{1/2}\left(C + \frac{G'(\eta)\xi}{1 + \Lambda}\right),$$

$$u_p^* = \frac{\xi^{1/2}(x^*)^{1/2}}{\varrho_p^*}\left(C + \frac{H'(\eta)\xi}{1 + \Lambda}\right), \qquad (10.105)$$

where $C = -C_1$, $G' = G_1'$. In this near-field region the particle slip will be large, because the particles have not had time to adjust to the local flow conditions. Thus, one would expect the velocity to depend on their initial conditions, $1 + \Lambda$, and on the integrated effect of their motion through the flow field, as indeed is apparent in Eq. (10.104).

In the far-field region, of small slip, the particulate phase velocity is found to be a function only of the local flow conditions.

It can also be shown that

$$\lim_{\eta\to\infty} g_1'(\eta) = -\frac{\Lambda}{1 + \Lambda}. \qquad (10.106)$$

Figure 10.6 shows $g_1'(\eta)$ and $g_1(\eta)$ versus η; $g_1'(\eta)$ gives, in effect, the particulate phase horizontal velocity profile. It is negative because the decelerating fluid phase exerts a drag force on the particulate phase.

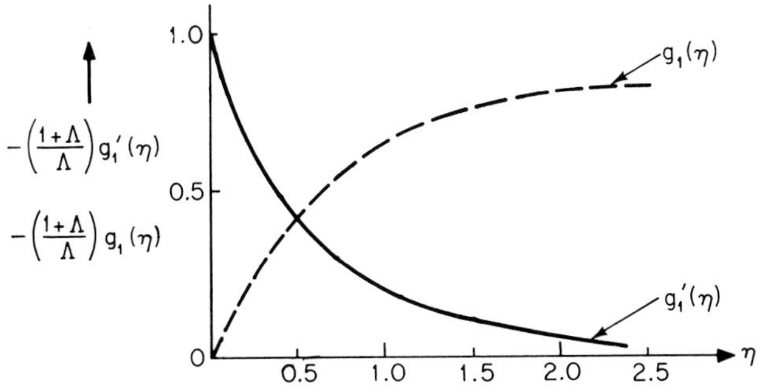

Fig. 10.6

There is also a drag force in the y direction, and the first-order particle velocity in the y direction is

$$v_{p1}^* = \xi^{1/2}\left(\frac{\eta}{2} g_1' - \frac{3}{2} g_1\right). \tag{10.107}$$

A graph of this is shown in Fig. 10.7. The effect of this particle transverse velocity is to move particles away from the clear fluid region.

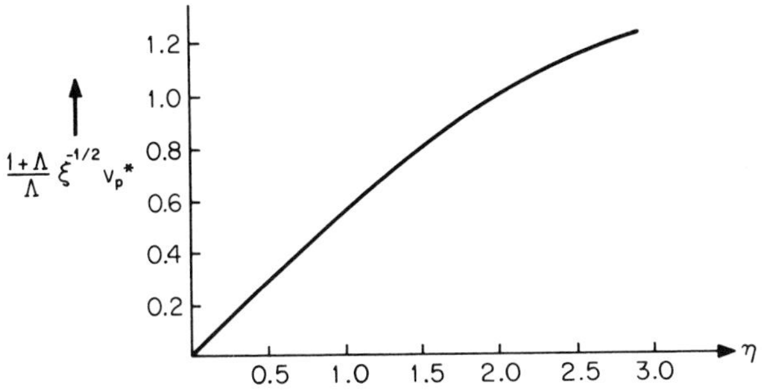

Fig. 10.7

10.6.2 Effect of Slip-Shear Forces

From Eq. (10.96), after substituting for $g_1'(\eta)$ and $g_1''(\eta)$, one obtains

$$\eta h_1' - 2h_1 = 0. \tag{10.108}$$

After integration,

$$h_1 = C_1\eta^2; \tag{10.109}$$

since $h_1(\infty) = 0$, therefore $C_1 = 0$, and so

$$h_1(\eta) = 0; \tag{10.110}$$

that is, at the start of the mixing region where $\xi \ll 1$, the drag effects do not produce a change in the particulate phase density. The increase in the density due to the slowing down of the phase in the x direction is just counterbalanced by the removal of particles from the boundary layer in the positive y direction.

The higher-order terms of Eqs. (10.78) and (10.79) are, respectively,

$$h_2' - \frac{5}{2\eta} h_2 = g_2'' - \frac{5}{2\eta} g_2', \tag{10.111}$$

$$\eta^2 g_2'' - 4\eta g_2' + \frac{21}{4} g_2 = \frac{Z}{1+A} (f_0')^{1/2}(g_0' - f_0'). \tag{10.112}$$

Integrating Eq. (10.111), one obtains

$$h_2 = g_2'(\eta), \tag{10.113}$$

since $h_2(\infty) = g_2'(\infty) = 0$. On defining

$$G_2 = \frac{g_2(1+A)}{ZA^{3/2}}, \tag{10.114}$$

Eq. (10.112) becomes

$$\eta^2 G_2'' - 4\eta G_2' + \frac{21}{4} G_2 = F(\eta), \tag{10.115}$$

where

$$F(\eta) = 4(g_0' - f_0')(f_0'')^{1/2} = \frac{4}{\pi^{1/4}} \left(1 - \operatorname{erf}\left(\frac{\eta}{2}\right)\right) \exp\left(-\frac{1}{8}\eta^2\right). \tag{10.116}$$

The solution to the homogeneous part of Eq. (10.115) is

$$G_2(\eta) = \eta^{3/2} A(\eta) + \eta^{7/2} B(\eta). \tag{10.117}$$

Using the variation of parameters again, one gets

$$A(\eta) = \frac{1}{2} \int_\eta^\infty \frac{F(x)}{x^{5/2}}\, dx, \tag{10.118}$$

$$B(\eta) = -\frac{1}{2} \int_\eta^\infty \frac{F(x)}{x^{9/2}}\, dx. \tag{10.119}$$

So, the *complete* solution is

$$G_2(\eta) = \frac{1}{2}\eta^{3/2}\int_\eta^\infty \frac{F(x)}{x^{5/2}}\,dx - \frac{1}{2}\eta^{7/2}\int_\eta^\infty \frac{F(x)}{x^{9/2}}\,dx + C_1\eta^{3/2} + C_2\eta^{7/2}; \tag{10.120}$$

therefore

$$G_2'(\eta) = \frac{3}{4}\eta^{1/2}\int_\eta^\infty \frac{F(x)}{x^{5/2}}\,dx - \frac{7}{4}\eta^{5/2}\int_\eta^\infty \frac{F(x)}{x^{9/2}}\,dx + \frac{3}{2}C_1\eta^{1/2} + \frac{7}{2}C_2\eta^{5/2}.$$

Since $C_2'(\infty) = 0$, therefore $C_1 = C_2 = 0$, and so

$$G_2'(\eta) = \frac{3}{4}\eta^{1/2}\int_\eta^\infty \frac{F(x)}{x^{5/2}}\,dx - \frac{7}{4}\eta^{5/2}\int_\eta^\infty \frac{F(x)}{x^{9/2}}\,dx. \tag{10.121}$$

Since $h_2 = g_2'(\eta)$,

$$\left(\frac{1+\Lambda}{\Gamma\Lambda^{3/2}}\right)h_2(\eta) = \frac{3}{4}\eta^{1/2}\int_\eta^\infty \frac{F(x)}{x^{5/2}}\,dx - \frac{7}{4}\eta^{5/2}\int_\eta^\infty \frac{F(x)}{x^{9/2}}\,dx. \tag{10.122}$$

Finally, the transverse (dimensionless) particle phase velocity due to the action of slip-shear forces is

$$v_{p2}^* = \xi^{3/4}\left(\frac{\eta}{2}g_2' - \frac{7}{4}g_2\right) \tag{10.123}$$

or

$$\frac{1+\Lambda}{Z\Lambda^{3/2}}v_{p2}^* = \xi^{3/4}\left(\frac{\eta}{2}G_2' - \frac{7}{4}G_2\right) \tag{10.124}$$

This is plotted in Fig. 10.8.

The effect of this velocity is to transport the particle phase into the clear fluid. The maximum of this velocity is at $\eta = 0$. Previously it was shown that due to drag forces the particles acquire a velocity away from the clear fluid, and that this is a minimum at $\eta = 0$. This produces a two-directional migration of the particles.

At a value of η given by the equation

$$\xi^{1/4} = \left|\frac{v_{p2}^*(\eta_0)}{v_{p1}^*(\eta_0)}\right|, \tag{10.125}$$

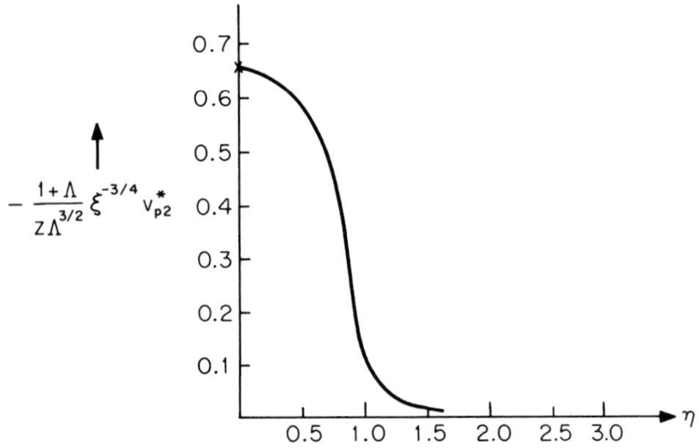

Fig. 10.8

those particles located at $\eta > \eta_0$ are moving toward the clear fluid. In fact, if there were no slip-shear effects, the particles would not penetrate the clear fluid region. Equation (10.122) gives the corresponding decrease in the particle phase density of the suspension. It can be shown that

$$\lim_{\eta \to 0} h_2 = \frac{8}{10} \lim_{\eta \to 0} \frac{F(\eta)}{\eta}.$$

Since $F(0) = 4/\pi^{1/4}$, h_2 is singular at $\eta = 0$. This implies that in the region $\eta = 0$, the basic assumption of a particle continuum would break down, due to particle removal; $h_2(\eta)$ is plotted in Fig. 10.9.

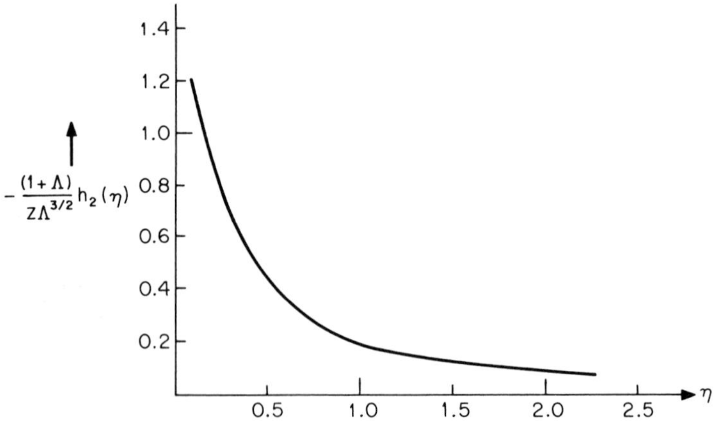

Fig. 10.9

It must be noted that

$$f_0(\eta), \quad \frac{f_1(\eta)}{\Lambda W}, \quad \frac{g_1'(\eta)}{\Lambda}, \quad \frac{(1+\Lambda)g_2(\eta)}{Z\Lambda^{3/2}}, \quad \text{and} \quad h_2(\eta)$$

are universal functions of η. Once these have been computed, the first-order solution for the problem of mixing of any solid particle/fluid suspension with a clear fluid is available.

As experimental results are virtually nonexistent, it is not yet possible to directly compare the above theoretical predictions with such results. In the experiments of Karmis *et al.* [21], however, observations were carried out on a suspension pumped through a tube. It was observed that where the solid phase lagged behind the fluid phase, a particle free zone developed near the wall. In these experiments the Saffman condition of $(\mathrm{Re})_k/(\mathrm{Re})_\mathrm{p}^2 \gg 1$ was not satisfied. The ratio varied from 10 to 100 for those experiments in which the free particle zone was observed. The clear zone would be compatible with the explanation that it is caused by the combined effect of the inertial wall effect and the slip-shear forces which would tend to move the particles toward the center of the tube. The Segre and Silberberg observations can also be explained by the opposing effects of slip-shear and wall forces.

In the treatment of particle fluid mechanics expounded here, it has been assumed that the particle effect on the fluid can be represented by means of body forces in the Navier–Stokes equations. It is unlikely that such a procedure gives complete information on the mechanism of momentum transfer from the particles to the fluid. The particles form interior boundary conditions for the flow and the effect of the particles on the fluid is dependent on the momentum distribution around each particle. This stage in the development of particle fluid mechanics is extremely difficult and not yet resolved. Its resolution may only be made possible by use of large digital computer simulations.

BIBLIOGRAPHY

1. G. G. Stokes, On the effect of internal friction of fluids on the motion of pendulums. "Mathematics and Physics Papers," Vol. 3, pp. 1–147. Cambridge Univ. Press, London and New York, 1891.
2. S. Goldstein, "Modern Developments in Fluid Dynamics," pp. 83, 492. Oxford Univ. Press (Clarendon), London and New York, 1938.
3. G. Segre and A. Silberberg, *Nature* **189**, 209 (1961).

4. G. Segre and A. Silberberg, *J. Fluid Mech.* **14**, 115 (1962).
5. S. Rubinow and J. Keller, The transverse force on a spinning sphere moving in a viscous fluid, *J. Fluid Mech.* **11**, 447–459 (1961).
6. P. G. Saffman, *J. Fluid Mech.* **22**, 385 (1965).
7. A. Tchernov, *Izv. Akad. Nauk Kaz. SSR, Ser. Energ.* No. 8 (1955).
8. R. Eichorn and S. Small, *J. Fluid Mech.* **20**, 513 (1965).
9. A. B. Basset, "Hydrodynamics," p. 270. Dover, New York, 1961.
10. R. Hughes and E. Gilliland, *Chem. Eng. Progr.* **48**, 497 (1952).
11. G. K. Batchelor, "An Introduction to Fluid Dynamics," p. 248. Cambridge Univ. Press, London and New York, 1967.
12. J. Happel and H. Brenner, "Low Reynolds Number Hydrodynamics." Prentice-Hall, Englewood Cliffs, New Jersey, 1965.
13. C. Sadron, *in* "Flow Properties of Disperse Systems" (J. J. Hermans, ed.). North-Holland Publ.,, Amsterdam, 1953.
14. H. Frisch and R. Simha, The Viscosity of Colloidal Suspensions and Macromolecular Solutions, *in* "Rheology" (F. R. Eirich, ed.), Vol. 1, Chapter 14. Academic Press, New York, 1956.
15. F. E. Marble, Dynamics of a gas containing small solid particles, *5th Agardograph Colloq.*, p. 175 (1963).
16. S. L. Soo, "Fluid Dynamics of Multiphase Systems." Ginn (Blaisdell), Boston, 1967.
17. G. O. Hinze, Momentum and mechanical energy balance for a flowing homogeneous suspension, *Appl. Sci. Res.* **11**, (1962).
18. F. E. Marble, *5th Agard. Combustion Colloq., Brauschweig*, 1962.
19. B. Otterman, Laminar Boundary Layer Flows of a Two-Phase Suspension. Ph. D. Thesis, Suny at Stonybrook (1968).
20. B. Otterman and S. Lee, Particle migrations in laminar mixing of a suspension with a clean fluid, *ZAMP* **20**, 730–749 (1969).
21. A. Karmis, H. Goldsmith, and S. Mason, The flow of suspensions through tubes, *Can. J. Chem. Eng.* pp. 181–193 (1966).
22. F. Zenz and D. Othmer, "Fluidization and Fluid–Particle Systems." Van Nostrand-Reinhold, Princeton, New Jersey, 1960.
23. R. S. Brodkey, "The Phenomena of Fluid Motions," Chapters 16–18. Addison-Wesley, Reading, Massachusetts, 1967.

CHAPTER ELEVEN *HYDRODYNAMICS OF SUPERFLUIDS*

As is well known, there are two stable isotopes of helium, ^{3}He and ^{4}He, of mass 3 and 4, respectively, in atomic units.

The ^{4}He isotope was first liquefied in 1908 [1]. The liquid form is commonly designated as helium I or II. The boiling point is 4.21°K and it has a critical point at a temperature of 5.2°K and a pressure of 2.26 atm.

About 1927 Keesom and Wolfke measured the dielectric constant of liquid helium as a function of temperature. A small discontinuity in the curve was found at a temperature of 2.18°K. Some sort of phase transition was indicated.

On a thermodynamic basis there are two kinds of phase transitions which commonly occur—the first-order phase transitions which have a latent heat and the second-order ones with no latent heat. The condensation of a saturated vapor into a liquid phase is a typical first-order transition. The transition from ferro- to paramagnetism on heating a piece of iron through its Curie temperature is a second-order type.

Keesom[2], in 1932, measured the specific heat versus temperature. The curve obtained is shown in Fig. 11.1 and shows a jump in the specific heat at the same temperature at which the discontinuity in the dielectric constant occurred.

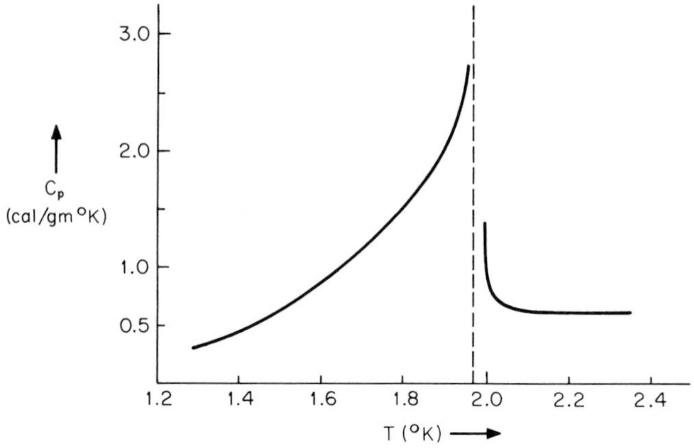

Fig. 11.1

The latent heat of the transition was found to be negligible and so the transition was second-order. This discontinuity is in fact a logarithmic singularity [2, 3].

The similarity of the shape of the specific heat curve to the Greek letter "lambda" led Keesom to call the transition the "λ point." The temperature region above this point is designated as He I, and below as He II.

The phase transition from He I to He II is also accompanied by a sharp maximum and a discontinuity in the slope of the density curve [4].

11.1 Fountain Effect

Allen and Jones [5] at Cambridge, in 1938, made the following remarkable discovery. Two variations of the "fountain-effect" experiment are shown in Fig. 11.2.

The first version (Fig. 11.2a) involved a capillary tube with a wide end, the latter being tightly packed with fine powder. Both ends are open and the latter end is immersed in He II. When a radiant heat source outside the cryostat falls on the powder, a liquid He fountain emerges from the capillary. This does not happen when He I is used.

The second version (Fig. 11.2b) involves electric heat, and an appreciable head of liquid can be produced in the wide end of the capillary.

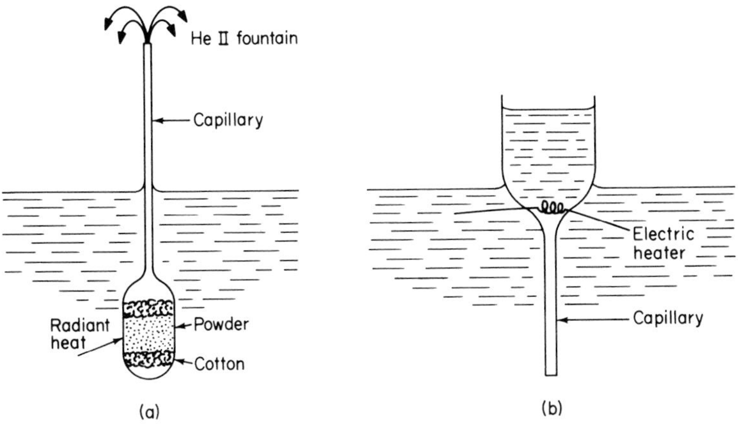

Fig. 11.2

11.2 Flow of Liquid Helium

By 1939, several measurements of the viscosity of He II had been made. These were based on three accepted methods:

(i) *Poiseuille flow through a narrow channel* under a steady pressure gradient. The well-known expression for the volumetric flow rate is

$$\dot{v} = \frac{\pi a^4}{8\eta} \frac{p_1 - p_2}{l} \quad \text{cm}^3/\text{sec},$$

where a is the capillary radius, $p_1 - p_2$ the pressure drop, and η the coefficient of viscosity of the liquid.

(ii) *Oscillating disk method* A thin disk suspended by a torsion wire executes simple harmonic oscillations while immersed in the liquid. The damping of the oscillations depends on the product of the density and viscosity of the liquid.

(iii) *Concentric cylinder viscometer* The inner cylinder is suspended by a torque wire, and the outer cylinder is rotated. The fluid fills the gap between the two cylinders and transmits a torque to the inner one which is measured by the deflection produced:

$$\text{torque} = 4\pi\eta L \frac{a_1^2 a_2^2}{a_2^2 - a_1^2} \omega,$$

where L is the length of the cylinder, and a_1, a_2 are the radii of the outer and inner cylinders, respectively.

For any "classical" liquid, all three methods should give the same value for the viscosity.

Kapitza [6], using methods (i) and (ii), found startling differences in the values for η in He II. He found that He II flowed easily through even the narrowest channel even under very small pressure heads ($\sim$100 dyn/cm^2). It was also found that for such narrow channels ($\sim$0.1 micron) the flow was independent of the pressure head. Now, for laminar flow, a viscous fluid flows with a volume flow proportional to Δp. Thus, method (i) predicted a virtually zero viscosity for He II.

Method (ii), however, at the same temperature, indicated considerable damping of the disk, which decreased sharply, however, as the temperature was reduced below the λ point. The value predicted for the viscosity by method (ii) was about 10^6 times that predicted by method (i).

The ability of He II to flow through extremely fine capillaries with apparently no viscous friction is lost when the speed of the liquid exceeds a certain critical velocity [7, 8].

11.3 Second Sound

Tisza [9] in 1940 and Landau [10] in 1941, developed theories of He II. Both theories predicted that small fluctuations in entropy (or temperature) are propagated as an undamped dispersionless wave in He II. The existence of this temperature wave, or second sound, was first experimentally confirmed by Peshkov [11] in 1944.

These strange properties of He II are known as "super" properties, and He II is called a *superfluid*.

11.4 Two-Fluid Model of He II

The "super," or peculiar, properties of He II can be largely accounted for on the basis of phenomenological two-fluid theory [9, 10].

A most striking characteristic of liquid helium is that, as far as is known, it exists in the liquid state down to the absolute zero temperature. This is due to two things:

(a) the van der Waals forces in helium are weak;
(b) the zero-point energy, due to the light mass, is large.

The atoms in the liquid are relatively far apart and considerable pressure is required (about 26 atm at 0°K) to force them into the solid state. Thus, the characteristics of liquid helium are governed by quantum effects and it is called a quantum liquid.

In fact, London [12], in 1954, suggested that below the λ point helium is a quantum fluid whose essential feature is the macroscopic occupation of a single quantum state. This implies long-range order in He II and is called *Bose condensation.* The "condensed" units are held together by exchange forces and hence the whole condensate has negligible fluctuations and moves as a whole.

In the ground state $|\,0\rangle$, a macroscopic number of particles n_0 is in the state with zero momentum; they form the condensate.

Thus, London attributed the λ transition to a Bose–Einstein condensation occurring in the liquid. There now seems little doubt of the correctness of the essential features of the macroscopic quantum effects first set out by London.

Starting from the general features of the Bose–Einstein condensation, Tisza in 1938 first proposed a two-fluid model to explain the phenomena occurring in He II. Landau [13], in 1947, developed a somewhat different version of the two-fluid model and has been remarkably successful in describing the thermal and hydrodynamic properties of He II.

The basic assumptions of the two-fluid model are as follows:

(a) He II consists of a kind of mixture of two components, a normal component and a superfluid component. The density of the fluid, ϱ, can thus be separated into a normal density ϱ_n and a superfluid density ϱ_s:

$$\varrho = \varrho_n + \varrho_s. \tag{11.1}$$

Similarly, the fluid motion, characterized by its local velocity $\mathbf{v}$, may also be considered to be due to the combined motions of the fluid components, so that

$$\mathbf{J} = \varrho \mathbf{v} = \varrho_n \mathbf{v}_n + \varrho_s \mathbf{v}_s, \tag{11.2}$$

where $\mathbf{v}_n$ and $\mathbf{v}_s$ are the velocities of the normal and superfluid components, respectively, and $\mathbf{J}$ is the mass current density.

(b) It is assumed that the entropy of the superfluid component is zero ($S_n = 0$), so that the total entropy of the liquid is

$$\varrho S = \varrho_n S_n.$$

(c) The superfluid component can move without friction as long as certain velocity limits are not exceeded. The normal component has viscosity.

The components of He II are not well defined and reveal themselves only collectively as a bulk fluid element. The reason we cannot isolate, in a classical sense, the particles of the normal and superfluid components is because He II is a fluid for which the quantum effects are important even on the macroscopic scale.

11.5 Critical Velocities

In the two-fluid model proposed by Landau, the normal component of the liquid is composed of thermal excitations—that is, the excitations have an effective mass and momentum attributable to the normal component. The superfluid background in which these excitations are embedded is the ground state liquid helium and hence is at zero temperature. It is not necessarily stationary in the macroscopic sense, for whether the bulk fluid is in motion or not is a relative matter. The motion of these macroscopic elements of "background" is represented by $\mathbf{v}_s$.

It is the excitation free superfluid which can move through capillaries without friction.

The excitations in liquid helium are very similar to those in solids and in fact are called *phonons*. There are also other excitations, called *rotons*. Based on this physical theory, we can obtain the macroscopic momenta and energy by the appropriate summation of the microscopic momenta and energies. But ϱ_n and ϱ_s are not basic quantities; they are derived.

In fact, there is still a certain amount of inconsistency with the macroscopic hydrodynamic theory of liquid helium.

Landau attributed the existence of a critical velocity in He II to the breakdown of the superfluid due to the creation of excitations.

Suppose the superfluid helium flows through a capillary with velocity v (in the ground state of energy). If this flow were accompanied by friction (as between the fluid and the capillary walls), then some of the kinetic energy is dissipated and transformed into thermal energy. If the helium is heated, then it makes transitions to excited states. But a quantum liquid cannot receive energy continuously. To get into the lowest excited state, an elementary excitation must be created. Let the energy of such an excitation be $\varepsilon(p)$, where p is the corresponding momentum, in the frame of reference moving with the liquid. Then, in a fixed frame of reference (that is, with respect to the walls), the energy of the system has changed by an amount

$$\varepsilon(p) + \mathbf{p} \cdot \mathbf{v}, \tag{11.3}$$

and this is the energy of the excitation ε'. Since the excitation can be created only by a decrease in the kinetic energy of the liquid, ε' must be negative to favor the transition

$$\varepsilon(p) + \mathbf{p} \cdot \mathbf{v} < 0. \tag{11.4}$$

The smallest value of $\mathbf{v}$ corresponding to this inequality is for $\mathbf{p}$ being directed opposite to $\mathbf{v}$ and so

$$\mathbf{v} > \frac{\varepsilon(p)}{p} \tag{11.5}$$

to produce the excitation; that is, the flow will remain frictionless if the velocity is less than the critical value v_c, where

$$v_c = \frac{\varepsilon}{p}.$$

In the energy spectrum (energy versus momentum) the lowest value for v_c will come at the minimum point in the curve, where ε/p is minimum.

Landau proposed that phonons and rotons are the two types of excitations which make up the normal fluid. Phonons are really the excitations of longitudinal sound waves; rotons, as per Landau, correspond to possible rotational modes of motion in the liquid.

For the phonons the energy–momentum relation is

$$\varepsilon = ap, \tag{11.6}$$

where a is the velocity of sound. For rotons, Landau proposed the following spectrum:

$$\varepsilon = \Delta + \frac{(p - p_0)^2}{2\mu}, \tag{11.7}$$

where p_0 is the value of the momentum at which ε has a minimum equal to Δ. These two types of excitations are part of a single energy–momentum curve as proposed by Landau to explain the experimental values obtained for the thermodynamic functions of liquid helium (Fig. 11.3).

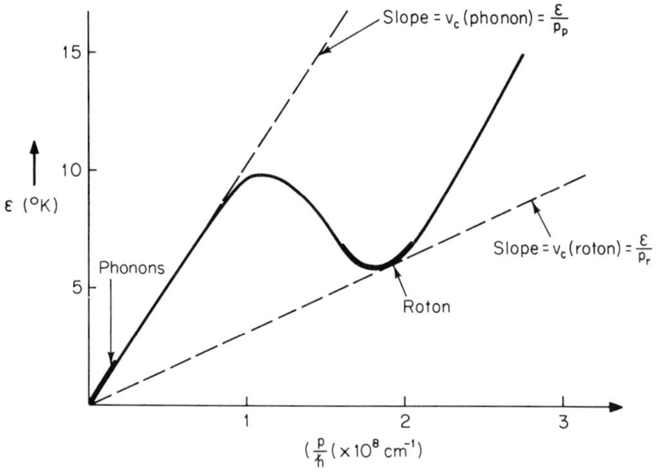

Fig. 11.3

Monochromatic neutrons emit or absorb elementary excitations in helium. By measuring the energies of neutrons scattered at given angles, one can determine the whole spectrum of elementary excitations [14,15]. Thus, the existence of this energy/momentum excitation curve was confirmed and the following values of the spectrum parameters were determined:

$$\frac{\Delta}{k} = 8.6°\text{K}, \qquad \frac{p_0}{k} = 1.91 A^{-1}, \qquad \mu = 0.16 m_{\text{He}}:$$

μ has the dimensions of mass and is commonly called the effective roton mass.

Using Landau's criterion ($v_c = \varepsilon/p$), it is found that for phonon creation

$$v_c = a = 2.4 \times 10^4 \quad \text{cm/sec},$$

and for roton creation

$$v_c = 6 \times 10^3 \quad \text{cm/sec}.$$

The critical velocities observed experimentally are two orders of magnitude less than these values, and they also depend on channel size. This discrepancy throws a shadow of doubt on Landau's theory and has yet to be satisfactorily resolved.

To study a more detailed exposition of excitations in liquid helium the reader is referred to the article by de Boer [16].

11.6 Quantized Vortices

The assertion that superfluid motion is irrotational has been verified experimentally. Both Landau and London had already arrived at the condition that

$$\nabla \times \mathbf{v}_s = 0.$$

It would follow then that if one rotates a cylindrical vessel containing superfluid helium, only the normal component would be carried along by the rotation. The superfluid component would remain at rest. The height of the meniscus formed in a rotating bucket of He II would be smaller by a factor ϱ_n/ϱ than the one formed in a classical liquid. But observations by Osborne [17] in 1950 showed that the free superfluid surface assumed the shape of a paraboloid, just as the classical liquid does.

On the basis of quantum mechanics, Onsager [18] in 1950 showed that there was a possible wave function for the liquid which would produce a motion analogous to classical vortex motion. In fact, he suggested the possible existence of quantized vortices in He II. Feynmann [19] in 1955 developed this conjecture into a theory which has been successful in explaining many of the strange phenomena observed in the superfluid. The argument runs as follows.

Suppose the ground state wavefunction for the superfluid at rest is ψ_0. Feynmann then postulated that the wave function which represents the flowing liquid would be of the form

$$\psi_{\text{flow}} = \psi_0 \exp i \sum_j W(x_j). \tag{11.8}$$

Then

$$\mathbf{v}_s = \frac{\hbar}{m} \nabla W, \tag{11.9}$$

where W is a function of position x_j, $W(x_j)$ is its value at the jth atom, and m is the mass of the helium atom. Equation (11.9) implies that

$$\nabla \times \mathbf{v}_s = 0.$$

For a simply connected volume of fluid, the circulation around any curve C is then zero:

$$\text{circulation} = \oint_C \mathbf{v} \cdot d\mathbf{l} = \iint_A (\nabla \times \mathbf{v}_s) \cdot \mathbf{n}\, dA = 0, \tag{11.10}$$

where A is any surface spanning the curve. For a multiply connected region this is not true, and

$$\oint \mathbf{v} \cdot d\mathbf{l} = \oint_C \frac{\hbar}{m} \nabla W \cdot d\mathbf{l} = \frac{\hbar}{m} \oint dW. \tag{11.11}$$

The circulation is equal to the change of W, the phase of the wave function, in passing around the closed curve C. The superfluid state wave function must be single-valued and so

$$\oint \mathbf{v}_s \cdot d\mathbf{l} = 2\pi n \frac{\hbar}{m}, \tag{11.12}$$

where $n = 0, 1, 2, \ldots$.

These singular lines around which the circulation is not zero are analogous to vortex lines in classical hydrodynamics.

Now a classical vortex line placed at the origin has a velocity field given by

$$\mathbf{v} = \frac{\varkappa}{2\pi r} \mathbf{e}_\vartheta, \tag{11.13}$$

where $\mathbf{e}_\vartheta$ is a unit tangential vector, $\varkappa$ a constant (the vortex strength), and r the distance from the origin. The streamlines are concentric circles about the vortex line and the vorticity $\boldsymbol{\zeta} = \nabla \times \mathbf{v}$ is zero everywhere except for the vortex line itself.

The circulation around a contour surrounding the vortex line is

$$\oint \mathbf{v} \cdot d\mathbf{l} = \oint \frac{\varkappa}{2\pi r} \mathbf{e}_\vartheta \cdot d\mathbf{l} = \varkappa. \tag{11.14}$$

For a vortex line in He II,

$$\oint \mathbf{v} \cdot d\mathbf{l} = \varkappa = \frac{nh}{m}. \tag{11.15}$$

So it is possible to have states of motion in the superfluid that are vortex lines, but with the constraint that the circulation about any of these lines must be an integral multiple of h/m. The circulation obeys a quantization condition.

These quantized vortex lines explain how there can be nonzero circulation in a singly connected vessel containing He II while there is zero vorticity throughout nearly all the liquid. This can be demonstrated by analyzing the motion of the superfluid in a rotating bucket.

The velocity field per unit length of a vortex parallel to the axis of rotation is

$$v_s = \frac{\varkappa}{2\pi} \times \frac{1}{r}, \qquad \varkappa = \frac{hn}{m}. \tag{11.16}$$

The kinetic energy per unit length is

$$E_k = \frac{1}{2} \varrho_s \int_{r_a}^{r_b} v_s^2 2\pi r \, dr = \varrho_s \frac{\varkappa^2}{4\pi} \ln \frac{r_b}{r_a}, \tag{11.17}$$

where r_a is the vortex core radius and r_b is some dimension external to the vortex. If the total number of vortices per unit surface area is N, then r_b is given by

$$\pi r_b^2 = \frac{1}{N}. \tag{11.18}$$

Using Stokes' theorem on Eq. (11.15), one obtains

$$N = \frac{|\operatorname{curl} \mathbf{v}_s|}{\varkappa}. \tag{11.19}$$

If F is the free energy of the moving liquid, M the angular momentum, and ω_0 the angular velocity of the vessel, then in the equilibrium state the quantity $F - M\omega_0$ will have a minimum value. The kinetic energy of the motion is

$$\varrho_s \frac{v_s^2}{2} + \varrho_n \frac{v_n^2}{2}.$$

Considering only the motion of the superfluid component, one can minimize the quantity $E_s - M_s\omega_0$. The normal component of the fluid will undergo rotation with a velocity $v_n = \omega_0 r$, just like a solid body.

The energy of the superfluid motion is given by

$$E = \frac{1}{2}\varrho_s \int v^2 2\pi r\,dr + \varrho_s \frac{\varkappa}{4\pi} \int |\operatorname{curl}\mathbf{v}| \ln\left(\frac{\varkappa^{1/2}}{\pi^{1/2}|\operatorname{curl}\mathbf{v}|^{1/2} r_a}\right) 2\pi r\,dr. \quad (11.20)$$

The second term is the energy of the vortices, which is obtained by using Eqs. (11.16)–(11.19).

Similarly, the angular momentum of the fluid is determined to be

$$M = \varrho_s \int vr 2\pi r\,dr + \varrho_s \frac{\varkappa}{2\pi} \int 2\pi r\,dr. \quad (11.21)$$

The difference $E - M\omega_0$ will now be varied with respect to δv:

$$\int \delta v \left\{ (v - \omega_0 r) + \frac{\varkappa}{8\pi} \frac{\dfrac{\partial}{\partial r}\dfrac{1}{r}\dfrac{\partial}{\partial r}(vr)}{\dfrac{1}{r}\dfrac{\partial}{\partial r}(vr)} \right\} 2\pi r\,dr + \frac{\varkappa}{4\pi} \int \frac{\partial}{\partial r} \left\{ r\,\delta v \ln \frac{\varkappa^{1/2}}{(\pi^{1/2}|\operatorname{curl}\mathbf{v}|^{1/2} r_a)} \right\} 2\pi r\,dr, \quad (11.22)$$

where

$$|\operatorname{curl}\mathbf{v}| = \frac{1}{r}\frac{\partial}{\partial r}(vr)$$

in cylindrical coordinates. The variation must be zero, and the first integral will be zero if

$$(v - \omega_0 r)\frac{1}{r}\frac{\partial}{\partial r}(vr) + \frac{\varkappa}{4\pi}\frac{\partial}{\partial r}\frac{1}{r}\frac{\partial}{\partial r}(vr) = 0. \quad (11.23)$$

The second integral is zero if δv is zero at the boundaries of the domain of integration. Hence Eq. (11.23) determines the superfluid velocity in a rotating bucket. It has the exact solutions

$$v = \omega_0 r, \tag{11.24}$$

$$v = \frac{A}{r}. \tag{11.25}$$

Equation (11.24) implies solid body rotation, and Eq. (11.25) implies irrotational motion with curl $\mathbf{v} = 0$.

It is now assumed that solid body rotation occurs in a region inside some radius R_c. Outside this region the motion is irrotational, as given by Eq. (11.25). It can be shown [20] that $R - R_c$ is given by

$$R - R_c = \frac{1}{2}\left(\frac{\varkappa}{\omega_0}\ln\frac{r_b}{er_a}\right)^{1/2}. \tag{11.26}$$

Thus the region of irrotational motion is small but observable. So, when a bucket of He II is rotated, the vortices formed cause solid body rotation in nearly the whole bucket. However, in a small region near the walls of the bucket, there are no vortices and the motion here is completely irrotational.

It is also obvious that the same explanation will hold for the shape of the free liquid surface observed.

In 1961 Vinen [21] measured the circulation about a single vortex. He observed that the average circulation of He II around a fine wire along the axis of a slowly rotating cylindrical container was more stable at the value h/m than at any other value. Hess and Fairbank [22] in 1966 have shown that the angular momentum of the equilibrium state of slowly rotating superfluid equals the angular momentum of the lowest free energy state allowed by the quantized vortex model.

It appears, then, that the quantized vortex concept is correct and accurately describes the states which He II can assume when rotated.

11.7 Quantized Vortices and Critical Velocities

Quantized vortex lines can also explain the critical velocity effects observed in flowing He II.

Now a vortex line has an energy per unit length and, as seen in Chapter 5, a vortex ring has an energy

$$E = \frac{\varrho\varkappa^2 r}{2}\left\{\ln\frac{8r}{r_a} - \frac{7}{4}\right\}, \tag{11.27}$$

where ϱ is the liquid density, $\varkappa$ the vortex strength, or circulation, r the ring radius, and r_a the core radius. The vortex ring also has an impulse

$$I = \pi\varrho\varkappa r^2. \tag{11.28}$$

Since quantized vortices can exist in the superfluid, they can be regarded as another type of excitation and one can apply Landau's criterion, $\varepsilon/p = v_c$. These excitations do not change the thermodynamic variables of the liquid, because their statistical weight is so small. For a tube of diameter d, the vortex ring would be expected to have a maximum radius $d/2$. Then

$$\frac{\varepsilon}{p} = v_c = \frac{\hbar}{mr} \ln\left(\frac{8r}{r_a} - \frac{7}{4}\right) \tag{11.29}$$

since $\varkappa = h/m$ for the quantized vortex ring.

The critical velocities predicted by this expression are much smaller than those derived from phonons and rotons, and are closer to the experimentally observed values.

There is considerable evidence for the creation of quantized vortex rings. Rayfield and Reif [23] in 1964 observed vortex rings with circulation h/m created by ions. The low energy ions after creating the vortex rings were trapped by them. By measuring the energy and velocity of the ions, Rayfield and Reif were able to show that the circulation was always h/m to within 3%, and that the core radius was 1.3 Å to within 10%. They directly measured the relationship between the energy and velocity of a vortex ring and established that, as usual, $v \propto 1/\varepsilon$. This is especially convincing evidence.

11.8 Jet Flow of Superfluid Helium

Consider the jet of helium emerging from an orifice, forming a vortex sheet along the boundary between the moving and the stationary fluid.

Inside the jet the velocity is v and outside the jet it is zero. The jet can be represented by an array of vortex rings (Fig. 11.4). The circulation (strength) per unit length of the jet is

$$\oint \mathbf{v} \cdot dl = v.$$

Now

$$\oint \mathbf{v} \cdot dl = N \frac{nh}{m},$$

Fig. 11.4

where N is the number of vortex lines per centimeter. It has been assumed that each vortex has a single quantum of circulation, $n = 1$. Therefore

$$N = \frac{mv}{h}. \tag{11.30}$$

The vortex rings move with the velocity of the fluid at their position, which is $v/2$. Therefore,

$$Nv = \frac{mv^2}{2h} = g \tag{11.31}$$

vortex rings per second are formed.

Energy is required to form these rings, which once formed start to dissipate as heat. Thus, a flow resistance appears. The energy required per second is

$$\varepsilon g = \frac{\varrho hr}{4m}\left(\ln\frac{8r}{a} - \frac{7}{4}\right)v^2. \tag{11.32}$$

The total kinetic energy per second available from the fluid is $\frac{1}{2}\pi r^2 \varrho v^3$. The critical velocity v_c is defined as that velocity for which there is just enough energy to form the rings. Hence,

$$\tfrac{1}{2}\pi r^2 \varrho v_c^3 = \frac{\varrho hr}{4m}\left(\ln\frac{8r}{a} - \frac{7}{4}\right)v_c^2,$$

and so

$$v_c = \frac{\hbar}{mr}\left(\ln\frac{8r}{a} - \frac{7}{4}\right). \tag{11.33}$$

This is the same result as was obtained by using the Landau relation [see Eq. (11.29)]. The theory as presented above is a very idealized picture of what may take place in the liquid. As Feymann has pointed out, the flow may be very complex and irregular at low velocities which are near the critical velocity. At high velocities one would expect the flow to be turbulent as in ordinary fluid flow.

11.9 Reversible Hydrodynamic Equations

As mentioned before, current macroscopic theories of liquid helium are formulated in terms of a two-fluid concept on the basis of Bose–Einstein condensation.

There are two modes of motion, each of which is associated with its own "effective mass." The sum of these masses equals the total true mass of the liquid.

To recapitulate the two-component model: ϱ_n is the density of the component, S_n the entropy per unit mass, and ϱ_s, S_s the corresponding quantities for the superfluid component. Each component has its own velocity field, $\mathbf{v}_n$ and $\mathbf{v}_s$, respectively, and

$$\varrho = \varrho_n + \varrho_s, \qquad \varrho S = \varrho_n S_n + \varrho_s S_s, \qquad \varrho \mathbf{v} = \varrho_n \mathbf{v}_n + \varrho_s \mathbf{v}_s. \tag{11.34}$$

It can be assumed that $S_s = 0$. The equations for the conservation of mass and entropy are

$$\frac{\partial \varrho}{\partial t} + \nabla \cdot [\Gamma \varrho \mathbf{v}_n + (1 - \Gamma)\varrho \mathbf{v}_s] = 0, \tag{11.35}$$

$$\frac{\partial(\varrho S)}{\partial t} + \nabla \cdot (\varrho S \mathbf{v}_n) = 0, \tag{11.36}$$

where $\Gamma = \varrho_n/\varrho$.

The derivation of the equations for $\mathbf{v}_n$ and $\mathbf{v}_s$ is very lengthy and the final form [10, 12, 24] is as follows:

$$\begin{aligned}\frac{\partial v_n}{\partial t} + (\mathbf{v}_n \cdot \nabla)\mathbf{v}_n = &-\frac{1}{\varrho}\nabla p - \frac{1-\Gamma}{\Gamma} S\,\nabla T - \nabla \Omega \\ &- \frac{1-\Gamma}{2}\nabla(\mathbf{v}_n - \mathbf{v}_s)^2 - (\mathbf{v}_n - \mathbf{v}_s)\frac{\lambda}{\Gamma\varrho},\end{aligned} \tag{11.37}$$

$$\frac{\partial v_s}{\partial t} + (\mathbf{v}_s \cdot \nabla)\mathbf{v}_s = -\frac{1}{\varrho}\nabla p + S\,\nabla T - \nabla\Omega + \frac{\Gamma}{2}\nabla(\mathbf{v}_n - \mathbf{v}_s)^2, \tag{11.38}$$

where Ω is the potential of the external force field, and

$$\lambda = \frac{\partial \varrho_n}{\partial t} + \nabla \cdot (\varrho_n \mathbf{v}_n) = -\left[\frac{\partial \varrho_s}{\partial t} + \nabla \cdot (\varrho_s \mathbf{v}_s)\right];$$

λ is the source density of the normal fluid. It must be noted that apart from the ideas of the two-fluid theory, the approaches and reasoning used in setting up the above equations are those normally used in the mechanics of continuum. The ultimate justification is the agreement with experimental observation. From this aspect, the theory is quite successful. The ability to flow through very narrow channels, the fountain effect, and the propagation of temperature waves (second sound) can be satisfactorily explained by the linearized version of the above set of equations. This is mainly due to the presence of the thermomechanical terms $S\,\nabla T$ in Eqs. (11.37) and (11.38).

In Eqs. (11.37) and (11.38) the term $-(\mathbf{v}_n - \mathbf{v}_s)(\lambda/\Gamma\varrho)$ can be interpreted as the average increment in $\mathbf{v}_n$ per second due to the interactions between normal and "superfluid" particles. The term $\frac{1}{2}(\mathbf{v}_n - \mathbf{v}_s)^2$ is $(\partial U/\partial\Gamma)_{\varrho,S}$, where $U(\varrho, S, \Gamma)$ is the internal energy density of the fluid [12]. Multiplying Eq. (11.37) by ϱ_n and (11.38) by ϱ_s and adding, one obtains

$$\frac{D\mathbf{v}}{Dt} + \frac{1}{\varrho}\nabla\cdot[\varrho_n\mathbf{v}_n\mathbf{v}_n + \varrho_s\mathbf{v}_s\mathbf{v}_s - \varrho\mathbf{v}\mathbf{v}] = -\frac{1}{\varrho}\nabla p - \nabla\Omega, \qquad (11.39)$$

where

$$\frac{D}{Dt} \Rightarrow \frac{\partial}{\partial t} + (\mathbf{v}\cdot\nabla).$$

If Eq. (11.37) is now multiplied by $\varrho_n\mathbf{v}_n$ and (11.38) by $\varrho_s\mathbf{v}_s$ and they are added, one gets

$$\begin{aligned}\frac{D}{Dt}\left[\Gamma\frac{v_n^2}{2} + \frac{1-\Gamma}{2}v_s^2 + U\right] &= -\frac{1}{\varrho}\nabla\cdot\left[\frac{\varrho_n v_n^2}{2}(\mathbf{v}_n - \mathbf{v}) + \frac{\varrho_s v_s^2}{2}(\mathbf{v}_s - \mathbf{v}) + p\mathbf{v}\right.\\ &\quad \left. + \varrho ST(\mathbf{v}_n - \mathbf{v}) + \varrho\Gamma(1-\Gamma)\left(\frac{\partial U}{\partial\Gamma}\right)(\mathbf{v}_n - \mathbf{v}_s)\right] - \mathbf{v}\cdot\nabla\Omega,\end{aligned} \qquad (11.40)$$

where

$$p = \varrho^2\left(\frac{\partial U}{\partial\varrho}\right)_{S,\Gamma}, \qquad T = \left(\frac{\partial U}{\partial S}\right)_{\varrho,\Gamma}. \qquad (11.41)$$

The divergence term on the left side of Eq. (11.39) is the apparent stress due to diffusive transfer of momentum, characteristic of all mixtures. The divergence term on the right of Eq. (11.40) is due to the flux of energy.

The advantage of Eqs. (11.39) and (11.40) over (11.37) and (11.38) is that in the former the terms with Γ and $\nabla(\mathbf{v}_n - \mathbf{v}_s)^2$ have been eliminated. These terms arise because of the assumption of a particular kind of interaction between the normal and superfluid "particles"; that is, the velocity is not a thermal average and so all superfluid particles should have the same velocity $\mathbf{v}_s$ and any change in $\mathbf{v}_s$ due to interactions would be a change for all "particles." This would be an unsubstantiated assumption. Thus Eqs. (11.39) and (11.40) are independent of the mechanism of interaction. They represent the conservation of momentum and energy.

Besides Eqs. (11.35)–(11.38), the condition of irrotationality is usually imposed on the superfluid component (following Landau):

$$\nabla \times \mathbf{v}_s = 0. \tag{11.42}$$

Two approaches have been used to justify this condition.

One starts from the variational principle of a macroscopic fluid [12, 24]. The other is based on a microscopic argument. Since ^{4}He are Bose particles, the superfluid component being a giant quantum system with concerted motion, should be irrotational [19]. Lin has disagreed with both these approaches. He pointed out that in the variational approach there should be an additional condition to ensure the definite identification of the fluid particles. The necessity for irrotationality of the superfluid component then is removed. But the point is not resolved yet and is still one of the central problems in the hydrodynamics of superfluid helium.

11.10 Irreversible Hydrodynamic Equations

The Landau equations discussed above are apparently adequate for the treatment of reversible processes. The behavior of superfluid helium is much more complex when irreversible processes are active.

Equation (11.36) must now be changed to

$$\frac{\partial(\varrho S)}{\partial t} + \nabla \cdot (\varrho S \mathbf{v}_n + \mathbf{q}) = Q, \tag{11.43}$$

where $\mathbf{q}$ represents the nonconvective heat flux and Q the entropy production due to normal thermal conduction and frictional dissipation.

To the right-hand side of Eqs. (11.37) and (11.38) one must add $(1/\varrho_n)(-\mathbf{F}_n + \mathbf{F}_{sn})$ and $(1/\varrho_s)(-\mathbf{F}_s - \mathbf{F}_{sn})$, respectively, where $\mathbf{F}_n$ is the frictional force term for the normal component, $\mathbf{F}_s$ the force term for the superfluid component, and $\mathbf{F}_{sn}$ is the mutual friction term.

The nature of these extra terms is still under investigation, but experimental observations dictate their presence. The $\mathbf{F}_n$ is assumed to arise from the ordinary viscous stresses; $\mathbf{q}$ involves normal thermal conduction.

It is natural to assume that $\mathbf{F}_s = 0$ due to the phenomenon of superfluidity. But, as pointed out by Lin, the absence of $\mathbf{F}_s$ may be due to irrotationality of the superfluid component rather than the absence of viscosity. The shear strains in this component could still exist and the momentum be transformed without energy dissipation.

The existence of a critical velocity for He II suggests the introduction of mutual friction. On the microscopic scale the critical velocity implies that besides phonons and rotons there must be other types of low energy excitation. It was this consideration which led to the introduction of quantized vortex lines [19], the basic features of which have now been supported by experimental results [23, 25].

The concept of quantized vortex lines implies that when the fluid flow is supercritical, the superfluid component is in essence rotational and frictional. It can serve as the basis of a kinetic theory to derive the macroscopic mutual friction between the normal and superfluid components. This is feasible only when the number of vortex lines or rings is small. When the number is large, it is difficult to calculate the interaction due to their entanglements. This has led to the suggestion of other forms of mutual friction derived on semiempirical bases.

There are three different recognized formulations.

11.10.1 The Gorter–Mellink Formula [26]

In this case the force terms are taken to have the following forms:

$$\mathbf{F}_s = 0, \tag{11.44}$$

$$(-F_n)_i = \frac{\partial}{\partial x_k}\left\{\eta\left(\frac{\partial v_{ni}}{\partial x_k} + \frac{\partial v_{nk}}{\partial x_i} - \frac{2}{3}\,\delta_{ik}\,\frac{\partial v_{nl}}{\partial x_l}\right) + \eta_B \delta_{ik}\,\frac{\partial v_{nl}}{\partial x_l}\right\}_{i,k=1,2,3}, \tag{11.45}$$

$$\mathbf{F}_{sn} = \alpha(1 - \Gamma)\Gamma\,|\,\mathbf{v}_s - \mathbf{v}_n\,|^2(\mathbf{v}_s - \mathbf{v}_n), \tag{11.46}$$

where η and η_B are the coefficients of shear and bulk viscosity, respectively, Γ the density ratio, and α the coefficient of mutual friction.

The nonconvective heat flux is given by

$$\mathbf{q} = \frac{\varkappa}{T} \nabla T, \tag{11.47}$$

and the entropy production by

$$Q = \frac{1}{T} \left\{ \frac{\varkappa}{T} (\nabla T)^2 + \eta_B (\nabla \cdot \mathbf{v}_n)^2 + \frac{\eta}{2} \left(\frac{\partial v_{ni}}{\partial x_k} + \frac{\partial v_{nk}}{\partial x_i} - \frac{2}{3} \delta_{ik} \frac{\partial v_{nl}}{\partial x_l} \right)^2 + \alpha \varrho \Gamma (1 - \Gamma)(\mathbf{v}_s - \mathbf{v}_n)^4 \right\}. \tag{11.48}$$

At the boundary over which the fluid flows, the conditions are as follows:

(a) the perpendicular component of the total mass flux is zero;
(b) the perpendicular component of the heat flux is continuous;
(c) the tangential component of the normal fluid velocity is zero.

For problems involving bulk flows, $\mathbf{q}$ and Q can normally be neglected because of the high efficiency of heat transfer by the internal convection.

11.10.2 The Lin Formulation

Lin [27] assumed that the viscous stresses are linearly dependent of the rate of strain. As there are two components of fluid motion, there must be exchange coefficients relating the stress acting on one component to the rate of strain of the other. Thus, there must be four exchange coefficients, defined as follows:

$$\begin{aligned} \tau_{ij}^{(n)} &= 2\eta^{(nn)} e_{ij}^{(n)} + 2\eta^{(ns)} e_{ij}^{(s)}, \\ \tau_{ij}^{(s)} &= 2\eta^{(sn)} e_{ij}^{(n)} + 2\eta^{(ss)} e_{ij}^{(s)}, \end{aligned} \tag{11.49}$$

where $\tau_{ij}^{(n)}$ is the stress tensor acting on the normal component and $e_{ij}^{(n)}$ is the rate of strain of the normal component:

$$e_{ij}^{(n)} = \frac{1}{2} \left\{ \frac{\partial v_{in}}{\partial x_j} + \frac{\partial v_{jn}}{\partial x_i} \right\}. \tag{11.50}$$

The other symbols have similar meanings. The formulation is for incompressible fluids and recognizes neither irrotationality or inviscidness in the superfluid component.

The values for the force contributions are as follows:

$$-\mathbf{F}_{\mathrm{n}} = \eta^{(\mathrm{nn})} \nabla^2 \mathbf{v}_{\mathrm{n}}, \tag{11.51}$$

$$-\mathbf{F}_{\mathrm{s}} = \eta^{(\mathrm{sn})} \nabla^2 \mathbf{v}_{\mathrm{n}}, \tag{11.52}$$

$$\mathbf{F}_{\mathrm{sn}} = \eta^{(\mathrm{ns})} \nabla^2 \mathbf{v}_{\mathrm{s}} + \varrho \Gamma (1 - \Gamma)(\boldsymbol{\nabla} \times \mathbf{v}_{\mathrm{s}}) \times (\mathbf{v}_{\mathrm{n}} - \mathbf{v}_{\mathrm{s}}). \tag{11.53}$$

Thus, the additional force per unit volume for the normal component is

$$-F_{\mathrm{n}i} + F_{\mathrm{sn}i} = \frac{\partial \tau_{ij}^{(\mathrm{n})}}{\partial x_i} = \eta^{(\mathrm{nn})} \nabla^2 \mathbf{v}_{i\mathrm{n}} + \eta^{(\mathrm{ns})} \nabla^2 v_{i\mathrm{s}}, \tag{11.54}$$

and for the supercomponent it is

$$-F_{\mathrm{s}i} - F_{\mathrm{sn}i} = \frac{\partial \tau_{ij}^{(\mathrm{s})}}{\partial x_j} = \eta^{(\mathrm{sn})} \nabla^2 v_{i\mathrm{n}} + \nabla^2 v_{i\mathrm{s}}. \tag{11.55}$$

The second term in Eq. (11.53) does not contribute to the dissipation and so is strictly not a mutual friction term. It is not included in Eq. (11.55), and is absent when the flow is irrotational.

It is assumed also that $\eta^{(\mathrm{ss})} + \eta^{(\mathrm{ns})} = 0$, so that the viscous effect of the normal component is present only in the equation governing the total fluid.

Because of Eq. (11.53), an additional boundary condition is needed to determine the velocity of the superfluid component.

If $\mathbf{w}_{\mathrm{s}}$ is the velocity of the superfluid relative to the boundary, which is impermeable to both components, and if there is no conduction of heat through the boundary, then the normal boundary conditions are

$$w_{\mathrm{n}i} n_i = 0, \qquad w_{\mathrm{s}i} n_{\mathrm{s}} = 0. \tag{11.56}$$

If there is a heat supply, then the perpendicular component of mass flux must be zero, while the perpendicular component of the heat flux must be continuous.

The extra condition replaces the no-slip condition which governs the normal component, and is that the tangential component of the shear stress vector will be directly related to the slip velocity:

$$\tau_\alpha = F_{\mathrm{s}}(w_{\mathrm{s}}^2) w_{\alpha \mathrm{s}} \tag{11.57}$$

when $\boldsymbol{\tau}$ is the stress vector at the boundary, with unit normal $\mathbf{n}$; that is,

$$\tau_{ij}^{(\mathrm{s})} = n_j\left\{\eta^{(\mathrm{sn})}\left(\frac{\partial v_{\mathrm{n}i}}{\partial x_j} + \frac{\partial v_{\mathrm{n}j}}{\partial x_i}\right) - \eta^{(\mathrm{ns})}\left(\frac{\partial v_{\mathrm{s}i}}{\partial x_j} + \frac{\partial v_{\mathrm{s}j}}{\partial x_i}\right)\right\}, \quad (11.58)$$

as defined in Eq. (11.49). The $\mathbf{w}_{\mathrm{s}}$ is the velocity of the superfluid component relative to the boundary, and the subscript α implies the tangential direction.

By a power series expansion of F_{s} and retaining only the first nonvanishing term, as an approximation, one can then write

$$\tau_{\alpha}^{(\mathrm{s})} = \beta \mid w_{\mathrm{s}} \mid^2 w_{\alpha\mathrm{s}}. \quad (11.59)$$

This is a valid approximation since the velocities of flow of He II are usually small with respect to the speed of sound.

The force functions F_{n} and F_{s} are expected to be different from each other. It is usual to assume that F_{n} is very large and that F_{s} is very small at small velocities. This implies that there is a complete absence of shear interaction between the superfluid and the boundary at low speeds. So, again, to a first approximation

$$w_{\alpha\mathrm{n}} = 0, \quad (11.60)$$

$$\tau_{\alpha}^{(\mathrm{s})} = 0. \quad (11.61)$$

In Lin's formulation the simplest possible situation has been considered. In the actual situation, it could be that $F_{\mathrm{s}} \equiv 0$ for $\mid w_{\mathrm{s}} \mid$ less than a certain critical value. But as the critical value is certainly very small, Eq. (11.57) is a valid approximation to the real behavior. Moreover, the coefficient relating the stresses and the rates of strain could be a function of the rates of strain.

11.10.3 The Hall, Vinen, Bekarevitch, and Khalatnikov (HVBK) Formulation

In the Lin formulation it was also assumed that the rotation of the supercomponent does not produce a significant change in the thermodynamical properties of the medium. However, the above authors [28, 29] have suggested a continuum theory in which the internal energy of the fluid depends on the magnitude of the superfluid component. Indeed, this concept results in ϱ_{s} being a tensor when the fluid is in rotation.

This formulation is the most complete of the three. Moreover, essentially the same equations can be derived from either a continuum approach or a microscopic model. The results obtained are as follows:

$$\mathbf{q} = \frac{\varkappa}{T} \nabla T, \tag{11.62}$$

$$-\mathbf{F}_s = -\boldsymbol{\omega} \times [\nabla \times \lambda \boldsymbol{\nu}], \tag{11.63}$$

$$(-F_n)_i = \frac{\partial}{\partial x_k} \left\{ \eta \left(\frac{\partial v_{ni}}{\partial x_k} + \frac{\partial v_{nk}}{\partial x_i} - \frac{2}{3} \delta_{ik} \frac{\partial v_{nl}}{\partial x_l} \right) \right.$$
$$\left. + \eta_{b1} \delta_{ik} \frac{\partial}{\partial x_l} \varrho_s (v_{sl} - v_{nl}) + \eta_{b2} \delta_{ik} \frac{\partial v_{nl}}{\partial x_l} \right\}, \tag{11.64}$$

$$\mathbf{F}_{sn} = -[\varrho_s \nabla \{\eta_{b3} \nabla \cdot \varrho_s (\mathbf{v}_s - \mathbf{v}_n) + \eta_{b1} \nabla \cdot \mathbf{v}_n\}]$$
$$+ [B_1 \boldsymbol{\omega} \times \boldsymbol{\xi} + B_2 \boldsymbol{\nu} \times (\boldsymbol{\omega} \times \boldsymbol{\xi}) - B_3 \boldsymbol{\nu} (\boldsymbol{\omega} \cdot \boldsymbol{\xi}), \tag{11.65}$$

$$Q = \frac{1}{T} \left\{ \varkappa \frac{(\nabla T)^2}{T} + \frac{\eta}{2} \left(\frac{\partial v_{ni}}{\partial x_k} + \frac{\partial v_{nk}}{\partial x_i} - \frac{2}{3} \delta_{ik} \frac{\partial v_{nl}}{\partial x_l} \right)^2 \right.$$
$$+ \eta_{b2} (\nabla \cdot \mathbf{v}_n)^2 + \eta_{b3} [\nabla \cdot \varrho_s (\mathbf{v}_s - \mathbf{v}_n)]^2$$
$$+ 2\eta_{b1} (\nabla \cdot \mathbf{v}_n) [\nabla \cdot \varrho_s (\mathbf{v}_s - \mathbf{v}_n)]$$
$$\left. + \frac{B_2}{|\boldsymbol{\omega}|} (\boldsymbol{\omega} \times \boldsymbol{\xi})^2 + \frac{B_3}{|\boldsymbol{\omega}|} |\boldsymbol{\omega} \cdot \boldsymbol{\xi}|^2 \right\}, \tag{11.66}$$

where

$$\boldsymbol{\omega} = \nabla \times \mathbf{v}_s, \qquad \boldsymbol{\xi} = \mathbf{v}_n - \mathbf{v}_s - \frac{1}{\varrho_s} \nabla \times \lambda \boldsymbol{\nu}, \qquad \boldsymbol{\nu} = \frac{\boldsymbol{\omega}}{|\boldsymbol{\omega}|}, \tag{11.67}$$

and $\varkappa$, η, η_{b2}, η_{b1}, η_{b3}, B_1, B_2, B_3, and λ are coefficients responsible for thermal conduction, viscosities, and mutual frictions.

If $\mathbf{N}$ is the unit normal at the boundary, and V the boundary velocity, then the boundary condition is

$$\left\{ \mathbf{v}_s - V + \frac{1}{\varrho_s} \nabla \times \lambda \boldsymbol{\nu} + \frac{B_1}{\varrho_s} \boldsymbol{\xi} + \frac{B_2}{\varrho_s} \boldsymbol{\nu} \times \boldsymbol{\xi} \right\} \times \boldsymbol{\omega}$$
$$= \varepsilon_b \mathbf{N} \times \boldsymbol{\omega} + \varepsilon_b' (\mathbf{N} \times \boldsymbol{\nu}) \times \boldsymbol{\omega}, \tag{11.68}$$

where ε_b and ε_b' are the boundary dissipation coefficients. In fact, Eq. (11.68) is derived from a consideration of the energy dissipation at the surface due to vortex slippage.

This of course is a set of very complex, highly nonlinear equations. It involves several undetermined physical coefficients and so it makes comparison with experimental very difficult.

In Eq. (11.65) the terms in the second bracket are present only when $\omega \neq 0$. They are the outcome of the rotation of the superfluid component or the quantized vortex lines. In the microscopic physical model, the force is transmitted from the normal component of the fluid through the collisions between rotons and vortex lines. The vortex lines have circulation and the force is transmitted to the superfluid component through the Magnus effect.

The B_3 term is longitudinal and the force is in the direction of vorticity, while the B_1 and B_2 terms are transverse. The B_1 term does not contribute to energy dissipation.

There is a fundamental difference between the HVBK and the Gorter–Mellink formulation. In the former case, irrotational flow is a permissible state of motion, while in the Gorter–Mellink formulation the superfluid component cannot be irrotational because of the mutual friction.

Lin's formulation is quite different from the other two, but on a theoretical continuum basis it is quite as legitimate. This indicates that the true nature of the nonlinear and irreversible aspects of superfluid helium flow is still not resolved.

11.11 Applications to Some Simple Flow Problems [30]

Consideration of the steady parallel flow of He II through a circular pipe can serve to show the different conclusions which result from the above three formulations.

In cylindrical coordinates (r, ϑ, z),

$$\mathbf{v}_\mathrm{n} = (0, 0, u(r)), \tag{11.69}$$

$$\mathbf{v}_\mathrm{s} = (0, 0, v(r)), \tag{11.70}$$

and

$$\nabla \cdot \mathbf{v}_\mathrm{n} = 0 = \nabla \cdot \mathbf{v}_\mathrm{s}. \tag{11.71}$$

Moreover,

$$\nabla \times \mathbf{v}_n = \left(0, -\frac{du}{dr}, 0\right), \tag{11.72}$$

$$\nabla \times \mathbf{v}_s = \left(0, -\frac{dv}{dr}, 0\right). \tag{11.73}$$

The analysis will be carried out for incompressible fluids with $T =$ constant all through the fluid (so that all physical parameters may be taken as constant).

Neglecting external force, the equations of motion for this system are as follows:

$$-\frac{1}{\varrho}\nabla p - \frac{1-\Gamma}{2}\nabla(\mathbf{v}_n - \mathbf{v}_s)^2 - \frac{1}{\varrho_n}\mathbf{F}_n + \frac{1}{\varrho_n}\mathbf{F}_{sn} = 0, \tag{11.74}$$

$$-\frac{1}{\varrho}\nabla p + \frac{\Gamma}{2}\nabla(\mathbf{v}_n - \mathbf{v}_s)^2 - \frac{1}{\varrho_s}\mathbf{F}_s - \frac{1}{\varrho_s}\mathbf{F}_{sn} = 0. \tag{11.75}$$

Multiplying Eq. (11.74) by ϱ_n and (11.75) by ϱ_s and adding, one obtains

$$0 = -\nabla p - \mathbf{F}_n - \mathbf{F}_s. \tag{11.76}$$

From this stage on the analysis will be continued for *each* of the above three formulations.

11.11.1 Gorter–Mellink

In this case, Eq. (11.76) becomes

$$\nabla p = -\eta \nabla \times (\nabla \times \mathbf{v}_n),$$

so that

$$p = p(z)$$

and

$$\frac{dp}{dz} = \eta \frac{1}{r}\frac{d}{dr}\left(r\frac{du}{dr}\right) = -C_1, \tag{11.77}$$

where C_1 is a constant.

Since u vanishes at the wall of the pipe, one gets

$$u = \frac{C_1}{4\eta}(a^2 - r^2). \tag{11.78}$$

Now Eq. (11.75) becomes

$$\frac{C_1}{\varrho} - \frac{1}{\varrho_s}\mathbf{F}_{sn} = 0,$$

and

$$F_{sn} = \alpha\varrho(1-\Gamma)\Gamma(v-u)^3, \qquad 1-\varepsilon = \frac{\varrho_s}{\varrho_n+\varrho_s} = \frac{\varrho_s}{\varrho},$$

therefore

$$\frac{C_1}{\varrho} = \alpha\Gamma(v-u)^3 \tag{11.79}$$

or

$$v = u + \left(\frac{C_1}{\alpha\varrho\Gamma}\right)^{1/3}. \tag{11.80}$$

11.11.2 Lin

Let $\eta = \eta^{(nn)} + \eta^{(sn)}$, and then Eq. (11.76) becomes

$$\nabla p = -\eta\nabla\times(\nabla\times\mathbf{v}_n) \tag{11.81}$$

using Eqs. (11.51) and (11.52). Again,

$$u = \frac{C_1}{4\eta}(a^2 - r^2).$$

In this formulation Eq. (11.75) becomes

$$-\frac{1}{\varrho}\nabla p + \frac{\Gamma}{2}\nabla(\mathbf{v}_n - \mathbf{v}_s)^2 - \frac{\eta^{(sn)}}{\varrho_s}\nabla\times(\nabla\times\mathbf{v}_n)$$
$$+\frac{\eta^{(ns)}}{\varrho_s}\nabla\times(\nabla\times\mathbf{v}_s) - \Gamma(\nabla\times\mathbf{v}_s)\times(\mathbf{v}_n-\mathbf{v}_s). \tag{11.82}$$

In component form, because the r and ϑ velocity components are zero in this case, one gets

$$0 = \frac{C_1}{\varrho} - \frac{\eta^{(sn)}}{\eta}\frac{C_1}{\varrho_s} - \frac{\eta^{(ns)}}{\varrho_s}\frac{1}{r}\frac{d}{dr}\left(r\frac{dv}{dr}\right), \tag{11.83}$$

because

$$A = -\frac{dp}{dz} = \eta\frac{1}{r}\frac{d}{dr}\left(r\frac{du}{dr}\right),$$

$$\nabla\times\mathbf{v}_s = \left(0, -\frac{dv}{dr}, 0\right),$$

and

$$\Gamma(u-v)\frac{d}{dr}(u-v)+\Gamma(u-v)\frac{dv}{dr}=0, \tag{11.84}$$

since the curl terms in this case are zero and

$$\frac{\Gamma}{2}\nabla(u-v)^2=\frac{\Gamma}{2}\frac{d}{dr}[(u-v)^2]=\Gamma\frac{d}{dr}(u-v), \qquad \text{etc.}$$

From Eq. (11.84), therefore,

$$\Gamma(u-v)\frac{du}{dr}=0 \qquad \text{or} \qquad u=v. \tag{11.85}$$

This is incompatible with the assumption that the tangential component of the shear stress vector is directly related to the slip velocity of the superfluid component used in the derivation of the expression for the tangential stress as in Eq. (11.59).

Also, from Eq. (11.83),

$$0=\frac{1}{\varrho}-\frac{\eta^{(\mathrm{sn})}}{\eta}\frac{1}{\varrho_{\mathrm{s}}}-\frac{1}{C_1}\left[\frac{\eta^{(\mathrm{ns})}}{\varrho_{\mathrm{s}}}\frac{1}{r}\frac{d}{dr}\left(r\frac{dv}{dr}\right)\right],$$

and since

$$C_1=-\eta\frac{1}{r}\frac{d}{dr}\left(\frac{1}{r}\frac{du}{dr}\right) \qquad \text{and} \qquad u=v,$$

thus

$$0=\frac{1}{\varrho}-\frac{1}{\varrho_{\mathrm{s}}}\left(\frac{\eta^{(\mathrm{sn})}}{\eta}-\frac{\eta^{(\mathrm{ns})}}{\eta}\right)$$

or

$$\frac{\eta^{(\mathrm{nn})}+\eta^{(\mathrm{sn})}}{\varrho}=\frac{\eta^{(\mathrm{sn})}-\eta^{(\mathrm{ns})}}{\varrho_{\mathrm{s}}}. \tag{11.86}$$

Solving Eq. (11.83) for v, one gets

$$v=\frac{C_1}{4\eta^{(\mathrm{ns})}}\left[(1-\Gamma)-\frac{\eta^{(\mathrm{ns})}}{\eta}\right]r^2+\text{constant}. \tag{11.87}$$

The equation for the tangential shear stress in the superfluid component given in Eq. (11.59) gives

$$\eta^{(\mathrm{sn})}\frac{du}{dr}-\eta^{(\mathrm{ns})}\frac{dv}{dr}=\beta v^3 \qquad \text{at} \quad r=a, \tag{11.88}$$

where v has been equated with w_s. Therefore

$$v = \frac{A}{4\eta^{(\mathrm{ns})}}\left[\frac{\eta^{(\mathrm{sn})}}{\eta} - (1 - \Gamma)\right](a^2 - r^2) - \left[\frac{(1 - \Gamma)aC_1}{2\beta}\right]^{1/3}. \qquad (11.89)$$

11.11.3 HVBK

In this case Eq. (11.76), using Eqs. (11.63) and (11.64), becomes

$$-\nabla p - \eta \nabla \times (\nabla \times \mathbf{v}_n) - \lambda(\nabla \times \mathbf{v}_s) \times (\nabla \times \boldsymbol{\nu}), \qquad (11.90)$$

and

$$\boldsymbol{\nu} = \frac{\nabla \times \mathbf{v}_s}{|\nabla \times \mathbf{v}_s|} = \left(0, \frac{-dv/dr}{|dv/dr|}, 0\right) = \left(0, -\operatorname{sgn}\frac{dv}{dr}, 0\right);$$

therefore

$$\nabla \times \boldsymbol{\nu} = \left(0, 0, -\frac{1}{r}\operatorname{sgn}\frac{dv}{dr}\right).$$

In component form, then, Eq. (11.90) becomes

$$\frac{\partial p}{dz} = \eta \frac{1}{r}\frac{d}{dr}\left(r\frac{du}{dr}\right), \qquad (11.91)$$

since the z component of $\nabla \times \mathbf{v}_s \equiv 0$ and

$$\frac{\partial p}{dr} = -\frac{\lambda}{r}\left(\frac{dv}{dr}\right)^2. \qquad (11.92)$$

From Eq. (11.92),

$$p = -C_1 z + f(r), \qquad (11.93)$$

and again

$$u = \frac{C_1}{4\eta}(a^2 - r^2). \qquad (11.94)$$

In this formulation, Eq. (11.75) assumes the form

$$-\frac{1}{\varrho}\nabla p + \frac{\Gamma}{2}\nabla(\mathbf{v}_n - \mathbf{v}_s)^2 - \frac{\lambda}{\varrho}\boldsymbol{\omega} \times [\nabla \times \boldsymbol{\nu}]$$
$$-\frac{1}{\varrho_s}[B_1\boldsymbol{\omega} \times \boldsymbol{\xi} + B_2\boldsymbol{\nu} \times (\boldsymbol{\omega} \times \boldsymbol{\xi}) - B_3\boldsymbol{\nu}(\boldsymbol{\omega} \cdot \boldsymbol{\xi})], \qquad (11.95)$$

and

$$\boldsymbol{\omega} \times \boldsymbol{\xi} = \left(-\frac{dv}{dr}[u-v] - \frac{\lambda}{\varrho_s}\frac{1}{r}\left|\frac{dv}{dr}\right|, 0, 0\right),$$

$$\mathbf{v} \times (\boldsymbol{\omega} \times \boldsymbol{\xi}) = \left(0, 0, -[u-v]\left|\frac{dv}{dr}\right| - \frac{\lambda}{\varrho_s}\frac{1}{r}\frac{dv}{dr}\right),$$

$$\boldsymbol{\omega} \cdot \boldsymbol{\xi} = 0.$$

Breaking Eq. (11.95) into components, one has

$$0 = \frac{C_1}{\varrho} + \frac{B_2}{\varrho_s}\left[(u-v)\left|\frac{dv}{dr}\right| + \frac{\lambda}{\varrho_s r}\frac{dv}{dr}\right], \tag{11.96}$$

$$0 = \left(\frac{1}{\varrho} - \frac{1}{\varrho_s}\right)\frac{\lambda}{r}\left|\frac{dv}{dr}\right| + \Gamma(u-v)\frac{d}{dr}(u-v)$$
$$+ \frac{B_1}{\varrho_s}\left[(u-v)\frac{dv}{dr} + \frac{\lambda}{\varrho_s r}\left|\frac{dv}{dr}\right|\right]. \tag{11.97}$$

Equations (11.96) and (11.97) are generally incompatible. From Eq. (11.96), dv/dr can vanish only at $r = 0$ and it certainly vanishes there. Since dv/dr does not change sign and if C_1 is taken to be positive, then $dv/dr \leq 0$. Thus, using Eq. (11.94), Eq. (11.96) becomes

$$\left[v + \frac{A}{4\eta}(r^2 - a^2) + \frac{\lambda}{\varrho_s}\frac{1}{r}\right]\frac{dv}{dr} + \frac{A}{B_2}\frac{\varrho_s}{\varrho} = 0. \tag{11.98}$$

A series solution can be obtained for Eq. (11.98) near $r = 0$, but normally v must be determined by the numerical integration techniques.

Summarizing, one sees that for all three formulations,

$$\frac{\partial p}{\partial z} = -C_1 \tag{11.99}$$

and

$$u = \frac{C_1}{4\eta}(a^2 - r^2). \tag{11.100}$$

But the solutions for $\mathbf{v}_s$ are different.

Gorter–Mellink:

$$v = \frac{C_1}{4\eta}(a^2 - r^2) + \left(\frac{C_1}{\alpha\varrho\Gamma}\right)^{1/3}; \tag{11.101}$$

Lin:

$$v = \frac{C_1}{4\eta^{(\mathrm{ns})}}\left[\frac{\eta^{(\mathrm{sn})}}{\eta} - (1 - \Gamma)\right](a^2 - r^2) - \left[\frac{(1 - \Gamma)aC_1}{2\beta}\right]^{1/3}; \qquad (11.102)$$

HVBK. The v term satisfies the equation

$$\left[v + \frac{C_1}{4\eta}(r^2 - a^2) + \frac{\lambda}{\varrho_\mathrm{s}}\frac{1}{r}\right]\frac{dv}{dr} + \frac{C_1(1 - \Gamma)}{B_2} = 0. \qquad (11.103)$$

Equations (11.101) and (11.102) represent parabolic profiles, but v from Eq. (11.103) is certainly not parabolic.

For the HVBK case

$$\lim_{r\to 0}\frac{d^2v}{dr^2} = -\frac{C_1\varrho_\mathrm{s}^2}{B_2\lambda\varrho}, \qquad (11.104)$$

$$\lim_{r\to 0}\frac{d^3v}{dr^3} = \frac{2C_1\varrho_\mathrm{s}^3}{B_2\varrho\lambda^2}\left[v(0) - \frac{C_1a^2}{4\eta}\right]. \qquad (11.105)$$

One way of testing the validity of these formulations could be to measure the curvature of the velocity profile in the central part of the pipe.

Another difference is that in the HVBK case the pressure varies with both r and z, but in the other cases it varies only with z.

The incompatibility in the Lin and HVBK formulations can be removed by replacing the constant temperature assumption by $T = T(r)$. This is not a problem in the Gorter–Mellink formulation.

Another point worth noting is the assumption in the Lin formulation that $\eta^{(\mathrm{ns})} + \eta^{(\mathrm{ss})} = 0$; that is, that the superfluid component cannot exert a net transfer of total momentum by shear forces, although it can affect the exchange of momentum between the two components. The assumption that $\eta^{(\mathrm{sn})} = 0$ again reflects the idea that the superfluid component is not easy to excite. Experimental evidence seems to support these contentions.

Of the three formulations, the Gorter–Mellink mutual friction one is the simplest and is semiempirical. It does lack a direct microscopic physical basis. It can serve as a useful starting point for the theoretical analysis of hydrodynamic problems in superfluid helium.

11.12 Applications of the Gorter–Mellink Formulation to Flow Problems

By considering only flows at low speeds, the condition of incompressibility can be imposed and the entropy density regarded as constant. The thermal conduction and dissipation terms can also be neglected, because the internal convection will be the predominant mechanism for heat transfer.

The basic equations now become

$$\nabla \cdot \mathbf{v}_s = 0, \tag{11.106}$$

$$\nabla \cdot \mathbf{v}_n = 0, \tag{11.107}$$

$$\begin{aligned} \frac{\partial \mathbf{v}_n}{\partial t} + (\mathbf{v}_n \cdot \nabla) = &-\frac{1}{\varrho} \nabla p - \frac{1-\Gamma}{\Gamma} S \nabla T - \nabla \Omega \\ &- \frac{1-\Gamma}{2} \nabla (\mathbf{v}_n - \mathbf{v}_s)^2 + \frac{\eta}{\Gamma \varrho} \nabla^2 \mathbf{v}_n \\ &+ \alpha(1-\Gamma)(\mathbf{v}_s - \mathbf{v}_n)^3, \end{aligned} \tag{11.108}$$

and

$$\begin{aligned} \frac{\partial \mathbf{v}_s}{\partial t} + (\mathbf{v}_s \cdot \nabla)\mathbf{v}_s = &-\frac{1}{\varrho} \nabla p + S \nabla T - \nabla \Omega + \frac{\Gamma}{2} \nabla (\mathbf{v}_n - \mathbf{v}_s)^2 \\ &- \alpha \Gamma (\mathbf{v}_s - \mathbf{v}_n)^3, \end{aligned} \tag{11.109}$$

where Ω is the potential of the external force field.

Some exact solutions based on the above set of equations have been obtained and are reported in the literature [31].

11.12.1 Steady Flow through a Pipe of Annular Section

The section can be represented by

$$b \leq r \leq \mathrm{a}.$$

Let z be the direction of flow. Also,

$$A = -\frac{d}{dz}(p + \varrho\Omega) = \text{constant}, \qquad B = \varrho S \frac{dT}{dz} = \text{constant},$$

$$\mathbf{v}_n = (0, 0, u), \qquad \mathbf{v}_s = (0, 0, v).$$

The expressions obtained for the velocity components and total discharge rate are as follows:

$$u = \frac{A}{4\eta}\left(b^2 - r^2 + \frac{a^2 - r^2}{\log(a/b)} \log \frac{r}{b}\right), \tag{11.110}$$

$$v = u + \left(\frac{A + B}{\alpha\Gamma\varrho}\right)^{1/3}, \tag{11.111}$$

$$Q = \pi(a^2 - b^2)\left[\frac{A}{8\eta}\left(a^2 + b^2 - \frac{a^2 - b^2}{\log(a/b)}\right) + (1 - \Gamma)\left(\frac{A + B}{\alpha\Gamma\varrho}\right)^{1/3}\right]. \tag{11.112}$$

11.12.2 Plane Couette Flow

Consider a channel $-d < y < +d$ with one wall $(y = d)$ moving with a velocity u_0 in the z direction. Then, using the same notation as in Section 11.12.1, it is found that

$$u = \frac{A}{2\eta}(d^2 - y^2) + \frac{u_0}{2}\left(1 + \frac{y}{d}\right), \tag{11.113}$$

$$v = u + \left(\frac{A + B}{\alpha\Gamma\varrho}\right)^{1/3}. \tag{11.114}$$

11.12.3 Boundary Layer Plow

A two-dimensional steady boundary layer flow over a flat surface is considered, with x designating the direction of flow and y the normal to the surface.

The velocities will be designated as

$$\mathbf{v}_\mathrm{n} = (u_x, u_y), \qquad \mathbf{v}_\mathrm{s} = (v_x, v_y),$$

the boundary conditions are

$$\mathbf{v}_\mathrm{n} = \mathbf{v}_\mathrm{s} = (U(x), 0),$$

and outside the boundary layer,

$$T = T_0, \tag{11.115}$$

$$p + \frac{\varrho}{2} U^2 = \text{constant}. \tag{11.116}$$

The boundary layer equations are

$$\frac{\partial u_x}{\partial x} + \frac{\partial u_y}{\partial y} = 0, \tag{11.117}$$

$$\frac{\partial v_x}{\partial x} + \frac{\partial v_y}{\partial y} = 0, \tag{11.118}$$

$$u_x \frac{\partial u_x}{\partial x} + u_y \frac{\partial u_y}{\partial y} = U \frac{dU}{dx} + \frac{\eta}{\Gamma \varrho} \frac{\partial^2 u_x}{\partial y^2} + \alpha(1 - \Gamma)(v_x - u_x)^3, \tag{11.119}$$

$$v_x \frac{\partial v_x}{\partial x} + v_y \frac{\partial v_x}{\partial y} = U \frac{dU}{dx} - \alpha\Gamma(v_x - u_x)^3. \tag{11.120}$$

Equation (11.116) gives the pressure in the boundary layer, and the temperature variation is given by

$$T = T_0 - \frac{\Gamma}{2S}(u_x - v_x)^2. \tag{11.121}$$

For similarity, solutions of the form

$$u_x = Uf'(\eta) \qquad \text{and} \qquad v_x = Ug'(\eta),$$

where $\eta = y/h(x)$, are sought. It turns out that the only permissible solutions are those for which

$$U = -\frac{C_1}{x}, \qquad h = C_2 x. \tag{11.122}$$

These correspond to flows in convergent and divergent channels.

As in normal fluids, separation of the boundary layer from the wall tends to occur when there is an adverse pressure gradient (Fig. 11.5a). The separation point is as usual determined by the condition $(\partial u_x/\partial y) = 0$. The appearance of this separation point is delayed, however, compared with the situation in normal fluids, as a result of the forward drag of the superfluid component caused by the action of mutual friction.

The superfluid component will slip over the wall surface with a velocity less than U due to the mutual friction (Fig. 11.5b). Somewhere downstream of the normal separation point, a minimum will appear in the velocity profile, and finally at some point the streamline will divide and backflow of the superfluid component will also occur beyond this point.

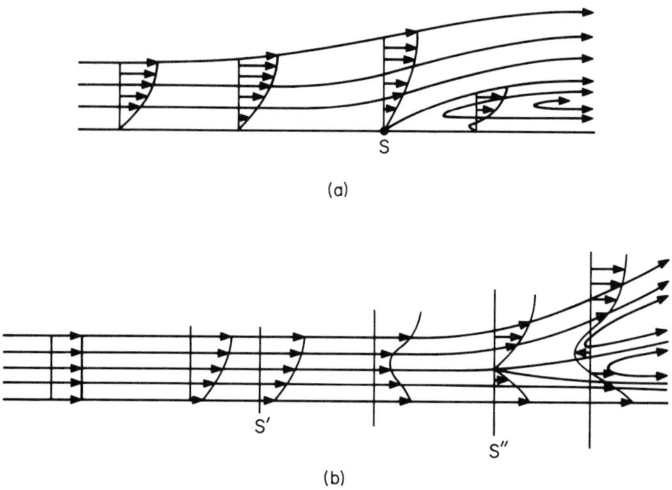

Fig. 11.5 (a) Normal separation. (b) Superfluid separation.

11.13 Molecular Basis of the Hydrodynamic Theory

The above purely phenomenological approach based on the concept of slip of the superfluid at the solid boundary leaves the molecular picture unresolved.

A quantum mechanical treatment of quasi-equilibrium phenomena involving long-range order (as in a superfluid) is not yet feasible. Hence, the reason for the adoption of the continuum theory.

In liquid helium the existence of cooperative phenomenon leads to a correlation length, which is the distance within which the momentum space ordering is very effective. For an ideal Bose gas this length is infinite. In the case of a dilute system of Bose particles with hard sphere interactions, Lee *et al.* [32] have shown that this correlation length should be given by

$$r_0 = (8\pi a n)^{-1/2},$$

where n is the particle number density and a the hard sphere radius. Taking $a = 10^{-8}$ cm and $n = 2 \times 10^{22}$ cm^{-3}, one obtains

$$r_0 = 1.5 \times 10^{-8} \quad \text{cm}.$$

The mean free path in liquid helium is much greater than this, and so the continuum concept appears to be justified. There is some experimental evidence, however, that suggests that r_0 has a value nearer 10^{-4} cm. If this is the case, then flow in films with a thickness of the order of 10^{-6} cm could not be handled by any hydrodynamic theory.

The correlation length increases as the temperature falls, and at or near 0°K it may be of the order of the dimension of the vessel. This would imply a macroscopic quantum mechanical behavior and the continuum treatment should be completely invalid.

In many of the hydrodynamic experiments, however, the continuum concept seems amply justified. Moreover, the prevailing hydrodynamic theory is not exclusively based on the continuum concept. The belief in the existence of quantized vortex lines of macroscopic dimensions is incorporated in most hydrodynamic problems.

The wave function used by Feynmann to represent a flowing liquid, quoted in Eq. (11.8), has been criticized by Lin on the basis that in general ψ_{flow} and ψ_0 cannot both be solutions of Schrödinger's equation, except in the trivial case of pure translation. He does not accept the fact that the existence of quantized irrotational motion has been definitely established, and he designates this as a major objective of theoretical physics. However, at present the most promising development theoretically has been the Onsager–Feynmann idea of a single unified theory (a macroscopic Schrödinger theory) that contains the Landau two-fluid along with the quantum conditions

$$\nabla \times \mathbf{v}_s = 0, \qquad \oint \mathbf{v}_s \cdot dl = \frac{nh}{m}, \qquad \psi \to 0 \quad \text{at a boundary.}$$

The clearest evidence for the quantization of circulation lies in the fundamental experiments of Rayfield and Reif [33], who measured the energy and velocity of free charged ions in He II. At low temperature they found the unexpected result that the greater the energy of the ion, the smaller the velocity. This behavior is just what one expects from a vortex ring of a given circulation. If Rayfield and Reif assumed that the ions became trapped on vortex rings, then the data they observed indicate that the circulation of the rings was h/m.

11.14 Concluding Remarks

The Landau formulation of the hydrodynamics of superfluid helium can satisfactorily explain the phenomena only up to a temperature below the λ point. If fails to account for the singular behavior in the λ transition.

This is probably due to a lack of understanding of the microscopic physics of liquid helium.

In the Landau theory the excitations corresponding to various regions of a single spectrum are identified with phonons and rotons.

Phonons represent sound waves (first sound), while the rotons generate a flow field very similar to that due to a very small classical vortex ring [34]. The questions then arise: Are rotons vortex rings and, if so, are they quantized?

Both phonons and rotons have their own energy spectrum. For phonons there is the dispersion relation

$$\varepsilon_{\mathrm{p}}(p) = ap, \tag{11.123}$$

where p is the momentum and a the speed of sound. As suggested by classical hydrodynamics, the corresponding dispersion relation for rotons can be taken as

$$\varepsilon_{\mathrm{r}}(p) = Ap^{1/2}, \tag{11.124}$$

where

$$A = \frac{C}{2}\left(\frac{\varrho\varkappa}{\pi}\right)^{1/2}, \tag{11.125}$$

$\varkappa$ is the circulation, and C is a constant of order unity, which depends on the size of the vortex ring and the physical nature of its core.

If the circulation is quantized, then for helium with atomic mass m,

$$\varkappa = \frac{h}{m} = 0.997 \times 10^{-3} \quad \mathrm{cm^2/sec}.$$

Presumably vortex rings have only one unit of circulation. To have two units the energy would have to increase eightfold.

For an assembly of large numbers of phonons and rotons, the state energy would be given as

$$E = E_{\mathrm{p}} + E_{\mathrm{r}} + E_{\mathrm{pr}}, \tag{11.126}$$

where E_{p} is the energy due to phonons alone, E_{r} that due to rotons alone, and E_{pr} the phonon–roton interaction energy.

To a first approximation, the exchange energy will be ignored. To the same approximation, if the phonon–phonon interactions are ignored, then

$$E_{\mathrm{p}} = \sum_i n_{i\mathrm{p}}\varepsilon_{i\mathrm{p}}(p) = \sum_i n_{i\mathrm{p}} a p_i, \tag{11.127}$$

where n_{ip} is the number of phonons with momentum p_i. Classical hydrodynamics shows that the energy of a system of vortex rings is

$$T = \sum_i \left[2p_i v_i' - \mathbf{r}_i \frac{dp_i}{dt} \right] + \frac{\varrho}{2} \iint_A V^2 \mathbf{r} \cdot \mathbf{n} \, dA, \quad (11.128)$$

where v_i' is the average velocity of the ith vortex ring in the direction normal to the plane of the ring, p_i the momentum of the ith vortex ring (assuming no vortex–vortex interaction), $\mathbf{r}_i$ the position vector of the center of the ith vortex ring, and $\mathbf{V}$ the fluid velocity.

The second term in Eq. (11.128) will produce a term like $\frac{1}{2}M\overline{V^2}$, where the bar implies the average of V^2 over the boundary and M is the total mass of the fluid. It is a constant and so can be omitted from T. The term $\sum_i \mathbf{r}_i \, d\mathbf{p}_i/dt$ is an interaction term and to a first approximation may also be dropped. Then

$$E_T = \sum_j 2p_i(v_j + w_j), \quad (11.129)$$

where v_j is the velocity of the jth roton, and w_j is the average velocity in the direction of v_j due to all the rest of the rings. Equation (11.129) can also be written

$$E_r = \sum_i n_{ir}[Ap^{1/2} + 2p_i u_i], \quad (11.130)$$

where n_{ir} is the number of rotons with momentum p_i, and u_i is the average of the w's over these n_{ir} rotons.

It thus appears that the energy of a state will depend not only on the number distribution of the rotons $\{n_{ir}\}$ but also on the positional and orientational distributions of the vortex rings. Thus,

$$E\{n_{ip}, n_{ir}, P\} = E_0 + \sum_{p_i} n_{ip} a p_i + \sum_{p_j} n_{jr} A p_j^{1/2} + \sum_{p_j} 2 n_{jr} p_j u_j(P). \quad (11.131)$$

The partition function Q is therefore

$$Q = \sum_{(n_{ip}, n_{jr}, P)} \exp\left[-\frac{E\{n_{ip}, n_{jr}, P\}}{kT} \right]. \quad (11.132)$$

Let

$$q = \sum_{\{P\}} \exp\left[-\frac{2 \sum_{p_j} n_{jr} p_j u_j(P)}{kT} \right], \quad (11.133)$$

where q normally depends on $\{n_{jr}\}$. If q is not very sensitive to $\{n_{jr}\}$ but depends on the total number of rotons present (which is directly related to the temperature and density of the fluid), then q can be factored out and Eq. (11.132) becomes

$$Q = qa^{-E_0/kT} \prod_i \left[1 - \exp\left(-\frac{ap_i}{kT}\right)\right]^{-1} \prod_j \left[1 - \exp\left(-\frac{Ap_j^{1/2}}{kT}\right)\right]^{-1}. \tag{11.134}$$

The information about the λ transition is contained in the expression for q.

One notes the resemblance of q to the partition function of the Ising problem.

In the two-dimensional Ising problem the energy of a state is given by

$$E(s_i) = -\sum_{\langle ij \rangle} \varepsilon s_i s_j, \tag{11.135}$$

where s_i can have the values ± 1, $\langle ij \rangle$ denotes a nearest-neighbor pair of spins, and $\varepsilon > 0$. It is found [35] that the specific heat of the lattice in the region of the transition temperature T_c is

$$C = -k\chi \ln | T - T_c |, \tag{11.136}$$

where

$$\chi = \begin{cases} 0.4781 & \text{for a hexagonal lattice,} \\ 0.4945 & \text{for a square lattice,} \\ 0.4991 & \text{for a triangular lattice.} \end{cases}$$

In the case of liquid helium, the specific heat per atom near the λ point [22] is given by

$$C = -0.63k \ln | T - T_\lambda |. \tag{11.137}$$

To calculate q is a much more difficult task, since it involves a three-dimensional problem with more than nearest-neighbor interaction. Furthermore, vortex rings can have various orientations.

Using the vortex-ring model, the phenomenon of superfluidity can be interpreted as follows. At temperatures below the λ point, there is long-range order among the positions and orientations of the vortex rings. Thus, the superfluid state exists for temperatures below the λ point. Above this point no long-range order is possible for the vortex rings (or large rotons). Superfluidity disappears.

The difficulty in calculating q has delayed the quantitative testing of this model. A very serious unexplored question is an understanding of the mechanism of generation of vortex rings.

The continuum theory of the hydrodynamics of He II as developed here can provide an adequate description for most macroscopic problems. Macroscopic phenomena which depend more heavily on the molecular interactions—for example, the attenuation of sound—cannot be fully accounted for by a simple macroscopic theory. It would appear that at the microscopic level there is a growing body of evidence for the existence of quantized vortex lines.

BIBLIOGRAPHY

1. H. Kammerlingh-Onnes, *Proc. Acad. Sci. Amsterdam* **11**, 168 (1908).
2. W. Keesom and K. Clusius, *Proc. Acad. Sci. Amsterdam* **35**, 307 (1932).
3. W. Fairbank, M. Buckingham, and C. Kellers, *Proc. Int. Conf. Low Temp. Phys., 5th* **50** (1957).
4. H. Kammerlingh-Onnes and G. Boks, *Comm. Phys. Lab. Univ. Leiden*, No. 1706 (1924).
5. J. Allen and H. Jones, *Nature* **141**, 243 (1938).
6. P. L. Kapitza, *Nature* **14**, 74 (1958).
7. J. Daunt and K. Mendelssohn, *Proc. Roy. Soc.* **A170**, 423, 467 (1939).
8. J. Allen and A. Misener, *Proc. Roy. Soc.* **A**173, 467 (1939).
9. L. Tisza, *Phys. Rad.* **1**, 165 (1940).
10. L. Landau, *J. Phys.* (*Moscow*) **5**, 71 (1941).
11. V. Peshkov, *J. Phys.* (*Moscow*) **8**, 131 (1944).
12. F. London, "Superfluids," Vol. 2. Wiley, New York, 1954.
13. L. Landau, *J. Phys.* (*Moscow*) **11**, 91 (1947).
14. D. Henshaw and A. Woods, *Phys. Rev.* **121**, 1266 (1961).
15. G. Arnold, G. Yamell, P. Bendt, and E. Kerr, *Phys. Rev.* **113**, 1379 (1959).
16. G. de Boer, *in Proc. Int. School of Phys.*, Course 21, "Liquid Helium" (G. Careri, ed.), Chapter 1. Academic Press, New York, 1963.
17. D. Osborne, *Proc. Phys. Soc.* **A63**, 909 (1950).
18. L. Onsager, *Nuovo Cimento* **6**, Suppl. 2, 249 (1949).
19. R. P. Feynmann, *in* "Progress in Low Temperature Physics" (C. J. Gorter, ed.), Vol. 1, Chapter 2. North-Holland Publ., Amsterdam, 1955.
20. J. Khalatnikov, "Introduction to the Theory of Superfluidity." Benjamin & Coy, New York, 1965.
21. W. Vinen, *Proc. Roy. Soc.* **A260**, 218 (1961).
22. G. Hess and W. Fairbank, *Proc. Int. Conf. Low Temp. Phys., 10th, Moscow* (1966).
23. G. Rayfield and F. Reif, *Phys. Rev.* **A136**, 1194 (1964).
24. P. Zisel, *Phys. Rev.* **79**, 309 (1950); **92**, 1106 (1953).
25. H. Hall and W. Vinen, *Proc. Roy. Soc.* **A238**, 204 (1956).
26. C. Gorter and G. Mellink, *Physica* **15**, 285 (1949).
27. C. C. Lin, "Liquid Helium." Academic Press, New York, 1963.
28. H. Hall, *Adv. Phys.* **9**, 89 (1960).
29. G. Bekarevitch and J. Khalatnikov, *Soviet Phys. JETP* **13**, 643 (1961).
30. D.-Y. Hsieh, Rep. No. 85–36, Div. of Eng. and Appl. Sci., Cal-Tech. (1966).
31. D.-Y. Hsieh, Rep. No. 327-3, Jet Propulsion Lab. Cal-Tech. (1966).
32. T. Lee, K. Huang, and C. Yang, *Phys. Rev.* **106**, 1135 (1957).
33. G. Rayfield and F. Reif, *Phys. Rev. Lett.* **11**, 305 (1963).
34. R. Feynman and M. Cohen, *Phys. Rev.* **102**, 1189 (1956).
35. R. Houtappel, *Physica* **16**, 425 (1950).

APPENDIX *VECTOR OPERATIONS AND IDENTITIES*

A1 Vector Operations

Only orthogonal coordinate systems in Euclidean 3-space will be considered. Only these are of practical interest since they allow separation of variables in the relevant partial differential equations (of the Laplace or Helmholtz type, for example).

Besides the rectangular Cartesian system, when analyzing 3-dimensional flows with axial symmetry, the use of circular cylindrical coordinates reduces the number of variables to two, and thus the boundary conditions are simplified. For conical flows a similar advantage is gained by studying them in spherical coordinates.

A1.1 Metric Coefficients

An orthogonal system (u^1, u^2, u^3) may be described by the metric coefficients g_{11}, g_{22}, g_{33}. An infinitesimal separation between two-points in this space is given by†

$$(ds)^2 = g_{11}(du^1)^2 + g_{22}(du^2)^2 + g_{33}(du^3)^2, \quad \text{(A.1)}$$

† G. Arfken, "Mathematical Methods for Physicists," 2nd ed. Academic Press, 1970.

where

$$g_{ii} = \left(\frac{\partial x^1}{\partial u^i}\right)^2 + \left(\frac{\partial x^2}{\partial u^i}\right)^2 + \left(\frac{\partial x^3}{\partial u^i}\right)^2 \tag{A.2}$$

and the x^i are rectangular coordinates.

Equation (A. 1) shows that the infinitesimal distances along the coordinate axes are

$$(g_{11})^{1/2}\, du^1, \qquad (g_{22})^{1/2}\, du^2, \qquad (g_{33})^{1/2}\, du^3.$$

Thus, an element of *area* on the u^1u^2 surface is

$$dA = [(g_{11})^{1/2}\, du^1][(g_{22})^{1/2}\, du^2] = (g_{11}g_{22})^{1/2}\, du^1\, du^2. \tag{A.3}$$

Similarly, an element of *volume* is

$$dV = (g_{11}g_{22}g_{33})^{1/2}\, du^1\, du^2\, du^3. \tag{A.4}$$

The *gradient* of a scalar function φ in orthogonal curvilinear coordinates (u^1, u^2, u^3) is

$$\text{grad}\, \varphi = \nabla\varphi = \frac{\mathbf{e}_1}{(g_{11})^{1/2}}\frac{\partial\varphi}{\partial u^1} + \frac{\mathbf{e}_2}{(g_{22})^{1/2}}\frac{\partial\varphi}{\partial u^2} + \frac{\mathbf{e}_3}{(g_{33})^{1/2}}\frac{\partial\varphi}{\partial u^3}, \tag{A.5}$$

where $\mathbf{e}_1$, $\mathbf{e}_2$, $\mathbf{e}_3$ are unit vectors. The *divergence* of a vector field $\mathbf{A}$ is expressed as

$$\text{div}\, \mathbf{A} = \nabla \cdot \mathbf{A} = g^{-1/2}\left\{\frac{\partial}{\partial u^1}\left[\left(\frac{g}{g_{11}}\right)^{1/2} A_1\right] + \frac{\partial}{\partial u^2}\left[\left(\frac{g}{g_{22}}\right)^{1/2} A_2\right] + \frac{\partial}{\partial u^3}\left[\left(\frac{g}{g_{33}}\right)^{1/2} A_3\right]\right\} \tag{A.6}$$

and

$$\text{curl}\, \mathbf{A} = \nabla \times \mathbf{A} = \begin{vmatrix} \mathbf{e}_1\left(\dfrac{g_{11}}{g}\right)^{1/2} & \mathbf{e}_2\left(\dfrac{g_{22}}{g}\right)^{1/2} & \mathbf{e}_3\left(\dfrac{g_{33}}{g}\right)^{1/2} \\ \dfrac{\partial}{\partial u^1} & \dfrac{\partial}{\partial u^2} & \dfrac{\partial}{\partial u^3} \\ (g_{11})^{1/2}A_1 & (g_{22})^{1/2}A_2 & (g_{33})^{1/2}A_3 \end{vmatrix}. \tag{A.7}$$

Now

$$\nabla^2 \varphi \equiv \text{div grad}\, \varphi$$

and, in orthogonal curvilinear coordinates,

$$\nabla^2\varphi = g^{-1/2} \sum_{i=1}^{3} \frac{\partial}{\partial u^i}\left[\frac{g^{1/2}}{g_{ii}} \frac{\partial \varphi}{\partial u^i}\right], \tag{A.8}$$

where $g \equiv g_{11}g_{22}g_{33}$.

A1.2 Rectangular Coordinates (x, y, z)

$$\begin{aligned} u^1 &= x, & -\infty < x < +\infty, \\ u^2 &= y, & -\infty < y < +\infty, \\ u^3 &= z, & -\infty < z < +\infty. \end{aligned} \tag{A.9}$$

Surfaces of constant x, y, or z are mutually orthogonal planes. The metric coefficients are

$$g_{11} = g_{22} = g_{33} = 1 = g^{1/2}. \tag{A.10}$$

Hence

$$\nabla\varphi = \mathbf{e}_x \frac{\partial \varphi}{\partial x} + \mathbf{e}_y \frac{\partial \varphi}{\partial y} + \mathbf{e}_z \frac{\partial \varphi}{\partial z}, \tag{A.11}$$

$$\nabla \cdot \mathbf{A} = \frac{\partial A_x}{\partial x} + \frac{\partial A_y}{\partial y} + \frac{\partial A_z}{\partial z}, \tag{A.12}$$

$$\nabla \times \mathbf{A} = \begin{vmatrix} \mathbf{e}_x & \mathbf{e}_y & \mathbf{e}_z \\ \dfrac{\partial}{\partial x} & \dfrac{\partial}{\partial y} & \dfrac{\partial}{\partial z} \\ A_x & A_y & A_z \end{vmatrix}, \tag{A.13}$$

$$\nabla^2\varphi = \frac{\partial^2 \varphi}{\partial x^2} + \frac{\partial^2 \varphi}{\partial y^2} + \frac{\partial^2 \varphi}{\partial z^2}. \tag{A.14}$$

A1.3 Circular Cylindrical Coordinates (r, ϑ, z)

Circular cylindrical coordinates are depicted in Fig. A.1.

$$\begin{aligned} u^1 &= r, & 0 \le r < +\infty, \\ u^2 &= \vartheta, & 0 \le \vartheta < 2\pi, \\ u^3 &= z, & -\infty < z < +\infty, \end{aligned} \tag{A.15}$$

$$x = r\cos\vartheta, \qquad y = r\sin\vartheta, \qquad z = z. \tag{A.16}$$

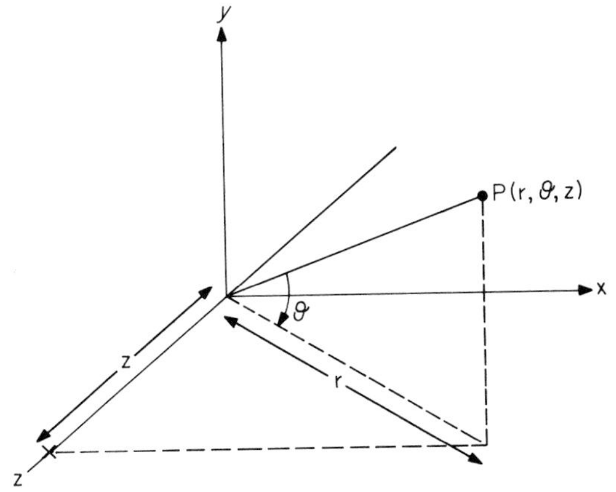

Fig. A.1

The metric coefficients are

$$g_{11} = 1, \qquad g_{22} = r^2, \qquad g_{33} = 1, \qquad g^{1/2} = r; \tag{A.17}$$

$$(ds)^2 = (dr)^2 + (r^2\, d\vartheta)^2 + dz^2, \tag{A.18}$$

$$\nabla\varphi = \mathbf{e}_r \frac{\partial\varphi}{\partial r} + \frac{\mathbf{e}_\vartheta}{r}\frac{\partial\varphi}{\partial\vartheta} + \mathbf{e}_z \frac{\partial\varphi}{\partial z}, \tag{A.19}$$

$$\nabla\cdot\mathbf{A} = \frac{\partial A_r}{\partial r} + \frac{A_r}{r} + \frac{1}{r}\frac{\partial A_\vartheta}{\partial\vartheta} + \frac{\partial A_z}{\partial z}, \tag{A.20}$$

$$\nabla\times\mathbf{A} = \frac{1}{r}\begin{vmatrix} \mathbf{e}_r & \mathbf{e}_\vartheta r & \mathbf{e}_z \\ \dfrac{\partial}{\partial r} & \dfrac{\partial}{\partial\vartheta} & \dfrac{\partial}{\partial z} \\ A_r & A_\vartheta r & A_z \end{vmatrix}, \tag{A.21}$$

$$\nabla^2\varphi = \frac{\partial^2\varphi}{\partial r^2} + \frac{1}{r}\frac{\partial\varphi}{\partial r} + \frac{1}{r^2}\frac{\partial^2\varphi}{\partial\vartheta^2} + \frac{\partial^2\varphi}{\partial z^2} \tag{A.22}$$

A1.4 Spherical Coordinates (r, ϑ, ψ)

Spherical coordinates are shown in Fig. A. 2.

$$\begin{aligned} u^1 &= r, & 0 &\le r < +\infty, \\ u^2 &= \vartheta, & 0 &\le \vartheta \le \pi, \\ u^3 &= \psi, & 0 &\le \psi < 2\pi, \end{aligned} \tag{A. 23}$$

$$x = r\sin\vartheta\cos\psi, \qquad y = r\sin\vartheta\sin\psi, \qquad z = r\cos\vartheta. \tag{A. 24}$$

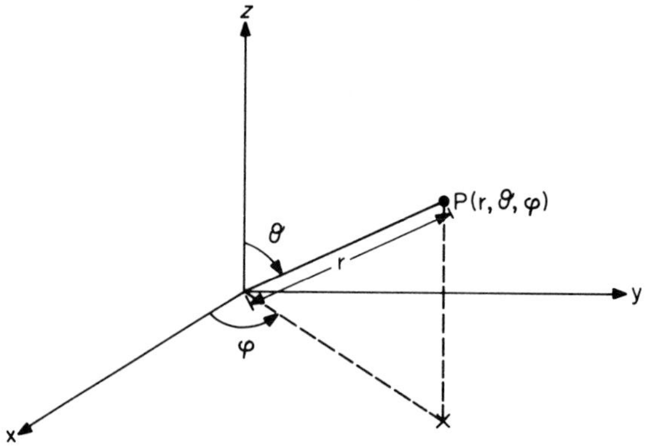

Fig. A.2

The metric coefficients are

$$g_{11} = 1, \qquad g_{22} = r^2, \qquad g_{33} = r^2 \sin^2 \vartheta, \qquad g^{1/2} = r^2 \sin \vartheta, \quad \text{(A.25)}$$

and

$$(ds)^2 = (dr)^2 + r^2(d\vartheta)^2 + r^2 \sin^2 \vartheta (d\psi)^2, \quad \text{(A.26)}$$

$$\nabla\varphi = \mathbf{e}_r \frac{\partial\varphi}{\partial r} + \frac{\mathbf{e}_\vartheta}{r} \frac{\partial\varphi}{\partial\vartheta} + \frac{\mathbf{e}_\psi}{r \sin \vartheta} \frac{\partial\varphi}{\partial\psi}, \quad \text{(A.27)}$$

$$\nabla \cdot \mathbf{A} = \frac{\partial A_r}{\partial r} + \frac{2}{r} A_r + \frac{1}{r} \frac{\partial A_\vartheta}{\partial\vartheta} + \frac{\cot \vartheta}{r} A_\vartheta + \frac{1}{r \sin \vartheta} \frac{\partial A_\psi}{\partial\psi}, \quad \text{(A.28)}$$

$$\nabla \times \mathbf{A} = \frac{1}{r^2 \sin \vartheta} \begin{vmatrix} \mathbf{e}_r & \mathbf{e}_\vartheta r & \mathbf{e}_\psi r \sin \vartheta \\ \frac{\partial}{\partial r} & \frac{\partial}{\partial\vartheta} & \frac{\partial}{\partial\psi} \\ A_r & A_\vartheta r & A_\psi r \sin \vartheta \end{vmatrix}, \quad \text{(A.29)}$$

$$\nabla^2\varphi = \frac{\partial^2\varphi}{\partial r^2} + \frac{2}{r} \frac{\partial\varphi}{\partial r} + \frac{1}{r^2} \frac{\partial^2\varphi}{\partial\vartheta^2} + \frac{\cot \vartheta}{r^2} \frac{\partial\varphi}{\partial\vartheta} + \frac{1}{r^2 \sin^2 \vartheta} \frac{\partial^2\varphi}{\partial\psi^2}. \quad \text{(A.30)}$$

A1.5 Indicial Notation

It has been pointed out that in the Cartesian coordinate system (x_1, x_2, x_3) the vector operations referred to above can be expressed very concisely as

$$\boldsymbol{\nabla}\varphi = \frac{\partial \varphi}{\partial x_i} = \varphi_{,i}, \tag{A.31}$$

$$\boldsymbol{\nabla} \cdot \mathbf{A} = \frac{\partial A_i}{\partial x_i} = A_{i,i}, \tag{A.32}$$

$$\boldsymbol{\nabla} \times \mathbf{A} = \varepsilon_{ijk} \frac{\partial A_k}{\partial x_j} = \varepsilon_{ijk} A_{k,j}. \tag{A.33}$$

A2 Vector Identities

For any two vectors **A** and **B** the following vector identities can be derived:

$$\boldsymbol{\nabla} \times (\mathbf{A} \times \mathbf{B}) = \mathbf{B} \cdot \boldsymbol{\nabla}\mathbf{A} - \mathbf{A} \cdot \boldsymbol{\nabla}\mathbf{B} + \mathbf{A}(\boldsymbol{\nabla} \cdot \mathbf{B}) - \mathbf{B}(\boldsymbol{\nabla} \cdot \mathbf{A}), \tag{A.34}$$

$$\boldsymbol{\nabla}(\mathbf{A} \cdot \mathbf{B}) = \mathbf{A} \cdot \boldsymbol{\nabla}\mathbf{B} + \mathbf{B} \cdot \boldsymbol{\nabla}\mathbf{A} + \mathbf{A} \times (\boldsymbol{\nabla} \times \mathbf{B}) + \mathbf{B} \times (\boldsymbol{\nabla} \times \mathbf{A}), \tag{A.35}$$

$$\boldsymbol{\nabla} \times (\boldsymbol{\nabla} \times \mathbf{A}) = \boldsymbol{\nabla}(\boldsymbol{\nabla} \cdot \mathbf{A}) - \nabla^2\mathbf{A}, \tag{A.36}$$

where

$$\nabla^2\mathbf{A} = \boldsymbol{\nabla} \cdot (\boldsymbol{\nabla}\mathbf{A})$$

In indicial notation this last identity is written as follows:

$$\varepsilon_{rsi}\varepsilon_{ijk}A_{k,js} = A_{s,sr} - A_{r,ss}, \tag{A.37}$$

$$\boldsymbol{\nabla} \cdot \boldsymbol{\nabla} \times \mathbf{A} = 0, \tag{A.38}$$

$$\boldsymbol{\nabla} \times (\varphi\mathbf{A}) = \boldsymbol{\nabla}\varphi \times \mathbf{A} + \varphi\boldsymbol{\nabla} \times \mathbf{A}, \tag{A.39}$$

$$\boldsymbol{\nabla} \cdot (\varphi\mathbf{A}) = \mathbf{A} \cdot \boldsymbol{\nabla}\varphi + \varphi\boldsymbol{\nabla} \cdot \mathbf{A}. \tag{A.40}$$

For two scalars A and B, there is the useful identity

$$\boldsymbol{\nabla} \cdot (A\,\boldsymbol{\nabla}B - B\,\boldsymbol{\nabla}A) = A\,\nabla^2 B - B\,\nabla^2 A \tag{A.41}$$

or, in indicial notation,

$$(AB_{,i} - BA_{,i})_{,i} = AB_{,ii} - BA_{,ii}. \tag{A.42}$$

SUBJECT INDEX

J

K

L

M

N

O

P

W